# Plant and Vegetation

Volume 14

**Series editor**
M.J.A. Werger, Utrecht, The Netherlands

Plant and Vegetation is a new Springer series comprising a series of books that present current knowledge and new perspectives on world vegetation. Examining the ecology of plants and vegetation at all scales – from plant to landscape – and covering key issues such as globalization, invasive species, climate change and the dynamics of plant biodiversity, this book series draws together a wide range of material of interest to plant ecologists, vegetation scientists, and geographers around the world. The series provides a valuable resource for both graduate students and researchers in environmental and biological sciences, as well as for landscape planners and policy makers involved in land-use and restoration projects at local, regional and international levels.

More information about this series at http://www.springer.com/series/7549

Milan Chytrý • Jiří Danihelka • Zdeněk Kaplan
Petr Pyšek
Editors

# Flora and Vegetation of the Czech Republic

Springer

*Editors*
Milan Chytrý
Department of Botany and Zoology
Masaryk University
Brno, Czech Republic

Jiří Danihelka
Department of Botany and Zoology
Masaryk University
Brno, Czech Republic

Zdeněk Kaplan
Institute of Botany
The Czech Academy of Sciences
Průhonice, Czech Republic

Petr Pyšek
Institute of Botany
The Czech Academy of Sciences
Průhonice, Czech Republic

ISSN 1875-1318 ISSN 1875-1326 (electronic)
Plant and Vegetation
ISBN 978-3-319-63180-6 ISBN 978-3-319-63181-3 (eBook)
DOI 10.1007/978-3-319-63181-3

Library of Congress Control Number: 2017954304



Printed on acid-free paper

This Springer imprint is published by Springer Nature
The registered company is Springer International Publishing AG
The registered company address is: Gewerbestrasse 11, 6330 Cham, Switzerland

# Preface

The Czech Republic, a country in the heart of Europe, has a long tradition of botanical research. More than two centuries of systematic studies on its flora and almost a century of vegetation research have resulted in a wealth of detailed botanical information. This country is perhaps the only one in the world for which detailed multivolume monographs of both of its flora (the last, ninth volume is under preparation) and vegetation (in four volumes) have recently been published. There are also large databases of records of plant distribution and vegetation plots, which are frequently used in basic and applied research.

Despite the existence of extensive botanical information, until quite recently general descriptions of the main patterns in this country's flora and vegetation were available only in Czech, mainly in the introductory chapters of botanical monographs. On the occasion of the centenary of the Czech Botanical Society in 2012, the society published a special issue of its journal *Preslia* that contained new checklists and Red Lists of the vascular plant flora and bryoflora of this country (Danihelka et al. 2012; Grulich 2012; Kučera et al. 2012) and review articles dealing with the history of botanical research (Krahulec 2012), general information about vegetation (Chytrý 2012), flora (Kaplan 2012), plant invasions (Pyšek et al. 2012) and basic overviews of bryoflora (Kučera et al. 2012) and lichen biota (Liška 2012). As these review articles were well received by botanists both within this country and internationally, we decided to use some of them as a basis for developing a reference book that would summarize the botanical information about the Czech Republic mainly for an international readership.

The current book focuses on the Czech flora, including the alien flora, bryoflora and lichen biota, and vegetation, including the history of the vegetation since the last glacial. It also deals with the history of botanical research and conservation of plant diversity in this country. The chapters on vascular flora (Chapter 3), bryoflora (Chapter 4), vegetation (Chapter 7) and plant invasions (Chapter 8) are based on considerably modified, updated and extended papers published in the 2012 special issue of *Preslia*. Chapters on the history of botanical research (Chapter 2), lichen biota (Chapter 5), vegetation history since the late Pleistocene (Chapter 6) and conservation of botanical diversity (Chapter 9) are entirely new. The chapter on this country's physical geography (Chapter 1), which provides basic information necessary for understanding botanical diversity in an environmental context, uses some materials originally included in the first part of the *Preslia* review paper on vegetation; however, also these materials have been considerably extended and revised.

This book is a publication of the Centre of Excellence Pladias (Plant Diversity Analysis and Synthesis), funded by a grant from the Czech Science Foundation in the years 2014–2018 (project no. 14-36079G). This centre includes the Department of Botany and Zoology of Masaryk University, Brno, Institute of Botany of The Czech Academy of Sciences and Department of Botany of the University of South Bohemia, České Budějovice. The Pladias researchers have created and critically revised extensive data sets on this country's flora and vegetation and integrated them in the Pladias database, which is now being used for various syntheses, research and applications. Some of the information presented in this book is based on data from this new database. Another of the results of the Pladias project is a new field flora, Key to the Flora of the Czech Republic (Kaplan et al. 2017), the manuscript of which was prepared shortly before the manuscript of this book. This field flora is used as the source of the nomenclature of the vascular plants throughout this book. The nomenclature of bryophytes follows Kučera et al. (2012) and that of lichens follows Liška et al. (2008).

We thank colleagues who helped us with the preparation of this book. Apart from the contributors to the Pladias database and authors of individual chapters, it was mainly Ondřej Hájek, who prepared nearly all the maps used in this book; Kryštof Chytrý, who prepared the plates with photographs and R scripts for some graphs; Pavel Dřevojan and Zuzana Sixtová, who checked references; authors of photographs, listed in the captions; Tony Dixon, who proofread the English texts; Martin Kočí, who prepared Index; and Irena Axmanová, who helped us with final proofreading. Various data were kindly provided by the Nature Conservation Agency of the Czech Republic. Special thanks go to Marinus Werger, the editor of the Springer book series Plant and Vegetation, for conceptual advice on the content of this book and specific comments on individual chapters.

We hope that this book will become a useful source of information for the international and national community of botanists, bryologists, lichenologists, palaeoecologists, vegetation scientists, invasion ecologists and conservationists, including those doing research in the Czech Republic, travelling to this country for excursions, or those who are searching for a reference on specific topics related to Central European plant diversity.

Brno, Czech Republic — Milan Chytrý
Brno, Czech Republic — Jiří Danihelka
Průhonice, Czech Republic — Zdeněk Kaplan
Průhonice, Czech Republic — Petr Pyšek
April 2017

## References

Chytrý M (2012) Vegetation of the Czech Republic: diversity, ecology, history and dynamics. Preslia 84:427–504
Danihelka J, Chrtek J Jr, Kaplan Z (2012) Checklist of vascular plants of the Czech Republic. Preslia 84:647–811
Grulich V (2012) Red List of vascular plants of the Czech Republic: 3rd edition. Preslia 84:631–645
Kaplan Z (2012) Flora and phytogeography of the Czech Republic. Preslia 84:505–574
Kaplan Z, Danihelka J, Chrtek J Jr, Kirschner J, Kubát K, Štech M, Štěpánek J (eds) (2017) Klíč ke květeně České republiky [Key to the Flora of the Czech Republic]. Academia, Praha
Krahulec F (2012) History of the studies on the flora and vegetation in the Czech Republic. Preslia 84:397–426
Kučera J, Váňa J, Hradílek Z (2012) Bryophyte flora of the Czech Republic: updated checklist and Red List (2012) and a brief analysis. Preslia 84: 813–850
Liška J (2012) Lichen flora of the Czech Republic. Preslia 84:851–862
Liška J, Palice Z, Slavíková Š (2008) Checklist and Red List of lichens of the Czech Republic. Preslia 80:151–182
Pyšek P, Chytrý M, Pergl J, Sádlo J, Wild J (2012) Plant invasions in the Czech Republic: current state, introduction dynamics, invasive species and invaded habitats. Preslia 84:575–629

# Contents

# Contributors

**Vojtěch Abraham** Department of Botany, Faculty of Science, Charles University, Praha, Czech Republic

**Milan Chytrý** Department of Botany and Zoology, Masaryk University, Brno, Czech Republic

**Jiří Danihelka** Department of Botany and Zoology, Masaryk University, Brno, Czech Republic

Institute of Botany, The Czech Academy of Sciences, Průhonice, Czech Republic

**Handrij Härtel** Bohemian Switzerland (České Švýcarsko) National Park Administration, Krásná Lípa, Czech Republic

**Zdeněk Kaplan** Institute of Botany, The Czech Academy of Sciences, Průhonice, Czech Republic

**Jan Kučera** Department of Botany, University of South Bohemia, České Budějovice, Czech Republic

**Petr Kuneš** Department of Botany, Faculty of Science, Charles University, Praha, Czech Republic

**Zdeněk Palice** Institute of Botany, The Czech Academy of Sciences, Průhonice, Czech Republic

Department of Botany, Faculty of Science, Charles University, Praha, Czech Republic

**Jan Pergl** Institute of Botany, The Czech Academy of Sciences, Průhonice, Czech Republic

**Petr Pyšek** Institute of Botany, The Czech Academy of Sciences, Průhonice, Czech Republic

Department of Ecology, Faculty of Science, Charles University, Praha, Czech Republic

**Jiří Sádlo** Institute of Botany, The Czech Academy of Sciences, Průhonice, Czech Republic

**Jan Wild** Institute of Botany, The Czech Academy of Sciences, Průhonice, Czech Republic

# Chapter 1
# Physical Geography of the Czech Republic

**Milan Chytrý**

**Abstract** Geodiversity, climate, hydrology and historical and recent effects of humans on land-cover in the Czech Republic are briefly described. The aim of this chapter is to provide basic information necessary for understanding current and historical patterns in the distribution of the flora and vegetation in this country, their ecology and dynamics.

## 1.1 Introduction

The Czech Republic (*Česká republika*, also called Czechia, *Česko*) is a land-locked country in Central Europe bordering Germany in the west and north-west, Austria in the south, Slovakia in the east and Poland in the north-east. It stretches from 48°33'09"N to 51°03'33"N and from 12°05'33"E to 18°51'33"E (Fig. 1.1). The country's south-north length is 278 km, west-east length 493 km and area 78,870 km$^2$ (Czech Statistical Office, www.czso.cz).

Historically the country was divided into three lands (Fig. 1.1, see also Chap. 2, this book). The largest land, Bohemia (*Čechy*), in the western part of the country, was the historical Kingdom of Bohemia with the capital city of Prague (*Praha*) in its centre. Moravia (*Morava*) in the east was a margraviate subject to the Kingdom of Bohemia. Silesia (*Slezsko*) was a duchy, which currently mostly belongs to Poland, with a small part extending to the north-east of present-day Czech Republic (sometimes referred to as Czech Silesia). Currently the country is divided into 14 administrative regions that do not respect the borders of the historical lands; however, these lands are still commonly used in botanical literature and will be referred to throughout this book.

M. Chytrý (✉)
Department of Botany and Zoology, Masaryk University,
Kotlářská 2, 611 37 Brno, Czech Republic
e-mail: chytry@sci.muni.cz



M. Chytrý et al. (eds.), *Flora and Vegetation of the Czech Republic*, Plant and Vegetation 14, DOI 10.1007/978-3-319-63181-3_1

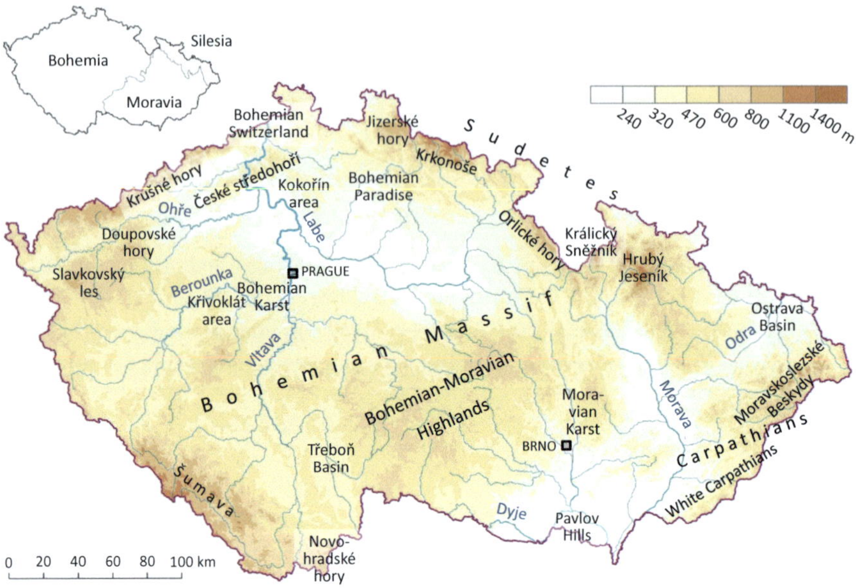


**Fig. 1.1** Topographic map of the Czech Republic. See Box 1.1 for a guide to Czech toponyms (all the maps in this chapter were prepared by O. Hájek)

**Box 1.1 Czech Topographic Names**

Czech belongs to the group of Western Slavic languages, together with Polish and Slovak. Many Czech names of topographic features (e.g. mountain ranges, basins or lowlands) are derived from the names of settlements, to which an adjectival ending is added. For example, the name *Pavlovské vrchy*, literally meaning *Pavlov Hills*, is derived from the village name *Pavlov*. Generic names of topographic features (*vrchy* in this case) are not capitalized in Czech even if they are inseparable part of the toponyms; an exception is the names of towns and villages.

Some better known topographic features in the Czech Republic have established English translations of their Czech names, but most Czech toponyms either have no English equivalent or the equivalent is rarely used and poorly known. In contrast, many Czech toponyms have German equivalents, which for most part are rarely used now, but were routinely used in older botanical literature. The following glossary provides several Czech words that are a common part of topographic names and a list of Czech topographic names frequently used in this book followed by their English and German equivalents. For each name the variant used in this book is set in italics. In most cases, the usage accepted in the recent book *Landscapes and landforms of the Czech Republic* (Pánek and Hradecký 2016) is followed.

(continued)

**Box 1.1** (continued)
**Frequent Words in Czech Topographic Names**

| | |
|---|---|
| český, -á, -é | 1. Czech, 2. Bohemian |
| dolní | lower |
| hora, -y | mountain, -s |
| horní | upper |
| jezero | lake |
| jižní | southern |
| kras | karst |
| les | forest, woodland, wood |
| malý, -á, -é | little, small |
| město | town, city |
| moravský, -á, -é | Moravian |
| nížina | lowland |
| nízký, -á, -é | low |
| pahorkatina | hilly (colline) landscape |
| pánev, pánve | basin |
| potok | brook |
| řeka | river |
| rybník | fishpond |
| severní | northern |
| skála, -y | rock, -s |
| slezský | Silesian |
| údolí | valley |
| velký, -á, -é | big, great, large |
| vrch, -y | hill, -s |
| vrchovina | highland |
| východní | eastern |
| západní | western |

### Historical Lands

| | | |
|---|---|---|
| Čechy | *Bohemia* | Böhmen |
| Morava | *Moravia* | Mähren |
| Slezsko | *Silesia* | Schlesien |

### Topographic Features

| | | |
|---|---|---|
| Adršpašsko-teplické skály | *Adršpach-Teplice Rocks* | Adersbach-Weckelsdorfer Felsenstadt |
| Bílé Karpaty | *White Carpathians* | Weiße Karpaten |
| *České středohoří* | n/a | Böhmisches Mittelgebirge |

(continued)

**Box 1.1** (continued)

| | | |
|---|---|---|
| České Švýcarsko | *Bohemian Switzerland* | Böhmische Schweiz |
| Českomoravská vrchovina | *Bohemian-Moravian Highlands* | Böhmisch-Mährische Höhe |
| Český kras | *Bohemian Karst* | Böhmischer Karst |
| Český ráj | *Bohemian Paradise* | Böhmisches Paradies |
| Český masiv | *Bohemian Massif* | Böhmische Masse |
| *Doupovské hory* | Doupov Mts | Duppauer Gebirge |
| Hradčanské stěny | *Hradčany Rocks* | Kummergebirge |
| *Hrubý Jeseník* | n/a | Altvatergebirge, Hohes Gesenke |
| *Jizerské hory* | Jizera Mts | Isergebirge |
| Karpaty | *Carpathians* | Karpaten |
| Kokořínsko | *Kokořín area* | Daubaer Schweiz |
| *Králický Sněžník* | n/a | Glatzer Schneeberg |
| *Krkonoše* | Giant Mountains | Riesengebirge |
| *Krušné hory* | Ore Mts | Erzgebirge |
| Křivoklátsko | *Křivoklát area* | Pürglitzer Wald |
| Labské pískovce | *Elbe Sandstones* | Elbsandsteingebirge |
| *Lužické hory* | Lusatian Mts | Lausitzer Gebirge |
| *Moravskoslezské Beskydy* | Moravian-Silesian Beskids | Mährisch-Schlesische Beskiden |
| Moravský kras | *Moravian Karst* | Mährischer Karst |
| *Novohradské hory* | Nové Hrady Mts | Gratzener Bergland |
| *Orlické hory* | Orlice Mts | Adlergebirge |
| Ostravská pánev | *Ostrava Basin* | Ostrauer Becken |
| Pavlovské vrchy | *Pavlov Hills* | Pollauer Berge |
| Praha | *Prague* | Prag |
| *Slavkovský les* | Slavkov Forest | Kaiserwald |
| Sudety | *Sudetes* | Sudeten |
| *Šumava* | Bohemian Forest (this term is usually used to include the Šumava Mts, Český les Mts and Bayerischer Wald Mts) | Böhmerwald |
| Třeboňská pánev | *Třeboň Basin* | Wittingauer Becken |
| Západní Karpaty | *Western Carpathians* | Westkarpaten |

**Rivers**

Czech names are used for all rivers throughout the text. Large rivers or those shared with other countries are internationally better known under their German names, which are usually used in English-language literature: Dyje – Thaya, Labe – Elbe, Morava – March, Odra – Oder, Ohře – Eger, Vltava – Moldau.

## 1.2 Geodiversity

The altitudinal range of the Czech Republic is 115–1603 m a.s.l., however, 67% of the area is at altitudes below 500 m and 99% below 1000 m. The mean altitude is 430 m (Czech Statistical Office, www.czso.cz).

The country comprises two major geological units with contrasting topography and landscape features: the Bohemian Massif (*Český masiv*) in the west and the Western Carpathians (*Západní Karpaty*) in the east. The geology of the country is summarized in geological maps (Cháb et al. 2007; mapy.geology.cz) and a reference handbook (Chlupáč et al. 2011), which are used as the main source of the geological information presented in this chapter. Descriptions of the different landscapes in this country are provided by Pánek and Hradecký (2016).

**The Bohemian Massif** occupies the whole of Bohemia and western and north-western parts of Moravia, extending also to eastern and south-eastern Germany, northern Austria and south-western Poland. It is an old mountain system created by the Variscan (Hercynian) orogeny in the Late Palaeozoic, during which two palaeo-continents, southern Gondwana and northern Laurussia, collided to form the super-continent Pangea. Tectonically the Bohemian Massif is related to other Variscan massifs in France, Belgium and Germany, such as the Armorican Massif, Massif Central, Vosges, Schwarzwald, Ardennes, Rheinisches Schiefergebirge and Harz. Since the Permian the Bohemian Massif has been subject to erosion, which erased its originally rugged mountainous topography and created its current gently undulating landscape (Fig. 1.2a).

The upland areas in the Bohemian Massif are formed mainly of hard, poorly-weathering igneous and metamorphic rocks of Proterozoic and Lower Palaeozoic age, such as granite, granodiorite, gneiss (Fig. 1.2b), schist and granulite, which in places are covered by younger sedimentary or volcanic rocks (Fig. 1.3). Predominant rocks are base-poor and acidic while only small areas of base-rich rocks such as limestone, erlane, amphibolite or serpentinite occur locally. A diverse complex of Proterozoic and Lower Palaeozoic sedimentary and volcanic rocks, called Barrandien, occurs in the area along the Berounka River between the cities of Prague and Plzeň. It is formed mainly of shale and greywacke, with patches of extremely base-poor silicite and quarzite, base-rich volcanic and subvolcanic rocks such as diabase and spilite (Fig. 1.2c), and areas of Silurian and Devonian limestone in the Bohemian Karst between Prague and Beroun. The fine mosaic of base-poor and base-rich, and poorly and easily weathered rocks, supports a highly diverse flora in this region. Another region with a predominance of sedimentary rocks in upland areas occurs along the eastern edge of the Bohemian Massif between the cities of Brno and Opava. Lower Carboniferous base-poor shale and greywacke prevail in this area, though smaller areas of Devonian limestone are also found there, most notably in the Moravian Karst north of Brno (Fig. 1.2d).

The sediments in the Bohemian Massif deposited from the Upper Carboniferous onwards were not influenced by any further orogeny, therefore they are not tilted or folded, still laying horizontally or near-horizontally. Red Permian claystones, sand-

**Fig. 1.2** Examples of landforms and bedrocks in the Czech Republic: (**a**) Gently undulating landscape typical of the crystallinic areas of the Bohemian Massif: Pohoří na Šumavě, southern Bohemia; (**b**) Gneiss outcrop at the upper edge of the deep valley of the Dyje River, Podyjí National Park, southern Moravia; (**c**) Outcrop of spilite, a type of Upper Proterozoic basalt, above the Berounka River in the Barrandien zone: Čertova skála near Hracholusky, central Bohemia; (**d**) Pustý žleb, a karst gorge in Devonian limestone in the Moravian Karst, southern Moravia; (**e**) An example of a landscape formed by weathering of Cretaceous sandstone: Prachovské skály, eastern Bohemia. Credits for all photos in this chapter: M. Chytrý

stones and conglomerates, in places calcareous, are present over large areas in central and eastern Bohemia and in the graben *Boskovická brázda* in western Moravia. Triassic and Jurassic rocks are very rare in the Bohemian Massif.

In the Cretaceous period extensive lowland areas in northern, central and eastern Bohemia were flooded by a sea, which left behind two contrasting types of sedi-

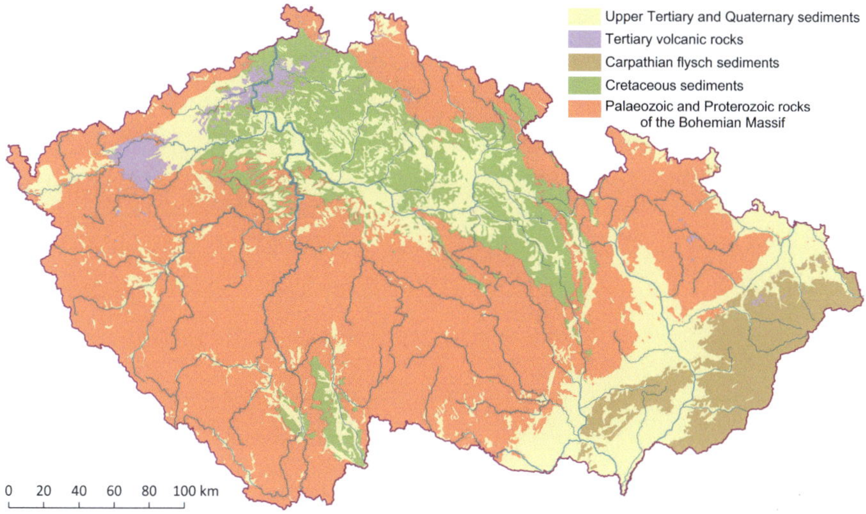


**Fig. 1.3** Simplified geological map of the Czech Republic (based on the digital geological map 1:500,000 – GEOČR500 of the Czech Geological Survey 1998)

ments. The first type is acidic siliceous sandstone, which locally forms sandstone pseudokarst areas with deep gorges and isolated rock towers, in Czech called *skalní města*, literally meaning 'rock cities' (Figs. 1.2e and 1.4; Härtel et al. 2007). The second type is a group of soft calcareous sediments, such as marl (Fig. 1.5a), marlite and fine calcareous sandstone, which give rise to gently undulating landforms with fertile base-rich soils. There is a strong contrast between the poor acidophilous flora of vascular plants on the former type and the rich basiphilous flora on the latter type of Cretaceous rocks. Cretaceous sediments also fill two large basins in southern Bohemia, České Budějovice Basin and Třeboň Basin, but here sedimentation occurred in freshwater conditions. The former basin is filled mainly with argillaceous sediments, while base-poor sand prevails in the latter, supporting acidophilous vegetation including peat bogs.

By the Tertiary the Bohemian Massif had been weathered to a gently undulating peneplain, which was broken into smaller tectonic units during the Alpine orogeny. Some of these units were lifted, forming mountain ranges especially along what are now the national borders with Austria, Germany and Poland. The highest of these ranges are the Sudetes along the Czech-Polish border, consisting of several more or less isolated mountain groups, most notably the Krkonoše (Karkonosze in Polish; 1603 m a.s.l.), Králický Sněžník (Śnieżnik Kłódzki in Polish; 1424 m) and Hrubý Jeseník (1491 m). Other prominent mountain ranges are the Krušné hory Mts (1244 m) on the border with Saxony and Šumava Mts (1456 m) on the border with Bavaria and Upper Austria. Since the upheaval of these mountains occurred with little folding, much of their summit areas are gently undulating plateaus, with peatlands in places. In the Pleistocene local glaciers developed on the eastern slopes of some mountain plateaus and formed cirques, especially in the Krkonoše, Hrubý

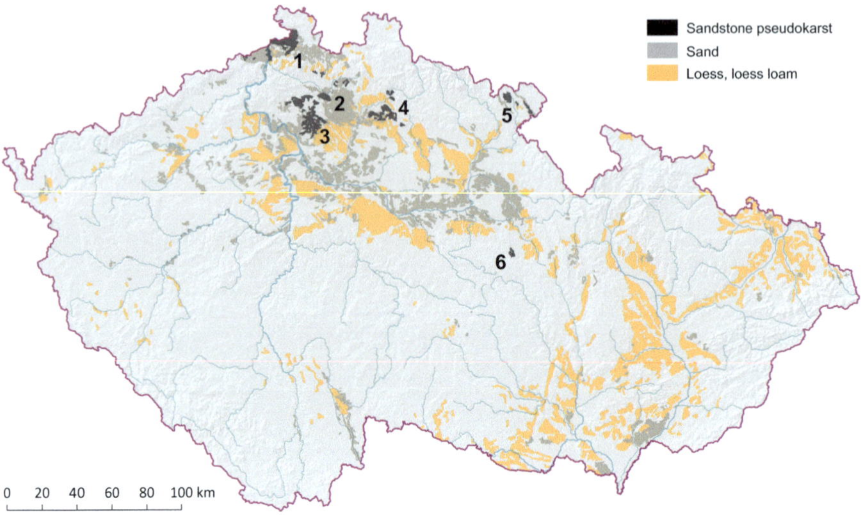


**Fig. 1.4** Areas of sandstone pseudokarst (based on topographic maps 1:50,000), sand deposits (based on interpretation of the soil map of the Czech Republic provided by the Czech University of Life Sciences to the Czech National Geoportal at http://geoportal.cenia.cz) and loess accumulations (based on Bajer and Houška 2009). (*1*) Elbe Sandstones (Labské pískovce, including Bohemian Switzerland), (*2*) Hradčany Rocks, (*3*) Kokořín area, (*4*) Bohemian Paradise (Český ráj), (*5*) Broumov area (including the Adršpach-Teplice Rocks), (*6*) Toulovcovy Maštale site

Jeseník and Šumava Mts (Fig. 1.5b). Two valley glaciers developed on the Czech side of the Krkonoše Mts, creating u-shaped elongated troughs of the Labský důl and Obří důl valleys.

In other areas in the Bohemian Massif, most notably in the southern foothills of the Krušné hory Mts, large land masses sank during the Alpine orogeny to form basins that were subsequently filled with Upper Tertiary sediments. Alpine orogeny also caused volcanic activity in the north-western and northern part of the Bohemian Massif, which gave rise to an extensive basalt stratovolcano of the Doupovské hory Mts and a number of isolated volcanic hills in the České středohoří Mts (Fig. 1.5c), which are formed of both base-rich (e.g. basalt) and acidic (e.g. phonolite and trachyte) rocks. These rocks are also typical of the volcanic mountain group of the Lužické hory Mts. Since the Upper Tertiary v-shaped river valleys were deeply cut in hard, poorly weathering rocks in the peneplain landscape in the Bohemian Massif after its upheaval due to the Alpine orogeny. Nowadays these valleys comprise high geodiversity with a broad variety of contrasting vegetation types (e.g. the Vltava, Otava, Lužnice, Berounka, Sázava, Dyje and Jihlava valleys; Fig. 1.6; see also Chap. 7, Sect. 7.4, this book).

Acidic bedrock prevails throughout the Bohemian Massif, especially in the submontane and montane areas. Therefore, local occurrences of limestone or other calcareous rocks (Fig. 1.7) markedly increase local diversity of flora and vegetation. Areas of hard, poorly weathering Silurian and Devonian limestone with well-developed karst features and rendzina soils occur especially in the Bohemian Karst in central Bohemia and Moravian Karst in southern Moravia. Small patches of metamor-

**Fig. 1.5** Examples of landforms and bedrocks in the Czech Republic: (**a**) Erosion-prone Cretaceous marl slope near Pokratice, northern Bohemia; (**b**) A glacial cirque in the Velká kotelní jáma in the Krkonoše Mts, eastern Bohemia; (**c**) Oblík, a solitary basalt hill formed as a result of Tertiary volcanism in the České středohoří Mts, northern Bohemia; (**d**) Serpentinite outcrop with the specialist fern *Notholaena marantae* on the slopes of the Jihlava River valley near Mohelno, southern Moravia; (**e**) Landscape in the basins in the foothills of the Krušné hory Mts transformed by open-cast coal mining: Jiří pit near Sokolov, north-western Bohemia

phic limestone (marble), occurring in association with siliceous metamorphic rocks, are found especially in south-western Bohemia (Šumava foothills), north-eastern Bohemia and western Moravia. Other calcareous rocks include sediments of Cretaceous age, which occur mainly in the lowlands in northern, central and eastern Bohemia. Another bedrock type with a strong effect on the local flora and diversity of vegetation is serpentinite, a metamorphic rock rich in magnesium and heavy metals, which is toxic for many plant species. Small patches of serpentinite occur in various areas in the Bohemian Massif, especially in the Slavkovský les Mts in western Bohemia, south-western Bohemia and the Bohemian-Moravian Highlands (Fig. 1.5d).

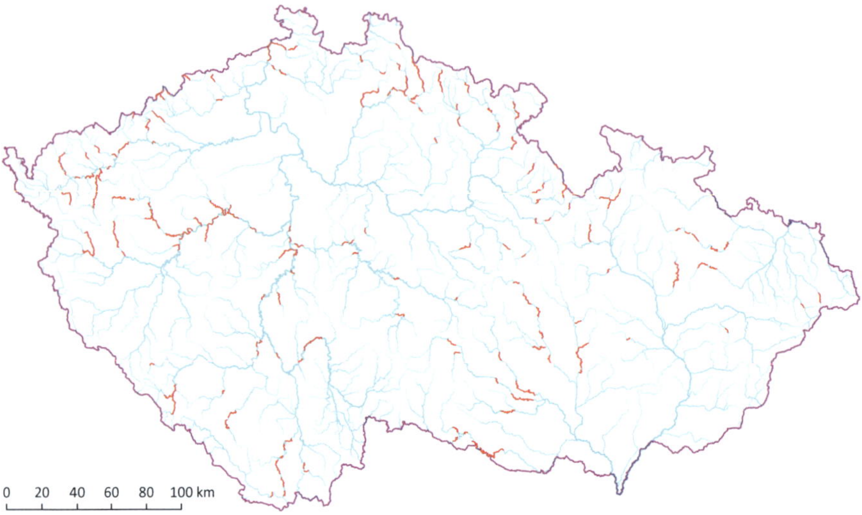


**Fig. 1.6** Deep river valleys, defined as areas with altitudinal difference of more than 60 m within 300 m on both sides of the river

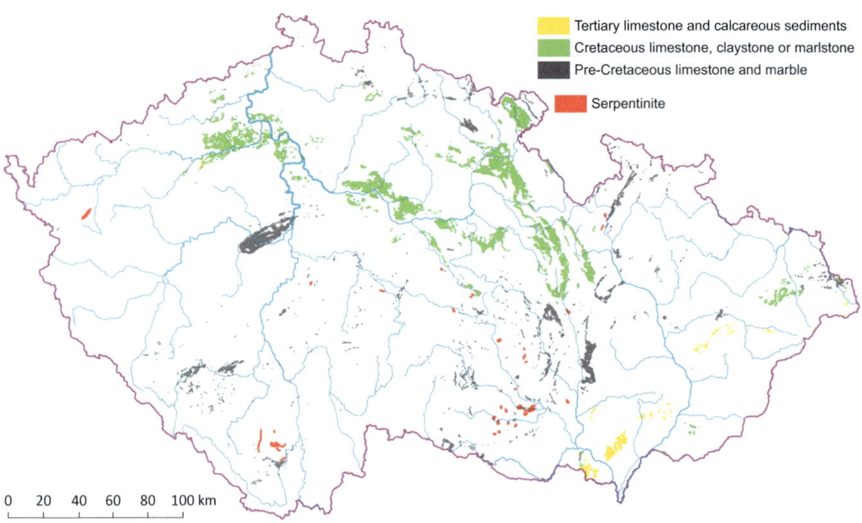


**Fig. 1.7** Areas of limestone, related calcareous bedrock and serpentinite (based on the digital geological map of the Czech Republic 1 : 50,000 of the Czech Geological Survey 2004). The map shows all limestone and serpentinite occurrences, including those covered by superficial deposits. (*1*) Bohemian Karst, (*2*) Moravian Karst

Ore has been mined in the Bohemian Massif since prehistory. A massive landscape transformation was caused by open-cast brown coal mining in the second half of the twentieth century, particularly in the 1970s–1980s in the basins at the foot of the Krušné hory Mts in northern and north-western Bohemia. Mining still occurs there, though with lower intensity, leaving large pits and waste dumps, partly rehabilitated, in an area larger than 1000 $km^2$ (Fig. 1.5e).

**The Western Carpathians** were formed by the closure of the Tethys Ocean during the Alpine orogeny in the Cretaceous and Tertiary, when oceanic crust was subducted under the African plate and the Western Carpathian rocks were thrust over the margin of the Eurasian plate. Only the Outer Western Carpathians, formed of flysch, are present in the Czech territory (in eastern and southern Moravia; Fig. 1.3). Flysch is a deep-ocean sedimentary sequence of alternating sandstones and claystones deposited by turbidity currents in the foreland basin of the developing Carpathian orogen in the Cretaceous and Palaeogene. With the progressive closure of the Tethys Ocean, the flysch layers were thrust over the margin of the Bohemian Massif, forming a series of nappes (thrust sheets). The highest flysch nappes in the Carpathian part of the Czech Republic are in the Moravskoslezské Beskydy Mts, reaching an altitude of 1323 m.

Flysch is a soft and erosion-prone rock, forming landscapes with gentle slopes and broad valleys. Rock outcrops are very rare in the flysch Carpathians, but fairly common in the Bohemian Massif. Still the topography of the flysch Carpathians, with their nappe structure forming deep valleys and relatively narrow crests (most of them running in a WSW–ENE direction) is more rugged at a coarser scale than that of the Bohemian Massif (Fig. 1.8a, b). Some flysch facies are calcareous (especially in the southern Moravian lowlands and upland fringes including the White Carpathians) while others are acidic (especially in the Moravskoslezské Beskydy Mts in north-eastern Moravia). As the sandstone layers of flysch are water-permeable while the claystone layers hold water, springs and small-scale landslides often occur on slopes of the Outer Western Carpathians. A combination of calcareous and relatively wet soils, otherwise rare in the Czech Republic, occurs on the calcareous flysch facies, supporting species-rich plant communities. Isolated remnant parts of eroded nappes, called klippes, occur at the outer edges of the Western Carpathians. The most prominent klippes are the Pavlov Hills in southern Moravia (Fig. 1.8c) and hills near the town of Štramberk in north-eastern Moravia, both formed of Jurassic and Lower Cretaceous limestone. These limestone klippes are important stepping stones for basiphilous and thermophilous flora.

The Western Carpathians are separated from the Bohemian Massif by the broad elongated Carpathian foreland basin, running in a SW–NE direction across Moravia, which is filled with molasse sediments of Neogene and Quaternary age, e.g. gravel, sand, loam and clay, both acidic and calcareous. The boundary between the Western Carpathians and the Eastern Alps is the Vienna Basin, also filled with Neogene and Quaternary sediments. While its main part is situated in eastern Austria and western Slovakia, it also reaches southern Moravia along the lower Dyje and Morava rivers. The soils developed in these lowland areas are among the most fertile in the country, therefore they were mostly converted to arable land.

**Fig. 1.8** Examples of landforms and bedrocks in the Czech Republic: (**a**) A typical topography at higher altitudes in the flysch Carpathians with long valleys and relatively narrow crests, some of them historically used as summer pastures: Mionší, Moravskoslezské Beskydy Mts, Silesia; (**b**) A typical topography at low altitudes in the flysch Carpathians with broad valleys and gentle slopes: Hustopeče, southern Moravia; (**c**) A klippe of Jurassic limestone in a marginal area of the Western Carpathians: Soutěska in the Pavlov Hills, southern Moravia; (**d**) Loess, aeolian sediment deposited in Pleistocene full glacials: Dolní Věstonice, southern Moravia; (**e**) Eroded chernozem below the Pavlov Hills in southern Moravia

Many areas in the Czech Republic below 300 m a.s.l. (or below 400 m in the driest regions) are covered by loess (Figs. 1.4 and 1.8d), a wind-blown calcareous sediment of Pleistocene age (Cilek 2001). Chernozems, black soils with a thick organo-mineral horizon containing charred organic matter, often develop on loess (Fig. 1.8e; Vysloužilová et al. 2014). Also these soils are very fertile and mostly used for agriculture.

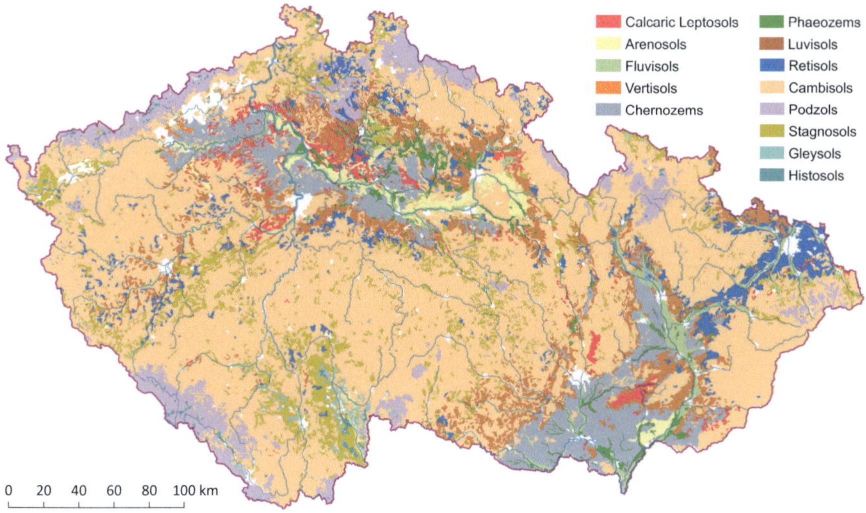


**Fig. 1.9** Soil types in the Czech Republic (based on the digital soil map provided by the Czech University of Life Sciences to the Czech National Geoportal at http://geoportal.cenia.cz). Soil types follow the typology and nomenclature of the IUSS Working Group WRB (2015)

The main soil types are chernozems and luvisols in the lowlands, cambisols at middle altitudes and podzols in the mountains (Kozák and Němeček 2010; Vavříček and Pancová Šimková 2014; Fig. 1.9). In general the fertility of soils depends on their richness in mineral nutrients, which decreases from chernozems on loess in the lowlands towards higher altitudes, being lowest in the mountain areas in the Bohemian Massif. In the flysch Carpathians nutrient-rich soils reach on average to higher altitudes than in the Bohemian Massif due to the high sorption capacity of soils developed on flysch sediments (Fig. 1.10; Bajer and Houška 2009). Poor soils also occur on lowland sandy plains, which develop especially along the Labe River in eastern and northern Bohemia, lower Vltava in central Bohemia, Lužnice in southern Bohemia and on the terraces of the Morava and Dyje rivers in southern Moravia (Fig. 1.4).

## 1.3 Climate

According to the Köppen-Geiger global classification, the climate in the Czech Republic is cold, with warm summers and no dry seasons (Peel et al. 2007). In the bioclimatic classification of Rivas-Martínez et al. (2004), this country has a transitional climate between temperate oceanic in the west (Bohemian Massif) and temperate continental in the east (Carpathians). The climate is seasonal with warm summers, cold winters and two distinct transitional seasons.

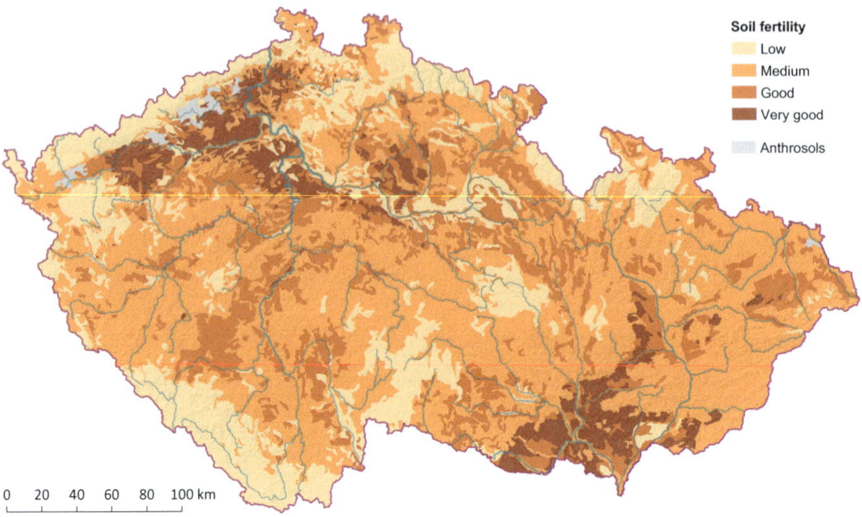


**Fig. 1.10** Soil fertility defined by the content of mineral nutrients (based on Bajer and Houška 2009)

Generally, temperature decreases and precipitation increases with altitude in the Czech Republic (Fig. 1.11). Summer and winter temperatures are positively correlated and so is summer and winter precipitation. Both temperature and precipitation peak in July. Lowlands are warm and dry, with a mean annual temperature of 8–9.5 °C (January mean −2 to 0 °C, July mean 18–20 °C) and annual precipitation of 400–600 mm. In contrast, the highest areas in the mountains have a mean annual temperature of about 1–2 °C (January mean about −7 to −6 °C, July mean about 8–10 °C) and annual precipitation of 1200–1400 mm (Tolasz et al. 2007).

Precipitation is brought predominantly by frontal systems originating over the northern part of the Atlantic Ocean. Cool summers and mild wet winters occur in years when there is a large difference in atmospheric pressure between the two major stable pressure areas in the North Atlantic Oscillation (the Icelandic Low and Azores High), which strengthens the westerly wind flow. In contrast, in the years with little difference in pressure westerlies are weaker, which results in hot and dry summers and winters with distinct frosty periods caused by the easy penetration of cold air from North-eastern Europe. These events account for the absence in the Czech Republic of several frost-sensitive species that are widespread at the same latitudes in Western Europe.

The average number of days with snow cover each year ranges from about 30 in the lowlands in Bohemia and southern Moravia up to more than 160 days on the summits of the highest mountains. The average seasonal maximum snow cover depth is less than 15 cm in the lowlands and more than 150 cm in the mountains (Tolasz et al. 2007). In the lowlands snow tends to melt soon after it falls, leaving the landscape without snow cover for most of the winter. Since the periods with temperatures below and above 0 °C alternate frequently in winter, plants in the lowlands are often not protected from frost by an insulating layer of snow.

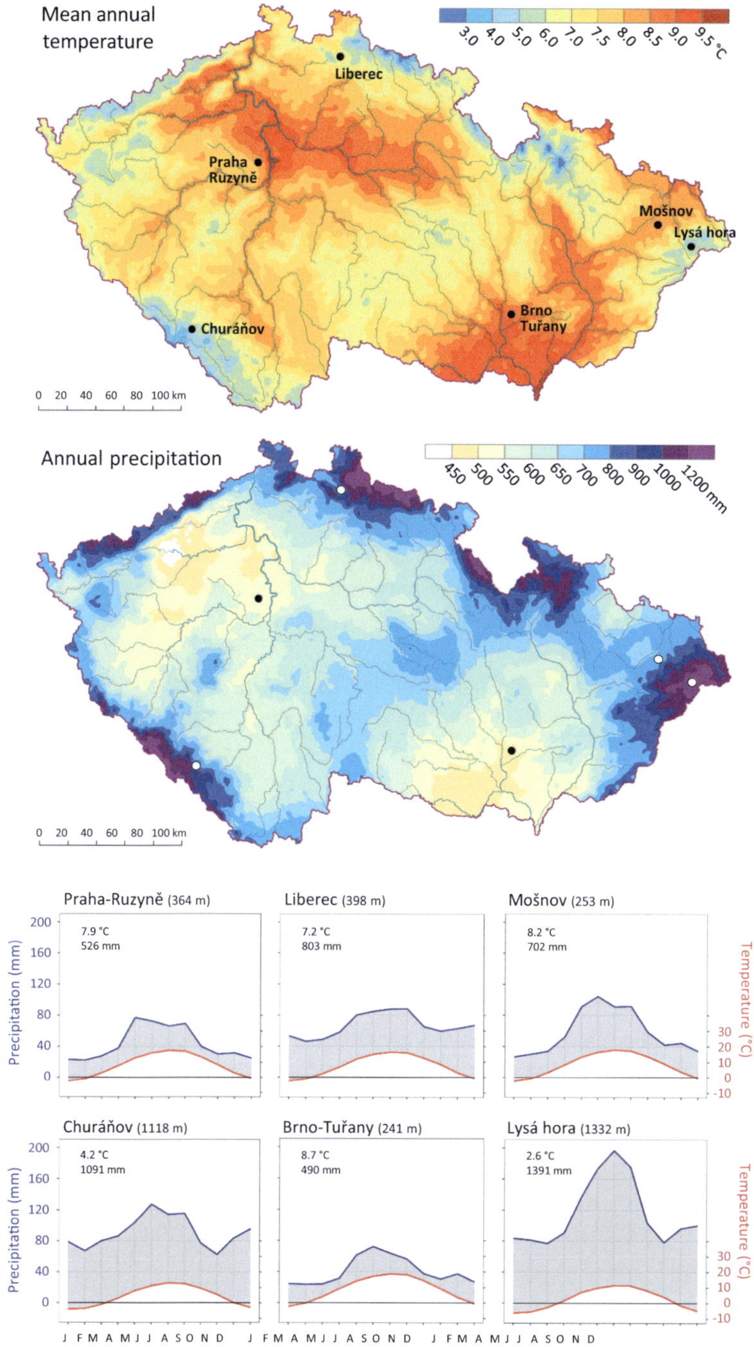


**Fig. 1.11** Mean annual temperature and annual precipitation in the Czech Republic (based on the source data for the Climate atlas of Czechia, Tolasz et al. 2007, provided by the Czech Hydrometeorological Institute) and climate diagrams for selected stations based on the measurements from 1961–1990 (data from www.chmu.cz). Locations of the climate stations are shown in the maps

The distribution of precipitation is strongly dependent on topography. The north-western slopes in the mountain systems, facing the moisture-laden westerlies, receive relatively high amounts of precipitation, whereas lowland areas in the lee of these mountain systems are dry due to the rain-shadow effect. The driest area in the Czech Republic is the middle Ohře valley in northern Bohemia, located in the lee of the Krušné hory Mts, which receives less than 450 mm of rain per year, which corresponds to a forest-steppe climate at this latitude. Relatively low amounts of precipitation (below 550 mm) are also typical of the lowland and low hilly areas adjacent to the south and south-east, up to the city of Plzeň in western Bohemia and Prague in central Bohemia (Fig. 1.11: Praha-Ruzyně). Another dry area (with less than 500 mm of precipitation per year) is the lowland part of southern Moravia (Fig. 1.11: Brno-Tuřany), situated to the south-east of the Bohemian-Moravian Highlands. This is not a very high (837 m a.s.l.), but extensive highland system, which also creates a rain-shadow effect. All of these dry areas host numerous species of the continental steppe, some of them at the western limit of their broad temperate Eurasian distribution.

Another lowland area in the Czech Republic, the Ostrava Basin in north-eastern Moravia and adjacent margins of the Silesian Lowlands (Fig. 1.11: Mošnov), is much wetter than the lowlands in northern Bohemia and southern Moravia because of its location at the foot of a windward front ridge in the Moravskoslezské Beskydy Mts. Being relatively warm, this area receives 650–800 mm of precipitation per year, which makes it an exception from the general negative correlation between temperature and precipitation across the remaining area of the Czech Republic.

## 1.4 Hydrology

The Czech Republic is situated on the continental divide of Europe, with the Labe (Elbe) River draining most of the area of Bohemia into the North Sea, the Odra (Oder) River draining Silesia and a part of northern Moravia into the Baltic Sea and the Morava (March) River draining most of Moravia through the Danube to the Black Sea. All the major Czech rivers originate in this country or in the borderland mountains. For most Czech rivers the peaks in discharge occur in March and April due to snow melt in the mountains, while the period of low discharge generally lasts from June to November. Only in the rivers belonging to the Odra catchment, which receives a higher mean precipitation than other catchment areas in the country, the period of relatively high discharge lasts until July (Tyl 2009; Vlnas 2009).

Natural lakes are very rare in the Czech Republic, except for small lakes in river floodplains and mire complexes (Vondrák et al. 2015). The general lack of lakes is in marked contrast to that in nearby areas that were glaciated in the Pleistocene including the Alps, their foothills, and northern German and Polish lowlands. Only five lakes of glacial origin occur in cirques in the Šumava Mts, the largest being

**Fig. 1.12** Examples of water bodies in the Czech Republic: (**a**) Plešné Lake in a glacial cirque in the Šumava Mts is one of six recently still existing glacial lakes in the Czech Republic. Natural spruce forest around this lake was damaged by a bark beetle outbreak; (**b**) The shore of Rožmberk, the largest Czech fishpond, built in the sixteenth century, is fringed with marshes and willow and alder carrs: Třeboň, southern Bohemia

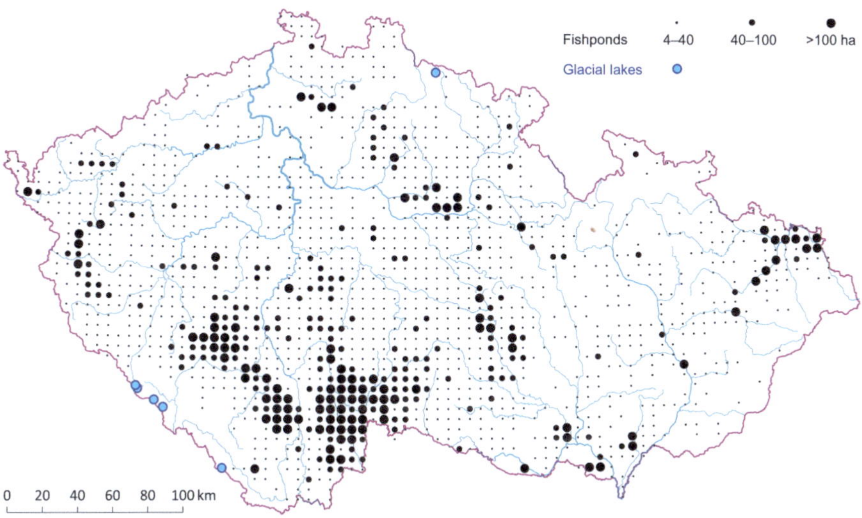


**Fig. 1.13** Distribution of fishponds and glacial lakes. Dot sizes for fishponds indicate summed hectarage of fishponds per grid square of 3 × 5 geographical minutes. Blue dots for glacial lakes indicate their exact position

Černé Lake (18 ha), followed by the lakes Čertovo, Plešné (Fig. 1.12a), Prášilské and Laka (Fig. 1.13). The Krkonoše Mts have large glacial lakes on the Polish side, whereas a large lake on the Czech side of these mountains disappeared due to terrestrialization in the Early Holocene (Engel et al. 2010). Currently only one very small glacial lake (~470 m$^2$) remains on the Czech side (Engel et al. 2003). Several natural lakes occurred also in the lowlands, most of them originating in the Late Glacial period. Saline lakes existed in southern Moravia near the villages Měnín,

Čejč and Kobylí. These lowland lakes vanished due to either natural terrestrialization in the Holocene or draining by humans, mainly in the nineteenth century (Břízová 2009; Jankovská and Pokorný 2013; Vondrák et al. 2015).

A typical feature of the Czech landscape is fishponds, shallow water reservoirs built from the eleventh century onwards for fish farming. The main species farmed has always been common carp (*Cyprinus carpio*), which requires warm and shallow water. Therefore, most of the fishponds were built in the lowlands or mid-altitude basins and are on average about 2 m deep. In the heydays of fish farming at the turn of the seventeenth century there were about 70,000 fishponds in the Bohemian lands. Later on, many of them, especially those in the lowlands, were drained to obtain agricultural land, and currently there are about 25,000 fishponds (Čítek et al. 1998). The largest extant fishpond, Rožmberk near Třeboň in southern Bohemia, has an area of almost 500 ha (Fig. 1.12b). The fishponds are concentrated mainly in the basins of southern Bohemia around Blatná, České Budějovice and Třeboň, in the Ostrava Basin in north-eastern Moravia and Silesia and in flat areas in the Bohemian-Moravian Highlands (Fig. 1.13). Further artificial lakes were built by damming deep river valleys, most notably on the Vltava River, during the twentieth century, for hydroelectricity production and flood protection. Unlike the fishponds, these water reservoirs are deep, with steep shores that provide few opportunities for the development of wetland habitats. A large system of shallow artificial lakes with a total area of 32 $km^2$ was built on the Dyje River in the lowland area of southern Moravia in the 1970s–1980s, destroying unique floodplain ecosystems.

## 1.5 Human Effects on Land-Cover

The territory of the current Czech Republic was inhabited by humans already in the Palaeolithic, but more significant effects of humans on land-cover probably only occurred in the Neolithic, which started here ~7600 years before present. Until the Early Middle Ages, human settlements and agriculture were concentrated in the warm and dry lowlands along the lower Ohře and Labe rivers in northern, central and eastern Bohemia, and in the Moravian lowlands along the Dyje, Jihlava, Svratka, Morava and Odra rivers. This old-settlement area occupied less than 30% of the total area of the country (Boháč 1987; Nováková 2009). In some periods, most notably in the Bronze and Iron Ages, isolated areas were settled also at lower altitudes in western or southern Bohemia, namely in the Plzeň region and the Vltava and Otava valleys (Čišecký and Dreslerová in Pokorný 2011). Boháč (1987) estimated the population in the territory of the current Czech Republic at ~100,000 people in the Roman Period and ~680,000 people in the Early Middle Ages (AD 1000).

Long-term agricultural activities and collection of firewood contributed to landscape openness in the old-settlement area, whereas higher altitudes both in the Bohemian Massif and the Carpathians were covered by pristine forest until the Early Middle Ages. A dramatic change occurred between the late twelfth and late fourteenth centuries, when many new villages and towns were founded and the mid-altitude areas were deforested and colonized, partly by a German-speaking population coming from other areas of Central Europe (Klápště 2012). By the end of fourteenth century, the population reached ~3 millions and the previously isolated deforested areas in Bohemia and Moravia were connected (Boháč 1987). Only the highest areas in the mountains remained continuously forested. However, because of wars and plague outbreaks, population numbers subsequently declined and remained below two million between the fifteenth and seventeenth centuries (Fialová 2007).

While the colonization of the higher altitudes on the topographically relatively gentle Bohemian Massif was proceeding by gradual extension of the deforested area, creating a matrix of open land with patches of woodland, in the more rugged landscapes of the higher altitudes of the flysch Carpathians the settlements remained concentrated in the lower parts of the valleys. During the so-called Wallachian colonization in the sixteenth and early seventeenth centuries, a livestock grazing system using mountain summer pastures was introduced in the Moravian Carpathians, resulting in deforestation of the summit areas, which were used as summer pastures for livestock (Futák 2008). However, forest was generally preserved in the belt between the valley bottoms and range summits. The highest mountain ranges in the Bohemian Massif were colonized even later. In the Krkonoše Mts summer or permanent chalets were established near or above the timberline in the seventeenth and eighteenth centuries, supporting mountain pastoralism and hay making (Krahulec et al. 1997). The Šumava Mts, forming the state border between Bohemia and Bavaria, had a very sparse network of settlements even in the Modern Period, with most of them along trade trails. Unlike other mountain areas in Central Europe, it escaped pastoral colonization, probably because of the lack of an alpine zone. Its central part was colonized only in the eighteenth century (Beneš 1996).

Rapid population growth in Bohemia and Moravia started after the end of Thirty Year's War in 1648, continuing until World War II. Increasing deforestation led Empress Maria Theresa to issue the forest regulation orders for Bohemia (1754), Moravia (1756) and Silesia (1769), which prevented forest grazing, leading to separation of forests from grasslands and reversing the trend of forest decline (Nožička 1957). Previous diffuse transitions between forests and grasslands were replaced by sharp borders in the first half of the nineteenth century when the so-called stable cadastre was established for the purpose of the calculation of land taxes, in which strictly defined land-use was assigned to each plot (Brůna et al. 2005). Forests were further markedly changed due the advent of plantation forestry at the turn of nineteenth century, based mainly on monocultures of spruce and pine (*Picea abies* and *Pinus sylvestris*, both native trees of this country; Nožička 1957). In spite of large qualitative changes in forests, their total area and spatial pattern in the country has been relatively stable from the mid-nineteenth century to the present, covering about one third of the territory (Bičík and Krupková 2009a; Fig. 1.14a). Grasslands also changed considerably between mid-eighteenth and mid-nineteenth century.

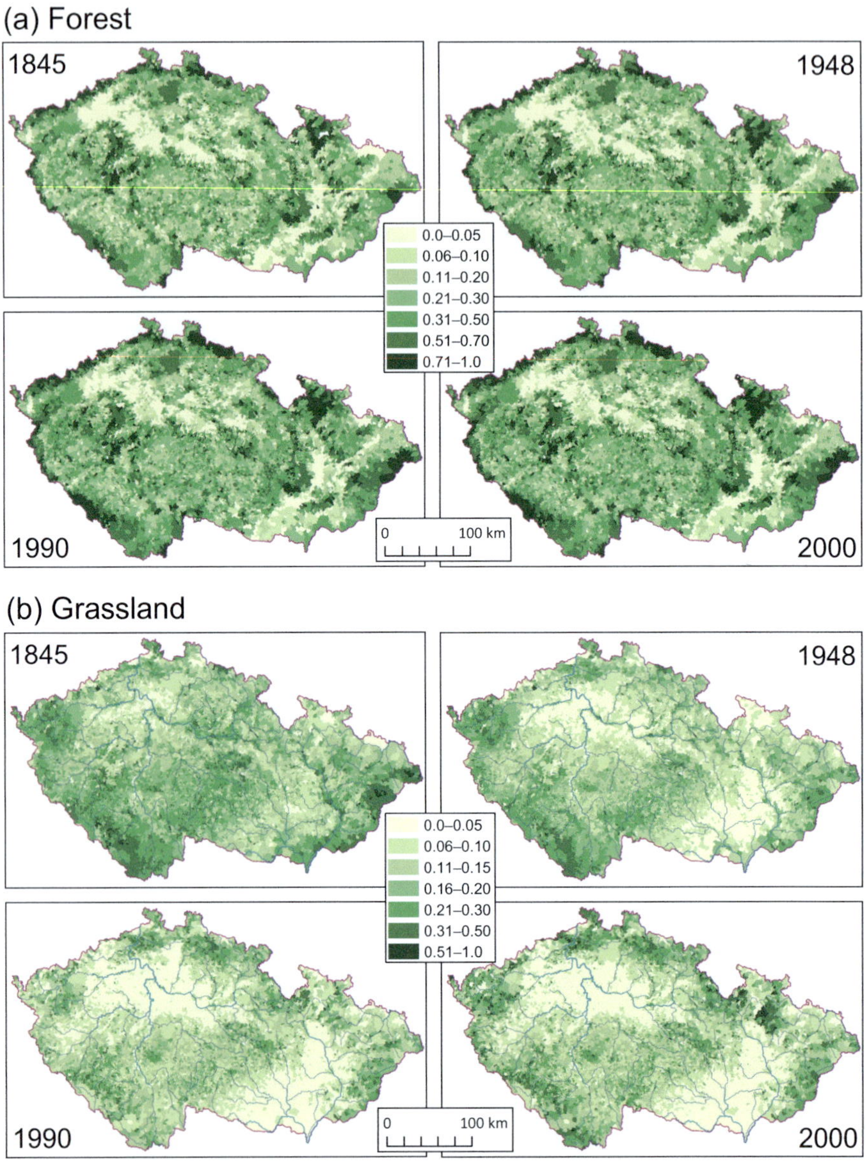


**Fig. 1.14** Changes in the area of forest and grassland between 1845 and 2000. Numbers are proportions of forest and grassland areas in territorial administrative units (based on the data provided by LUCC Czechia, Charles University; http://web.natur.cuni.cz/ksgrrsek/lucc)

The growing population required an increase in agricultural production, which was achieved partly by extension of arable land, partly by intensification through the application of farmyard manure. Livestock was therefore increasingly moved from free ranging to enclosures, and required additional feeding by hay. As a result, grasslands were separated into pastures and hay meadows (Hejcman et al. 2013).

In 1930 the area of the current Czech Republic reached a population of 10.7 million (Czech Statistical Office, http://bit.ly/2pn3jDp), corresponding to 135 people per $km^2$. This was historically the highest count of all population censuses (the 2011 census reported 10.4 million people). The decline to 8.9 million in 1950 was partly due to war casualties and partly due to forced emigration of citizens of German nationality after World War II. Before the war, ethnic Germans were the prevailing nationality in the areas adjacent to the national borders with Germany (including German Silesia, currently in Poland) and Austria. Some of these areas were repopulated by Czechs, others were depopulated and abandoned. Some of them were converted into military training areas, the largest one (332 $km^2$) being in the Doupovské hory Mts in north-western Bohemia. Others became a part of the so-called Iron Curtain, a prohibited strip of land on the border between communist Czechoslovakia and Western Germany or Austria. Large-scale spontaneous succession towards woodland occurred on former arable land, grasslands and even at the sites of abandoned villages in these areas (Kopecký and Vojta 2009).

Huge changes in agricultural landscapes occurred across the whole country in the 1950s as a result of the introduction of collective farming enforced by the communist government. Small private fields were united, forming large uniform areas of arable land that replaced the heterogeneous environment of previous farmland. Agricultural intensification with heavy machinery, artificial fertilizers and pesticides was supported by government subsidies to the cooperative farms. Intensive management was increasingly applied not only to arable land, but also to grasslands. In contrast, remote, poorly accessible or less productive areas of agricultural land were gradually abandoned and left to spontaneous succession towards scrub and forest. Post-war agricultural intensification was associated with a decline in the number of people involved in agriculture (in 2015 less than 3% of Czech employees worked in agriculture and forestry; Czech Statistical Office, http://bit.ly/2pDiLum). Consequently, the heterogeneous agricultural landscape typical of the period between 1850 and 1950, which was maintained by various activities of numerous small land owners, was replaced by a coarse mosaic consisting of large patches with homogenized management. This landscape transformation is associated with the loss of valuable natural or semi-natural habitats, especially grasslands (Bičík and Krupková 2009b; Fig. 1.14b), and a decline in quality and biodiversity of most habitat types. Further changes in land-cover occurred after the break-up of the communist regime in 1989, when both arable farming and animal husbandry became unprofitable in many areas after the cessation of governmental subsidies to cooperative and state-run farms. Large areas of arable land, especially at higher altitudes, were converted into grasslands, but at the same time many grasslands were abandoned because of the decrease in demand for pasture land and hay (Bičík and Krupková 2009b; Fig. 1.14b).

**Acknowledgements** I thank Jiří Danihelka, Handrij Härtel, Zdeněk Kaplan, Zdeněk Palice and Marinus Werger for comments on this text, Ondřej Hájek for preparing the maps, Kryštof Chytrý for preparing an R script for climate diagrams and technical processing of photographic plates, Pavel Dřevojan and Zuzana Sixtová for editing the list of references, Tony Dixon for English proofreading and Irena Axmanová for her help with proof correcting. The work was supported by the Czech Science Foundation (project no. 14-36079G, Centre of Excellence Pladias).

# References

Bajer A, Houška J (2009) Minerální síla půd [Mineral power of soils]. In: Hrnčiarová T, Mackovčin P, Zvara I (eds) Atlas krajiny České republiky [Landscape atlas of the Czech Republic]. Ministerstvo životního prostředí České republiky, Praha and Výzkumný ústav Silva Taroucy pro krajinu a okrasné zahradnictví, Průhonice, p 139

Beneš J (1996) The synantropic landscape history of the Šumava Mountains (Czech side). Silva Gabreta 1:237–241

Bičík I, Krupková L (2009a) Změna výměry trvalých travních porostů (1845–2000) [Change of permanent grassland area (1845–2000)]. In: Hrnčiarová T, Mackovčin P, Zvara I (eds) Atlas krajiny České republiky [Landscape atlas of the Czech Republic]. Ministerstvo životního prostředíČeské republiky, Praha and Výzkumný ústav Silva Taroucy pro krajinu a okrasné zahradnictví, Průhonice, p 88

Bičík I, Krupková L (2009b) Změna výměry lesních ploch (1845–2000) [Change of forest land area (1845–2000)]. In: Hrnčiarová T, Mackovčin P, Zvara I (eds) Atlas krajiny České republiky [Landscape atlas of the Czech Republic]. Ministerstvo životního prostředí České republiky, Praha and Výzkumný ústav Silva Taroucy pro krajinu a okrasné zahradnictví, Průhonice, p 89

Boháč Z (1987) Postup osídlení a demografický vývoj českých zemí do 15. století [Settlement expansion and demographic development of the Czech Lands until the 15th century]. Historická demografie 12:59–88

Břízová E (2009) Quaternary environmental history of the Čejčské Lake (S. Moravia, Czech Republic). Bulletin of Geosciences 84:637–652

Brůna V, Křováková K, Nedbal V (2005) Stabilní katastr jako zdroj informací o krajině [Stable cadastre as a source of landscape information]. Historická geografie 33:397–409

Cháb J, Stráník Z, Eliáš M (2007) Geologická mapa České republiky 1: 500 000 [Geological map of the Czech Republic 1: 500 000]. Česká geologická služba, Praha

Chlupáč I, Brzobohatý R, Kovanda J, Stráník Z (2011) Geologická minulost České republiky [Geological history of the Czech Republic], 2nd edn. Academia, Praha

Cilek V (2001) The loess deposits of the Bohemian Massif: silt provenance, palaeometeorology and loessification processes. Quaternary International 76–77:123–128

Čítek J, Krupauer V, Kubů F (1998) Rybnikářství [Fishpond management], 2nd edn. Informatorium, Praha

Engel Z, Šobr M, Křížek M (2003) Mechové jezírko v Krkonoších – fyzickogeografická studie [Mechové jezírko Lake in the Krkonoše Mts – a physico-geographical study]. Opera Corcontica 40:201–207

Engel Z, Nývlt D, Křížek M, Treml V, Jankovská V, Lisá L (2010) Sedimentary evidence of landscape and climate history since the end of MIS 3 in the Krkonoše Mountains, Czech Republic. Quaternary Science Reviews 29:913–927

Fialová L (2007) Obyvatelstvo České republiky v dlouhodobé perspektivě (1. díl) [Population of the Czech Republic in a long-term perspective]. Geografické rozhledy 17:26–27

Futák P (2008) Vývoj osídlení [History of settlement]. In: Jongepierová I (ed) Louky Bílých Karpat [Grasslands of the White Carpathian mountains]. ZO ČSOP Bílé Karpaty, Veselí nad Moravou, pp 29–37

Härtel H, Cílek V, Herben T, Jackson A, Williams R (eds) (2007) Sandstone landscapes. Academia, Praha
Hejcman M, Hejcmanová P, Pavlů V, Beneš J (2013) Origin and history of grasslands in Central Europe – a review. Grass and Forage Science 68:345–363
IUSS Working Group WRB (2015) World reference base for soil resources 2014, update 2015. International soil classification system for naming soils and creating legends for soil maps. FAO, Rome
Jankovská V, Pokorný P (2013) Reevaluation of the palaeoenvironmental record of the former Komořanské jezero lake: late-glacial and Holocene palaeolimnology and vegetation development in north-western Bohemia, Czech Republic. Preslia 85:265–287
Klápště J (2012) The Czech Lands in medieval transformation. Brill, Leiden
Kopecký M, Vojta J (2009) Land use legacies in post-agricultural forests in the Doupovské Mountains, Czech Republic. Applied Vegetation Science 12:251–260
Kozák J, Němeček J (2010) Soil atlas of the Czech Republic. Czech University of Life Sciences, Prague
Krahulec F, Blažková D, Balátová-Tuláčková E, Štursa J, Pecháčková S, Fabšičová M (1997) Louky Krkonoš: rostlinná společenstva a jejich dynamika [Grasslands of the Krkonoše Mountains: plant communities and their dynamics]. Opera Corcontica 33(1996):1–250
Nováková B (2009) Krajina ve 12. století [Landscape in the 12th century]. In: Hrnčiarová T, Mackovčin P, Zvara I (eds) Atlas krajiny České republiky [Landscape atlas of the Czech Republic]. Ministerstvo životního prostředí České republiky, Praha and Výzkumný ústav Silva Taroucy pro krajinu a okrasné zahradnictví, Průhonice, pp 74–75
Nožička J (1957) Přehled vývoje našich lesů [Historical overview of our forests]. Státní zemědělské nakladatelství, Praha
Pánek T, Hradecký J (eds) (2016) Landscapes and landforms of the Czech Republic. Springer, Cham
Peel MC, Finlayson BL, McMahon TA (2007) Updated world map of the Köppen-Geiger climate classification. Hydrology and Earth System Sciences 11:1633–1644
Pokorný P (2011) Neklidné časy. Kapitoly ze společných dějin přírody a lidí [Unstable times. Chapters from the common history of nature and humans]. Dokořán, Praha
Rivas-Martínez S, Penas Á, Díaz TE (2004) Bioclimatic map of Europe. Bioclimates. University of León, León
Tolasz R, Míková T, Valeriánová A, Voženílek V (eds) (2007) Atlas podnebí Česka [Climate atlas of Czechia]. Český hydrometeorologický ústav/Univerzita Palackého v Olomouci, Praha/ Olomouc
Tyl R (2009) Režim průměrných měsíčních průtoků (1961–2005) [Regime of average monthly discharges (1961–2005)]. In: Hrnčiarová T, Mackovčin P, Zvara I (eds) Atlas krajiny České republiky [Landscape atlas of the Czech Republic]. Ministerstvo životního prostředí České republiky, Praha and Výzkumný ústav Silva Taroucy pro krajinu a okrasné zahradnictví, Průhonice, p 128
Vavříček D, Pancová Šimková P (2014) Atlas lesních půd ČR [Atlas of forest soils of the Czech Republic]. Mendelova univerzita v Brně, Brno
Vlnas R (2009) Hydrologická bilance (1980–2005) [Hydrological balance (1980–2005)]. In: Hrnčiarová T, Mackovčin P, Zvara I (eds) Atlas krajiny České republiky [Landscape atlas of the Czech Republic]. Ministerstvo životního prostředí České republiky, Praha and Výzkumný ústav Silva Taroucy pro krajinu a okrasné zahradnictví, Průhonice, pp 126–127
Vondrák D, Prach J, Houfková P (2015) Sedimenty postglaciálních jezer v ČR – unikátní přírodní archívy (českou) limnologií přehlížené [Sediments of postglacial lakes in the Czech Republic – unique natural archives overlooked by (Czech) limnology]. In: Rádková V, Bojková J (eds) XVII. konference České limnologické společnosti a Slovenskej limnologickej spoločnosti Voda – věc veřejná – Sborník příspěvků. Masarykova univerzita, Brno, pp 162–167
Vysloužilová B, Danková L, Ertlen D, Novák J, Schwartz D, Šefrna L, Delhon C, Berger JF (2014) Vegetation history of chernozems in the Czech Republic. Vegetation History and Archaeobotany 23(Suppl 1):S97–S108

# Chapter 2
# History of Botanical Research in the Czech Republic

**Jiří Danihelka, Milan Chytrý, Jan Kučera, and Zdeněk Palice**

**Abstract** The history of non-experimental scientific botany, including research on the flora of vascular plants, bryophytes, lichens and plant communities, is reviewed for the territory that is now the Czech Republic. This review is organized chronologically, starting from the Middle Ages. For each period it describes the development of universities, research institutions and scientific societies and outlines the developments in various branches of botany.

## 2.1 Introduction

There are several literature sources that describe the history of botany in the current territory of the Czech Republic, mainly in Czech or German, starting with the early attempts by d'Elvert (1868) for Moravia and Maiwald (1904) for Bohemia. There is an excellent overview of the history of botany in the context of other life sciences in the monograph Life Sciences in the Czech Lands 1750–1850 (Janko 1997). The history of field botany for the whole country in Czech, including brief biosketches

J. Danihelka (✉)
Department of Botany and Zoology, Masaryk University,
Kotlářská 2, 611 37 Brno, Czech Republic

Institute of Botany, The Czech Academy of Sciences, 252 43 Průhonice, Czech Republic
e-mail: danihel@sci.muni.cz

M. Chytrý
Department of Botany and Zoology, Masaryk University,
Kotlářská 2, 611 37 Brno, Czech Republic
e-mail: chytry@sci.muni.cz

J. Kučera
Department of Botany, University of South Bohemia,
Branišovská 1760, 370 05 České Budějovice, Czech Republic
e-mail: kucera@prf.jcu.cz

Z. Palice
Institute of Botany, The Czech Academy of Sciences, 252 43 Průhonice, Czech Republic

Department of Botany, Faculty of Science, Charles University, Benátská 2, 128 01 Praha 2, Czech Republic
e-mail: zdenek.palice@ibot.cas.cz



M. Chytrý et al. (eds.), *Flora and Vegetation of the Czech Republic*, Plant and Vegetation 14, DOI 10.1007/978-3-319-63181-3_2

of many botanists, is provided by Klášterský et al. (1982) and a review in English by Krahulec (2012). The history of life sciences within the Czech Academy of Sciences and Arts and then in the Czechoslovak Academy of Sciences was recently outlined by Franc (2015). A description of botany at Prague University up to the 1830s is provided by Krombholz (1837), while the history of botany at Masaryk University in Brno in the twentieth century is described by Vacek and Bureš (2001). The history of the Czech Botanical Society is outlined by Novák (1962) and Hejný et al. (1982). Biographies of important botanists have been regularly published in the journals of the Czech Botanical Society. The history of lichen research in the former Czechoslovakia and the Czech Republic is reviewed by Liška (1992, 2012), while detailed portraits of many lichenologists working in the present territory of the Czech Republic are presented by Halda (2009). An important source of information for the period up to 1952 is the botanical bibliography compiled by Futák and Domin (1960), followed by *Bibliographia botanica čechoslovaca* and for the period 1993–2000 *Bibliographia botanica čechica*. A lichenological bibliography up to 1999 was compiled by Vězda and Liška (1999) and a syntaxonomic bibliography up to 1970 (*Bibliographia syntaxonomica čechoslovaca ad annum 1970*) by a team of the Institute of Botany in Průhonice and published in 20 volumes between 1983 and 1992. This chapter is largely based on these sources.

This review is organized chronologically. Description of each period starts with a brief introduction of the political history followed by information on the development of institutions and societies and an outline of the developments in various branches of botany. Up to the establishment of Czechoslovakia in 1918, the developments in Bohemia, Moravia and Silesia are explained separately as there was no formal administrative or political connection between these historical lands apart from being part of the Habsburg Monarchy or Austrian Empire. For completeness, we describe the history up to the present, but we are aware that it is too early to make any serious conclusions about the historical importance of the developments in recent decades.

## 2.2 The Middle Ages and Renaissance

The present territory of the Czech Republic consists of three historical lands (see Fig. 1.1 in Chap. 1, this book): Kingdom of Bohemia (*Království české* or *Čechy* in Czech, *Böhmen* in German) in the west with Prague (*Praha*, *Prag*) as its capital city, Margraviate of Moravia (*Morava*, *Mähren*) in the east with Olomouc (*Olmütz*) and from the seventeenth century Brno (*Brünn*) as its capital, and Silesia (*Slezsko*, *Schlesien*, *Śląsk* in Polish) in the north-east with the capital in Wrocław (*Breslau*; now in Poland) and later a local administrative centre in Opava (*Troppau*). All these three lands, together with Upper and Lower Lusatia (now in Germany and Poland), constituted the Lands of the Bohemian Crown (Bohemian lands or less precisely Czech lands), a political entity established in the mid-fourteenth century by Holy Roman Emperor Charles IV. At that time, Bohemia and Moravia had a

Czech-speaking majority. However, as early as the late twelfth century Bohemian kings invited German-speaking colonists to settle in peripheral, less populated or previously unsettled parts of their lands and also in towns. In 1347–1349 Charles University (*Universitas Pragensis* or *Carolina*), the first university in Central Europe, was established in Prague following the model of the universities of Paris and Bologna. For more than four centuries this university remained the only scientific institution in the kingdom. However, it lost its international importance in the fifteenth century after becoming the chief doctrinal authority of the Hussite movement.

In 1526, the Bohemian lands became part of the Habsburg Monarchy, i.e. an assemblage of countries and provinces ruled by the junior branch of the House of Habsburg and between 1780 and 1918 by the House of Habsburg-Lorraine. Emperor Ferdinand invited the Jesuits to Prague, who opened their own academy, the Clementinum (in 1562), and another public university in Olomouc, then the capital of Moravia (in 1573). Under Rudolf II (Emperor 1576–1612) Prague replaced Vienna as the capital of the Habsburg Monarchy and the city became an important centre of arts and sciences. The Thirty Years' War started in 1618 with the rebellion of Bohemian non-Catholic estates against King Ferdinand II. It was defeated two years later and non-Catholic nobility, clergy and many educated town dwellers had either to convert to Catholicism or to leave Bohemia and Moravia. German was introduced as the language of administration with equal rights with (but effectively superior to) Czech. In 1654, the Carolinum and Clementinum were merged by an imperial decree to create the Charles-Ferdinand University (*Universitas Carolo-Ferdinandea*).

The Renaissance period witnessed a boom in interest in natural history including botany. The earliest printed herbal in Czech, compiled by Jan Černý, a physician in Litomyšl, was published in Nuremberg in 1517. It contains, among others, a record of *Angelica archangelica* from the Krkonoše Mts. Pietro Andrea Mattioli (1501–1577), born in Siena, arrived in Prague in 1554, where he served as a physician at the Habsburg court. In 1563 he travelled to the Krkonoše Mts, where he discovered *Allium victorialis* and *Geum montanum* (Fig. 2.1a; Matthiolus 1565). His famous herbal, *Petri Andreae Matthioli Senensis medici, Commentarii in sex libros Pedacii Dioscoridis Anazarbei de Medica materia*, was translated into Czech and adapted to a Central European flora by the Prague physician Tadeáš Hájek z Hájku. This book, illustrated with numerous woodcuts, was printed in Prague in 1562. An augmented edition followed in 1596.

Herbarium specimens were being produced already in the sixteenth century, which can be documented by the fact that the description of *Cortusa matthioli* in one of the editions of Mattioli's herbal is based on specimens sent to him to Prague from Italy. Probably the earliest herbarium specimens from Bohemia that currently still exist were collected by Joachim Burser (1583–1639), who worked in 1615–1624 as a physician in the town of Annaberg on the Saxon side of the Krušné hory Mts. He made trips to north-western Bohemia and collected plants for instance in the surroundings of Jáchymov, Kadaň, Žatec and Karlovy Vary. His *Hortus siccus*, now at Uppsala University, was often consulted by Linnaeus when working on *Species plantarum* (Speta 2000).

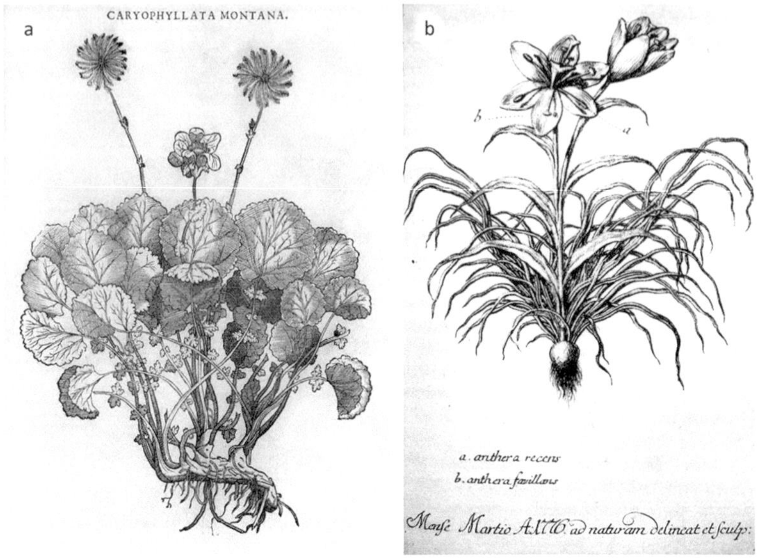


**Fig. 2.1** Examples of early botanical illustrations from Bohemia: (**a**) *Geum montanum* (*Caryophyllata montana*) discovered by Pietro A. Mattioli on his trip to the Krkonoše Mts in 1563 (Matthiolus 1565); (**b**) *Gagea bohemica* (*Ornithogalum bohemicum*) as depicted by Johann Z. Zauschner in the protologue (Zauschner 1776)

## 2.3 The Age of Enlightenment: Second Half of the Eighteenth Century

The reign of Empress Maria Theresa (1740–1780) was affected by the war of the Austrian succession, in which most of the Duchies of Silesia were lost to Prussia in 1742. The empress and her son Joseph II (1780–1790) introduced various administrative and economic reforms, including the abolition of serfdom, religious freedom for non-Catholic Christians, more rights for Jews, compulsory education for children and other improvements in the education system. One of the main goals of university studies in the Habsburg Monarchy became the education of efficient bureaucrats for various positions in the state administration.

### *2.3.1 Bohemia*

Until the mid-eighteenth century newly appointed professors at the Faculty of Medicine of Charles-Ferdinand University started as teachers of theoretical disciplines and advanced to teaching practical subjects with resignations or deaths of their older colleagues. This system required all professors to be polymaths but made any research difficult. By the decision of the government in Vienna this system was replaced in 1747 by specialized professorships. Two years later Joseph J. Scotti von Compostella (1722–1794) was appointed as the first specialist professor of botany and *materia medica*, but nothing is known about his achievements in botany; he resigned in 1762. He was followed by Johann B. Bohadsch (Jan Boháč; 1724–1768), who gathered a large collection of minerals, plants and other objects when travelling across Bohemia, but his manuscript on the natural history of the kingdom was not published after his untimely death. The professorship in botany was combined with that in chemistry and in 1775 granted to Joseph G. Mikan (1743–1814), a student of Nicolaus J. Jacquin. He initiated the establishment of a botanical garden in Smíchov near Prague (now part of the city). With two specialized professorships in natural history established at the Faculty of Medicine in 1752 and the Faculty of Arts in 1784, Charles-Ferdinand University had three professional natural historians at the end of the eighteenth century. In contrast, the small University in Olomouc remained rather unimportant in natural history research.

The Scholar Society (*Gelehrte Gesellschaft*, *Společnost učená*) was formed in Prague about 1770 as an informal private association of scholars and scientists, supported by high-ranking members of the kingdom's administration. In 1790 Emperor Leopold II granted the society public status as the Royal Bohemian Society of Sciences (*Königliche böhmische Gesellschaft der Wissenschaften*, *Královská česká společnost nauk*). The society published a journal, *Abhandlungen einer Privat-Gesellschaft in Böhmen*, and established a natural history collection and library. Approximately at the same time the Patriotic-economic societies (*Patriotisch-oekonomische Gesellschaft*, *Vlastenecká hospodářská společnost*) were established in the individual crown lands by governmental decisions aiming at an improvement in agricultural production. Despite their applied focus, these societies also improved the knowledge of natural history, especially in Moravia. In Bohemia this society was headed by Count Joseph E. Malabaila von Canal (1745–1826), who established a private park and botanical garden with greenhouses and a lecture hall in Prague. In 1791–1830, lectures in 'philosophical botany' and natural history were regularly given there by, for instance, Franz W. Schmidt (see Schmidt 1790–1792), Johann C. Mikan and Ignaz F. Tausch.

The research into this country's flora by the members of the Scholar Society started in the 1770s. In 1776 Johann B. Zauschner, later a professor of natural history at the Faculty of Arts, published in the society's journal a description of *Ornithogalum bohemicum* (now *Gagea bohemica*; Fig. 2.1b; see also Kirschner et al. 2007). To our knowledge, this is the first description of a species native to Bohemia by a Bohemian botanist. In 1786 the society organized a research expedition to the Krkonoše Mts (Jeník 1986). One of the four expedition members was the botanist Thaddaeus (Tadeáš) P. X. Haenke (1761–1816/1817; Fig. 2.2a). The scientific report published five years later in Dresden (Haenke 1791) includes, among others, the description of *Poa laxa* as a species new to science (Kirschner et al. 2007) and some of the earliest scientific records of bryophytes and lichens from Bohemia. Several records of lichens from Bohemia in the doctoral thesis of the German naturalist Alexander von Humboldt (Humboldt 1793) also belong to this period. Perhaps the first lichen record in a taxonomic study concerns the rare cyanolichen *Polychidium muscicola* collected near Karlovy Vary (Bernhardi 1799).

**Fig. 2.2** Remarkable Bohemian botanists from the late eighteenth to the early twentieth century: (**a**) Thaddaeus P. X. Haenke (1761–1816/1817); (**b**) Jan S. Presl (1791–1849); (**c**) Karel B. Presl (1794–1852); (**d**) Philipp M. Opiz (1787–1858); (**e**) Josephine Kablik (also Kablíková; 1787–1863); (**f**) Ladislav J. Čelakovský (1834–1902); (**g**) Josef Velenovský (1858–1949); (**h**) Josef Rohlena (1874–1944); (**i**) Karel Domin (1882–1953). Sources: Viniklář (1931; a–d), Pluskal (1849; e), Polívka (1904; f), Časopis Národního musea (1926; g) and archive of the Czech Botanical Society (h, i)

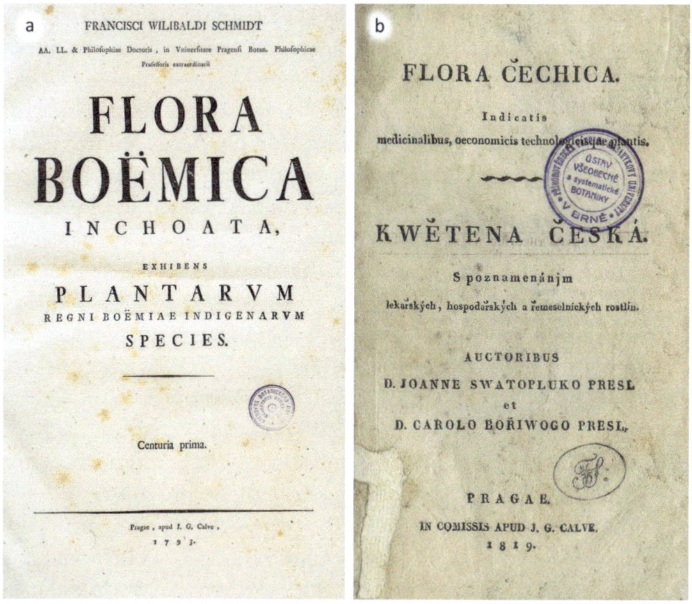


**Fig. 2.3** Title pages of early Bohemian floras: (**a**) The first centuria of *Flora boëmica inchoata* (Schmidt 1793): due to untimely death of its author only four hundred species were described in four volumes published in 1793 and 1794; (**b**) *Flora Čechica* (Presl and Presl 1819), the first complete flora of Bohemia

Franz W. Schmidt (1764–1796), a botanist and lecturer at Count Canal's garden, started publishing the first kingdom's flora in the mid-1790s. Each of the four published volumes of the *Flora Boëmica inchoata, exhibens plantarum regni Boëmiae indigenarum species* (Schmidt 1793, 1794; Fig. 2.3a) contains 100 species regardless of the Linnaean system. Schmidt, born to a family of painters and himself an excellent draughtsman, prepared 244 plates for centuriae 1 and 2, but these were never published. Unfortunately, he died at the age of 32 and Bohemia entered the new century without a complete flora.

## 2.3.2 Moravia and Silesia

In the 1780s a private natural history society was also established in Brno, involving high-ranking members of the Moravian provincial administration. In 1794 the society was reorganized and renamed *Mährische Gesellschaft der Natur- und Vaterlandskunde*. Its main tasks included research, developing a natural history collection and organizing meetings with lectures by its members.

From 1763 the surgeon Norbert Boccius (1729–1806), member of the Order of Hospitallers, lived in Valtice (*Feldsberg*) in southern Moravia (at that time a part of Lower Austria). Apart from his professional interest in medicinal plants, he initiated the production of fourteen volumes of *Liber regni vegetabilis*, known also as the Codex Liechtenstein (Lack 2000). It contains 2748 plates of around 3100 species of plants, including also many native plants of southern Moravia. The plates were painted by the famous botanical illustrators Franz A. Bauer and Ferdinand Bauer, during their early careers. Boccius also discovered *Crambe tataria* near the village of Kurdějov in southern Moravia and sent it to the author of its scientific description (Sebeók 1779).

Probably the earliest preserved herbarium specimens with standard labels from Moravia are those collected by Heinrich Schott (1759–1819), who worked as a gardener of Count J. B. Mitrovský from about 1786 to 1800. Several dozen specimens, collected mainly near Brno, are now preserved in the herbarium of the Moravian Museum in Brno (acronym BRNM) and in Budapest (herbarium BP; Sutorý 1995). However, there was no attempt to produce a flora of the province.

The situation in Silesia was different. The province had a university founded in 1702 in Wrocław as a school of Philosophy and Catholic theology and reorganized in 1811 as a regular university with five faculties. Although Austria had to cede most of the province to Prussia in 1742, the natural history research organized from Wrocław dealt with the whole of Silesia, including the small part that remained in the Habsburg Monarchy. In 1776 the *Flora silesiaca* was published by Count Heinrich G. Mattuschka (1734–1779), followed by an updated version published under a different name three years later (Mattuschka 1776, 1779). Also the four-volume *Flora Silesiae renovata* (Krocker 1787–1823) by the Wrocław physician Anton J. Krocker (1743–1823) was started in this period. Therefore, the smallest province of the Bohemian lands entered the nineteenth century with two complete floras, which was in a marked contrast with the situation in Bohemia and Moravia.

## 2.4 The First Half of the Nineteenth Century

The first years of the century were marked by the Napoleonic wars, bringing frequent changes not only in government policies, but also in educational and scientific institutions. The ideas of the Enlightenment were gradually replaced by policies of Prince Klemens W. Metternich, from 1809 Foreign Minister and in 1821–1848 State Chancellor of the Austrian Empire (formally established in 1804), aimed at fighting liberalism and nationalist movements. In spite of that, the cultural movement of the Czech National Revival started, followed by a gradual improvement of the economic situation of the Czech-speaking population. Initially this cultural and economic emancipation found some support also among the German-speaking compatriots, which lasted until 1848. The conservative polices also affected the education system: in 1818, for instance, natural history was removed from the curricula of secondary schools as a concession to the Catholic Church.

### 2.4.1 Bohemia

At Charles-Ferdinand University the system of professorships remained unchanged (see also Svojtka 2016). At the Faculty of Medicine lectures in botany were given by Johann C. Mikan (until 1826), who took part in the Austrian expedition to Brazil, and Vinzenz F. Kosteletzky. Jan S. Presl was appointed as a professor of natural history in 1820. Lectures in natural history at the Faculty of Arts were given by Karel B. Presl. However, the university's importance in terms of its research into the flora of Bohemia remained small.

A discussion about the need for institutional support for natural history research resulted in 1818 in the establishment of the *Museum Regni Bohemiae*, inspired by a similar institution in France and the recently founded *Joanneum* in Graz. The museum was a private institution supported by Bohemian estates including the head of the kingdom's administration, Count Franz A. Kolowrat. The botanist Count Kaspar M. Sternberg (1761–1838), founder of modern palaeobotany (Sternberg 1820), became the first President of the Society of the Patriotic Museum, which also included some other botanists. The museum focused on assembling diverse collections, thus becoming the first biological record centre in the kingdom. The society started publishing a journal in German in 1824 (closed in 1832) and another journal in Czech in 1827, named *Časopis Společnosti Vlastenského museum v Čechách* (Journal of the Patriotic Museum in Bohemia) and nowadays published as the Journal of the National Museum (Prague). The society purchased a part of T. Haenke's collection from America, and K. B. Presl, curator of the natural history collections, participated together with his younger brother J. S. Presl and Count Sternberg in its taxonomic treatment (Presl 1825–1835). Karel B. Presl later turned his attention mainly to the taxonomy of *Lobeliaceae* and pteridophytes with a world-wide scope. Thus, the contribution of the museum to the knowledge of the kingdom's flora was minor during the first decades of its existence.

Another attempt to produce a flora of Bohemia was Johann E. Pohl's *Tentamen florae Bohemiae* (Pohl 1809, 1814). Pohl used the records of his predecessors and information from botanical amateurs working in various parts of this country. The flora is slightly more critical than Schmidt's work but many records are unreliable. In 1817 Pohl left with an Austrian expedition for Brazil and the flora remained unfinished. In 1821 he returned to Vienna, where he worked mainly with the botanical collections from Brazil.

The first complete flora of the kingdom, *Flora Čechica/Kwětena česká* (Presl and Presl 1819; Fig. 2.3b), was published by the Presl brothers, Jan S. Presl (1791–1849; Fig. 2.2b) and Karel B. Presl (1794–1852; Figs. 2.2c and 2.4a). Written in Latin, it has a parallel title and foreword in Czech and includes also Czech plant names, mainly based on vernacular ones. This book includes accounts of 1498 species of phanerogams, and although being far from perfect, it is definitely more critical than earlier attempts. The authors explicitly say that they consider many records published by earlier writers as erroneous. Specimens of most of the species' records still exist in the Charles University herbarium (PRC). The Presls published a supple-

**Fig. 2.4** Herbarium labels and handwritings of some Bohemian botanists: (**a**) Karel B. Presl (1794–1852); (**b**) Philipp M. Opiz (1787–1858); (**c**) Ladislav J. Čelakovský (1834–1902); (**d**) Antonín Weidmann (1850–1915); (**e**) Josef Velenovský (1858–1949); (**f**) Václav Kuťák (1876–1956); (**g**) Karel Domin (1882–1953); (**h**) Günther Beck (1856–1931); (**i**) Karl Preis (1913–1941); (**j**) Josef Rohlena (1874–1944); (**k**) Ivan Klášterský (1901–1979). Credits: Herbaria of Charles University (a, b, e, g, h, i), the Department of Botany of the National Museum in Prague (c, d, j, k) and the Institute of Botany of The Czech Academy of Sciences, Průhonice (f)

ment to their flora three years later, and this was their last contribution to research on the Bohemian flora. Later they turned their attention to their work in the National Museum and on taxonomy. Jan S. Presl also compiled several natural history books in Czech, introducing Czech terminology in biology and chemistry. Some of these books are still important in botanical nomenclature because of the priority principle.

The first identification key for the Bohemian flora, *Clavis analytica in floram Bohemiae phanerogamicam* (Kosteletzky 1824), was written by its author as a doctoral dissertation. However, already at that time Latin was a major obstacle to many potential users. The author later became the first professor of botany at Charles-Ferdinand University.

Another important Bohemian botanist at that time was Ignaz F. Tausch (1793–1848), who worked for a long time as a lecturer in the botanical garden of Count Malabailla (see Tausch 1823, 1825). He travelled to various parts of Bohemia and collected plants, producing several exsiccatae, which he offered for sale, with the largest being the *Herbarium florae bohemicae*. The specimens, marked with handwritten numbers and accompanied by labels written by someone else, are found in many European herbaria. Tausch did not write a flora but published numerous articles in the journal *Flora* (then published in Nuremberg). Tausch's exsiccatae of phanerogams were actually not the earliest in Bohemia, being predated by the collection *Flora bohemica* of Franz X. Sieber distributed in 1814.

In the 1820s Philipp M. Opiz (1787–1858; Figs. 2.2d and 2.4b) became the leading personality doing research on the Bohemian flora. He was neither a university professor or medical doctor or pharmacist as were many of the contemporary botanists, but he served as an official in the administration of state-owned forests in Prague. He travelled to the Krkonoše Mts and some parts of central Bohemia to explore their flora. In 1819 he established a *Pflanzentausch-Anstalt*, a society for the exchange of herbarium specimens, seeds and insects, which continued until his death. This exchange scheme grew to 856 members in 1857, the year before his death. The number of specimens that passed through the exchange society is estimated at more than three million, which gave Opiz an unprecedented knowledge of the Central European flora. In 1823–1828[–1830] Opiz issued a catalogue of specimens offered for exchange, named *Naturalientausch*, which was later modified to a short-lived natural history journal; its 12 volumes are an important source of information on the flora of Bohemia and Central Europe. Numerous taxa were described as new to science in this journal. Opiz's concept of species was very narrow, therefore many taxa described by him are no longer recognized, although some of them were "resurrected" in the twentieth century. Keeping always one specimen for himself, Opiz established a large private herbarium, which is now in the National Museum in Prague (PR). He published only two books, in recent terminology check-lists. The first one, Bohemian phanerogamic and cryptogamic plants (*Böheims phanerogamische und kryptogamische Gewächse*; Opiz 1823) was in German, but with Czech ("provincial") names. The second, a List of plants of the Bohemian flora (*Seznam rostlin květeny české*; Opiz 1852), was in Czech. The first edition is actually a reprint of the botanical chapter in a "statistical-topographical" description of the Kingdom of Bohemia published one year earlier (Ponfikl 1822). Both editions also include lichens and fungi, and were followed by additions published in the journals *Flora* and *Lotos*, respectively. Opiz also contributed many taxonomic parts of the unfinished *Oekonomisch-technische Flora Böhmens* (Berchtold and Seidl 1836–1843). However, his major contribution to the research on the Bohemian flora was the creation of a dense network of collaborators in different parts of the country, including priests, civil servants, lawyers, medical doctors and pharmacists who collected plants in different places, substantially improving the knowledge of the kingdom's flora.

One of the most diligent contributors to Opiz's exchange scheme was the first female Bohemian botanist Josephine Kablik (also Kablíková; 1787–1863; Fig. 2.2e), a wife of a pharmacist in Vrchlabí at the foot of the Krkonoše Mts (Pluskal 1849). She is reported to have collected altogether about 200,000 specimens for Opiz' exchange scheme, mainly in the Krkonoše Mts. She was a member of several societies, including the Society of the Patriotic Museum and the Botanical Society of Regensburg (Ilg 2012).

Opiz, who described several dozen taxa of fungi, lichens and bryophytes as new to science, was not the only botanist working on Bohemian cryptogams in that period. The dissertation of the physician Václav (Wenzel) Mann (1825) is the first book exclusively dealing with the lichens in Bohemia. Exsiccatae of lichens were issued by the Presl brothers in 1812 and by P. M. Opiz as *Flora cryptogamica Boëmiae/Böheims cryptogamische Gewächse* in 1828 and 1829. However, probably the earliest exsiccatae of lichens and bryophytes from Bohemia (and Polish Silesia) were collected already in 1799–1806 in the Krkonoše Mts (*Kryptogamische Gewächse des Riesengebirges*) by Carl Ludwig, a castle gardener in Unięcice (*Meffersdorf*) near Pobiedna in Lower Silesia. The cryptogam flora in the Krkonoše Mts was explored also by the local hobby botanists Karl Mosig and Johann Weigel, both operating from the Silesian side. The earliest validly published lichen name based unequivocally on a specimen from Bohemia (*ad rupes Quartzosas Bohemiae*) is *Lecanora falsaria* (now *Fuscidea cyathoides*), published by E. Acharius in 1810; the specimen was collected by K. Mosig. Several of the type specimens of lichen species that were collected in the Krkonoše Mts are included in the exsiccatae *Lichenen vorzüglich in Schlesien, der Mark und Pommern gesammelt* issued in 1829 by Julius von Flotow, a military officer and amateur lichenologist. A specialized exsiccatae series *Musci bohemici* was issued by A. Poech in 1845. Further records of bryophytes, mainly from the Krkonoše Mts, may be found in the works of Christian G. Nees von Esenbeck. Several dozen species of bryophytes, mostly liverworts, were described and skilfully illustrated by August J. Corda (1809–1849; Corda 1829), from 1835 curator of the zoological collections at the National Museum, otherwise famous as a creator of exquisite mycological illustrations.

### 2.4.2 *Moravia and Silesia*

A provincial museum was established in Brno in 1816–1818, which was also a private institution of the Moravian Estates, supported by their most prominent members. To honour Emperor Francis, the museum was named *Franzens-Museum* (*Františkovo museum*) and had this name until 1900, when it legally became a possession of the Land of Moravia and was renamed the Moravian Museum (*Moravské zemské museum, Mährisches Landesmuseum*). Ten years after its foundation the museum's collections included 7605 herbarium specimens. A small museum affiliated with the local secondary school, *Gymnasialmuseum*, was established in 1814 in Opava, the administrative centre of Austrian Silesia. The current Silesian Museum (*Slezské zemské museum*) in Opava is its direct successor. This museum also had botanical collections from the very beginning.

The Patriotic-economic societies continued to exist, being repeatedly reorganized by the central government. However, only the Moravian *Ackerbaugesellschaft* contributed significantly to the research into the province's flora. Its journal *Mittheilungen der k. k. Mährisch-schlesischen Gesellschaft* was the only journal dealing with natural history in Moravia up until the early 1860s. The most important for the natural history research turned out to be the *Schlesische Gesellschaft für vaterländische Cultur*, established in 1803 in Wrocław, from 1809 known under this name and from 1820 with a section of natural history. Since 1824 the society issued a journal, *Jahresbericht der Schlesischen Gesellschaft für vaterländische Cultur*, in which many botanical reports were published. The society also assembled a large herbarium collection and a library, both now at the University of Wrocław.

Botanical research in Moravia started about two decades later than in Bohemia. Alois Carl (1765–1831), a physician in Brno and Uherské Hradiště, collected plants mainly in south-eastern Moravia from about 1806. However, the undisputable pioneer of Moravian botany was Christian F. Hochstetter (1778–1860; Fig. 2.5a). Born in Stuttgart and educated in theology, he served as a Lutheran minister in Brno from 1817–1824. During this period he visited various places, mainly in the southern half of the province. He produced exsiccatae *Gewächse des Brünner Kreises*, containing probably 500 species. He also collected some lichens for *Lichenes exsiccati* issued by L. Reichenbach and C. Schubert in 1822–1826. Hochstetter established a herbarium of southern Moravian flora and in 1823 issued a list of the plants of Moravia and Austrian Silesia comprising more than 1200 species (Danihelka 2008). After returning to Germany, he published the first summary article on the Moravian flora (Hochstetter 1825), which marks the beginning of modern botanical research in this province. It includes also some records of cryptogams.

The first Moravian flora, *Vorarbeiten zu einer Flora des Mährischen Gouvernements* (Rohrer and Mayer 1835; Fig. 2.6a), was compiled by two amateur botanists, Rudolph Rohrer, an owner of a printing office in Brno, and August Mayer, an administrator of an estate in Silesia. It contains 1346 species of seed plants from Moravia and 1172 from Silesia. It is based on records by earlier authors (mainly Hochstetter and Carl) supplemented by Mayer's observations from Silesia and

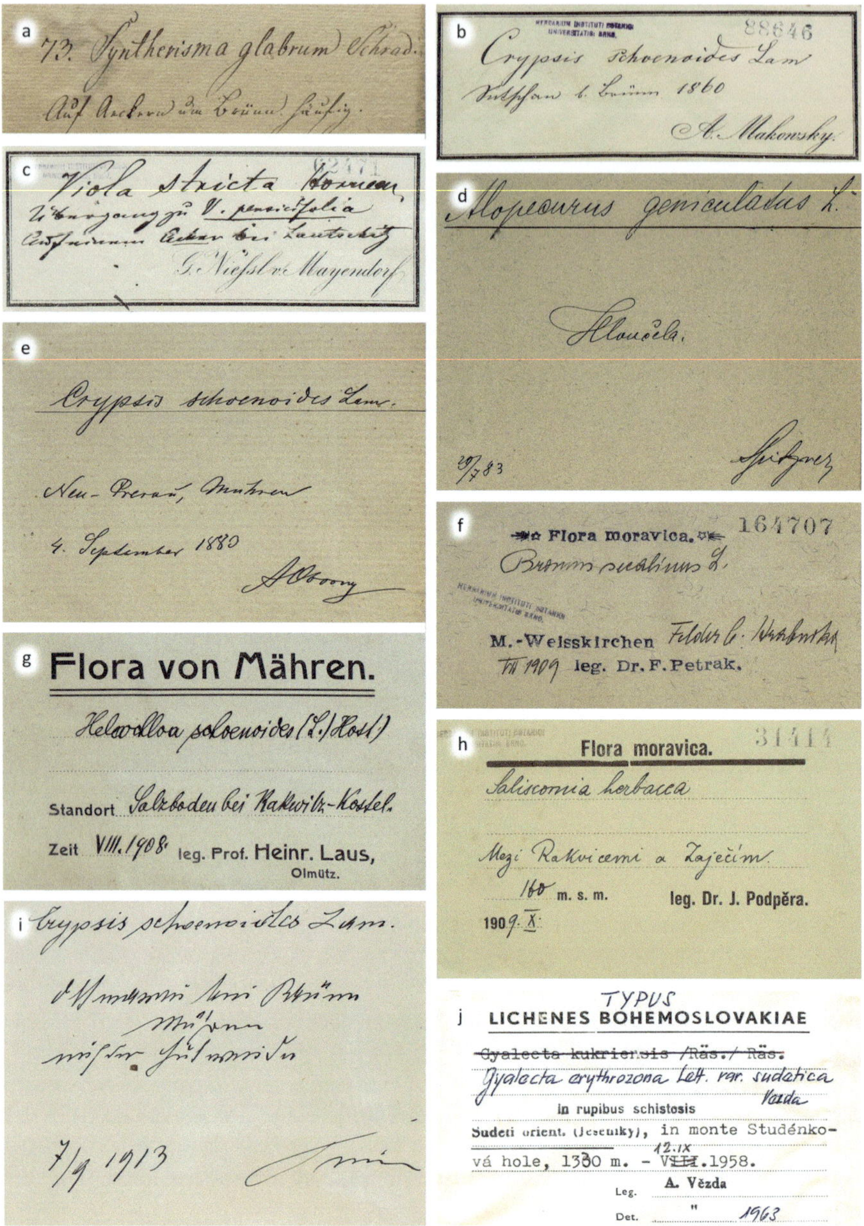


**Fig. 2.5** Herbarium labels and handwritings of some Moravian botanists: (**a**) Christian F. Hochstetter (1778–1860); (**b**) Alexander Makowsky (1833–1908); (**c**) Gustav Niessl von Mayendorf (1839–1919); (**d**) Václav Spitzner (1852–1907); (**e**) Adolf Oborny (1840–1924); (**f**) Franz Petrak (1886–1973); (**g**) Heinrich Laus (1872–1941); (**h**) Josef Podpěra (1878–1954); (**i**) Otto Thenius (dates unknown); (**j**) Antonín Vězda (1920–2008). Credits: Herbaria of Charles University (a), Masaryk University (b–i) and the Institute of Botany of The Czech Academy of Sciences, Průhonice (j)

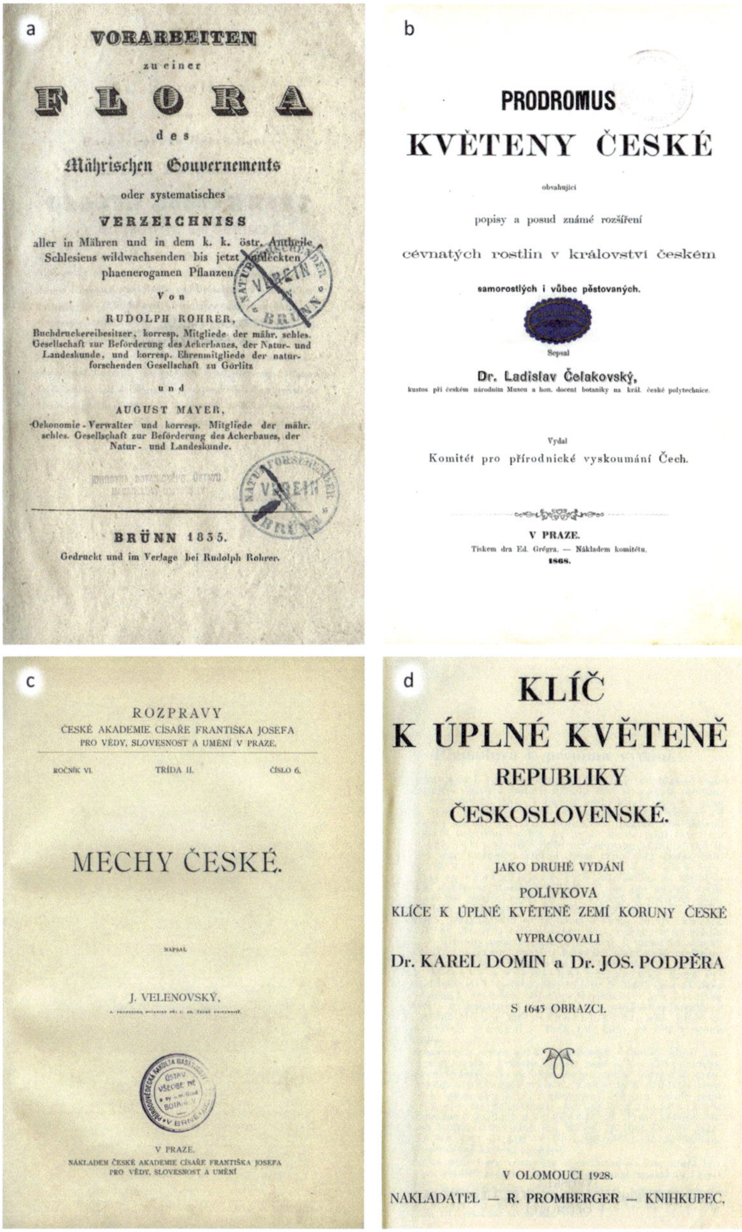


**Fig. 2.6** Title pages of the nineteenth and early twentieth century floras: (**a**) *Vorarbeiten zu einer Flora des Mährischen Gouvernements*, the earliest flora of Moravia and Austrian Silesia (Rohrer and Mayer 1835); (**b**) Czech edition of the first volume of the Prodromus of the Bohemian Flora (*Prodromus květeny české*; Čelakovský 1868), published by the Commission for the Exploration of Natural History of Bohemia; (**c**) Mosses of Bohemia (*Mechy české*; Velenovský 1897), published by the Emperor Francis Joseph's Czech Academy of Sciences, Literature and Arts; (**d**) Key to the Complete Flora of the Czechoslovak Republic (*Klíč k úplné květeně Republiky československé*; Domin and Podpěra 1928)

Rohrer's from the surroundings of Brno. Herbarium specimens for most of the records in the flora may still be found in the Charles University herbarium. Numerous records from the mountains of northern Moravia and Silesia, many of them provided by Franz Mückusch, are erroneous. The publication of this flora stimulated new interest in the flora of both provinces and resulted in additions published mainly by Siegfried Reissek, which increased within five years the number of species known in Moravia up to 1471. The first key to the Moravian flora was written by Joseph C. Schlosser (Schlosser 1843), a Moravian-born physician who graduated from the University of Vienna. The book contains many erroneous records, probably due to the lack of experience of its young author. It was issued when Schlosser was already in Croatia and no longer participated in the exploration of the Moravian flora.

The best explored part of the Bohemian lands was once again the small Austrian Silesia. Apart from the floras mentioned above, *Flora Silesiae* in two volumes was published by Wimmer and Grabowski (1827–1829). Christian F. H. Wimmer (1803–1868), a secondary school teacher in Wrocław, studied in detail mainly the genera *Salix* and *Hieracium*, and prepared three editions of another Silesian flora written in German (Wimmer 1840–1857). Heinrich (also Henryk) E. Grabowski, a pharmacist in Opole, later compiled a flora for Upper Silesia and the Hrubý Jeseník Mts (Grabowski 1843). These Silesian floras contain many important records from the Hrubý Jeseník and Krkonoše Mts.

## 2.5 From the 1848 Revolution to the Foundation of Czechoslovakia in 1918

The revolutionary year 1848 strongly affected the situation in the Austrian Empire. It became clear that the interests of the German- and Czech-speaking communities in Bohemia and Moravia were different. While German-speaking political and intellectual elites wanted a united Germany with a liberal constitution, the newly formed Czech elite preferred social and nationalist demands. The Austro-Hungarian compromise in 1867 created Austria-Hungary, a constitutional union of the Austrian Empire and the Kingdom of Hungary, with two national parliaments and a reduced central government. Although the Bohemian lands did not receive similar autonomy as Hungary, the social-economic and cultural advancement of the Czech community continued, manifested, for instance, by the establishment of numerous secondary schools with Czech as the only language of instruction from the 1860s onwards. After the failed attempt at a Czech-Austrian compromise in 1871 the relationship between Czech and German communities further deteriorated. A strong decrease in the loyalty of Czech community to the imperial dynasty and Austria occurred during WWI. Defections of Czech soldiers were particularly frequent on the Russian front, and Czech and Slovak prisoners of war formed military units that joined the French and Russian armies. The Austrian power in Bohemia and Moravia collapsed on 28 October 1918, when an independent state of Czechoslovakia was proclaimed in Prague.

### 2.5.1 *Institutions and Societies*

The development of life sciences in the Austrian Empire was supported by reforms of universities and secondary schools shaped by Minister Count Leo Thun-Hohenstein in 1849, which effectively introduced academic freedom. Following these reforms, research became an integral part of the activity not only of university professors, but also of secondary school teachers. In the Faculty of Arts of Charles-Ferdinand University the professorship of natural history was replaced by specialized professorships in mineralogy, botany and zoology. Also the importance of the Prague Polytechnical Institute (now Czech Technical University in Prague), formed in 1806, increased with the introduction of studies specializing in forestry and agriculture, and the establishment of professorships in zoology and botany. Olomouc University was dissolved in 1860 but this loss was partly compensated for by the establishment of a technical college in Brno in 1850, which was transformed into a technical university in 1873. The German-speaking amateur botanists Alexander Makowsky and Gustav Niessl von Mayendorf (Fig. 2.5b, c) were affiliated with the latter. The division of Charles-Ferdinand University in 1882 into two parallel German and Czech institutions was finally supported also by the German-speaking politicians, who felt endangered by the growing influence of the Czech majority. This division considerably supported the development of a scientific infrastructure, suddenly almost doubling the number of professional scientists in Bohemia. However, the gap between the Czech and German intellectual elite grew. For example, the herbarium remained at the German institute and because of mutual animosities, it was not used by Czech botanists.

From 1875 the Royal Bohemian Society of Sciences had exclusively Czech-speaking scholars as presidents. It continued to publish a journal, *Sitzungsberichte der Königlichen böhmischen Gesellschaft der Wissenschaften in Prag*, from 1886 with the main title in Czech as *Věstník Královské české společnosti nauk*, and a series of scientific monographs, *Abhandlungen…* (from 1886 with the main title in Czech as *Rozpravy…*), both of which also included botanical studies. In 1890 the Emperor Francis Joseph's Czech Academy of Sciences, Literature and Arts (*Česká Akademie císaře Františka Josefa pro vědy, slovesnost a umění*) was founded as an institution of the Czech-speaking scholars and scientists, with the formal protection of the Emperor. This institution, whose main task was deciding on research, travel and publication grants, was financially supported by Josef Hlávka (1831–1908), a Czech architect, builder and philanthropist. The first 19 ordinary members appointed by the Emperor included one botanist, Ladislav Čelakovský, while another botanist, Josef Velenovský, was elected an extraordinary member. This society supported the publication of numerous scientific monographs, one of the earliest being *Flora bulgarica* (Velenovský 1891), and published a specialist journal, *Rozpravy České akademie císaře Františka Josefa pro vědy, slovesnost a umění*, which contained many botanical studies. In 1891 a similar society, the Association for Fostering German Science, Arts and Literature in Bohemia (*Gesellschaft zur Förderung deutscher Wissenschaft, Kunst und Literatur in Böhmen*), was founded by German-speaking scholars and scientists. This society supported, for instance, the publication of a monograph on the genus *Euphrasia* written by R. Wettstein.

The natural history society *Lotos* was established in Prague in 1848; three years later it started a journal, named also *Lotos*, which later became the flagship journal of the Prague German-speaking university professors and scientists. A similar society, *Naturforschender Verein in Brünn*, was founded in Brno in 1861. It immediately established its own herbarium (now incorporated into the herbarium of Masaryk University; BRNU) and started publishing a journal, *Verhandlungen des Naturforschenden Vereines in Brünn*. This society later became famous because of the membership of Gregor J. Mendel (1822–1884), the discoverer of the rules of heredity, which were published in this journal. Both societies later profiled themselves as exclusively German. Czech natural history societies emerged about two decades later, the first being the Natural History Club of Prague (*Klub přírodovědecký v Praze*), founded in 1869 (Viniklář 1931). Roughly in the same period almost ten local museums were formed in various Bohemian and Moravian cities. The Czech Museum Society in Olomouc and its museum were founded in 1883; this society also published a journal. The natural history museum, *Josefo-Ferdinandeum*, was established there in 1896. Other Czech natural history societies in Moravia were established rather late: in 1898 in Prostějov in central Moravia and in 1904 in Brno, then still with a German-speaking majority. Both societies immediately started publishing their annual bulletins, which also contained local botanical studies, while some of the museums became important centres of local research by amateur naturalists. With a growing specialization in different branches of natural history, a need for a specialized botanical society became clear in the years shortly before WWI. The Czech Botanical Society was established in 1912, and Professor Velenovský was elected its first president. Two years later he resigned, being replaced by young and ambitious Karel Domin. In the same year the Society published the first volume of its yearbook *Preslia*, named in honour of the Presl brothers. The Prague natural history journals *Živa*, established by the prominent physiologist Jan E. Purkyně in 1853, and *Vesmír* (Universe), launched about two decades later, both published up to now with interruptions, were important not only because of publishing some botanical notes and studies in Czech, but also as a platform for discussion among scientists, teachers and amateur researches living in remote parts of this country. An important publication platform became the series of monographs published by the Commission for the Exploration of Natural History in Bohemia (*Komitét pro přírodovědné proskoumání Čech*), which was created by the National Museum and the Patriotic Economic Society in 1864. The monographs were usually issued in both languages.

## *2.5.2 Bohemia*

The most important Czech botanist in the second half of the nineteenth century was Ladislav J. Čelakovský (1834–1902; Figs. 2.2f and 2.4c). After the death of his father František L. Čelakovský, a Czech poet and scholar, he was raised by Professor Purkyně's family and in 1860 he replaced Purkyně's eldest son as curator of botanical collections in the National Museum. With Čelakovský's appointment the museum became the centre of research into the flora of Bohemia. His first major work was the Prodromus of the Bohemian flora (*Prodromusder Flora von Böhmen*), a four-volume flora of Bohemia (three basic volumes and one volume with additions; Čelakovský 1867–1881). Each volume of this second complete Bohemian flora was printed first in German and one or two years later in a modified and improved version also in Czech (*Prodromus květeny české*; Čelakovský 1868–1883; Fig. 2.6b). The Prodromus, whose scientific quality equalled that of contemporary Austrian or German floras, was followed by a series of additions (*Resultate der botanischen Durchforschung Böhmens im Jahre…*), first published in 1882 (Čelakovský 1882) and then almost annually until 1894. Čelakovský also served as the Bohemian officer for the Flora of Germany prepared by P. Ascherson and P. Graebner (Čelakovský 1885). He used a broad concept of species in marked contrast to that used previously by Opiz and his contemporaries. He was a good and critical taxonomist, who removed many erroneous records from the Czech botanical literature. A considerable part of the plant records in the Prodromus is based on herbarium specimens that are still stored in the herbarium of the National Museum, in which he established a Bohemian collection, separated from the general collection. As the Prodromus lacked identification keys, Čelakovský prepared an Analytical flora of Bohemia (*Analytická květena česká*; Čelakovský 1879), published in three editions up to 1897. The geographic scope of the second edition (1887) was enlarged to include Moravia and Austrian Silesia. This booklet was the standard work for Czech-speaking botanists for more than three decades. Čelakovský revived interest in the Bohemian flora not only by publishing these standard handbooks but also by recruiting local contributors in various parts of the country, usually amateur botanists, who sent him plant specimens for revision and provided him with plant records. Some of these collaborators later produced valuable local floras.

Čelakovský became associate professor of botany at the Prague Technical University in 1866. Five years later he became extraordinary professor and in 1880 he was appointed as an assistant to Professor H. Moritz Willkomm at Charles-Ferdinand University. In 1882, when this university was divided, he became the first professor of botany in the Faculty of Arts of the Czech University and held this position for the next 20 years. At the university he focused on comparative plant morphology.

Čelakovský had several talented students, the most prominent being Josef Velenovský, Karel Vandas, Bohumil Němec and Edvin Bayer. However, none of them continued Čelakovský's studies on the flora of Bohemia, preferring to study cryptogams, Balkan flora, or even different branches of biology. It is possible that

Čelakovský felt endangered by the professional growth of his former students (Klášterský et al. 1982; Janko 1997). The rivalry between the professor and his ambitious students became almost a rule in botany in Prague, and the generation change was never smooth. The small size of the scientific community, with only one prestigious professorship and one or two additional second-rank positions at the Technical University, may have been the main reason.

Josef Velenovský (1858–1949; Figs. 2.2g and 2.4e) replaced the deceased Čelakovský as professor of botany in 1902. His research interests included palaeontology, flora of Bulgaria, bryology, mycology and comparative plant morphology, which is also the title of the four-volume monograph published in parallel in German as *Vergleichende Morphologie der Pflanzen* in 1905–1913 and in Czech. He also established a new herbarium at the Czech University.

Botany at the Prague German University had a brilliant head in Richard Wettstein (1863–1931), who was appointed as a professor of botany in 1892 and remained in Prague until 1898 when he replaced his deceased father-in law, Anton Kerner, as a professor of botany at the University of Vienna. His most significant work written in Prague was the Basics of the Geographical-morphological Method of Plant Systematics (*Grundzüge der geographisch-morphologischen Methode der Pflanzensystematik*), a pioneering work of botanical taxonomy (Wettstein 1898). Using this taxonomic approach and applying the concept of seasonal dimorphism, he wrote a monograph on the genus *Euphrasia* (Wettstein 1896). The next professor of botany at the German University in Prague was Günther Beck (1856–1931; Fig. 2.4h), who remained active until 1921. Already in the 1890s he wrote a detailed and critical flora for Lower Austria (Beck 1890–1893). In Prague he continued his studies on the flora of the western Balkan Peninsula, publishing a monograph on the vegetation in this area and then in three volumes the Flora of Bosnia and Hercegovina and the District of Novi Pazar (*Flora Bosne, Hercegovine i Novopazarskog sandžaka*; Beck 1903–1927). However, the participation of the German University in the research on the flora of Bohemia was very limited.

A good example of the many local floras written under Čelakovský's influence is the flora of the Cidlina and Mrlina river basins, which deals with a large part of eastern-central Bohemia, by the secondary school teacher Eduard Pospíchal (1838–1905). This flora was published both in Czech and German (Pospíchal 1881, 1882), and received an award in a competition organized by the Opiz Foundation.

In the twentieth century the centre of floristic research moved from the museum to the university where it was organized mainly by Karel Domin (1882–1953; Figs. 2.2i and 2.4g), since 1904 assistant to Professor Velenovský. Domin coined a new kind of geobotanical study of various parts of Bohemia, based on the principles of phytogeography introduced by the German botanist Oscar Drude. The most representative of these studies is probably the description of the České středohoří Mts (Domin 1904). The author classified the study area into seven districts with different climates and floras, distinguishing 23 vegetation formations. The second part is a travelogue describing various landscapes in the area and listing records of plants noted during the author's field trips or extracted from the literature. Josef Podpěra, together with K. Domin the most influential Czech botanist in the first half of the twentieth century, tried to explain the historical development of the flora of the Bohemian lands (Podpěra 1906), based on the latest knowledge about glacial periods and the phytogeographic spectrum of the flora. Although many of his opinions and premises are outdated, it was a reasonable attempt at a phytogeographic analysis.

### 2.5.3 *Moravia and Silesia*

In Moravia a handful of local floras were published in the 1850s and 1860s. Research was stimulated by the establishment of the Brno Natural History Society, and Makowsky's *Flora des Brünner Kreises* (Makowsky 1863) was published already in the first volume of its journal. Alexander Makowsky (1833–1908; Fig. 2.5b) was a professor at the technical college in Brno. His flora is a careful and critical synthesis of plant records from the beginning of botanical research in this area. It was soon followed by a pocket field guide with the same geographic scope (Haslinger 1869), which was revised and reissued after a decade.

Makovsky's flora may have served as a model for the *Flora des Znaimer Kreises* (Oborny 1879), dealing mainly with the flora of south-western Moravia. Its author, Adolf Oborny (1840–1924; Fig. 2.5e), a German-speaking secondary school teacher in Znojmo, became the leading Moravian botanist in the late nineteenth century. The two-volume *Flora von Mähren und österr. Schlesien* (Oborny 1883–1886) is a synthesis of crucial importance, being precise and critical, and therefore a respectable counterpart of Čelakovský's *Prodromus* or Fiek's *Flora von Schlesien*. Starting from 1884, Oborny annually summarized records of vascular plants from Moravia and Austrian Silesia (Oborny 1885). Throughout his botanical life he also studied *Hieracium* and wrote a monograph of this difficult genus in both provinces (Oborny 1905, 1906). A less critical two-volume flora of Moravia and Austrian Silesia was written in Czech and appeared soon after Oborny's flora (Formánek 1887–1897). Its author, Eduard Formánek (1845–1900), a secondary school teacher in Brno, also studied the flora of the Balkan Peninsula. His Moravian flora is very similar to Oborny's book, in places resembling a translation, but it contains some additional records. Local floras were published also in Czech: a valuable flora covering two districts in central Moravia was written by Václav Spitzner (1852–1907; Fig. 2.5d), a secondary school teacher in Prostějov (Spitzner 1887). A few studies on the Moravian flora were published by Franz Petrak (1886–1973; Fig. 2.5f), a student of Professor Richard Wettstein at the University of Vienna. He started his first exsiccatae *Flora Bohemiae et Moraviae exsiccata* already as a student in 1907 (Petrak 1910). It includes 1600 numbers of vascular plants collected by the joint efforts of members of the Petrak family and 13 other mainly German-speaking botanists from various parts of Bohemia and Moravia. It also contains specimens of numerous rare plants.

Moravia was the place where the first botanical identification key written in Czech was compiled (Sloboda 1852). The book, covering the flora of the whole Austrian Empire, was written by a Lutheran minister from Rusava in eastern Moravia and published by the Patriotic Museum in Prague. It is a compilation based on the 1840 edition of P. F. Cürie's pocket flora of central and northern Germany and the *Österreichs Flora* by J. A. Schultes from 1814. Because it was written in Czech it became crucially important in the dissemination of botanical knowledge among Czech-speaking students and local botanists. However, many workers used German floras published by H. G. L. Reichenbach and W. D. J. Koch because of their scientific quality, and later also the handbook published by G. Lorinser, born in Mimoň in northern Bohemia, of which five editions were produced between 1854 and 1883 (e.g. Lorinser 1854).

A pocket flora for Moravia and Austrian Silesia in German was written by Gustav Merker (1910). Its author was a teacher at the Forest Academy in Hranice in northern Moravia, and his book is the last complete provincial flora published in German. In contrast, the importance of the botanical handbooks written by František Polívka (1860–1923), another Moravian secondary school teacher, was also greatly appreciated well beyond the province's borders. His four-volume Illustrated Flora of the Lands of the Bohemian Crown (*Názorná květena zemí koruny české*; Polívka 1900–1904) of 2490 pages with identification keys, species descriptions and 3159 illustrations remained an indispensable identification aid for half a century and his pocket flora with the same geographic scope (Polívka 1912) was used until 1928. Twenty editions of his pocket flora for students containing more than 1000 species and 650 drawings, were printed, the first in 1904 (Polívka 1904) and last in 1976.

Geobotanical studies in the contemporary style were also published in Moravia. Heinrich Laus (1872–1941; Fig. 2.5g), from 1902 a professor at the academy for the education of German-speaking teachers in Olomouc and from 1908 curator of the natural history collection in the Olomouc museum, wrote a monograph on the halophilous vegetation in southern Moravia and another on the Moravian weed and ruderal flora (Laus 1907, 1908a). He also published two editions of a student's flora for the Bohemian lands (Laus 1908b). He was an extremely diligent collector who exchanged plants with many botanists abroad: there are thousands of his specimens in many herbaria all over Europe. Probably the best contemporary local flora is that of Podpěra (1911), which covers central Moravia. This book combines descriptions of vegetation formations, geographical range analyses and traditional list of localities for particular species, even with indications of the collectors and herbarium specimens; it includes also specialized chapters on algae, diatoms, micromycetes, macromycetes, lichens and bryophytes written partly by Podpěra and partly by his friends.

Though the plants of Austrian Silesia were included in Moravian floras, the research into this province's flora continued to be organized mainly from Wrocław. A new critical flora of the whole of Silesia was written by Emil Fiek, assisted by Rudolf Uechtritz (Fiek 1881). During the 1890s Theodor Schube (1860–1934), a secondary school teacher in Wrocław, became the leading personality in Silesian botany. The most prominent among his works is the Flora of Silesia (*Flora von*

*Schlesien, preussischen und österreichischen Anteils*; Schube 1904), which remained the standard work on Silesian botany for many decades. Starting from 1896, he also compiled annual reports on the progress of the research into Silesian flora (Schube 1897). A geobotanical description of the Silesian flora, accompanied by numerous photographs, was written by Ferdinand Pax, professor of botany at the University of Wrocław (Pax 1915).

### *2.5.4 Bryology and Lichenology*

The bryological exploration of the Bohemian lands was initially done by renowned German-speaking bryologists: the Sudetes Mts were explored by Julius Milde (1824–1871) and Karl G. Limpricht (1834–1902), both secondary school teachers in Wrocław. The most important of the Bohemian German-speaking bryologists was Viktor F. Schiffner (1862–1944), born in Česká Lípa in northern Bohemia. He was an assistant to Professor Willkomm and later Professor at the German University, before leaving for Vienna in 1902, where he focused on the taxonomy of exotic hepatics. Schiffner, assisted by Franz Matouschek, explored mainly the bryoflora in northern Bohemia. The first bryological flora written in Czech, Liverworts of the Bohemian Flora (*Mechy jatrovkovité (Hepaticae) květeny české*), was written by a secondary school teacher, Josef Dědeček. It was published first in Czech and then in German (Dědeček 1883, 1886). Ernst Bauer, a lawyer and public servant, studied bryophytes in central and western Bohemia and curated the *Musci europaei exsiccati*. The Bohemian mosses were comprehensively covered by the primary school teacher Antonín Weidmann (Fig. 2.4d) in his Prodromus of Bohemian Leafy Mosses (*Prodromus českých mechů listnatých*; Weidmann 1895). Professor Josef Velenovský focused on bryology only for about a decade between 1893 and 1903. He became acquainted with Bohemian bryoflora to such an extent that he was able to produce the floras, Mosses of Bohemia (*Mechy české*; Velenovský 1897; Fig. 2.6c) and Liverworts of Bohemia (*Jatrovky české*; Velenovský 1901–1903). Velenovský's student Karel Kavina, during his short bryological career, wrote monographs on the peat-mosses (1911) and frondose liverworts (1915). Perhaps the only attempt to describe the cryptogamic flora in Moravia in a systematic manner was made by Jakob Kalmus (1834–1870), a physician in Brno. He published a moss flora in 1867 and a liverwort flora posthumously in 1871 (edited by Gustav Niessl), both in the *Verhandlungen des Naturforschenden Vereines in Brünn*.

In the mid-nineteenth century the Bohemian lands had no lichenologist, but the Sudetes Mts were explored by German-speaking Silesian researchers. The lichenological handbooks published by them (Koerber 1855; Körber 1859–1865; Stein 1879) were also important beyond Central Europe. The popular handbook by Rabenhorst (1870) dealt with lichens in Saxony, Thuringia, Upper Lusatia and northern Bohemia. The lichenofloristic reports from various parts of Moravia published by V. Spitzner in the 1890s were based mainly on records of other workers, including J. Kalmus, G. Niessl and F. Kovář. An identification key to Bohemian macrolichens and an up-to-date list of all lichens known from Bohemia was compiled by Josef Novák both in Czech and German (Novák 1888, 1893). The most important field lichenologist around the turn of the century was Filip Kovář (1863–1925), a worker in a shoe-making factory and later museum curator. He wrote several reports on the lichen biota in the surroundings of Žďár nad Sázavou and other parts of Moravia, containing many new records for the province. He also produced small monographs on the species of the genera *Peltigera* and *Cladonia* in Moravia (Kovář 1909, 1912). Another excellent field lichenologist, Václav Kuťák (1876–1956; Fig. 2.4f), worked mainly in north-eastern Bohemia. He became well-known because of his wide-scale exchange of herbarium specimens and collections, *Lichenes Bohemiae*, present in many herbaria worldwide. Eugen Eitner (1839–1921), a retired merchant working in Wrocław, also collected lichens in the Sudetes Mts. He described several dozen taxa of lichens, some of them based on collections from the Bohemian side of the Krkonoše Mts. Unfortunately, his herbarium was destroyed in WWII, and only duplicates have survived. Josef Anders (1863–1936), a teacher in Česká Lípa, influenced his contemporaries with his Shrubby and Leafy Lichens of Northern Bohemia (*Die Strauch- und Blattflechten Nordböhmens*; Anders 1906), later enlarged to include Central Europe (Anders 1928).

### 2.5.5 World War I

There was a general decline in botanical activity during WWI, although some botanists did not forget their beloved hobby in spite of their direct involvement in the war. Josef Podpěra was mobilized in the summer of 1914; he was one of 119,000 troops that surrendered to Russian forces in March 1915 and spent the rest of the war in Russian captivity, working in the provincial museum of Ufa in south-eastern European Russia and later as botany lecturer at the University of Tomsk. In 1917 he joined the Czechoslovak legion and returned to his homeland via Vladivostok and North America as late as in August 1920. Wherever he was, he probably used every moment to study plants. Albert Latzel, a medical doctor serving in the Austro-Hungarian Navy in Dubrovnik and Boka Kotorska, collected plants and mosses during the war in Dalmatia and Ukraine. Also Heinrich Laus, who was with an Austrian military unit stationed in the Bosnian capital Sarajevo, collected plants until autumn 1918 and was able to transport numerous herbarium specimens to his homeland despite the chaos at the end of the war. However, botanical research in Bohemia and Moravia was greatly restricted during the war years, as is documented by the small number of specimens from this period in herbaria.

## 2.6 The First Czechoslovak Republic (1918–1939)

The end of WWI brought the ultimate collapse of Austria-Hungary, and the Bohemian lands became part of the newly established Czechoslovak Republic, which included the Bohemian lands in their historical boundaries, Slovakia and Carpathian Ruthenia (or Carpatho-Ukraine, *Podkarpatská Rus*, now the *Zakarpattia Oblast* of Ukraine). Several minorities were included in the population of this new country: almost 31% of the citizens in the Bohemian lands were German-speaking and in Slovakia almost 22% were Hungarians. These minorities, which actually constituted the majority in many towns and villages, were to a large extent hostile to the new republic, as their political elites lost power. Czechoslovakia was a parliamentary democracy, the most liberal and after 1935 the only one in Central Europe. The "Czechoslovak nation", including Czechs and Slovaks, was the main constituent of the new state, but the German minority had a system of educational and cultural institutions fully equal to the Czechs. However, the Republic failed in granting political autonomy to ethnic minorities, which later proved to be fatal. The Czechoslovak economy quickly recovered from WWI and the country ranked among the most developed in the world, but the economic crisis in the early 1930s most strongly affected the industrial areas with a German majority. The Sudeten German Party, supported by Nazi Germany, became the strongest of all the parties in the country in the mid-1930s. At the Munich Conference in September 1938 Czechoslovakia had to cede the areas with a predominant German population to Germany, and in March 1939 the Bohemian lands were occupied by the Nazis, creating the Protectorate of Bohemia and Moravia as an autonomous Nazi-administered territory, while a puppet state was established in Slovakia and Carpathian Ruthenia annexed by Hungary.

### *2.6.1 Institutions and Societies*

The establishment of Czechoslovakia brought a sudden improvement in the research infrastructure. Masaryk University, established in Brno in January 1919 as the second Czech university in the country, had a Faculty of Science from the very beginning. In 1921 Josef Podpěra (1878–1954; Figs. 2.5h and 2.7a) was appointed as the first professor of botany and head of the botanical institute. The University of Agriculture in Brno was also founded in 1919; from 1922 it included the Institute of Dendrology, much later with a full professorship. At Charles University, which dropped Emperor Ferdinand from its name, the Faculty of Science was separated from the Faculty of Arts, but Professor Velenovský remained the head of the Institute of Botany until 1927. In 1922 Professor Domin created the Institute of Pharmaceutical Botany in order to satisfy his personal ambitions, and fused both institutes after Velenovský's retirement in 1927.

**Fig. 2.7** Remarkable Czech botanists of the twentieth century: (**a**) Josef Podpěra (1878–1954) in his office in Brno in 1941; (**b**) František A. Novák (1892–1964); (**c**) Josef Dostál (1903–1999); (**d**) Antonín Vězda (1920–2008). Photo credits: Archive of the Department of Botany and Zoology, Masaryk University (a), Archive of the Czech Botanical Society (b, c) and H. Michalcová (d)

At the Institute of Botany of the German University in Prague Professor Beck retired in 1921 and Fritz (Friedrich J.) Knoll (1883–1981), specialized in flower biology, was appointed as a new professor. As a prominent member of the Nazi party he was appointed Rector (chancellor) of the University of Vienna in 1938 and was personally responsible for the persecution of professors and students for racial and political reasons. His successor at the German University, Adolf A. Pascher (1881–1945), from 1927 professor and from 1933 head of the institute, was a prominent algologist and served for two decades as the editor of the journal *Beihefte zum Botanischen Centralblatt*. Pascher's achievements made the institute probably the most important centre of algological research in the world. He substantially reshaped the classification of algae, but probably most famous are his monographs published in *Die Süsswasser-Flora Deutschlands, Österreichs und der Schweiz* (e.g. Pascher 1915, 1927), which he founded. He died in May 1945 in Prague or Doksy (northern Bohemia) under circumstances that remain unclear.

Both the Royal Bohemian Society of Sciences and the Czech Academy of Sciences and Arts, which dropped Emperor Francis Joseph from its name, continued their activities, and so did the renamed German Society of Sciences and Arts for the Czechoslovak Republic (*Deutsche Gesellschaft der Wissenschaften und Künste für die Tschechoslowakische Republik*). In 1924 the Moravian Natural History Society (*Moravská přírodovědecká společnost*) was established in Brno. It immediately started a journal, *Práce Moravské přírodovědecké společnosti*, which became the most important publication venue for natural sciences in Moravia. Also other natural history societies, for instance in Třebíč and Moravská Ostrava, published journals. The faculties of science, both in Prague and Brno, launched in the early 1920s their scientific journals *Spisy vydávané přírodovědeckou fakultou*... (with a parallel title in French, *Publications de la Faculté des sciences*...). Valuable taxonomic studies appeared mainly in the Brno series (e.g. Nábělek 1923). Professor Domin (until 1928 together with J. Podpěra) edited the journal *Acta botanica bohemica*, which published plant geographical and taxonomical articles, in the 1930s and 1940s written mainly by the editor. There was a handful of popular journals, e.g. *Věda přírodní* (Natural Science), published in cooperation with the Natural History Club and Czechoslovak Botanical Society, edited also by Domin, and *Příroda* (Nature); both journals also published short research reports, mainly descriptions of flora and vegetation. Among the journals published in German, botany was best represented in *Lotos* and the *Verhandlungen des Naturforschenden Vereines in Brünn*, published in Prag and Brno, respectively. In 1930 they were joined by the journal *Natur und Heimat*, published by the *Deutscher Verband für Heimatforschung und Heimatbildung* in Ústí nad Labem, which focused on flora, fauna and nature conservation.

In 1927 the Czechoslovak state purchased the chateau in Průhonice and the surrounding park of 250 hectares. The park was founded in 1885 by Count Ernst E. Silva-Tarouca (1860–1936), a politician and self-educated dendrologist, who co-authored (together with Camillo Schneider) several editions of a popular dendrological and horticultural handbook. The state established the Research and Breeding Institute for Ornamental Gardening (nowadays the Silva Tarouca Research Institute for Landscape and Ornamental Gardening) in the Průhonice chateau. In 1950 the Botany Department and herbarium of the National Museum were moved to the chateau, where they remained for more than half a century. Since the 1960s this chateau has been the headquarters of the Institute of Botany of the Czechoslovak (now Czech) Academy of Sciences.

In 1934 the National Museum in Prague, until then privately owned by the Society of the Patriotic Museum in Bohemia, became the property of the Land of Bohemia. Together with numerous state subsidies provided to societies, this change documented the growing role of the state in the support of scientific infrastructure. The activities of the museum's botanical department were shaped by Ivan Klášterský (1901–1979; Fig. 2.4k), who was appointed as department head in 1933. He focused on re-identifying, sorting and incorporating the valuable historical collections acquired by the museum, including those of Haenke, Opiz and Kablik.

**Fig. 2.8** Participants of the Fifth Phytogeographical Excursion on the serpentinite slopes near Mohelno in south-western Moravia on 13 July 1928. Photo credit: Archive of the Department of Botany and Zoology, Masaryk University

The Czech Botanical Society enlarged its geographic scope to include the whole of the new country, changing its name to Czechoslovak Botanical Society. Its main activities remained excursions, meetings and lectures. In October 1921 the Society organized the first Congress of Czechoslovak botanists. As there was one participant from the Kingdom of Serbs, Croats and Slovenes, Professor Vale Vouk, the event was also declared the 1st Congress of Slavic Botanists. The next congress, both Czechoslovak and Slavic, was held in 1928 to celebrate the 10th anniversary of Czechoslovakia. It hosted numerous representatives from Poland and Yugoslavia, and also a few from Bulgaria and the Soviet Union. The Society also co-organized the 5th International Phytogeographical Excursion to Czechoslovakia and Poland in 1928 (Fig. 2.8). In 1923, after a long break due to the war, the second volume of the Society's journal *Preslia* appeared, followed by further volumes at more regular intervals. As the established journals were firmly in the hands of the older generation and provided only limited publication space for young botanists, about two dozen young botanists formed the Club of Czechoslovak Botanists in 1937 with the only purpose of publishing at their own expense. Their journal, *Studia botanica čechoslovaca*, edited by Alfred Hilitzer, published 12 volumes, before being fused with *Preslia* in 1951. In parallel, monographic studies were issued in the series *Opera botanica čechica*, in which a valuable modern monograph on the genus *Sesleria* (Deyl 1946) was published.

### *2.6.2 Studies on Vascular Flora*

Both leading personalities in Czech botany at that time, K. Domin and J. Podpěra, concluded an agreement demarcating their spheres of interest within Czechoslovakia: Professor Domin claimed Bohemia, Slovakia and Carpathian Ruthenia for himself and his collaborators, while Moravia was allotted to Professor Podpěra and his assistants. This agreement held until the mid-1930s, when Vladimír Krist, Podpěra's assistant, started research on the halophilous flora in southern Slovakia. Actually, K. Domin was mainly interested in the Carpathians, while Bohemia was left to amateur botanists. Accounts of local species were compiled mainly for natural history monographs for administrative districts in various parts of the country (e.g. Klement 1930). Domin himself returned to local floristics during WWII when he could not travel to Slovakia, producing instead a comprehensive flora of the Kokořín area in central Bohemia (Domin 1942). In the meantime, the research into the flora of the whole of Bohemia was organized by Josef Rohlena (1874–1944; Figs. 2.2h and 2.4j), a primary school teacher and amateur botanist, who from 1895 lived in Prague. He helped L. J. Čelakovský as a voluntary museum assistant in the late 1890s. Later he shifted his attention to the flora of Montenegro, where he spent most of his summer vacations. He returned to the flora of his homeland in the early 1920s, when he published a series of reports on new floristic records for Bohemia (e.g. Rohlena 1922). He organized an informal network of local contributors, who sent him their field records and herbarium specimens for identification. Of the 13 reports published in 1922–1938, the last three were prepared together with Josef Dostál. However, Rohlena's lifework is undoubtedly *Conspectus Florae Montenegrinae*, edited by Dostál and published as a double-volume of Preslia (Rohlena 1942).

The newly established Institute of Botany at Masaryk University in Brno became a centre of botanical research in Moravia. Its herbarium was formed by fusing the personal collections of Professor Podpěra with the purchased herbaria of H. Laus, F. Petrak and the *Naturforschender Verein in Brünn*. This Institute also purchased the library of this society. In 1931, i.e. 10 years after its establishment, the institute's herbarium had about 225,000 specimens. Podpěra soon started working on a new flora for Moravia, of which three volumes were finished (Podpěra 1926–1930), including ferns, gymnosperms and graminoids. The title of this flora may be translated into English as Flora of Moravia in a Systematic and Geobotanical Context, which reflects its unusual format: for many species the taxonomical and chorological information is supplemented by detailed descriptions of their distributions and habitats, particularly in Eastern Europe and Siberia, based on the author's personal observations and Russian literature. The complicated hierarchy of infraspecific taxa resembles that of the *Synopsis der mitteleuropäischen Flora* by Ascherson and Graebner, sometimes with two parallel infraspecific classifications within a single species.

Local floristic studies appeared less frequently in this period. A nice example of a local flora is that for the town of Štramberk in northern Moravia (Otruba 1930), once again a work of an amateur botanist. Roughly at the same time Heinrich Laus described the flora on Mt. Petrovy kameny (Petersteine) in the Hrubý Jeseník Mts (Laus 1927).

Young members of the botanical institutes studied the taxonomy of various genera, including *Dianthus* (e.g. Novák 1924), *Festuca* (e.g. Krajina 1930) and *Centaurea* (Dostál 1938). Particularly valuable are the monographs on *Fabaceae* written by Grigorij Ivanovič Širjaev (1882–1954). He was born in Kharkiv (Ukraine), where he also studied botany. During the Russian civil war, he joined the White Army, in 1920 leaving Crimea for Turkey together with the remnants of General Wrangel's army. In 1923 he started working at Masaryk University as the first herbarium curator. In Brno he published monographs on the genera *Onobrychis* (e.g. Širjaev 1925–1937), *Trigonella* and *Ononis*, as well as taxonomic studies on some species of *Lathyrus* and *Astragalus*.

The interwar period was a golden age of exsiccatae. Professor Domin with his assistant Vladimír Krajina, supported by numerous collaborators, prepared four centuries of *Flora čechoslovenica exsiccata*, issued from 1929 to 1936. The Brno series, *Flora exsiccata Reipublicae bohemicae slovenicae*, was started in 1925 and instalments were issued almost annually until 1938. The series was continued until 1979 under modified names, reaching altogether 17 centuries. Both collections were produced in numerous copies and are present in many herbaria all over the world.

There was a need for a flora of Czechoslovakia as the already outdated floras published before WWI by F. Polívka covered only Bohemia and Moravia, while information about Slovak and Carpatho-Ruthenian floras had to be extracted from Hungarian sources. A pragmatic solution was to update and extend the time-proven Polívka's pocket flora of 1912 (Domin and Podpěra 1928; Fig. 2.6d). A year later a flora of Bohemia, written in German, the *Bestimmungsbuch der Flora von Böhmen* (Tannich 1929) was published by Anton Tannich, a forester and amateur botanist expelled to Austria in 1945. In the early 1930s an attempt of Jaromír Klika to launch a new flora failed at the planning phase: Klika was a vegetation scientist at the Czech Technical University in Prague and failed to obtain the support of taxonomists. Instead, Domin hastily compiled a list of Czechoslovak vascular plants (Domin 1935), which was probably intended as a taxonomic backbone of the future flora edited by him. This publication was followed by a thorough survey of the literature related to the flora of Czechoslovakia, much later published as a comprehensive bibliography (Futák and Domin 1960), and by a systematic extraction of plant occurrence records on card files.

### 2.6.3 *Bryology and Lichenology*

The most influential Czechoslovak bryologist in the interwar period was Professor Josef Podpěra, a student of Velenovský at the Czech University in Prague and founder of scientific bryology at Masaryk University in Brno. He is best known for his studies and a monograph on the genus *Bryum* (e.g. Podpěra 1942), which were published throughout his career, and for a compilation of a modern and comprehensive index of European mosses, *Conspectus muscorum europaeorum* (Podpěra 1954). He also authored a handful of bryofloristic studies during his early years in Moravia.

The interwar lichenological research in Czechoslovakia was dominated by two scientists. Alfred Hilitzer (1899–1940), associate professor at the Czech Technical University in Prague, was a field botanist specializing in lichenology. His most important work was a book on epiphytic cryptogam communities, *Étude sur la végetation épiphyte de la Bohême* (Hilitzer 1925), inspired by the Scandinavian phytosociogical approach. The most productive lichenologist of that time was Jindřich Suza (1890–1951), from 1921 assistant to Podpěra at Masaryk University in Brno and from 1932 professor at Charles University in Prague. He focused on phytogeographic and geobotanical aspects of lichenology, working mainly in Moravia and the Western Carpathians. He published more than 100 papers on lichenology initially focusing mainly on lichenofloristics; later he also included geographical and ecological aspects. This resulted in a synthesis of the Moravian lichen biota (Suza 1925). Josef Anders continued his studies on lichens in northern Bohemia, while Oskar Klement (after 1945 expelled to Bavaria) studied lichens mainly in the Krušné hory Mts.

### 2.6.4 *Vegetation Studies*

In the 1920s the research interest of young botanists turned towards the emerging field of geobotany or phytosociology. The first attempts to study Czech vegetation using a phytosociological approach were made by Professor František Schustler (1893–1925) at Charles University in Prague. Unfortunately, his untimely death prevented him from continuing his vegetation research in the Krkonoše Mts and Bohemian dry grasslands, which were continued by his student Alois Zlatník (1902–1979; Fig. 2.9a). Professors Domin and Podpěra were also interested in vegetation studies, although their main interests were in phytogeography and plant taxonomy. Domin published a book on phytosociological methods (Domin 1923) and introduced the internationally well-known Domin scale for estimating cover in vegetation-plots. Podpěra published insightful descriptions of vegetation at various sites, especially in Moravia, framed in a broad phytogeographic context.

**Fig. 2.9** Botanical events and remarkable botanists of the twentieth century: (**a**) *from left*: Witold H. Paryski, Jaromír Klika (1888–1957), Alois Zlatník (1902–1979), Vladimir N. Sukachev, Rezső Soó, Mrs. Szafer, Władysław Szafer and Bálint Zólyomi in the High Tatras during the Congress of the Czechoslovak Botanical Society in 1955; (**b**) Josef Holub (1930–1999; *left*) and Stuart M. Walters in a field trip organized during the Congress of the Czechoslovak Botanical Society in 1962. Photo credits: Archive of the Czech Botanical Society

An important event in the development of vegetation science in Czechoslovakia was the Fifth International Phytogeographical Excursion in Czechoslovakia and Poland in the summer of 1928 (Rübel 1930). This excursion, organized for the Czechoslovak part under the leadership of Karel Domin, Josef Podpěra and Karl Rudolph, was attended by leading vegetation scientists of that time and pioneers of phytosociology, including Josias Braun-Blanquet, Heinrich Brockmann-Jerosch, Helmut Gams, Henry Gleason, Rolf Nordhagen, Eduard Rübel, Arthur Tansley and Heinrich Walter (Fig. 2.8). For this event, Czech botanists prepared pioneering vegetation studies of various excursion sites (e.g. Domin 1928; Firbas and Sigmond 1928; Klika 1928; Podpěra 1928a, b, c; Rudolph 1928; Rudolph et al. 1928; Suza 1928; Zlatník 1928a, b). The establishment of international contacts and research motivated by this excursion had a great influence on further developments of vegetation science in Czechoslovakia.

In the 1930s, vegetation scientists from Prague and Brno continued surveying various parts of this country, though many of them were attracted mainly by the relatively more natural and less explored Carpathian landscapes in Slovakia (e.g. Josef Dostál, Vladimír Krajina, Rudolf Mikyška and Pavel Sillinger) and Carpathian Ruthenia (e.g. Miloš Deyl and Alois Zlatník). A long-term systematic survey of vegetation in various parts of Bohemia, Moravia and Slovakia was undertaken by Jaromír Klika (1888–1957; Fig. 2.9a), professor at the Czech Technical University in Prague and from 1951 professor at Charles University. Klika was in close contact with the international community of vegetation scientists cooperating with Braun-Blanquet's institute SIGMA in Montpellier. Influenced by Braun-Blanquet attempts at creating an international hierarchical classification of plant communities, Klika started preparing a hierarchical overview of the Czechoslovak vegetation types, which was published in four editions (Klika 1941, 1948, 1955; Klika and Hadač 1944) containing first descriptions of several high-ranking syntaxa, which are still widely accepted internationally. Three of these editions were published in handbooks and textbooks of vegetation science written in Czech and edited by Klika, which contributed to a national standardization of concepts and methods. The 1944 edition of the overview also contained a pioneering attempt by Emil Hadač at creating a syntaxonomic system for communities of bacteria, fungi, algae, lichens and bryophytes.

Modern palaeobotanical research was started at the German University in Prague by Karl Rudolph (1881–1937) already during WWI. In the 1920s he was joined by Franz Firbas (1902–1964) and several students who carried out palynological studies on mires in various areas of Czechoslovakia. Unfortunately, the successful work of this group was discontinued by the expulsion of Germans from Czechoslovakia after WWII. Firbas resumed his research at the University of Göttingen, where he prepared a monograph summarizing the research results of this group (Firbas 1949–1952).

## 2.7 World War II and Postwar Turbulence (1939–1948)

Although the ultimate goal of the Nazi occupation was the full Germanization of the Bohemian lands, the repressions were initially aimed mainly at the Czech intellectual elite, politicians, former military officers and people classified as Jews. After anti-Nazi demonstrations in autumn 1939, all Czech universities were closed for three years, nine student leaders were executed and more than 1000 students and many professors sent to concentration camps. After the assassination in 1942 of the Deputy Reich Protector Reinhard Heydrich by Czechoslovak soldiers airlifted into the country by the British Airforce, the Nazi terror against the Czech population intensified. The military defeat of Nazi Germany in May 1945 was followed by the expulsion of the German-speaking population to Germany or Austria soon after the end of the war. The political regime in liberated Czechoslovakia was basically democratic, however, the Communist Party won the 1946 elections and seized absolute power in a coup d'etat in February 1948.

### *2.7.1 Botany in the Occupied Country*

The closure of Czech universities in November 1939 immediately stopped all teaching and most scientific research at these institutions. In Prague, Viktor Czurda, an excellent algologist but also a convinced Nazi, was appointed as the commissioner for the Czech Faculty of Science. Professor Domin barely managed to save his private library. With the expectations of bombing raids by the Allies, the herbarium was later moved to relative safety in a castle in south-western Bohemia, together with the collection of the German institute. Both collections were partly damaged because of the lack of care and unsuitable conditions. In Brno the situation was somewhat better since Professor Franz Frimmel (1888–1957) was appointed as the commissioner for the Institute of Botany. He was a prominent plant breeder and professor of agriculture at the German Technical University in Brno. Having nothing in common with Nazi ideology, he was also on good terms with Professor Podpěra. The institute was allowed to work as a kind of a research facility until autumn 1941. Then Professor Podpěra retired and only Širjaev remained as the last member in charge of the herbarium of the dissolved institute. In April 1945 Širjaev managed to escape to Bavaria to avoid his arrest by the Soviet secret police, but he was unable to find an adequate job and died nine years later in poverty in New York.

Although the situation in the occupied country was difficult, botanical research did not stop completely. Professor Domin and a handful of his assistants found a refuge in the Commission for Collecting Medicinal Plants, established by the Protectorate Ministry of Health. It was a small institute of botany, and its employees continued compiling literature records for the planned flora of Czechoslovakia. In early 1940 Domin published a call and instructions for research into the country's flora on a regional basis, which was an invitation to amateur botanists (Domin 1940). The research, referred to as the Floristic Action, was supported by the Czech Botanical Society. Already by the end of 1940 there were about 700 participants. This recording scheme enhanced the collecting of plants, which is reflected in a large number of herbarium specimens of that date in the Charles University herbarium. Although meetings and other activities of natural history societies were restricted by the Nazi authorities, the Czech Botanical Society was able to issue three volumes of Preslia, one of them containing Rohlena's *Conspectus Florae Montenegrinae* (Rohlena 1942).

## 2.7.2 War Victims and Heroes

Some Czech botanists became actively involved in the anti-Nazi resistance movement. Vladimír Krist (1905–1942) joined a local cell of Defending the Nation, a resistance group organized mainly by former officers of the Czechoslovak Army. This group, whose members sometimes met in the former building of the Institute of Botany at Masaryk University in Brno, was disclosed and Krist died in the Mauthausen concentration camp. Further victims of the Nazi regime include Professor August Bayer of the University of Agriculture in Brno and several amateur botanists. There were victims also on the other side: for instance, Karl Preis (*1913; Fig. 2.4i) became a German army soldier and was killed near Leningrad (now Saint Petersburg) in September 1941 (Pohl 1941).

Professor Vladimír Krajina (1905–1993) joined the Czech resistance movement immediately after the Nazi occupation. He was one of the leading members at the Political Headquaters of Homeland Resistance, being in charge of the encrypted radio messages sent from the Protectorate to the exiled Czechoslovak politicians in London. Krajina was finally captured by the Gestapo in 1943 and after a failed suicide attempt held in prison till May 1945 as a prominent prisoner of the Nazi authorities (Drabek 2012).

### *2.7.3 Consequences of the War*

The defeat of Nazi Germany was followed by the expulsion of Germans from the country, sanctioned by the Potsdam Agreement of 1945, which ended the long and sometimes difficult coexistence of Czechs and Germans in the Bohemian lands. Only a small fraction of the German-speaking citizens was allowed to stay in the country, including professional botanists Professors Erich Daumann, a former assistant to Professor Knoll, and Franz Frimmel. The German University in Prague and other German educational institutions were dissolved. Many herbarium specimens collected by the German-speaking compatriots were transferred to public herbaria (e.g. collection of Otto Thenius to the Masaryk University herbarium; Fig. 2.5i) after their owners had to leave the country with a few personal items only.

The first task after the country's liberation in May 1945 was to rebuild the institutions. At Charles University in Prague, Professor František A. Novák (1892–1964; Fig. 2.7b) was appointed as a head of the Institute of Botany in 1945, while Professor Domin was accused of collaboration with the Nazis and could not return to his former job. In 1946 he was imprisoned but a year later he was expunged of all allegations by the National Tribunal. However, this court decision did not help him reverse his exclusion from the faculty, and he was ultimately retired by decision of the first communist Minister of Education in spring 1949. He blamed his former assistants for his situation. It is now clear that the collaboration charges were fabricated. Domin is said to have been an honest person, a Czech nationalist with right-wing political views and strong political ambitions, but with a poor sense of politics. The reasons for his exclusion from scientific life were partly political as left-wing views prevailed in the postwar period and partly may have been personal, reflecting the reluctance of his former students and colleagues to accept the continuation of his dominance and autocratic style. The Institute of Botany at Masaryk University in Brno was ruined in terms of staff as only Professor Podpěra returned, resuming his work as a head of the institute. A new university was (re-)established in Olomouc in 1946 and named Palacký University. Initially it did not have a Faculty of Science, but a small Institute of Botany was established later at the short-lived University of Education, which was merged with Palacký University in the early 1960s.

The postwar developments forced Professor Krajina to accept the job of Secretary General of the Czechoslovak National Socialist Party in 1946. In spite of its name it was a democratic party never associated with Nazi ideology, which in the postwar period was the main opponent of the Communists. In February 1948, immediately after the communist coup, Krajina was arrested and released only with the help of President Beneš. He finally managed to flee from the country skiing over the Šumava

Mts. He started a new professional career at the University of British Columbia in Vancouver, where he taught plant ecology for 24 years. There he developed an ecologically-based vegetation classification now widely used as the basis for forest management, and made a major contribution to the establishment of a network of ecological reserves in British Columbia (Drabek 2012).

## 2.8 The Communist Period (1948–1989)

The Communists introduced a Soviet style one-party political regime and a centrally planned economy. The first years of communist rule were marked by harsh repressions not only against its active opponents but also against any assumed dissenters. Economic difficulties in the early 1960s indicated an urgent need of reforms, which were prepared by liberal economists and intellectuals within the Communist Party. These reforms were accompanied by pressure from Slovak representatives asking for more autonomy, resulting in the federalization instituted in 1968. The conservative Communist Party leadership was removed from office in January 1968, which opened the way for an eight-month period of democratic reforms, usually referred to as the Prague Spring. This short period of liberalization was terminated by the Warsaw Pact invasion, led by the Soviets, in August 1968, followed by the restoration of a conservative communist regime, which was less severe than that of the 1950s, but still oppressive of any opposition movement. In the late 1980s there was a crisis in the country's planned economy, growing domestic opposition, and change in the Soviet policies towards other communist countries, which ultimately resulted in the collapse of the communist regime in the fall of 1989.

### *2.8.1 Life and Work of Botanists Under the Communist Regime*

The communist regime had a profound influence on all spheres of Czechoslovak society, including the educational system and research. It immediately suppressed the academic freedoms that had existed since the late 1840s. Officially Soviet science became the only model to follow, including a shift towards applied research. Research was also impeded by international isolation: travelling to western countries and communication with their institutions and workers were severely restricted. Exchange and purchase of literature was also substantially reduced. The other devastating factor were the persecutions of real or assumed political opponents. The Communists almost immediately removed many professors, students and scientists from universities and other institutions. For example, Josef Svoboda, later an Arctic plant ecologist in Canada, was arrested in 1949, when he was a student at Masaryk University, and kept in work camps and prison until 1958 (Svoboda 2011). Even the

Czechoslovak Botanical Society terminated the membership of Professor Domin, its former President, Miroslav Pulchart, who testified in Domin's favour at the National Tribunal, and Vladimír Krajina, a hero of non-Communist anti-Nazi resistance. Another wave of cleansing of "untrustworthy" persons followed in 1958, when numerous scientists had to leave their institutions or were removed from important positions. For instance, Antonín Vězda, later a prominent lichenologist, had to leave the Faculty of Forestry at the University of Agriculture in Brno. Jan Šmarda, a bryologist and vegetation scientist, was expelled from Masaryk University in 1961. Probably the greatest loss of intellectual capacity came after the defeat of the 1968 reform movement. During the early 1970s several people were expelled from universities, such as the vegetation scientists Jan Jeník and Jarmila Kubíková at Charles University, and Jiří Vicherek at Masaryk University (then renamed J. E. Purkyně University), while others were given lowly jobs. Some of them were replaced by Communist Party members or conformists with a lack of expertise in the field. As the state, completely controlled by the Communist Party, was the universal employer, only few of these people were fortunate enough to find adequate employment outside universities. Others left the country in order to escape political oppression or find better working conditions in a democratic society. For instance, Adolf Češka, who left almost immediately after the 1968 Warsaw Pact invasion, worked later as a museum curator and vegetation scientist in British Columbia, Canada. The former political prisoner Josef Svoboda left the country in the same year and later did his PhD in Canada and became a professor at the University of Toronto. The plant ecologist Marcel Rejmánek left the country via Yugoslavia in 1983 and became a professor at the University of California, Davis, in 1993; he made fundamental contributions to the ecology of plant invasions. It is indisputable that this brain drain caused by political oppression had an adverse effect on the quality of research in all fields of science.

The communist system gave the members of the Communist Party in leading positions in scientific institutions extraordinary powers, with the result being dependent on their personal qualities. For example, Slavomil Hejný, the long-serving Director of the Institute of Botany of the Czechoslovak Academy of Sciences in Průhonice, had a balanced personality and dared to support good scientists expelled from universities or otherwise in trouble with the regime as long as it did not endanger his own position. He even employed former political prisoners such as Josef Svoboda and Josef Petr Ondok, which undoubtedly was risky. The Institute of Botany consequently entered the post-communist period in a relatively good condition. In contrast, Radovan Hendrych, the head of botany at Charles University in Prague, made the climate in the department substantially worse than the general political situation required, which had a considerable negative effect on the department's teaching and research (Hrouda 2005).

### 2.8.2 Institutions

The traditional research infrastructure was profoundly changed by the establishment of the Czechoslovak Academy of Sciences (CAS) in October 1952, following the Soviet example. Formally it succeeded both the Royal Bohemian Society of Sciences and the Czech Academy of Sciences and Arts, but the main difference was that the Academy fulfilled not only the role of a body of the most eminent scientists, but also ran numerous institutes specialized in various fields of basic research. From the very beginning, the research topics were to a large extent prescribed by state authorities as part of a Soviet-type planned economy, nevertheless, it started to have an important role in many fields of research, which it still has.

The Academy of Sciences founded the Geobotanical Laboratories in Průhonice near Prague in 1954 and in Brno in 1955. The Průhonice laboratory was headed by Rudolf Mikyška (1901–1971), assisted by Jaroslav Moravec, while the Brno laboratory was headed by Jan Šmarda (1904–1968). The Institute of Botany came into existence by merging these two laboratories in January 1962. Slavomil Hejný (1924–2001), a plant ecologist who specialized in aquatic botany and ruderal vegetation, became the first director of this institute, keeping this position for 29 years. In 1971, a new Hydrobotanical Department of the Institute of Botany was established in Třeboň in southern Bohemia.

The restoration of the conservative communist order after the 1968 Warsaw Pact invasion adversely affected botany especially in universities. As there was no suitable Communist Party member available for the headship of the Department of Botany at Masaryk University from among staff members, the department was merged in 1971 with the Department of Plant Physiology and its staff gradually reduced. In contrast, the Department of Botany at Charles University was temporarily divided into the Department of Botany of Higher Plants in 1978, led by Radovan Hendrych, and Department of Botany of Lower Plants, headed by Zdeněk Urban, in order to satisfy the personal ambitions of the two influential party members.

As the communist system aimed at absolute control of many human activities, voluntary associations and societies lost their right to publish books and periodicals. Also local museums, in many cases previously backed and owned by voluntary associations, were taken over by the state. Despite many negative effects, gradually professional curators of the collections in various museums were appointed. This is one of the reasons for the existence of numerous public herbaria in the Czech Republic. In February 2017 the Index Herbariorum listed 51 Czech herbaria with an international code, though some of them are actually defunct or very small. Still, apart from the large university and museum herbaria in Prague and Brno, there are further 12 herbaria housing from 47,000 to 209,000 specimens each, some of them perfectly curated. The total number of herbarium specimens in the collections is estimated at about 8 million.

In 1966 the Institute of Botany of the Czechoslovak Academy of Sciences started publishing its own journal, *Folia Geobotanica et Phytotaxonomica Bohemoslovaca*. This journal, now published by Springer as *Folia Geobotanica*, was intended to be international, and it was published mainly in German and English. In addition to these national and to some extent international journals, there were numerous periodicals, journals and yearbooks issued by local museums, which published natural history papers of local or regional importance, such as *Severočeskou přírodou* (published in Litoměřice), *Sborník Jihočeského muzea v Českých Budějovicích* (České Budějovice) and *Vlastivědný sborník Vysočiny* (Jihlava).

### 2.8.3 Czechoslovak Botanical Society

The activity of the Czechoslovak Botanical Society increased rather slowly, probably because of the workload of the leading personalities during postwar reconstruction, frequent changes in Society's governance and its division in three branches based in Prague, Brno and Bratislava. Professor Domin resigned as President in June 1945 and was replaced by Silvestr Prát (1895–1990), a prominent plant physiologist, who was successively replaced by Josef Podpěra, Jaromír Klika and Alois Zlatník. The situation stabilized only after 1956 with the election of Professor Karel Hrubý (1910–1962), an excellent geneticist, botanist and lepidopterologist, who was, however, in disgrace because of his firm adherence to Mendelian genetics and opposition to politically enforced pseudoscientific Lysenkoism. A law passed in 1951 virtually abolished the freedom of association, and the Society was affiliated by an administrative decision of state authorities with the Biological Section of the Czechoslovak Academy of Sciences. It was reshaped into an "exclusive scientific society", in which only professional botanists could become members. Consequently, it lost some dedicated amateur members. This was later partially remedied by introducing extraordinary memberships. These unfortunate state interventions were partly compensated by regular subsidies, which enabled the Society to employ an officer and a librarian at its Prague headquarters.

The activities of the Society gradually expanded. Several local branches were established, often informally associated with museums and co-organized by botanists employed in these facilities. The most active were probably those in northern Bohemia (Ústí nad Labem), western Bohemia (Plzeň) and southern Bohemia (České Budějovice). These branches focused mainly on doing research on the local flora, field trips, lectures, plant identification workshops and publication of local botanical bulletins. In parallel, specialized sections were established, including the Algological Section, Dendrological Section, Floristic Section, Section of Plant Physiology, Geobotanical Section and Section of Synanthropic Flora and Vegetation. Some of them were short-lived, some later split off as separate societies, while others remained important parts of the Society. The most popular was the Floristic Section, established in 1956, which carried out field surveys of flora across the country. Its most important activity was the organization of Summer Schools of Field Botany. The first Summer School was held in 1957 in the České středohoří Mts and since 1964 up to now these one-week field meetings have been organized annually. The number of participants grew from about 30 in the mid-1960s to about 150 in the

1980s. These meetings became the Society's flagship activity with a range of purposes, including dissemination of botanical knowledge and research on the flora of less known areas. Plant records collected during these courses are systematically summarized and published. Equally important is their social function as an informal platform for meetings between professional and amateur botanists, secondary school teachers and students. For many former attendants this encounter turned out to be decisive for their professional carriers or life-long hobbies.

In the early 1970s the Society launched a series of two-day conferences or workshops on various topics. In the first decade, the topics of the conferences included, for instance, the phytogeographic land division of the country, Czech plant names, taxonomically difficult groups of vascular plants, vegetation classification and biology of selected species. The 1976 workshop, focused on vanishing flora and vegetation, resulted in the preparation of the first Red List of the country's vascular flora (Holub et al. 1979). The number of conference participants sometimes exceeded 200, which clearly indicated the hunger for information of a scientific and expert community prevented from exchanging information with non-communist countries. The conference lectures were usually published as special issues in the *Materiály* series of the Society's bulletin. This series of annual conferences continues until the present.

In the summer of 1958 the Society organized the 12th International Phytogeographical Excursion across Czechoslovakia. Its itinerary, which included the most important sites of botanical interest in the country, was prepared by Josef Dostál, Ján Futák and Alois Zlatník. The excursion lasted for more than a month and was attended by 60 botanists from Czechoslovakia and 58 from abroad.

The Society continued the tradition of congresses. The fourth society's congress, already with international attendance (Fig. 2.9a), took place in July 1955 in the High Tatra Mts in Slovakia. However, the most important congress, organized to celebrate the Society's 50th anniversary, was convened in summer 1962. It was a window to the botanical world, something like a miracle at that time. There were 180 participants from Czechoslovakia and 66 from abroad. The foreigners came not only from the Soviet Union and its satellite countries, but also from Canada, Denmark, Federal Republic of Germany, the UK and the USA. Of key importance was the presence of the British botanists David H. Valentine and Stuart M. Walters (Fig. 2.9b), whose valuable reviews appeared in the Preslia double-issue published on the occasion of the Society's anniversary (Valentine 1962; Walters 1962). The atmosphere at the congress and personal encounters with eminent foreign botanists had a marked effect on the future of Czechoslovak plant taxonomy and facilitated the participation of Czechoslovak botanists in the Flora Europaea project. A comparison with the next congress organized in 1972 reflects the political climate of the 1970s: this congress, held in České Budějovice, had only 18 participants from abroad, all from the Soviet Union.

The Society's situation stabilized during the 1960s. After Karel Hrubý, who died in 1962, František A. Novák (until 1964) and Zdeněk Černohorský, both highly respected botanists, successively chaired the Society. In 1965 the Society heralded the future federalization of the country by creating the Slovak Botanical Society, initially with 155 members. The chairman of the Slovak Botanical Society became automatically vice-chairman of the Czechoslovak Botanical Society and vice versa. As in numerous similar cases, the structure remained asymmetric since the maternal society retained its name (Czechoslovak) and a corresponding Czech society was not created. The election of Slavomil Hejný as President of the Czechoslovak Botanical Society in 1976 was a reasonable decision with positive consequences. Hejný was a Director of the Institute of Botany, a Corresponding Member of the Czechoslovak Academy of Sciences and an influential member of the Communist Party, which enabled him to protect the Society from the worst of the arbitrary acts of the state. As he was fully occupied with the management of the Institute of Botany and his memberships in numerous bodies, the Society's Scientific Secretary Josef Holub was virtually given a free hand in shaping the Society's activities.

The Society resumed publishing *Preslia* as a yearbook in 1948. In 1951 the journal was taken over by a state-owned publishing house and published quarterly, which has continued up to now. A new journal, *Zprávy Československé botanické společnosti* (Bulletin of the Czechoslovak Botanical Society), was launched in 1966. It contained mainly flora surveys, phytogeographic and taxonomic studies and botanical news. Its two parallel series, *Přílohy* (Supplements) and *Materiály* (Materials), published mainly plant records from Summer Schools and proceedings of the Society's conferences, respectively. The youngest Society's journal, *Bryonora*, was launched in 1988 as a bulletin of the very active Bryological-Lichenological Section.

### 2.8.4 Towards a New Flora

A new illustrated flora of Czechoslovakia was published in two volumes soon after WWII (Dostál 1948–1950). This handbook was edited by Josef Dostál (1903–1999; Fig. 2.7c), who invited almost 30 botanists of different ages as collaborators (e.g. Slavomil Hejný, Radovan Hendrych, Josef Holub, Václav Jirásek, Jaroslav Moravec and František A. Novák). As the distribution records compiled by Professor Domin and his collaborators were not available to them, information about the distributions of the species was based on information in the more important earlier floras and specimens collected during WWII or shortly afterwards during the Floristic Action. An abridged version of this flora was issued as a one-volume field guide *Klíč k úplné květeně ČSR* (Key to the Complete Flora of the Czechoslovak Republic) in two editions (Dostál 1954, 1958). The flora and both editions of the key were criticised for outdated taxonomy, numerous mistakes and inconsistences, and the option of publishing a third edition was explicitly rejected at a public meeting (e.g. Holub 1959). However, altogether about 35,000 copies of the three books were quickly sold out and they remained standard reference handbooks for more than three decades, but were unavailable to botanical beginners in the 1970s and 1980s. As they were becoming rare and outdated, many botanists also used Rothmaler's pocket floras for Germany as an identification aid. For Moravia and Silesia, a nomenclatural update with keys to difficult and newly recognized taxa was published as a handbook for students by Miroslav Smejkal (Smejkal 1980), a taxonomist at Masaryk University. However, it was again Professor Dostál, already in his eighties, who tried to meet the demand for a new flora. His two-volume handbook *Nová květena ČSSR* (New Flora of Czechoslovakia; Dostál 1989) is partly based on earlier floras, while some keys resemble those in Flora Europaea. A dominant feature of the flora is a narrow generic concept, roughly the same as the one used in the Check-list of the Czechoslovak Flora published earlier (Dostál 1982). Taxonomic splitting was reflected also in Czech taxon names, which was quite confusing for the users of this flora.

A new multi-volume flora of Czechoslovakia was a recurrent topic within the Czech Botanical Society. An editorial committee was formed in 1954 and some preparatory work was done, of which probably the most important was the complete bibliography of the Czechoslovak botanical literature up to 1952 (Futák and Domin 1960). However, the work was discontinued for various reasons, and only two mycological volumes and the treatment of *Sphagnaceae* (Pilous 1971) were published. The publication of the nine-volume *Květena České republiky* (Flora of the Czech Republic), of which the first volume appeared in 1988 (Hejný et al. 1988), is an undisputable great achievement of Hejný, who championed the idea in various committees of the Czechoslovak Academy of Sciences and achieved its inclusion into the state plan of basic research. The Institute of Botany automatically took over the leadership of this task, and the necessary preparations started around 1975. The editorial work, from the beginning, was organized mainly by Bohumil Slavík (1935–2004) and backed by Hejný, who also co-authored treatments of some genera. Although the flora was edited at the Institute of Botany, many chapters were contributed by taxonomists at Charles University, Masaryk University and elsewhere. The Institute of Botany in 1967 also joined the international project Floristic Mapping of Central Europe. Slavík started working on a distribution atlas of the Czech flora in 1969 and the first volume of the Phytocartographical Syntheses of the Czech Socialist Republic (*Fytokartografické syntézy ČSR*) appeared 17 years later (Slavík 1986). It contained grid maps with a resolution of 6 × 10 min for 316 species. This mapping project was discontinued as the editor died, and the fourth volume was finished much later by Jitka Štěpánková. The four volumes of this atlas include about a half of the Czech flora.

### 2.8.5 Taxonomic and Floristic Research on Vascular Plants

The Czech plant taxonomy was in a difficult position after WWII. Most of the prewar leading taxonomists adopted a morphological or morphological-geographical approach, the weak points of which are obvious for example in the monographs of Karel Domin. New trends, indicated in the studies by Miloš Deyl and Alois Zlatník, were discontinued due to political developments. The recovery of the Czechoslovak taxonomy was encouraged by the 1962 congress of the Czechoslovak Botanical Society, which was attended by leading taxonomists from abroad.

The postwar taxonomic studies may be roughly divided in two groups based on the methods used (Hejný et al. 1978). The first group includes classical studies based mainly on herbarium specimens and field observations, including for instance the studies by Jindřich Chrtek sen., either alone (*Polygonum aviculare* agg.) or together with Václav Jirásek (*Corynephorus* and *Poa*), by Radovan Hendrych (*Thesium* and *Trifolium*), Josef Holub (*Avenula* s.l.), Anna Skalická (*Corothamnus*, i.e. *Cytisus*), Vladimír Skalický (*Agrimonia*) and Miroslav Smejkal (*Camelina*, *Euphrasia* and *Scleranthus*). The taxonomic studies of Jan O. Martinovský, a retired secondary school teacher, substantially improved the level of knowledge on European species of the genus *Stipa*. Jindřich Chmelař, based at the University of Agriculture in Brno, became one of the few worldwide specialists in the genus *Salix*; his best-known achievement is a brief monograph on European willows aimed at a broader public (Chmelař and Meusel 1976). The second group of taxonomic studies employed biosystematic approaches, most frequently chromosome counting, but also experimental crossing and studies of breeding systems. These include the studies by Ivan Klášterský and Irena Novotná (*Arabis hirsuta* agg.), Irena Klášterská (*Rosa*), Anna Chrtková (née Žertová; *Lotus corniculatus* agg. and *Vicia cracca* agg.), Miloslav Kovanda (*Campanula rotundifolia* agg.), Jiří Soják (*Potentilla*), Pavel Tomšovic (*Rorippa*), Václav Zelený (*Leucanthemum*) and later also by Jan Kirschner (*Luzula*). Some of these authors contributed taxonomic treatments to large international projects such as Flora Europaea, Flora Iranica and Flora of Turkey. From the mid-1970s most of the experts were involved in preparing the Flora of the Czech Republic, which limited the number of biosystematic studies.

Several Czech plant taxonomists were interested in apomictic groups. Numerous microspecies in *Alchemilla* were described by Alexander Plocek during the 1980s and early 1990s. Miloslav Kovanda described the first apomictic species of *Sorbus* (*S. bohemica*) in 1961 and continued studies on this genus until the 1990s; however, some of his conclusions were recently refuted. While the studies of J. Holub on *Rubus* were primarily descriptive, based on field observations and detailed morphological comparisons, the studies on *Taraxacum* by Jan Kirschner and Jan Štěpánek, started in the early 1980s, soon adopted a range of biosystematic methods. Both researchers have become leading worldwide specialists in this genus.

The four postwar decades also brought some local floristic reports and floras, published as books or in journals. Examples are the flora of the Pardubice District by Jan Hadač and Emil Hadač, written during WWII, flora of the Plzeň District by Emil Hadač and collaborators, flora of the Mladá Boleslav District by Čeněk Novotný, flora of the surroundings of Horažďovice in south-western Bohemia by Josef Vaněček and ruderal flora of Brno by František Grüll. Probably the last systematic contribution of a German-speaking botanist to research on the Bohemian flora was the flora of the surroundings of Šluknov in northern Bohemia written by the amateur botanist Hans Marschner (1912–1989). He was allowed to remain in the country after WWII and continued exploring the local flora before moving to Bavaria in 1969. His manuscript was translated into Czech by his friends and published when he was already abroad. A very valuable flora was written by Josef Šourek, a former officer in the Czechoslovak Army who moved to the Krkonoše Mts after his retirement in 1947 and spent the last 20 years of his life studying the local flora (Šourek 1969).

### 2.8.6 *Bryology and Lichenology*

Bryology in postwar Czechoslovakia was shaped by two students of Professor Podpěra. Jan Šmarda, who succeeded Podpěra at Masaryk University in 1952, studied especially bryophyte communities in the Slovak Carpathians and published a bryoflora of the Hrubý Jeseník Mts (Šmarda 1952). Josef Duda (1925–2012) spent his professional career in the Silesian Museum in Opava, where he studied the taxonomy of liverworts. He also initiated and carried out a detailed mapping of the distribution of liverworts in Czechoslovakia, which he co-authored with Jiří Váňa. The most notable amateur bryologist in this period was Zdeněk Pilous (1912–2000), a primary and later secondary school teacher and local museum worker, who is perhaps best known in the Czech Republic for compiling the last two complete bryofloras. The first one included only mosses (Pilous 1948), and the second, written together with J. Duda, comprised all bryophytes (Pilous and Duda 1960). Professor Jiří Váňa at Charles University in Prague is most renowned for his taxonomic studies on the liverwort family *Jungermanniaceae* and many other groups. He also participated in the compilation of the *Index Hepaticarum*, the World checklist of liverworts.

The leading personality in Czech academic lichenology was Zdeněk Černohorský (1910–2001), who was a Professor at Charles University in 1959–1977 and the first lichenologist to recognize the importance of UV fluorescence in lichen taxonomy (Černohorský 1950). He was the main editor of the only comprehensive key to the macrolichens of former Czechoslovakia (Černohorský et al. 1956). The two co-authors of this book, Miroslav Servít (1886–1959) and Josef Nádvorník (1906–1977), were lichen taxonomists. The most important of their taxonomic studies are those on *Verrucariaceae* (e.g. Servít 1954) and *Physciaceae* (e.g. Nádvorník 1940, 1947), but both of them also made local lichenofloristic surveys. The most influential Czech lichenologist was Antonín Vězda (1920–2008; Figs. 2.5j and 2.7d), who published more than 300 scientific papers. He published numerous taxonomic papers and monographic studies mainly on gyalectoid lichens (e.g. Vězda 1958), taxonomic-chorological papers related to the former Czechoslovakia (e.g. Vězda 1970, 1978) and co-authored two supplements to the identification key for European lichens (Poelt and Vězda 1977, 1981). His lichen exsiccatae, *Lichenes Bohemoslovakiae exsiccati*, *Lichenes rariores exsiccati* and *Lichenes selecti exsiccati,* became renowned; the latter, issued in 1960–1991, reached 2500 in number. In the late 1970s Vězda turned his attention to tropical foliicolous lichens. Ten genera described by Vězda, sometimes with co-authors, are accepted in the latest Czech lichen checklist (Liška et al. 2008). From the 1970s onwards attention was also paid to lichens as bioindicators of air pollution in the studies by Jiří Liška and collaborators.

## 2.8.7 Vegetation Studies

Several members of the postwar generation of botanists engaged in vegetation research, which was supported by newly established institutions. The main aim of the Geobotanical Laboratories of the Czechoslovak Academy of Sciences, established in the 1950s, was the preparation of a vegetation map for this country, a project led by an experienced field ecologist, Rudolf Mikyška, and involving several young researchers such as Miroslava Husová, Jaroslav Moravec, Robert Neuhäusl and Zdenka Neuhäuslová. The final product was published as a map of reconstructed natural vegetation (conceptually similar to potential natural vegetation of Tüxen 1956) at a scale of 1: 200,000 (Mikyška et al. 1968–1972). The extensive field work done during the preparation of this map yielded hundreds of vegetation-plot records, many of them with detailed soil analyses, and resulted in the consolidation of knowledge on vegetation diversity of Czech forests and other vegetation types. Results were systematically published in journals, especially *Preslia* and *Folia Geobotanica et Phytotaxonomica*, and in the book series *Vegetace ČSSR*, of which 15 volumes were published from 1965 to 1986. This accumulated knowledge was summarized in a new overview of vegetation classes, orders and alliances in Czechoslovakia, prepared by a team at the Institute of Botany in Průhonice (Holub et al. 1967).

Parallel research on forest vegetation was done by foresters of the Forest Management Institute, where the leading personalities were Alois Mezera, Karel Mráz and Věroslav Samek. The experience of extensive field sampling by forest site researchers conducted in the 1950s–1960s was summarized in the national classification system of forest site types. Another system of forest site types, reflecting mainly forest diversity in the Carpathians, was created by Alois Zlatník at the University of Agriculture in Brno (now Mendel University). Both systems were united by Karel Plíva and Eduard Průša in 1971 and the resulting system was used for forest site type mapping in the whole country at a scale 1: 10,000. Eduard Průša also carried out a very detailed survey of old-growth forests (Průša 1985). Forest history was studied especially by Josef Nožička (Nožička 1957).

Department of Botany at Charles University also contributed significantly to vegetation research in the 1950s–1980s. Jan Jeník, the successor of Jaromír Klika at the Department of Botany of Charles University, conducted studies on the diversity and ecology of alpine and subalpine vegetation in the Sudetes (Jeník 1961) and together with Jarmila Kubíková, Jiřina Slavíková and other colleagues also focused on dry grassland vegetation in Bohemia. Besides traditional phytosociological studies aimed at classifying vegetation, plant ecologists at this department were pioneering, at a national scale, a new approach to vegetation research that applied methods of quantitative ecology, involving detailed measurements of the characteristics of vegetation and environmental factors, which were analysed statistically. The most influential proponents of this new approach were young scientists such as Marcel Rejmánek in the 1970s and Jan Lepš and Karel Prach in the 1980s. Results of the studies conducted by several staff members and students are mainly summarized in a monograph on vegetation-environment relationships at Oblík hill in the forest-steppe region of northern Bohemia (Slavíková et al. 1983) and in a monograph on succession in abandoned fields in the Bohemian Karst (Osbornová et al. 1990). Unfortunately, the leading personalities were forced to leave Charles University due to political repressions after the 1968 Warsaw Pact invasion: J. Jeník and J. Kubíková left in 1971 and M. Rejmánek in 1977. Jan Jeník obtained a job at the Institute of Botany, Czechoslovak Academy of Sciences, J. Kubíková in state nature conservation institutions, and M. Rejmánek also at an institute of the Czechoslovak Academy of Sciences before ultimately leaving for the USA in 1983. In spite of these devastating losses of personnel, vegetation research continued at Charles University, with Jiřina Slavíková, the leader of the geobotanical group, trying to give as much support to students and colleagues as possible under the generally oppressive conditions imposed by the political establishment and the Department's head R. Hendrych.

Vegetation studies also continued at Masaryk University (in 1960–1990 renamed J. E. Purkyně University), but here the botanical research was even more adversely affected by the communist regime than at Charles University. Jan Šmarda, successor of J. Podpěra as the Department head in the 1950s, studied mainly mountain vegetation in the Hrubý Jeseník Mountains and bryophyte communities in the Tatras. He was forced to leave the university in 1961, being subsequently employed in Brno institutes of the Czechoslovak Academy of Sciences. A new vegetation research programme was developed by Jiří Vicherek in the 1960s, which focused on phytosociology of Pontic-Pannonian vegetation including dry grasslands, sandy grasslands and saline vegetation at a broad scale ranging from Moravia to Eastern and South-eastern Europe (e.g. Vicherek 1973). In 1975 Vicherek was also expelled from the university and his applications to obtain an alternative job in science were declined; he had to accept a position as a technician on a cooperative farm. Consequently, very few vegetation studies were conducted at Masaryk University in the late 1970s and 1980s, all of which were vegetation descriptions of local significance.

In spite of the worsened political situation after 1968, vegetation research successfully continued at the Institute of Botany of the Czechoslovak Academy of Sciences. Vegetation scientists at the Institute of Botany were able to keep in contact with the international community even in these difficult times, especially through the symposia of the International Association for Vegetation Science (IAVS) organized by Professor Reinhold Tüxen in Germany. After Tüxen died in 1980, the format of these symposia was changed and the first symposium of the new series was organized by Jaroslav Moravec and Robert Neuhäusl in Prague in 1982. This event was crucial for the transformation of the IAVS into a truly international English-speaking organization with symposium venues in different countries. Jaroslav Moravec (1929–2006) made a significant contribution to the development of the (International) Code of Phytosociological Nomenclature, first published in 1976 (Barkman et al. 1976). He also led the compilation of more detailed and updated phytosociological overviews of Czech vegetation types at the level of associations (Moravec et al. 1983, 1995). Robert Neuhäusl (1930–1991) started and coordinated the international project of the Map of Natural Vegetation of Europe, but after his untimely death the project coordination was taken over by Udo Bohn in Germany (Bohn et al. 2000–2003). Since the 1960s researchers at the Institute of Botany also systematically studied the flora and vegetation of man-made habitats (Slavomil Hejný, Karel Kopecký, Zdeněk Kropáč and others) and related studies were also conducted by researchers not affiliated to the Institute of Botany, e.g. Antonín Pyšek in western Bohemia and František Grüll in Brno. Recent successful research on plant invasions at the Institute of Botany is partly based on this tradition.

Ecosystem research on grassland and wetland ecosystems, to some extent within the framework of the International Biological Programme, was carried out at the Brno branch of the Institute of Botany under the leadership of Milena Rychnovská (Rychnovská 1993). Her team included, among others, the grassland specialist Emilie Balátová-Tuláčková. Ecosystem studies were also successfully developed at the Hydrobotanical Department of the Institute of Botany in Třeboň, which focused mainly on southern Bohemian fishponds and related wetlands and was led by Slavomil Hejný, Dagmar Dykyjová and Jan Květ (Dykyjová and Květ 1978).

The palaeoecological group at the Brno branch of the Institute of Botany, in particular Kamil Rybníček, Eliška Rybníčková, Vlasta Jankovská and Helena Svobodová, carried out palaeobotanical studies throughout the former Czechoslovakia, and contributed significantly to the knowledge of Late Pleistocene and Holocene vegetation (Kuneš et al. 2009), interpretation of different vegetation types for vegetation mapping, and the ecology and classification of mire vegetation.

## 2.9 Recent Period (Since 1989)

The collapse of the communist regime in November 1989 ('Velvet Revolution') started a transition towards liberal democracy and a market economy. This transition was accompanied by tensions between the Czech and Slovak political elites, which affected a large part of the Czech and Slovak populations. The 1992 election results offered only limited options for creating a stable Czecho-Slovak federal government. Therefore, following a parliamentary decision, on 1 January 1993 Czechoslovakia was peacefully separated into two independent countries, the Czech Republic and Slovak Republic. The tensions between Czechs and Slovaks gradually decreased, and the current relationships between the two countries are very friendly and cooperative. The Czech Republic joined NATO in 1999 and the European Union in 2004.

### 2.9.1 Institutions

The 1990s were extremely difficult for the Czechoslovak Academy of Sciences, as its budget was substantially reduced due austerity measures introduced by the government and even its existence was challenged by some politicians and the media. The Academy lost one third of its employees and some institutes were dissolved. With the end of Czechoslovakia, it was transformed and renamed to the Academy of Sciences of the Czech Republic, later renamed as The Czech Academy of Sciences. It continues to exist as a network of research institutes. The long-serving director of the Institute of Botany Slavomil Hejný was replaced by Robert Neuhäusl in 1990 and, after Neuhäusl's untimely death in 1991, by Jan Štěpánek, who led the institute through its most difficult period. Since the early 1990s the institutional funding of basic research has been combined with a system of grants from the newly established Czech Science Foundation. The scientific infrastructure and research support substantially improved with the country's accession to the European Union.

The devastated botany departments of universities needed to be restored in the early 1990s. At Charles University Professor Jan Jeník and his assistant Jarmila Kubíková returned to their former positions in 1990 and new staff members came mainly from the Institute of Botany in Průhonice. Professor Jiří Vicherek was able to resume his work at the restored botanical department at Masaryk University in Brno, and his natural personal authority facilitated the department's rehabilitation. The present Department of Botany and Zoology came into existence by the merge of the former Department of Botany and Department of Zoology and Ecology in 2006. Also at Palacký University the quality of the field of botany was substantially improved during the 1990s, based on the research activities of young staff members. Several new universities were established in the 1990s, of which the most important, at least in biological sciences, is the University of South Bohemia in České Budějovice. Its Faculty of Biology, now reorganized and renamed to Faculty of Science, which includes the Department of Botany, quickly reached the scientific level of the botanical departments at the traditional universities. The key to this success was the involvement of scientists from several biological research institutes of the Czech(oslovak) Academy of Sciences that the government in the 1980s decided to move from Prague to České Budějovice.

### 2.9.2 Czech Botanical Society

The Czechoslovak Botanical Society was also affected by the changes in Czech society after the collapse of the communist regime. In 1990 Slavomil Hejný was replaced as the Society's President by its former Scientific Secretary Josef Holub (1930–1999; Fig. 2.9b), who served in this position until his death in 1999. Vladimír Řehořek, Lubomír Hrouda and Karel Prach served as the next presidents. The Society got back its status of an independent legal entity but lost some members due to the separating off of the Plant Physiology Section to form the Czech Society of Experimental Plant Biology in 1992, and the Algology Section to form the Czech Phycological Society in 2003. Already in 1992 the Society returned to its original name, Czech Botanical Society. However, the relationship with the sister society in Slovakia remained friendly and close: once every three years since 1991 the Summer School of Field Botany takes place in Slovakia as a joint event of both societies. The Society's Bryological-Lichenological Section started a series of workshops called the Autumn Bryological-Lichenological Days in 1988 at Svätý Júr near Bratislava, which since then have been held annually, and Spring Meetings of the Bryological-Lichenological Section started 6 years later.

The tradition of the Society's decennial congresses continued with events held in Olomouc (in 1992), Lednice in southern Moravia (2002) and Prague (2012). For the 2012 congress, organized to celebrate the Society's 100th anniversary, a special issue of Preslia was published, including reviews of Czech flora and vegetation and updated checklists and Red Lists of vascular plants and the bryophyte flora (Danihelka et al. 2012; Grulich 2012; Kučera et al. 2012). Regardless of the scientific topics, the main message of the last two congresses was that the Czech botanists had reintegrated themselves into the Central European botanical community, which was documented by the participation of many botanists from neighbouring countries.

The journals issued by the Society have remained the same since the 1980s, only the title of the Society's bulletin was changed to reflect the new name of the Society. The main journal, Preslia, with Josef Holub (1990–1999) and later Petr Pyšek (since 1999) as Editors in Chief, was profiled as a leading journal with geographical focus on Central Europe. Its inclusion in the Web of Science database in 2003 and the growing impact factor have had a stimulating effect on Czech botany (Pyšek et al. 2014).

### 2.9.3 *Studies on Vascular Flora*

The publication of the Flora of the Czech Republic continued, although with increasingly longer intervals between consecutive volumes. Bohumil Slavík, the main editor of volumes 4–6, died in 2004, and the following volumes were edited by Jitka Štěpánková, assisted mainly by Jindřich Chrtek Jr. and Zdeněk Kaplan. The eight volumes published so far contain altogether 5005 pages and include accounts of 2952 species authored by 68 botanists. The plates with plant drawings, from the first volume onwards drawn by Anna Skoumalová-Hadačová and Eva Smrčinová, are based mainly on fresh plants collected in the field. A field guide under preparation by Josef Holub since the 1960s unfortunately never advanced beyond the preparatory stage. The urgent need for a pocket guide was met by the publication of the Key to the Flora of the Czech Republic (*Klíč ke květeně České republiky*; Kubát et al. 2002). Texts for the new pocket guide were prepared by 45 botanists, coordinated by seven editors, with Karel Kubát as editor in chief. The second, extensively revised and updated edition was prepared by a team led by Zdeněk Kaplan and submitted to the publisher in 2016.

Although the work on Slavík's distribution atlas of the Czech flora was discontinued, the research into the country's flora intensified. Already in the early 1990s Holub started to prepare the second version of the Red List of vascular plants: he organized a workshop and produced a draft list in 1995. The second edition of the Red List was finalized after Holub's death under the editorship of František Procházka (Holub and Procházka 2000). A year earlier the Red Data Book, authored by botanists from the Czech Republic and Slovakia and including both countries, was published (Čeřovský et al. 1999). These two publications stimulated further research on red-listed species.

An important achievement was a comprehensive inventory of alien plants, Catalogue of Alien Plants of the Czech Republic, produced by a team led by Petr Pyšek and published in two editions (Pyšek et al. 2002, 2012). It immediately became a basis for numerous ecological and biogeographical analyses and applications at the national scale and subsequently also continental and global scales. Currently the Department of Ecology of the Institute of Botany in Průhonice, led by Petr Pyšek, is a globally recognized centre of excellence for research on plant invasions, which has published many studies on alien plants both with a national and international focus.

Since 2002 records of red-listed, rare or recently introduced species provided by numerous botanists have been summarized annually in the *Additamenta ad floram Reipublicae Bohemicae* (Hadinec et al. 2002), with 15 instalments produced so far. This systematic registration of the records of rare and red-listed plants facilitated the work on the second edition of the Catalogue of Alien Plants of the Czech Republic (Pyšek et al. 2012) and the third edition of the Red List (Grulich 2012).

Flora mapping was made easier by the introduction of electronic databases. At the Institute of Botany in Průhonice the Database of the Distribution of Vascular Plants in the Czech Republic (*FLDOK*, *Floristická dokumentace*) was started in 1992 for storing published plant records. Several regional projects of flora mapping realized in the 1990s were designed to collect records of plant occurrences in grids that could be integrated into future distribution atlases. Examples include the mapping of the flora in the Podyjí National Park in south-western Moravia by Vít Grulich, Křivoklátsko Protected Landscape Area in central Bohemia by a team led by Jiří Kolbek and the White Carpathians by Jan W. Jongepier and collaborators. Some other recording schemes remain unpublished but are stored in databases. A regional database of plant records was established by the South Bohemian Branch of the Czech Botanical Society. In parallel, the Species Occurrence Database of the Nature Conservation Agency of the Czech Republic was developed, primarily for conservation purposes; it is by far the largest database of biological records (mainly of vascular plants), based mainly on field surveys of natural habitats, nature reserves and other protected areas. Large numbers of species occurrence records are also stored in the Czech National Phytosociological Database established in 1996. In 2009 the website florabase.cz established by Jiří Danihelka and collaborators made almost nine million plant records from six databases available to the public. This opened the way for a new mapping project launched by a team headed by Zdeněk Kaplan in 2014, based on a distribution record database and mapping tools integrated within the newly established Pladias database of Czech flora and vegetation (www.pladias.cz). Apart from a large collection of data on biological traits and ecological characteristics of the Czech flora, this database includes records of species distributions from seven large or middle-sized databases that represent all the main national data sources (Kaplan et al. 2015). This project is based on the co-operation of specialists from several institutions and dozens of amateur botanists. The database records were in most cases rigorously checked against herbarium records by taxonomic specialists. By November 2017 maps in ~5 × 5.5 km grid were published for 458 taxa and others are being prepared.

The taxonomic research combined traditional methods with multivariate morphometrics, chromosome counting, ploidy surveys across large areas using flow cytometry, studies of breeding systems and molecular methods. The studies on apomictic genera employing traditional methods focused mainly on the genus *Rubus* (Josef Holub, Bohumil Trávníček and others). The knowledge on species of *Taraxacum* has greatly improved, including descriptions of new species (Jan Kirschner, Jan Štěpánek, Bohumil Trávníček, Radim Vašut and others). Czech apomictic species of *Sorbus* were revised, using multivariate morphometrics, flow cytometry and molecular methods (Martin Lepší and collaborators), which resulted in a substantial correction of earlier opinions. The studies on the genera *Hieracium* and *Pilosella* included cultivation, chromosome counting and analyses of breeding systems (Jindřich Chrtek Jr., František Krahulec, Anna Krahulcová and Olga Rotreklová). Special attention was paid to endemic plants, such as *Cerastium alsinifolium* and *Dianthus arenarius* subsp. *bohemicus* (Petr Vít, Jan Suda and collaboratos). Other taxonomically difficult genera and polyploid complexes were studied in detail for example by Petr Bureš, Jiří Danihelka, Libor Ekrt, Zdeněk Kaplan, Filip Kolář, Petr Koutecký, Petr Šmarda, Milan Štech and Jan Suda. Many taxonomic studies involved collaborators from abroad and had a broader geographic scope than just the Czech Republic (see also Pyšek et al. 2014).

### *2.9.4 Bryology and Lichenology*

In bryological research the main focus has shifted from classical taxonomy and phytogeography towards molecular systematics, population genetics, and ecology including molecular ecology, ecophysiology and conservation biology. The most active centres of bryological research and education in the post-2000 period became the University of South Bohemia, Masaryk University and the University of Ostrava. Several active bryologists are also in local and regional museums. Significant improvements in the knowledge of the Czech bryoflora have been reflected in several updates of checklists and Red Lists of Czech bryophytes, the most recent published by Kučera et al. (2012).

The current knowledge on the taxonomy, sensitivity, endangerment and distribution of lichens is summarized in the Checklist and Red List of lichens of the Czech Republic (Liška et al. 2008). Lichenological research continues mainly at the Institute of Botany in Průhonice, Charles University, the Czech University of Life Sciences in Prague, University of South Bohemia, National Museum, West Bohemian Museum in Plzeň and several regional museums. Current research is diverse in including studies on lichen photobionts (ecology, mycobiont-photobiont specificity and taxonomy), lichenicolous fungi (taxonomy and distribution) and lichenized fungi (ecology, diversity, chorology, taxonomy and molecular phylogeny). Lichenofloristic research has greatly intensified in the new millennium, resulting in the addition of more than 100 new lichen taxa to the latest checklist of Czech lichens (Liška and Palice 2010).

### 2.9.5 Vegetation Studies

The change in the political regime in 1989 was also followed by a boom in vegetation research, which was resumed at both Charles University and Masaryk University with the return of Jan Jeník and Jiří Vicherek, respectively, who became professors, and arrival of new young researchers. Vegetation research also started at the new University of South Bohemia in České Budějovice. At this university, the group of Jan Lepš is studying various general aspects of plant community assembly and that of Karel Prach deals mainly with vegetation succession and its application in restoration ecology. The 40th Symposium of the International Association for Vegetation Science was hosted by this university in 1997. Vegetation research also continues at the Institute of Botany of The Czech Academy of Sciences, where Tomáš Herben and František Krahulec started a research programme focused on spatiotemporal processes in grassland vegetation, Zdenka Neuhäuslová and colleagues updated the vegetation map of the Czech Republic (Neuhäuslová et al. 1998) and Jaroslav Moravec started a book series aiming at more detailed descriptions of all plant communities of the country, but only four volumes devoted to forest vegetation were published and the series has remained unfinished (Moravec 1998–2003).

The first attempts to create a national database of vegetation plots were made at the Institute of Botany in Průhonice already in the 1980s (Neuhäuslová and Kolbek 1982), however, the database was developed only after John Rodwell of Lancaster University (UK) offered to train a group of young Czech vegetation ecologists in vegetation database management and Stephan Hennekens of Wageningen University (the Netherlands) provided his software Turboveg for this purpose. The Czech National Phytosociological Database was established in 1996 by Milan Chytrý as a national collaborative effort coordinated at Masaryk University (Chytrý and Rafajová 2003). This database and the analytical software Juice developed by Lubomír Tichý (Tichý 2002) provided the infrastructure for the project Vegetation of the Czech Republic, in which a team from Masaryk University, Institute of Botany of The Czech Academy of Sciences and other institutions used more than

139,000 vegetation plots to critically evaluate the concepts of all the plant associations occurring in this country, created a computer expert system for their identification and prepared descriptions, distribution maps and comparative synoptic tables of the floristic composition of these vegetation types (Chytrý 2007–2013). An applied by-product of these synthetic vegetation studies was the Habitat Catalogue of the Czech Republic (Chytrý et al. 2001, 2010), which provided a simplified vegetation classification for the purpose of habitat mapping. An ambitious project of habitat mapping at a scale of 1: 10,000 based on a legend defined in this catalogue was realized in 2001–2004 across this country, supervised by the Nature Conservation Agency of the Czech Republic. Since 2006 this mapping has been regularly updated by field surveys (Härtel et al. 2009). The results of these projects were presented to the international scientific community at the 17th Meeting of European Vegetation Survey (Brno 2008) and the 58th Symposium of the International Association for Vegetation Science (Brno 2015).

Research into forest vegetation dynamics has also flourished since the 1990s. The group of Tomáš Vrška in Brno (now at The Silva Tarouca Research Institute for Landscape and Ornamental Gardening) started studies on natural forest dynamics, including resurveying of forest reserves sampled by Eduard Průša in the 1970s. Later on, the research programme on forest dynamics was also established by the group of Miroslav Svoboda at the Czech University of Life Sciences in Prague. Forest history and changes in species diversity of forest vegetation have been studied at the Brno branch of the Institute of Botany of The Czech Academy of Sciences by the team of Radim Hédl and Péter Szabó. The Forest Management Institute digitized field records collected during the forest site type research running since the 1950s and created the second largest vegetation-plot database in this country (Zouhar 2012).

A new palaeoecological group established at Charles University under the leadership of Petr Kuneš created the Czech Quaternary Palynological Database (Kuneš et al. 2009). The analysis of this database using the Landscape Reconstruction Algorithm resulted in a new reconstruction of the history of the Holocene vegetation in the Czech Republic (Abraham et al. 2016). Important studies on vegetation history in the 2000s and 2010s were also carried out by Petr Pokorný and palaeoecologists based at Masaryk University, University of South Bohemia and Institute of Botany of The Czech Academy of Sciences.

Other remarkable research programmes have recently been developed for example on the ecology of clonal plants and grassland diversity by Leoš Klimeš and Jitka Klimešová in the Třeboň branch of the Institute of Botany, on mire ecology by Michal Hájek at Masaryk University, on population and metapopulation ecology of rare plants by Zuzana Münzbergová at Charles University and on ecology and management of grasslands by Michal Hejcman at the Czech University of Life Sciences in Prague and Vilém Pavlů in the Crop Research Institute in Liberec.

In summary, currently Czech botany is capitalizing on a strong tradition of field surveys and a wealth of data and herbarium material on the country's flora. At the same time current botanists are using the most up-to-date research methods ranging from laboratory analyses of plant genomes to statistical evaluation of large databases of plant occurrence and vegetation records. This work is benefitting from international scientific cooperation. The state of the art of current knowledge is presented in the chapters of this book.

**Acknowledgements** We thank Zdeněk Kaplan and Petr Pyšek for comments on the manuscript of this chapter, Lubomír Hrouda, Jan Kirschner and Jan Štěpánek for advice on diverse topic of botanical history, Kryštof Chytrý for preparation of the figure plates, Tony Dixon for English proofreading and Irena Axmanová for her help with proof correcting. Karel Sutorý provided us the portrait of J. Kablik, Jan Kirschner the photograph of the drawing of *Gagea bohemica* and H. Michalcová the photograph of her father, Antonín Vězda. We also thank the curators of the herbaria of Charles University, National Museum in Prague and Institute of Botany of The Czech Academy of Sciences for permissions to reproduce herbarium labels from their collections. This work was supported by the Czech Science Foundation (project no. 14-36079G, Centre of Excellence Pladias).

## References

Abraham V, Kuneš P, Petr L, Svitavská Svobodová H, Kozáková R, Jamrichová E, Švarcová MG, Pokorný P (2016) A pollen-based quantitative reconstruction of the Holocene vegetation updates a perspective on the natural vegetation in the Czech Republic and Slovakia. Preslia 88:409–434

Anders J (1906) Die Strauch- und Blattflechten Nordböhmens. Anleitung zum leichten und sicheren Bestimmen der in Nordböhmen vorkommenden Strauch- und Laubflechten. Böhm. Leipa

Anders J (1928) Die Strauch- und Laubflechten Mitteleuropas. Anleitung zum Bestimmen der in Mitteleuropa vorkommenden Strauch- und Laubflechten. G. Fischer, Jena

Barkman JJ, Moravec J, Rauschert S (1976) Code of phytosociological nomenclature. Code der pflanzensoziologischen Nomenklatur. Code de nomenclature phytosociologique. Vegetatio 32:131–185

Beck G (1890, 1893) Flora von Nieder-Österreich 1, 2. Carl Gerold's Sohn, Wien

Beck G (1903–1927) Flora Bosne, Hercegovine i novopazarskog sandžaka 1–3. In: Glasnik zemaljskog muzeja u Bosni i Hercegovini 16–33 (Vols 1, 2), Sonderausgaben der Kgl. Serbischen Akademie 63(15) (Vol 3)

Berchtold F, Seidl WB (onwards Berchtold F, Opiz PM) (1836–1843) Oekonomisch-technische Flora Böhmens 1–3. Prag

Bernhardi JJ (1799) Lichenum gelatinosorum illustratio. Journal für die Botanik (Schrader) [1]:1–27

Bohn U, Neuhäusl R, Gollub G, Hettwer C, Neuhäuslová Z, Raus T, Schlüter H, Weber H (eds) (2000–2003) Karte der natürlichen Vegetation Europas. Maßstab 1: 2 500 000. Teil 1–3/Map of the natural vegetation of Europe. Scale 1: 2 500 000. Part 1–3. Landwirtschaftsverlag, Münster

Čelakovský L (1867–1881) Prodromus der Flora von Böhmen. I.–IV. In: Archiv für die naturwissenschaftliche Landesdurchforschung von Böhmen, sect. 3a, 1–4: 1–955. Prag

Čelakovský L (1868–1883) Prodromus květeny české. I–IV [Prodromus of the Bohemian flora. I–IV]. In: Archiv pro přírodovědecký výzkum Čech, sect. 3a, 1–4:1–944. Praha

Čelakovský L (1879) Analytická květena česká. F. Tempský, Praha [2nd edn in 1887, 3rd edn in 1897]

Čelakovský L (1882) Resultate der botanischen Durchforschung Böhmens im J. 1881. Sitzungsberichte der Königl. böhm. Gesellschaft der Wissenschaften in Prag 1881:360–395

Čelakovský L (1885) Bericht der Commission für die Flora von Deutschland 1884. X. Böhmen. Berichte der Deutschen Botanischen Gesellschaft 3:CXXXIX–CXLIII

Černohorský Z (1950) Fluorescence of lichens in ultra-violet light. Genus *Parmelia* Ach. Studia Botanica Čechoslovaca 11:98–100

Černohorský Z, Nádvorník J, Servít M (1956) Klíč k určování lišejníků ČSR [Key for identification of lichens of Czechoslovakia] 1. Nakladatelství Československé akademie věd, Praha

Čeřovský J, Feráková V, Holub J, Maglocký Š, Procházka F (eds) (1999) Červená kniha ohrožených a vzácných druhů rostlin a živočichů ČR a SR 5. Vyšší rostliny [Red data book of threatened and rare plants and animals of the Czech Republic and Slovak Republic 5. Higher plants]. Príroda, Bratislava

Chmelař J, Meusel W (1976) Die Weiden Europas: die Gattung *Salix*. A. Ziemsen Verlag, Wittenberg-Lutherstadt [re-editions in 1979, 1986 and 2004]

Chytrý M (ed) (2007–2013) Vegetace České republiky 1–4 [Vegetation of the Czech Republic 1–4]. Academia, Praha

Chytrý M, Rafajová M (2003) Czech National Phytosociological Database: basic statistics of the available vegetation-plot data. Preslia 75:1–15

Chytrý M, Kučera T, Kočí M (eds) (2001) Katalog biotopů České republiky [Habitat Catalogue of the Czech Republic]. Agentura ochrany přírody a krajiny ČR, Praha

Chytrý M, Kučera T, Kočí M, Grulich V, Lustyk P (eds) (2010) Katalog biotopů České republiky [Habitat Catalogue of the Czech Republic], 2nd edn. Agentura ochrany přírody a krajiny ČR, Praha

Corda AJC (1829) Genera Hepaticarum. Die Gattungen der Lebermoose. In: Opiz PM (ed), Beiträge zur Naturgeschichte. [Forsetzung des] Naturalientausch[es] 12:643–655, Prag [reprint Annales Bryologici 10:9–15, 1937]

d'Elvert C (1868) Zur Geschichte der Pflege der Naturwissenschaften in Mähren und Schlesien. A. Ritsch, Brünn

Danihelka J (2008) Hvězdnice sivá (*Aster canus*), Christian Ferdinand Hochstetter a dva málo známé prameny ke květeně Moravy [*Aster canus*, Christian Ferdinand Hochstetter and two less known sources of information on the flora of Moravia]. Zprávy České botanické společnosti 43:1–16

Danihelka J, Chrtek J Jr, Kaplan Z (2012) Checklist of vascular plants of the Czech Republic. Preslia 84:647–811

Dědeček J (1883) Mechy jatrovkovité (*Hepaticae*) květeny české [Liverworts of the Bohemian flora]. In: Archiv pro přírodovědecké proskoumání Čech, Botanické oddělení 5(4):1–79. Praha

Dědeček J (1886) Die Lebermoose (*Hepaticae*) Böhmens. In: Archiv für naturwissenschaftliche Landesdurchforschung von Böhmen, Botanische Abtheilung 5(4):1–71. Prag

Deyl M (1946) Study of the genus *Sesleria*. Opera Botanica Čechica 3:1–256

Domin K (1904) České středohoří. Studie fytogeografická [České středohoří Mts. A phytogeographic study]. Praha

Domin K (1923) Problémy a metody rostlinné sociologie a jejich použití pro výzkum lučních a pastvinných porostů republiky Československé [Problems and methods of plant sociology and their application in the research on meadows and pastures of the Czechoslovak Republic]. Ministerstvo zemědělství republiky Československé, Praha

Domin K (1928) The plant associations of the valley of Radotín. Preslia 7:3–68

Domin K (1935) Plantarum Čechoslovakiae enumeratio. Preslia 13–15:1–305

Domin K (1940) Program vlastivědného výzkumu domácí květeny na podkladě regionalistickém [A programme of research into homeland's flora on a regional base]. Věda přírodní 20:2–8

Domin K (1942) Prodromus květeny mšenské [A prodromus of the flora of the Mšeno area]. Acta Botanica Bohemica 13(1941):1–236

Domin K, Podpěra J (1928) Klíč k úplné květeně republiky Československé [Key to the complete flora of the Czechoslovak Republic]. R. Promberger, Olomouc

Dostál J (1938) *Centaurea scabiosa* v Československu [*Centaurea scabiosa* in Czechoslovakia]. Spisy vydávané přírodovědeckou fakultou Karlovy university 160:1–43

Dostál J (1948–1950) Květena ČSR [Flora of Czechoslovakia] 1–2. Praha

Dostál J (1954, 1958) Klíč k úplné květeně ČSR [A key to the complete flora of the Czechoslovak Republic]. Edn 1, 2. Nakladatelství Československé akademie věd, Praha

Dostál J (1982) Seznam cévnatých rostlin květeny československé [Checklist of vascular plants of the Czechoslovak flora]. Pražská botanická zahrada, Praha-Troja

Dostál J (1989) Nová květena ČSSR [A new flora of the Czechoslovak Socialist Republic]. Academia, Praha

Drabek J (2012) Vladimir Krajina: World War II hero and ecology pioneer. Ronsdale Press, Vancouver

Dykyjová D, Květ J (eds) (1978) Pond littoral ecosystems: structure and functioning. Springer, Berlin

Fiek E (1881) Flora von Schlesien preussischen und österreichischen Antheils, enthaltend die wildwachsenden, verwilderten und angebauten Phanerogamen und Gefäss-Cryptogamen. J. U. Kern's Verlag, Breslau

Firbas F (1949, 1952) Spät- und nacheiszeitliche Waldgeschichte Mitteleuropas nördlich der Alpen 1, 2. G. Fischer, Jena

Firbas F, Sigmond H (1928) Vegetationsstudien auf dem Donnersberge im Böhmischen Mittelgebirge. Lotos 76:113–172

Formánek E (1887, 1897) Květena Moravy a rakouského Slezska [Flora of Moravia and Austrian Silesia] 1, 2. Brno [Vol 1], Praha [Vol 2]

Franc M (2015) Ke 125. výročí vzniku České akademie císaře Františka Josefa pro vědy, slovesnost a umění. I. Vznik a vědy o živé přírodě. [On the 125th anniversary of the Emperor Franz Joseph Czech Academy of Sciences, Letters and Arts. I. Establishment and life sciences]. Živa 2015:50–52

Futák J, Domin K (1960) Bibliografia k flóre ČSR do r. 1952 [Bibliography to the flora of Czechoslovakia until 1952]. Vydavateľstvo Slovenskej akadémie vied, Bratislava

Grabowski H (1843) Flora von Ober-Schlesien und dem Gesenke. A. Gosohorsky, Breslau

Grulich V (2012) Red list of vascular plants of the Czech Republic: 3rd edition. Preslia 84:631–645

Hadinec J, Lustyk P, Procházka F (eds) (2002) Additamenta ad floram Reipublicae Bohemicae. I. Zprávy České botanické společnosti 37:51–105

Haenke T (1791) Die botanischen Beobachtungen auf der Reise nach dem Böhmischen Riesengebirge. In: Jirasek J, Gruber T, Haenke T, Gerstner F (eds) Beobachtungen auf Reisen nach dem Riesengebirge. Waltherische Hofbuchhandlung, Dresden, pp 31–159

Halda JP (2009) Historie lichenologického bádání v Orlických horách a na Králickém Sněžníku [History of lichenological research in the Orlické hory Mts and Mt. Králický Sněžník]. Orlické hory a Podorlicko 15(2008):141–164

Härtel H, Lončáková J, Hošek M (eds) (2009) Mapování biotopů v České republice. Východiska, výsledky, perspektivy [Habitat mapping in the Czech Republic. Background, results and perspectives]. Agentura ochrany přírody a krajiny ČR, Praha

Haslinger F (1869) Botanisches Excursionsbuch für den Brünner Kreis und das angrenzende Gebiet. Buschak und Irrgang, Brünn [2nd edn in 1880]

Hejný S, Moravec J, Rychnovská M, Holub J (1978) Advances in geobotany, plant ecology, taxonomy and geography in the Czech Socialist Republic during last 30 years. Folia Geobotanica et Phytotaxonomica 13:337–349

Hejný S, Slavík B, Chrtek J, Tomšovic P, Kovanda M (eds) (1988) Květena České socialistické republiky 1 [Flora of the Czech Socialist Republic 1]. Academia, Praha

Hilitzer A (1925) Étude sur la végetation épiphyte de la Bohême. Spisy vydávané přírodovědeckou fakultou Karlovy university 41:1–202

Hochstetter G [recte C] F (1825) Uebersicht des Merkwürdigsten aus Mährens Flora. Flora (Regensburg) 8:513–525, 529–537

Holub J (1959) Několik kritických poznámek k 2. vydání Dostálova "Klíče k úplné květeně ČSR" [Some critical remarks on the 2nd edition of Dostál's "Key to the complete flora of the Czechoslovak Republic"]. Preslia 31:432–440

Hejný S, Holub J, Neuhäuslová Z, Toman J (1982) 70 let Československé botanické společnosti [70 years of the Czechoslovak Botanical Society]. Zprávy Československé botanické společnosti, Příloha 1982(1):1–146

Holub J, Procházka F (2000) Red List of vascular plants of the Czech Republic – 2000. Preslia 72:187–230
Holub J, Hejný S, Moravec J, Neuhäusl R (1967) Übersicht der höheren Vegetationseinheiten der Tschechoslowakei. Rozpravy Československé akademie věd, Řada matematických a přírodních věd 77(3):1–75
Holub J, Procházka F, Čeřovský J (1979) Seznam vyhynulých, endemických a ohrožených taxonů vyšších rostlin květeny ČSR (1. verze) [The list of extinct, endemic and threatened taxa of vascular plants of the Czech Socialistic Republic (1st version)]. Preslia 51:213–237
Hrouda L (2005) Radovan Hendrych a pražská univerzitní botanika [Radovan Hendrych and the Prague university botany]. Preslia 77:333–337
Humboldt FA (1793) Florae Fribergensis Specimen. Plantas Cryptogamicas praesertim subterraneas exhibens. H. A. Rottmann, Berolini
Ilg W (2012) Josephine Kablik (1787–1863). Erforscherin der Flora des Riesengebirges. Hoppea 73:7–26
Janko J (1997) Vědy o životě v českých zemích 1750–1950 [Life Sciences in the Czech Lands 1750–1950]. Archiv Akademie věd České republiky, Praha
Jeník J (1961) Alpinská vegetace Krkonoš, Králického Sněžníku a Hrubého Jeseníku [Alpine vegetation of the Krkonoše, Králický Sněžník and Hrubý Jeseník Mts]. Nakladatelství Československé akademie věd, Praha
Jeník J (1986) Dvousté výročí výpravy České společnosti nauk do Krkonoš [Bicentenary of the Czech Science Society's Expedition to the Krkonoše Mountains]. Opera Corcontica 23:53–76
Kaplan Z, Danihelka J, Štěpánková J, Bureš P, Zázvorka J, Hroudová Z, Ducháček M, Grulich V, Řepka R, ... Wild J (2015) Distributions of vascular plants in the Czech Republic. Part 1. Preslia 87:417–500
Kirschner J, Kirschnerová L, Štěpánek J (2007) Generally accepted plant names based on material from the Czech Republic and published in 1753–1820. Preslia 79:323–365
Klášterský I, Hrabětová-Uhrová A, Duda J (1982) Dějiny floristického výzkumu v Čechách, na Moravě a ve Slezsku I [History of floristic research in Bohemia, Moravia and Silesia I]. Severočeskou přírodou, Příloha 1982(1):1–242
Klement O (1930) Die Pflanzendecke unserer Heimat. Heimatkunde Bezirkes Komotau 1(5):1–128. Komotau
Klika J (1928) Une étude géobotanique sur la végétation de Velká Hora près de Karlštejn. Bulletin international, Academie Tchèque des Sciences, Classe des sciences mathématiques, naturelles et de la médecine 29:17–42
Klika J (1941) Přehled společenstev, svazů, řádů a tříd středoevropského jevnosnubného rostlinstva [An overwiev of communities, alliances, orders and classes of Central European phanerogamic vegetation]. In: Klika J, Novák V (eds) Praktikum rostlinné sociologie, půdoznalství, klimatologie a ekologie [Handbook of plant sociology, soil science, climatology and ecology]. Melantrich, Praha, pp 53–70
Klika J (1948) Rostlinná sociologie (fytocenologie) [Plant sociology (phytocoenology)]. Melantrich, Praha
Klika J (1955) Nauka o rostlinných společenstvech (fytocenologie) [Science of plant communities (phytocoenology)]. Nakladatelství Československé akademie věd, Praha
Klika J, Hadač E (1944) Rostlinná společenstva střední Evropy [Plant communities of Central Europe]. Příroda 36:249–259, 281–295
Koerber GW (1855) Systema lichenum Germaniae. Trewendt & Granier, Breslau
Körber GW (1859–1865) Parerga lichenologica. Trewendt, Breslau
Kosteletzky VF (1824) Clavis analytica in floram Bohemiae phanerogamicam. Typis Sommerianis, Pragae
Kovář F (1909) Moravské hávnatky [Moravian species of *Peltigera*]. *Peltigera* Willd. – hávnatka. Časopis Vlasteneckého spolku musejního v Olomouci 26:13–18, 80–86
Kovář F (1912) Moravské druhy rodu *Cladonia* [Moravian species of *Cladonia*]. Věstník Klubu přírodovědeckého v Prostějově 15:85–190, 193–199

Krahulec F (2012) History of the studies on the flora and vegetation in the Czech Republic. Preslia 84:397–426

Krajina V (1930) Monografická studie druhů *Festuca varia* (Haenke) a *Festuca versicolor* (Tausch) Krajina [A monographic study on *Festuca varia* and *Festuca versicolor*]. Spisy vydávané přírodovědeckou fakultou Karlovy university 106:1–46

Krocker AI (1787–1823) Flora silesiaca renovata… 1. Sumptibus Guilielmi Theophili Kornii, Vratislaviae

Krombholz JV (1837) Topographisches Taschenbuch von Prag zunächst für Naturforscher und Aerzte. Gottlieb Haase Söhne, Prag

Kubát K, Hrouda L, Chrtek J Jr, Kaplan Z, Kirschner J, Štěpánek J (eds) (2002) Klíč ke květeně České republiky [Key to the flora of the Czech Republic]. Academia, Praha

Kučera J, Váňa J, Hradílek Z (2012) Bryophyte flora of the Czech Republic: updated checklist and Red List and a brief analysis. Preslia 84:813–850

Kuneš P, Abraham V, Kovářík O, Kopecký M, PALYCZ contributors (2009) Czech Quaternary Palynological Database – PALYCZ: review and basic statistics of the data. Preslia 81:209–238

Lack HW (2000) A garden for eternity: the codex Liechtenstein. Benteli Verlag, Berne

Laus H (1907) Die Halophytenvegetation des südlichen Mährens und ihre Beziehungen zur Flora der Nachbargebiete. Mitteilungen der Kommission zur naturwissenschaftlichen Durchforschung Mährens, Botanische Abteilung 3:1–67. Brünn

Laus H (1908a) Mährens Ackerunkräuter und Ruderalpflanzen. Zugleich ein Beitrag zur Phytogeographie des Landes. Mitteilungen der Kommission zur naturwissenschflichen Durchforschung Mährens, Land- und forstwirtschafliche Abteilung 2:1–269. Brünn

Laus H (1908b) Schulflora der Sudetenländer mit besonderer Rücksicht auf Mähren, mit pflanzengeographischen und biologischen Hinweisen zum Schulgebräuche und für botanische Exkursionen. F. Irrgang, Brünn [2nd edn in 1911]

Laus H (1927) Květena Petrštýna ve Vysokých Jesenících se zvláštním zřetelem k rozšíření našich arkticko-alpských druhů vrb [Flora of Mt. Petrovy kameny in the Hrubý Jeseník Mts, with focus on the distribution of our Arctic-alpine willows]. Časopis Vlasteneckého spolku muzejního v Olomouci 39:27–52

Liška J (1992) Two hundred years of lichenology in Czechoslovakia. Bryonora 9:20–25

Liška J (2012) Lichen flora of the Czech Republic. Preslia 84:851–862

Liška J, Palice Z (2010) Červený seznam lišejníků České republiky (verze 1.1) [Red List of lichens of the Czech Republic (version 1.1)]. Příroda 29:3–66

Liška J, Palice Z, Slavíková Š (2008) Checklist and Red List of lichens of the Czech Republic. Preslia 80:151–182

Lorinser G (1854) Botanisches Excursionsbuch für die deutsch-österreichischen Kronländer und das angrenzende Gebiet. Tendler & comp, Wien

Maiwald V (1904) Geschichte der Botanik in Böhmen. C. Fromme, Wien

Makowsky A (1863) Die Flora des Brünner Kreises. Verhandlungen des Naturforschenden Vereines in Brünn 1(1862):45–210

Mann W (1825) Lichenum in Bohemia observatorum dispositio succinctaque descriptio. Typis Sommerianis, Pragae

Matthiolus PA (1565) Commentarii in sex libros Pedacii Dioscoridis Anazarbei de Medica materia. Ex Officina Valgrisiana, Venetiis

Mattuschka HG (1776, 1777) Flora silesiaca 1, 2. Wilhelm Gottlieb Korn, Leipzig

Mattuschka HG (1779) Enumeratio stirpium in Silesia sponte crescentium in usum herborisantium ... Sumptibus Guilielmi Theophili Kornii, Vratislaviae

Merker G (1910) Exkursionsflora für Mähren u. Österreichisch-Schlesien, Mährisch-Weißkirchen

Mikyška R, Deyl M, Holub J, Husová M, Moravec J, Neuhäusl R, Neuhäuslová-Novotná Z (1968–1972) Geobotanická mapa ČSSR 1. České země [Geobotanical map of the CSSR 1. Bohemian lands]. Vegetace ČSSR, Ser A, 2:1–204. Academia, Praha. [Explanatory text 1968, map sheets 1968–1972]

Moravec J (ed) (1998–2003) Přehled vegetace České republiky. Svazek 1–4 [Vegetation survey of the Czech Republic. Volume 1–4]. Academia, Praha

Moravec J, Balátová-Tuláčková E, Hadač E, Hejný S, Jeník J, Kolbek J, Kopecký K, Krahulec F, Kropáč Z, ... Vicherek J (1983) Rostlinná společenstva České socialistické republiky a jejich ohrožení [Red List of plant communities of the Czech Republic and their endangerment]. Severočeskou přírodou, Příloha 1983:1–110

Moravec J, Balátová-Tuláčková E, Blažková D, Hadač E, Hejný S, Husák Š, Jeník J, Kolbek J, Krahulec F, ... Vicherek J (1995a) Rostlinná společenstva České republiky a jejich ohrožení [Red List of plant communities of the Czech Republic and their endangerment], 2nd edn. Severočeskou přírodou, Příloha 1995:1–206

Nábělek F (1923) Iter Turcico-Persicum. Pars 1. Plantarum collectarum enumeratio (*Ranunculaceae–Dipsacaceae*). Spisy vydávané přírodovědeckou fakultou Masarykovy university 35:1–144

Nádvorník J (1940) Systematische Übersicht der mitteleuropäischen Arten der Flechtenfamilie *Caliciaceae*. I. Preslia 18–19:113–129

Nádvorník J (1947) *Physciaceae* Tchécoslovaques. Studia Botanica Čechoslovaca 8:69–124

Neuhäuslová Z, Kolbek J (1982) Seznam vyšších rostlin, mechorostů a lišejníků střední Evropy užitých v bance geobotanických dat BÚ ČSAV [List of vascular plants, bryophytes and lichens of Central Europe used in the geobotanic database of the Institute of Botany of the Czechoslovak Academy of Sciences]. Botanický ústav ČSAV, Průhonice

Neuhäuslová Z, Blažková D, Grulich V, Husová M, Chytrý M, Jeník J, Jirásek J, Kolbek J, Kropáč Z, ... Sádlo J (1998) Mapa potenciální přirozené vegetace České republiky. Textová část [Map of potential natural vegetation of the Czech Republic. Explanatory text]. Academia, Praha

Novák FA (1924) Monografická studie o *Dianthus monspessulanus* (L.) s.l. [A monographic study on *Dianthus monspessulanus* L. s.l.] Spisy vydávané přírodovědeckou fakultou Karlovy university 21:1–50

Novák FA (1962) Padesát let Československé botanické společnosti [Fifty years of the Czechoslovak Botanical Society]. Preslia 34:1–30

Novák J (1888) Lišejníky okolí německobrodského a seznam lišejníků v Čechách objevených, které scházejí v okolí německobrodském [Lichens occurring in the surroundings of Německý Brod and a list of lichens found in Bohemia but missing from the Německý Brod surroundings]. In: Archiv pro přírodovědecký výzkum Čech, Oddíl botanický 7(1):1–65. Praha

Novák J (1893) Die Flechten der Umgebung von Deutschbrod nebst einem Verzeichnis der überhaupt in Böhmen entdecken Arten. In: Archiv für naturwissenschaftliche Landesdurchforschung von Böhmen, Botanische Abtheilung 7(1):1–66. Prag

Nožička J (1957) Přehled vývoje našich lesů [An overview of the development of our forests]. Státní zemědělské nakladatelství, Praha

Oborny A (1879) Die Flora des Znaimer Kreises. Verhandlungen des Naturforschenden Vereines in Brünn 17(1878):105–304

Oborny A (1883–1886) Flora von Mähren und österr. Schlesien. Pars 1–4. Verhandlungen des Naturforschenden Vereines in Brünn 21(1882):1–268, 22(1883):269–636, 23(1884):637–888, 24(1885):889–1285

Oborny A (1885) Bericht der Commission für die Flora von Deutschland 1884. XI. Mähren (excl. des Höhenrückens des mähr. Gesenkes). Berichte der Deutschen Botanischen Gesellschaft 3:CXLIV–CXLVIII

Oborny A (1905, 1906) Die Hieracien aus Mähren und österr. Schlesien. Verhandlungen des Naturforschenden Vereines in Brünn 43(1904):135–276, 44(1905):1–79

Opiz FM (1852) Seznam rostlin květeny české [List of plants of the Bohemian flora]. In: Spisů musejních no 44:1–216. F. Řivnáč, Praha

Opiz PM (1823) Böheims phänerogamische und cryptogamische Gewächse. C. W. Enders, Prag

Osbornová J, Kovářová M, Lepš J, Prach K (eds) (1990) Succession in abandoned fields. Kluwer, Dordrecht

Otruba J (1930) Květena Štramberka [Flora of Štramberk]. Štramberk

Pascher A (1915) Die Süsswasser-Flora Deutschlands, Österreichs und der Schweiz. Heft 5: *Chlorophyceae* II. G. Fischer, Jena

Pascher A (1927) Die Süsswasser-Flora Deutschlands, Österreichs und der Schweiz. Heft 4: *Volvocales* = *Phytomonadinae*. *Flagellatae* IV. = *Chlorophyceae* I. G. Fischer, Jena

Pax F (1915) Schlesiens Pflanzenwelt. Eine pflanzengeographische Schilderung der Provinz. G. Fischer, Jena

Petrak F (1910) Beiträge zur Flora von Mähren. Allgemeine botanische Zeitschrift für Systematik, Floristik, Pflanzengeographie etc. 16:4–6, 20–23

Pilous Z (1948) Naše mechy. Ilustrovaný klíč k určování mechů československých [Our mosses. An illustrated key for the identification of Czechoslovak mosses]. Československá botanická společnost, Praha

Pilous Z (1971) *Bryophyta*. Mechorosty. *Sphagnidae* – Mechy rašeliníkové. Flora ČSSR, Ser C, 1:1–412. Academia, Praha

Pilous Z, Duda J (1960) Klíč k určování mechorostů ČSR [Key for the identification of Czechoslovak bryophytes]. Nakladatelství Československé akademie věd, Praha

Pluskal FS (1849) Biographie der berühmten, jetzt lebenden Pflanzenforscherin Oesterreich's, Frau Josephine Kablik. F. Gastl, Brünn

Podpěra J (1906) Vývoj a zeměpisné rozšíření květeny zemí českých ve srovnání s poměry evropskými [Development and geographical distribution of the flora of the Bohemian lands in the European context]. Mor. Ostrava

Podpěra J (1911) Květena Hané. Základy zeměpisného rozšíření rostlinstva na Horním úvalu moravském [Flora of the Haná area. Basic patterns of geographic distribution of vegetation in the upper Morava valley]. Brno

Podpěra J (1926–1930) Květena Moravy ve vztazích systematických a geobotanických [Flora of Moravia in systematic and geobotanical context] Vols 1, 6(2), 6(3). Práce Moravské přírodovědecké společnosti 1(1924)(10):393–618, 1926; 2(1925)(10):271–782, 1927; 5(1928)(5):57–415, 1930

Podpěra J (1928a) Steppe und Waldsteppe des Hutberges oberhalb Pouzdřany (Pausram). Preslia 7:153–167

Podpěra J (1928b) Die Vegetationsverhältnisse der Pollauer Berge. Acta Botanica Bohemica 6–7:77–131

Podpěra J (1928c) Die Vegetationsverhältnisse im Gebiete des Mährischen Karstes im Vergleiche mit der nächsten Umgebung: ein Beitrag zur Pflanzengeographie Mährens. Moravská musejní společnost, Brno

Podpěra J (1942) *Bryum* generis monographiae Prodromus. Pars 1. Práce Moravské přírodovědecké společnosti 14(2):1–58

Podpěra J (1954) Conspectus muscorum Europaeorum. Nakladatelství Československé akademie věd, Praha

Poelt J, Vězda A (1977, 1981) Bestimmungsschlüssel europäischer Flechten. Ergänzungsheft I, II. Bibliotheca Lichenologica 9:1–258, 16:1– 390

Pohl JE (1809, 1814) Tentamen florae Bohemiae 1, 2. Prag

Pohl F (1941) Dr. Karl Preis: im großen deutschen Daseinskampfe gefallen. Lotos 88:159–163

Polívka F (1900–1904) Názorná květena zemí koruny české [An illustrated flora of the lands of the Bohemian Crown] 1–4. R. Promberger, Olomouc

Polívka F (1904) Klíč k určování rostlin v naší květeně nejčastěji se vyskytujících [A key to the most common plants of our flora]. R. Promberger, Olomouc

Polívka F (1912) Klíč k úplné květeně zemí koruny české [A key to the complete flora of the lands of the Bohemian Crown]. R. Promberger, Olomouc

Ponfikl JE (1822) Vollständiger Umriß einer statistischen Topographie des Königreiches Böhmen, vol 1. Fürst-Erzbischöfliche Buchdruckerey bei Josepha verwitteten Vetterle von Wildenbrunn, Prag

Pospíchal E (1881) Flora des Flussgebietes der Cidlina und Mrlina. In: Archiv für die naturwissenschaftliche Landesdurchforschung von Böhmen, Botanische Abtheilung 4(5):1–103. Prag

Pospíchal E (1882) Květena poříčí Cidliny a Mrliny. In: Archiv pro přírodovědecké proskoumání Čech, Botanické oddělení 4(5):1–99. Praha

Presl JS, Presl KB (1819) Flora Čechica. Kwětena česká. J. G, Calve, Pragae
Presl KB (1825–1835) Reliquiae haenkeanae 1, 2. J. G. Calve, Pragae
Průša E (1985) Die böhmischen und mährischen Urwälder – ihre Struktur und Ökologie. Vegetace ČSSR, Ser A, 15:1–577. Academia, Praha
Pyšek P, Sádlo J, Mandák B (2002) Catalogue of alien plants of the Czech Republic. Preslia 74:97–186
Pyšek P, Danihelka J, Sádlo J, Chrtek J Jr, Chytrý M, Jarošík V, Kaplan Z, Krahulec F, Moravcová L, ... Tichý L (2012) Catalogue of alien plants of the Czech Republic (2nd edition): checklist update, taxonomic diversity and invasion patterns. Preslia 84:155–255
Pyšek P, Kaplan Z, Chytrý M, Hrouda L (2014) Preslia: hundred years of exploring central-European flora and vegetation. Preslia 86:1–11
Rabenhorst L (1870) Kryptogamen-Flora von Sachsen, der Ober-Lausitz, Thüringen und Nordböhmen mit Berücksichtigung der benachbarten Länder. E. Kummer, Leipzig
Rohlena J (1922) Příspěvky k floristickému výzkumu Čech [Contributions to the floristic research in Bohemia]. Časopis Musea království českého, Oddíl přírodovědný 96:54–66
Rohlena J (1942) Conspectus Florae Montenegrinae. Preslia 20–21:1–506
Rohrer R, Mayer A (1835) Vorarbeiten zu einer Flora des Mährischen Gouvernements. R. Rohrer, Brünn
Rübel E (1930) Chronik der Fünften I. P. E. Veröffentlichungen des Geobotanischen Institutes Rübel in Zürich 6:7–29
Rudolph K (1928) Die bisherigen Ergebnisse der botanischen Mooruntersuchungen in Böhmen. I. Die Moortypen Böhmens. Beihefte zum Botanischen Centralblatt, Abteilung B 45:1–64
Rychnovská M (ed) (1993) Structure and functioning of seminatural meadows. Academia, Praha
Schlosser JC (1843) Anleitung, die im Mährischen Gouvernement wildwachsenden… R. Rohrer's sel. Wittwe, Brünn
Schmidt FW (1790–1792) Hortus Canalius, continens pictiones plantarum tum exoticarum, quum Bohemiae indigenarum, quae in horto illustrissimi Josephi comitis Malabaila de Canal coluntur, ad naturam delineatae... Centuria I–IV. Pragae
Schmidt FW (1793, 1794) Flora Boëmica inchoata, exhibens plantarum regni Boëmiae indigenarum species. Cent. 1:iv + 1–86(1793); Cent. 2:1–97(1793); Cent. 3:1–112(1794); Cent. 4:1–96(1794) [Plates to Cent. 1 and Cent. 2, with only two printed copies, are deposited in the National Library, Prague (16 AA 20)]
Schube T (1897) Ergebnisse der Durchforschung der schlesischen Phanerogamenflora im Jahre 1896. Jahresbericht der Schlesischen Gesellschaft für vaterländische Cultur, Abteilung 2b, 74:39–64
Schube T (1904) Flora von Schlesien, preussischen und österreichischen Anteils. Korn, Breslau
Sebeók A (1779) Dissertatio inauguralis medico-botanica de Tataria Hungarica. Typis Mathiae Andreae Schmidt, Viennae
Servít M (1954) Československé lišejníky čeledi *Verrucariaceae* [Czechoslovak lichens of the family *Verrucariaceae*]. Nakladatelství Československé akademie věd, Praha
Širjaev GI (1925–1937) *Onobrychis* generis revisio critica. Partes 1–3. Suppl 2. Spisy vydávané přírodovědeckou fakultou Masarykovy university 56:1–196, 76:1–165, 242:1–14
Slavík B (1986) Fytokartografické syntézy ČSR 1 [Phytocartographical syntheses of the Czech Socialist Republic 1]. Botanický ústav ČSAV, Průhonice
Slavíková J, Molíková M, Rejmánek M, Rydlo J, Studnička M, Studničková I, Suchara I, Štolcová-Březinová J (1983) Ecological and vegetational differentiation of solitary conic hill. Vegetace ČSSR, Ser A, 13:1–221. Academia, Praha
Sloboda D (1852) Rostlinnictví [Botany]. In: Spisů musejních no 42:1–733. F. Řivnáč, Praha
Šmarda J (1952) Mechorosty Hrubého Jeseníku [Bryophytes of the Hrubý Jeseník Mts]. Přírodovědecký sborník Ostravského kraje 13:447–488
Smejkal M (1980) Komentovaný katalog moravské flóry [Annotated catalogue of the Moravian flora]. Univerzita J. E. Purkyně v Brně, Brno [2nd edn in 1981]

Šourek J (1969) Květena Krkonoš. Český a polský Krkonošský národní park [Flora of the Krkonoše Mts. The Czech and Polish Krkonoše National Park]. Academia, Praha

Speta F (2000) Joachim Burser (1583–1639) als Pflanzensammler in Böhmen. Severočeskou přírodou 32:7–28

Spitzner V (1887) Květena okresu prostějovského a plumlovského [Flora of the Prostějov and Plumlov Districts]. Prostějov

Stein B (1879) Flechten. In: Cohn F (ed) Kryptogamen-Flora von Schlesien 2(2):1–400. Kern's Verlag, Breslau

Sternberg C (1820) Versuch einer geognostisch-botanischen Darstellung der Flora der Vorwelt. Heft 1. Leipzig, Prag

Sutorý K (1995) Nejstarší datované herbářové doklady z Moravy [The oldest dated plant specimens from Moravia]. Acta Musei Moraviae, Scientiae biologicae 79:73–79

Suza J (1925) Nástin zeměpisného rozšíření lišejníků na Moravě vzhledem k poměrům evropským [A sketch of geographic distribution of lichens in Moravia in the European context]. Spisy vydávané přírodovědeckou fakultou Masarykovy university 55:1–151

Suza J (1928) Guide géobotanique pour le terrain serpentineux près de Mohelno dans la Moravie du sud-ouest (Tchécoslovaquie). Bulletin international, Academie Tchèque des Sciences, Classe des sciences mathématiques, naturelles et de la médicine 29:313–346

Svoboda J (2011) Tři inkarnace [Three incarnations]. Radioservis, Praha

Svojtka M (2016) Das Fach Naturgeschichte und seine Fachvertreter an der Universität Prag von 1749 bis 1849. Berichte der Geologischen Bundesanstalt 118:95–105

Tannich A (1929) Bestimmungsbuch der Flora von Böhmen. H. Korkisch, Prag

Tausch IF (1823, 1825) Hortus canalius. Decas 1, 2. T. Haase, Pragae

Tichý L (2002) JUICE, software for vegetation classification. Journal of Vegetation Science 13:451–453

Tüxen R (1956) Die heutige potentielle natürliche Vegetation als Gegenstand der Vegetationskartierung. Zentralstelle für Vegetationskartierung, Stolzenau

Vacek V, Bureš P (2001) Botanika. Dějiny oboru na Masarykově univerzitě v Brně [Botany. History of the discipline at Masaryk University in Brno]. Folia Historica 70:1–101

Valentine DH (1962) Variation and evolution in the genus *Viola*. Preslia 34:190–206

Velenovský J (1891) Flora bulgarica. F. Řivnáč, Pragae

Velenovský J (1897) Mechy české [Bohemian mosses]. Rozpravy České akademie císaře Františka Josefa pro vědy, slovesnost a umění, Cl 2, 6(6):1–352

Velenovský J (1901–1903) Jatrovky české [Bohemian liverworts]. Vols 1–3. Rozpravy České akademie císaře Františka Josefa pro vědy, slovesnost a umění, Cl 2, 10(12):1–49, 11(3):1–24, 12(4):1–38

Vězda A (1958) Československé druhy rodu *Gyalecta* a *Pachyphiale* s klíčem a přehledem evropských druhů [Czechoslovak species of the genera *Gyalecta* and *Pachyphiale* with a key and an overview of the European species]. Sborník Vysoké školy zemědělské a lesnické v Brně, řada C 1958(1):21–56

Vězda A (1970, 1978) Neue oder wenig bekannte Flechten in der Tschechoslowakei. I, II. Folia Geobotanica et Phytotaxonomica 5:307–337, 13:397–420

Vězda A, Liška J (1999) Katalog lišejníků České republiky [A catalogue of lichens of the Czech Republic]. Institute of Botany, Academy of Sciences of the Czech Republic, Průhonice

Vicherek J (1973) Die Pflanzengesellschaften der Halophyten- und Subhalophytenvegetation der Tschechoslowakei. Vegetace ČSSR, Ser A, 5:1–200. Academia, Praha

Viniklář L (ed) (1931) Vývoj české přírodovědy [Development of Czech natural science]. Přírodovědecký klub, Praha

Walters SM (1962) Generic and specific concepts and the European flora. Preslia 34:207–226

Weidmann A (1895) Prodromus českých mechů listnatých [Prodromus of Bohemian leafy mosses] 1, 2. A. Wiesner, Praha

Wettstein R (1896) Monographie der Gattung *Euphrasia*. Wilhelm Engelmann, Leipzig

Wettstein R (1898) Grundzüge der geographisch-morphologischen Methode der Pflanzensystematik. G. Fischer, Jena

Wimmer F (1840) Flora von Schlesien. F. Kist, Breslau etc. [2nd edn in 1844, 3rd edn in 1857]
Wimmer F, Grabowski H (1827–1829) Flora Silesiae 1–3. G. T. Korn, Vratislaviae
Zauschner J (1776) Charaktere des Ornithogali Bohemici und der Erucae tenuifoliae perennis, flore luteo, Joannis Bauhini. Abhandlungen einer Privatgesellschaft in Böhmen 2:119–127
Zlatník A (1928a) Aperçu de la végétation des Krkonoše (Riesengebirge). Preslia 7:94–152
Zlatník A (1928b) Lesy a skalní stepi v Milešovském středohoří (Typologická a synekologická studie československých lesů) [Forests and rocky steppes in the Milešovské středohoří Mts (A typological and synecological study of Czechoslovak forests)]. Lesnická práce 7:65–80, 151–160, 209–228
Zouhar V (2012) Database of Czech Forest Classification System. Biodiversity & Ecology 4:346

# Chapter 3
# Flora and Phytogeography of the Czech Republic

**Zdeněk Kaplan**

**Abstract** An overview of the diversity of vascular plants in the Czech Republic is presented. This country is situated at the intersection of several assumed important European migration routes. Consequently, the flora is composed of almost all the floristic elements that occur in Central Europe, of which the Central European geo-element is dominant. The occurrence of various sorts of relicts is discussed in the context of the changes in vegetation caused by Pleistocene climatic fluctuations. An account of Czech endemics includes 82 species and subspecies, which is 2.2% of the total vascular plant diversity in this country. Patterns in the distribution and occurrence of endemics in different habitat types are described. Groups of species with similar ecogeographical features within the Czech Republic are distinguished as regional types of distribution. Phytogeographical division of the country is described and the phytogeographical units distinguished are shown on a map.

## 3.1 Introduction

The present-day flora in the Czech Republic reflects the country's geographical position, climate, vegetation history (particularly its glacial and postglacial development), diversity of bedrocks, topography and habitats, and effect of human activities on the landscape. The geographical position roughly in the centre of the European continent means that the flora in this country includes plant species from the cold north and warm south as well as the oceanic west and continental east. The great changes in climate that occurred during the Pleistocene, with many cold (glacial, stadial) and warm (interglacial, interstadial) periods, had major effects on the flora (Ložek 1973; Ložek in Hejný et al. 1988: 31–35). The distributions of species repeatedly contracted and expanded, or shifted up and down mountains.

The country is covered by a heterogeneous mosaic of cultural landscapes with arable fields, deciduous, mixed and coniferous forests, meadows, pastures and human settlements. The dominant type of natural vegetation is forest. Natural

Z. Kaplan (✉)
Institute of Botany, The Czech Academy of Sciences, 252 43 Průhonice, Czech Republic
e-mail: kaplan@ibot.cas.cz



M. Chytrý et al. (eds.), *Flora and Vegetation of the Czech Republic*, Plant and Vegetation 14, DOI 10.1007/978-3-319-63181-3_3

treeless vegetation includes alpine and subalpine grasslands, steep rocky slopes, steppe, peat bogs and natural water bodies.

This chapter is an updated and adjusted version of the paper by Kaplan (2012). It provides a basic overview of the Czech flora, its taxonomic diversity, remarks on glacial and postglacial relicts, accounts of endemic species and subspecies, most important geoelements and regional phytochorotypes and an outline of phytogeographical divisions.

Taxonomy and nomenclature mostly follow Kaplan et al. (2017a), which largely corresponds to a previous inventory of the Czech flora (Danihelka et al. 2012). For taxa that are not listed in that overview taxonomy and nomenclature follow Flora Europaea (Tutin et al. 1964–1980) or the literature cited.

## 3.2 Basic Overview of the Flora of the Czech Republic

The latest statistics on the Czech flora (Danihelka 2013) includes 148 families of vascular plants, 925 genera, 3754 species and subspecies and 618 hybrids. Genera with 30 or more species include *Taraxacum* (221 species), *Rubus* (127), *Carex* (85), *Hieracium* (59), *Pilosella* (59), *Veronica* (35) and *Trifolium* (34), four of which include agamospermous species, which accounts for the high diversity. Families richest in species are the *Asteraceae* (666 species), *Rosaceae* (315), *Poaceae* (273), *Fabaceae* (171), *Brassicaceae* (148), *Cyperaceae* (127), *Lamiaceae* (112), *Caryophyllaceae* (108) and *Apiaceae* (99).

Most of the taxa (83%) are native; aliens include more neophytes (77% of all aliens) than archaeophytes (Danihelka 2013). Within the Raunkiær system of life forms, the most frequently represented among Czech plants included in the Key to the Flora of the Czech Republic (Kaplan et al. 2017a) are hemicryptophytes (47%), followed by therophytes (22%), geophytes (12%), nanophanerophytes (7%), macrophanerophytes (5%), chamaephytes (4%) and hydrophytes (3%).

Human activity affects vegetation and results in considerable changes in the structure of floras. Of the vascular plants recorded in the Czech Republic since the beginning of floristic research, 156 species or subspecies (4% of the total Czech flora and 5% of the native flora) became extinct or vanished (e.g. *Anacamptis coriophora*, *Betula humilis*, *Galatella cana*, *Geranium bohemicum*, *Gymnadenia odoratissima*, *Herminium monorchis*, *Iris spuria*, *Linnaea borealis*, *Oenanthe fistulosa*, *Pedicularis sceptrum-carolinum*, *Peucedanum arenarium*, *Plantago altissima*, *Potamogeton compressus*, *Selaginella helvetica* and *Woodsia alpina*), while many others are endangered. The latest Red List of the Czech flora (Grulich 2012) includes a total of 1720 taxa, which is 46% of all vascular plants and 59% of the native taxa and archaeophytes. Habitats with the highest percentage of endangered plants are still water bodies, bottoms of drained ponds, wet sandy sites, fens, mires, wet meadows, saline habitats, dry grasslands, sand dunes and open woodlands. In contrast, the Czech flora is being enriched by introduced plants. The alien flora is discussed in detail by Pyšek et al. (Chap. 8, this book).

In addition to common and widespread Central European species, the Czech flora includes also diverse geoelements, boundary and outlying elements, glacial and postglacial relicts and local endemics. These are discussed in detail below.

## 3.3 Relicts

Because of the climate changes that occurred in glacial and interglacial phases and the associated great changes in the Central European landscape during the Pleistocene, no Tertiary relicts are thought to have survived in the area of the present-day Czech Republic and all relicts that persisted here are of Quaternary age or, probably more precisely, Late Quaternary. During the last glacial period, the landscape was predominantly treeless in Central Europe, dominated by steppe-tundra (West 2000; Stewart and Lister 2001; Granoszewski 2003; Müller et al. 2003; Ložek 2011; Binney et al. 2017), but small patches of boreal woodland or even more extensive forests existed especially in continental areas and wet meadows were found at favourable sites (Willis and van Andel 2004; Deffontaine et al. 2005; Jankovská and Pokorný 2008; Kuneš et al. 2008; Kramp et al. 2009; Magyari et al. 2010; Pokorný 2011; Hošek et al. 2014; Petr and Novák 2014; Petr et al. 2014; Rybníčková and Rybníček 2014). Many arctic, alpine, boreal and steppe species were more widespread in Central Europe during the last glacial period and in the Early Holocene (e.g. Szafer 1912; Tralau 1963; Lang 1994; Ložek 2007, 2009b; Birks and Willis 2008; Ehrich et al. 2008). Some of them later disappeared when there was a marked change in climate and forests became more widely distributed (Ložek in Hejný et al. 1988: 31–35; Birks and Willis 2008; Ložek 2011). Others had markedly restricted ranges but survived in small and fragmented populations in suitable habitats (refugia), which resulted in disjunct distributions (e.g. Rull 2008, 2010; Stewart et al. 2010). These relicts now occur mostly in a few small areas in the Czech Republic, in some cases even at a single or a few localities. Depending on the species requirements, their refugia are rock outcrops (mainly in deep river valleys), screes, erosion-prone steep slopes with landslides, serpentinite outcrops (particularly those covered with open pine forests), fens, mires, peat bogs, avalanche tracks in glacial cirques or natural subalpine to alpine grasslands above the timberline.

In contrast, human activity during the Middle and Late Holocene, particularly that now considered as "traditional landscape management", contributed to the local spread of several rare (and presumably relict) species of alpine and dry grasslands, fens and other treeless communities. For example, subalpine to upper-montane anthropogenic meadows in the Krkonoše Mts now host relict subalpine and alpine taxa such as *Viola lutea* subsp. *sudetica* (Fig. 3.1) and *Campanula bohemica*, but these meadows are below the natural timberline.

Relicts as discussed above were formerly widespread taxa that are currently restricted to refugia that provide a suitable combination of long-term stable ecological conditions. However, our understanding of the exact history and distribution of

**Fig. 3.1** Examples of mountain species in the Czech Republic: (**a**) *Doronicum austriacum*, Frantoly near Mičovice, southern Bohemia; (**b**) *Anemonastrum narcissiflorum*, Rýchorská bouda challet, Rýchory ridge, Krkonoše Mts; (**c**) *Delphinium elatum*, Malá Morávka, Hrubý Jeseník Mts; (**d**) *Listera cordata*, between Mt. Praděd and Mt. Petrovy kameny, Hrubý Jeseník Mts; (**e**) *Viola lutea* subsp. *sudetica*, Mt. Praděd, Hrubý Jeseník Mts. Photo credits: J. Lepš (a), M. Pospíšil (b), M. Mejstřík (c) and L. Bureš (d, e)

the populations that survived throughout the Holocene is fragmentary due to the lack of palaeoecological records. Continuous occurrence of species throughout the Holocene can only be directly confirmed for species with easily determinable and well-preserved fossils. The precise evaluation and unequivocal proof of relict status is difficult and there is no good evidence for the supposed relict status of many species. The relict status of a species is usually inferred from indirect evidence such as knowledge of its current ecological requirements, dynamics of its recent habitats, the pattern of its present distribution and the general knowledge of the Holocene vegetation history in the area studied. Consequently, many relict species are reported for the Czech Republic. For example, assumed arctic or alpine glacial relicts include *Rubus chamaemorus* (Holub in Slavík et al. 1995: 54–206; Slavík in Slavík et al. 1995: 20–40; Ložek 2009b), *Pedicularis sudetica* (Hendrych and Hendrychová 1988), *Conioselinum tataricum* (Holub in Slavík et al. 1997: 370–371), *Pinus mugo* (Skalický in Hejný et al. 1988: 291–298 and 308), *Luzula spicata* (Kirschner 1989) and *Gentiana pannonica* (Holub et al. 1970). Another group, including *Helictotrichon desertorum* (Holub 1962; Holub in Čeřovský et al. 1999: 179), some species of *Stipa* (Martinovský 1965, 1975), *Allium strictum* (Martinovský 1969; Holub et al. 1970; Krahulec and Duchoslav in Štěpánková et al. 2010: 647–677), *Orobanche coerulescens* (Zázvorka 1984) and *Bassia prostrata* (Tomšovic 1989), are considered to be glacial relicts of continental steppes.

However, there are a few well-documented relict occurrences in the Czech flora. Pollen records indicate that *Betula nana* was widespread during the last glacial and Early Holocene (e.g. Pokorný 2002; Svobodová et al. 2002; Jankovská 2007; Kuneš et al. 2008), but is now restricted to a few sites at old peat bogs in high mountains. Based on palaeobotanical finds (fossil fruits), Pokorný et al. (2010) report that metapopulations of *Cladium mariscus* survived in fens along the Labe River throughout the entire Holocene, but currently this species' distribution in Bohemia (Kaplan et al. 2015) is more restricted than the fossil record, which corroborates the previously assumed status of *C. mariscus* as a relict from at least the Early Holocene (Sádlo 2000). Additional data came from analyses of the genetic structure of populations. Hensen et al. (2010) report low genetic diversity within and high genetic differentiation between populations of *Stipa capillata*, which is thought to be a relict that experienced strong bottlenecks in Central Europe, enhanced by isolation and selfing. A strong genetic differentiation between populations of *Saxifraga paniculata* is attributed to genetic drift in isolated populations and interpreted by Reisch et al. (2003) as evidence of this species being a glacial relict in Central Europe. Kajtoch et al. (2016) list several examples of populations of steppe species that show moderate to high genetic distinctiveness in eastern Central Europe, which indicates they survived in this area during the glacial maxima.

Plant macrofossils and pollen records indicate that many arctic species were widespread in tundra at mid-altitudes in Central Europe during the last glacial period (Birks and Willis 2008). The rapid climatic amelioration that occurred at the Pleistocene/Holocene boundary and the associated expansion of forest eliminated much of the arctic flora, remnants of which occur in Northern Europe or at the highest altitudes. Some of the arctic species, however, survived in alpine grasslands or at similar treeless sites above the timberline or habitats ecologically similar to tundra, such as peat bogs and mires. These glacial arctic relicts include *Andromeda polifolia*, *Betula nana*, *Carex bigelowii*, *C. capillaris*, *C. limosa*, *C. rupestris*, *C. vaginata*, *Eriophorum vaginatum*, *Oreojuncus trifidus*, *Pedicularis sudetica* (Fig. 3.2),

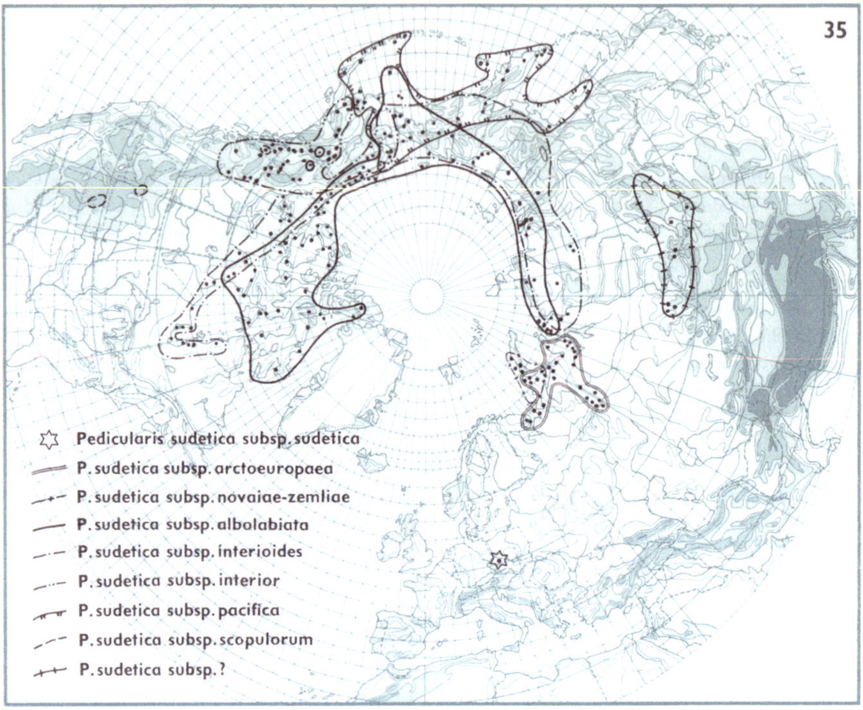


**Fig. 3.2** An isolated relict occurrence of *Pedicularis sudetica* subsp. *sudetica* in the Krkonoše Mts, along the Czech/Polish border, indicated by an asterisk with a dot in the middle. This is a very remote occurrence isolated from the main part of the distribution range of the *P. sudetica* complex, which has the highest diversity in the arctic and northern boreal areas (reproduced from Hultén 1971 with permission from Koeltz Scientific Books)

*Rhodiola rosea*, *Rubus chamaemorus* (Fig. 3.3), *Salix herbacea*, *S. lapponum*, *Saxifraga oppositifolia* and *Swertia perennis*.

While the arctic species spread from Northern Europe southwards and back with the waxing and waning of continental glaciers, those species that originated in Central European high mountains (particularly in the Alps and Carpathians) underwent shifts in altitude. After the retreat of the glaciers, these alpines recolonized their original mountain ranges but some of them persisted also in mountains in the Czech Republic. These Central European mountain relicts include *Adenostyles alliariae*, *Gentiana asclepiadea*, *Homogyne alpina*, *Hypochaeris uniflora*, *Pinus mugo* and *Primula minima*. Species distributed mainly in the Alps and Carpathians and now occurring in the Czech Republic in predominantly lowland refugia are, for instance, *Calamagrostis varia* and *Tofieldia calyculata*.

Another group includes continental plants that presumably colonized Central Europe via cold periglacial steppes (particularly the loess steppe, see Ložek 2009b) along the Sarmatian migration route north of the Carpathians (Martinovský 1969, 1971). During the Last Glacial Maximum, cold and arid steppe was the dominant

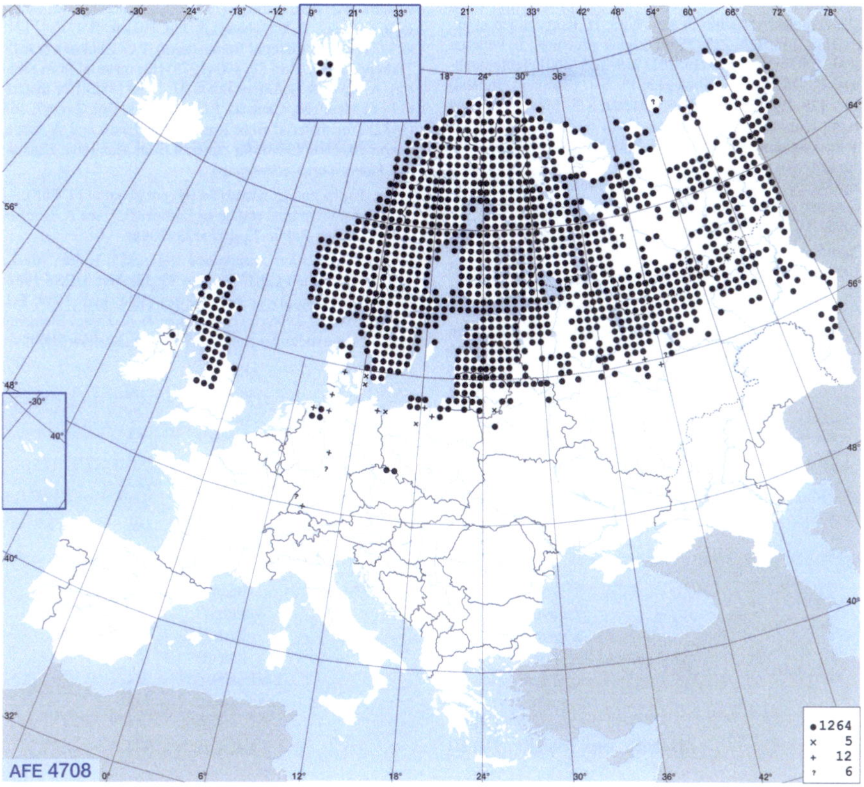


**Fig. 3.3** Distribution of *Rubus chamaemorus* in Europe. This species, belonging to the Sub-Arctic element, is widespread in Northern Europe, but it is found also in an isolated area in Central Europe, namely in peat bogs in the Krkonoše Mts, where it occurs as a glacial relict (reproduced from Kurtto et al. 2010 with permission of the Committee for Mapping the Flora of Europe, and Societas Biologica Fennica Vanamo)

type of vegetation across northern Eurasia south of approximately 57°N (Tarasov et al. 2000; Binney et al. 2017). This type of vegetation, sometimes called "steppe-tundra", included elements of modern steppic grassland and northern tundra. It does not have exact current analogues (Stewart and Lister 2001), but recent studies show that it was very similar to the cold continental landscapes of the southern Siberian mountain systems (Horsák et al. 2015; Chytrý et al. 2017). These cold steppes were widespread south and east of the Fennoscandian ice-sheet (Birks and Willis 2008) including the Czech lowlands, as indicated by fossil pollen (e.g. Kuneš et al. 2008). The species of continental steppe that are thought to have colonized Central Europe via the Sarmatian migration route in the Late Pleistocene are now often referred to as thermophilous species, but they are actually drought-adapted and eurythermic (cold-tolerant). Spreading westwards north of the Carpathians, they reached Bohemia and later survived in the warmest lowlands or adjacent hilly landscapes in its north-central part (see the distribution map of *Astragalus arenarius* in Kaplan et al. 2016b); less frequently they occur also in southern Moravia (see distribution maps of *Astragalus danicus* and *Helictotrichon desertorum* in Kaplan et al. 2016b). Other glacial relicts associated with the Sarmatian migration route are *Allium strictum* and *Jurinea cyanoides*.

Other species of continental steppe probably spread along the Pannonian migration route south of the Carpathians and reached southern Moravia, but not Bohemia. *Agropyron pectinatum*, *Bassia prostrata*, *Crambe tataria* (Fig. 3.4), *Jurinea mollis*, *Prunus tenella* (Fig. 3.4) and *Taraxacum serotinum* are the best examples, and *Crepis pannonica*, *Onosma arenaria*, *Phelipanche caesia*, *Phlomis tuberosa*, *Scorzonera austriaca* and *Trinia ucrainica* apparently also belong to this group. The Pannonian endemic *Artemisia pancicii* is at the north-western limit of its distribution in southern Moravia.

In addition to low temperatures, the climate in the lowlands in northern Eurasia was, in general, extremely dry during the last glacial period (Hubberten et al. 2004). Saline meadows and salt marshes were therefore probably more frequent at suitable sites in Central Europe than they are today (Janská et al. 2017). This vegetation largely disappeared later but saline soils locally remained treeless, which enabled light-demanding halophytes to survive there throughout the Holocene (Magyari et al. 2010). Putative glacial relicts associated with this kind of vegetation include *Glaux maritima*, *Plantago maritima*, *Salicornia perennans*, *Suaeda prostrata* and *Taraxacum bessarabicum*.

Climate warming at the end of the last glaciation and the Pleistocene-Holocene transition triggered major shifts in the distributions of species. Steppe persisted from the full glacial to the Early Holocene but cold steppe was gradually replaced by grass-rich steppe and forest-steppe, which occupied vast parts of Eastern and Central Europe (Magyari et al. 2010). Thermophilous species that during the Pleistocene retreated to more temperate refugia recolonized Central Europe. Many light-demanding species rapidly colonized open habitats where there was little competition. The thermophilous immigrants became mixed with eurythermic species, which were already present on cold steppes during the last glaciation.

Thermophilous plants migrated along several dispersal routes (e.g. Slavík in Slavík et al. 1995: 20–40; Taberlet et al. 1998; Hewitt 1999; Petit et al. 2002; Sádlo 2007; Parisod 2008; Ložek 2009a, b, 2011; Magyari et al. 2010; Windmaißer et al. 2016). The main migration routes for the re-colonization of the area of present-day Czech Republic passed through southern Moravia. The Sub-Mediterranean element spread from the south along the Illyrian-Noric migration route that passes along the eastern edge of the Alps. The West-Sub-Mediterranean element migrated along the Rhône-Rhine pathway (following the main valleys of these rivers) and the Upper Danube Corridor north of the Alps into the western part of Central Europe and reached Bohemia from the south-west. Regardless of their origin and the period of their migration, some of the previously widespread thermophilous or eurythermic species now occur in refugia and are considered to be relicts. These perhaps include *Astragalus exscapus* (see the distribution map in Kaplan et al. 2016b), *Linum flavum*, *Pedicularis exaltata* and *Viola ambigua*.

**Fig. 3.4** Examples of thermophilous species in the Czech Republic: (**a**) *Dianthus gratianopolitanus*, Malý Bezděz hill near Doksy, northern Bohemia; (**b**) *Vicia pannonica* subsp. *pannonica*, Podhradí near Jičín, north-eastern Bohemia; (**c**) *Astragalus exscapus*, Raná hill, České středohoří Mts, northern Bohemia; (**d**) *Melampyrum arvense*, Čestice near Častolovice, eastern Bohemia; (**e**) *Prunus tenella*, botanical garden, Prague; (**f**) *Crambe tataria*, Janská hora hill, Dunajovické kopce hills, southern Moravia. Photo credits: Z. Kaplan (a–d) and L. Hrouda (e, f)

Fens (calciphilous and eutrophic mires) were widely distributed in the open landscape of the Early Holocene (Sádlo 2000) but were later mostly eliminated as a consequence of forest expansion. Sádlo (2000) suggests that species of open fens may have occurred in Central Europe throughout the entire Holocene in shifting gaps in alder carrs. Cyclic alternations of open fens and alder carrs in lowland floodplains (Pokorný et al. 2000; Pokorný 2011) may have created small refugia surrounded by woodlands. These might have provided conditions for uninterrupted survival of open fens in the wooded phases of the Holocene. Similar hypotheses were proposed to account for the survival of relict plant populations in a dynamic mosaic of open habitats in mountain floodplains (Sádlo and Bufková 2002) and for the long-term survival of fen species at the meta-population level in river basins (Pokorný et al. 2010). The best example of a relict species from the glacial and early postglacial calciphilous and eutrophic mires is *Cladium mariscus*, while *Betula humilis*, *Calamagrostis stricta*, *Carex buxbaumii*, *Eriophorum gracile*, *Ligularia sibirica*, *Schoenus ferrugineus* and *S. nigricans* may also belong to this group.

Increasing humidity in the late-glacial and early postglacial periods also enabled the expansion of boreal taiga. Species that are now typical of boreal coniferous forests were previously probably more widespread in Central Europe. Examples of assumed postglacial relicts of boreal taiga include *Actaea europaea*, *Linnaea borealis*, *Rhododendron tomentosum*, *Rubus saxatilis*, *Stellaria longifolia* and *Trientalis europaea*.

Steppe grasslands, boreal taiga and other habitats characteristic of the Early Holocene were later increasingly replaced by deciduous forests. No relicts from more recent periods are distinguished.

## 3.4 Endemics

### *3.4.1 Definition of Concepts*

The distribution of endemics is usually defined in terms of conspicuous topographical features, such as a mountain range or an island, or other natural elements, such as a specific habitat of relict character. In Central Europe, many endemics occur in alpine and subalpine habitats. As the Czech national borders often coincide with peaks and ridges in summit areas, a majority of narrow-range endemics occurring in these border areas are included in the floras of two countries. Two of three highest mountain groups in the Sudetes, Krkonoše Mts (Karkonosze in Polish) and the Králický Sněžník Mts (Śnieżnik Kłódzki in Polish), are situated on the border with Poland. Many of the species endemic to them occur both in the Czech Republic and Poland, although their entire range covers an area of a few square kilometres. None of these endemics would be formally designated as endemic if only endemism within political borders was considered. Because of their very restricted ranges, also species and subspecies with distributions that only slightly exceed the borders of this country (generally by less than 1 km) and for which the majority of localities are in the Czech Republic, are also listed as endemics in this review. In addition, two *Sorbus* endemics confined to a small area in the valley of the Dyje River along the border between Lower Austria and Moravia are reviewed here. All these taxa may be formally designated as Czech subendemics.

In terms of the extent of their distributions, most of the endemics discussed here may be classified as stenoendemics, i.e. endemics confined to a small geographical area. Some of them are even confined to a single locality such as a glacial cirque or a rocky summit.

### 3.4.2 Changes in the Understanding of Czech Endemics

There have been several attempts to compile a list of plants endemic to the Czech Republic. The first treatment of endemic species (as currently defined) that aimed to be complete and critically analyzed was provided by Hadač (1977), who listed 118 Czech endemics and subendemics (including apomictic taxa). He compiled records previously scattered in the literature and, in contrast, excluded many previous putative endemics on the basis of a re-evaluation of their taxonomy and distribution.

A revised list by Holub et al. (1979) appeared soon afterwards. It included 43 endemic taxa completely or almost completely confined to the Czech Republic. Of the latter, only species with a slight transgression beyond the state borders were admitted (similar approach to that adopted in this treatment). In addition to these Czech endemics, the authors listed an additional 16 endemics whose distributions distinctly extend from the Czech Republic into bordering countries (subendemics) and 32 mostly Central European endemics that mainly occur outside the Czech Republic but with a few localities in this country. These endemics, of which the Pannonian endemic *Artemisia pancicii* occurring in southern Moravia, eastern Austria and north-eastern Serbia (Danihelka 1995; Danihelka and Marhold 2003) is a good example, were not included in this review.

Czech (and Slovak) endemics were discussed also by Hendrych (1981a) who provided the shortest list. He intentionally omitted all species known to be apomictic and listed only six supposedly Czech endemics (*Campanula bohemica*, *C. gelida*, *Cerastium alsinifolium*, *Dianthus carthusianorum* subsp. *sudeticus*, *Melampyrum bohemicum* and *Poa riphaea*). Eleven endemics were mentioned in the phytogeographical characteristics of the Czech Republic by Slavík (in Hejný et al. 1988: 65–102). Gerža (2009) listed 29 sexually reproducing and 40 apomictic Czech endemics. Krahulec (2006) reviewed the endemics in the Krkonoše Mts (including their Polish part) and also mentioned several endemics that occur in other areas in the Czech Republic.

Two detailed reviews of Czech endemics appeared recently. Kaplan (2012) listed and annotated 74 species and subspecies endemic to the Czech Republic and adjacent border areas. Suda and Kaplan (2018) provided a comprehensive monograph of the Czech endemics, which includes detailed treatments of taxonomy, descriptions of habitats, biology, evolution, threats and conservation measures. These publications serve as a basis for the review of Czech endemism presented here.

The considerable differences between these publications reflect not only the different approaches of the authors but also the changes in the state-of-the-art of taxonomy and knowledge of plant distribution that have developed over the last four decades. Some of the species that were previously considered to be Czech (or Czechoslovak) endemics were later recorded in neighbouring countries. For example, *Epipactis albensis* was described from floodplain forests along the middle Labe River in central Bohemia (Nováková and Rydlo 1978) but later discovered in several countries in Central Europe (Delforge 2006; Batoušek in Štěpánková et al. 2010: 439–464). *Tephroseris longifolia* subsp. *moravica* was described from a small area in the White Carpathians (Holub 1979) but later recorded also in the northwestern part of the Slovenské stredohorie Mts in Slovakia (Kochjarová 1997, 1998; Kochjarová and Hrouda in Slavík et al. 2004: 300–306). Similarly, *Taraxacum bohemicum* was described from a small area in eastern Bohemia (Kirschner and Štěpánek 1986) and considered as a Czech endemic (Kirschner and Štěpánek 1994) but later found at one site in westernmost Slovakia (Kirschner and Štěpánek 1998) and therefore had to be reclassified as a subendemic. *Potentilla lindackeri* was long known only from rocky slopes in river valleys and similar habitats in central Bohemia (e.g. Hadač 1977; Dostál 1989) but later also found in Saxony, Germany (Soják in Slavík et al. 1995: 283–314; Gerstberger 2003; Gregor and Müller 2005; Soják 2009a). Both *Aconitum plicatum* and *Gentianella praecox* subsp. *bohemica* (Fig. 3.5), although distributed mainly in the mountains of the Bohemian Massif in the Czech Republic, occur also in adjacent areas in Austria, Germany and Poland (Skalický in Hejný et al. 1988: 392–403; Procházka and Skalický in Čeřovský et al. 1999: 170; Kirschner and Kirschnerová in Slavík et al. 2000: 82–98) and are better classified as subendemics rather than Czech endemics. *Knautia serpentinicola*, a relict restricted to serpentinite outcrops, is endemic to the Bohemian Massif with most localities in the Czech Republic, but it also occurs at one site in Germany (Kaplan 1998; Kolář et al. 2015). The endemic status of *Symphytum bohemicum* depends on the taxonomic delimitation of this species. This diploid member of the *S. officinale* group was described from central Bohemia and Hadač (1977) and Kubát (in Slavík et al. 2000: 202–210) recognized it in the narrow sense as a Czech (Bohemian) endemic microspecies. However, very similar diploid plants, nowadays mostly considered conspecific with the Bohemian taxon, are more widespread in Central Europe and occur westwards in eastern England and southwards in northern Italy (Kaplan et al. 2016a). *Spergularia kurkae* was described as a hybrid from southern Bohemia but was recently revised as a stabilized allopolyploid species, which, however, also occurs along the Elbe River in Germany (Kúr et al. 2016). *Rubus josholubii* was described as endemic to Bohemia (Weber 2000), but recently was discovered in adjacent parts of Poland (Trávníček in Suda and Kaplan 2018). *Aconitum plicatum* subsp. *sudeticum* was described as an endemic of the Králický Sněžník Mts and the Hrubý Jeseník Mts (Mitka 2003), but it is apparently of hybrid origin, resulting from introgression of *A. firmum* subsp. *moravicum* into *A. plicatum*. Its evolutionary history and taxonomic validity should be tested using molecular methods.

Other taxa are excluded from the list of endemics presented here following expert taxonomic re-evaluations of the given groups. For example, *Melampyrum bohemicum* was long considered a Czechoslovak endemic species (e.g. Hadač 1966, 1977; Hendrych 1981a), which evolved from *M. subalpinum*, which is now confined

**Fig. 3.5** Examples of Czech endemics and subendemics: (**a**) *Pedicularis sudetica* subsp. *sudetica*, subalpine grassland near the Luční bouda chalet, Krkonoše Mts; (**b**) *Aconitum plicatum*, Velká kotlina cirque, Hrubý Jeseník Mts; (**c**) *Gentianella praecox* subsp. *bohemica*, mountain grassland near Olešnice v Orlických horách, Orlické hory Mts, eastern Bohemia; (**d**) *Dianthus arenarius* subsp. *bohemicus*, open sandy grassland at Kleneč, northern Bohemia. Photo credits: M. Štech (a), L. Bureš (b), J. Kučera (c) and J. Vítová (d)

to eastern Austria. However, in the revised delimitation by Štech (Štech in Slavík et al. 2000: 412–429; Štech 2006) the Czech populations and those in western Slovakia are considered to be conspecific with the Austrian populations traditionally called *M. angustissimum*. *Crepis mollis* subsp. *velenovskyi* was described from a deciduous forest at Sadská in the Labe River basin in central Bohemia (Domin 1904) and was still listed as a Czech endemic by Hadač (1977). The type specimen is indeed morphologically peculiar and very distinct from all other plants of *C. mollis*. However, it is only an aberrant phenotype as other collections from the type locality fall within the variation of *C. mollis* subsp. *succisifolia*. That is why the plant originally described as subsp. *velenovskyi* is currently not ascribed any rank (Kaplan and Kirschner in Slavík et al. 2004: 509–536). Previously indicated endemic Bohemian subspecies of *Iris aphylla* are no longer accepted taxonomically (Hrouda and Grulich in Štěpánková et al. 2010: 565–581). The level of morphological differentiation of *Dianthus lumnitzeri* subsp. *palaviensis* in the Pavlov Hills is rather low and does not warrant subspecific rank (Kovanda in Hejný et al. 1990: 200–213). *Sorbus hardeggensis* from the Dyje River valley in south-western Moravia was described as a hybridogenous apomictic species (Kovanda 1996b, 1999), but recent studies indicate that it is a recurrently formed population of primary and backcrossing hybrids *S. aria* × *S. torminalis* (Šefl 2007; Kaplan et al. 2016a; Lepší and Lepší in Kaplan et al. 2017a; Lepší and Lepší in Suda and Kaplan 2018). *Sorbus quernea*, an assumed Czech endemic described from two places in Prague, is a naturalized population of the alien *S. mougeotii* (Lepší et al. 2013b). *Euphrasia corcontica* was thought to be endemic to Mt. Sněžka in the Krkonoše Mts, having originated from the hybridization of *E. micrantha* and *E. minima* (Smejkal 1963; Smejkal and Dvořáková in Slavík et al. 2000: 430–449). This phenotype was repeatedly recorded during the nineteenth century but it is extinct now and no modern biosystematic investigation of its origin is therefore possible. However, recent studies show that these plants fall within the variation range of *E. coerulea* and may be a taxonomically unimportant aberration rather than a separate endemic species (Štech in Kaplan et al. 2017a; Štech et al. in Suda and Kaplan 2018). *Hieracium melanocephalum* described from the Krkonoše Mts is little known but apparently not endemic and doubtfully a separate species (Chrtek in Suda and Kaplan 2018). Previously distinguished endemic subspecies such as *H. saxifragum* subsp. *celakovskianum*, *H. schmidtii* subsp. *winkleri* and *H. schmidtii* subsp. *diversifolium* (Chrtek in Slavík et al. 2004: 504–701) are no longer recognized in the recent treatment of *Hieracium* (Chrtek in Kaplan et al. 2017a). *Hieracium chamaedenium*, described from the valley of the Dyje River in southern Moravia and known only from this type locality, has not been observed for a long time, is probably extinct and remains taxonomically unclear (Chrtek in Slavík et al. 2004: 504–701; Chrtek in Kaplan et al. 2017a; Chrtek in Suda and Kaplan 2018).

The decision of whether or not a unique local form of *Pilosella* of hybrid origin is an endemic depends on the evaluation of its evolutionary history and modes of reproduction (Krahulec and Chrtek in Suda and Kaplan 2018). The diversity of *Pilosella* includes the so called (1) basic species, (2) stabilized hybridogenous species and (3) recent hybrids (e.g. von Nägeli and Peter 1885; Zahn 1921–1923; Chrtek in Slavík et al. 2004: 504–701; Rotreklová et al. 2005; Krahulec et al. 2008). Both the latter groups include plants of hybrid origin, of which the hybridogenous species are stabilized and often occur in the absence of their parental species, unlike the F1 hybrids or very early subsequent generations, which may be produced repeatedly. Three *Pilosella* taxa known only from the Czech Republic appear to be of hybrid origin: *P. callimorphoides* is probably a hybrid between *P. longiscapa* and *P. officinarum* (formula *P. longiscapa* < *P. officinarum*), *P. pseudocalodon* the result of a cross between *P. calodon* and *P. setigera*, whereas *P. tephrophyton* is a hybrid between *P. bifurca* and *P. lactucella* (Chrtek in Slavík et al. 2004: 504–701; Chrtek in Kaplan et al. 2017a). However, these plants are each known from only a single or two sites and currently are considered to be hybrids and not stabilized species; consequently, they are not treated as endemics (Krahulec and Chrtek in Suda and Kaplan 2018). All these forms are now extirpated, thus their evolutionary history and breeding systems cannot be investigated.

In contrast, recent taxonomic revisions of critical groups revealed several new endemics. Multidisciplinary revisions of Bohemian *Sorbus* (Lepší et al. 2008, 2009b, 2013a, 2015; Velebil 2012; Vít et al. 2012) yielded 11 additional endemic species. A few of the regional species of *Rubus* discovered and described during the past two decades (Holub 1991, 1992; Weber 2000) are known only from the Czech Republic. Taxonomic and biosystematic revisions of subalpine *Hieracium* species (Chrtek 1995, 1997; Chrtek and Marhold 1998; Kaplan et al. 2016a) refined our understanding of the diversity and distribution within this group rich in endemics. An endemic sedge that occurs in the Krkonoše Mts, which was recorded in the literature under the provisional names *Carex oederi* subsp. *pseudoscandinavica* (Holub et al. 1979) or *C. viridula* subsp. *pseudoscandinavica* (Holub in Čeřovský et al. 1999: 83), has been thoroughly evaluated and validly described as *C. derelicta* (Štěpánková 2008). Other recent additions to the list of Czech endemics include, among others, *Carlina biebersteinii* subsp. *sudetica* (Kovanda 2002), *Dactylorhiza bohemica* (Businský 1989), *Minuartia corcontica* (Dvořáková 1999), *M. smejkalii* (Dvořáková 1988) and *Scilla bifolia* subsp. *rara* (Trávníček et al. 2010).

An extraordinary case of endemism is *Oenothera moravica*, which evolved as a result of hybridization between two alien species, *O. fallax* and *O. victorinii*, which only rarely co-occur (Jehlík and Rostański 1995). Its origin is similar to that of two *Tragopogon* allopolyploids, *T. mirus* and *T. miscellus*, which evolved from three diploids (*T. dubius*, *T. pratensis* and *T. porrifolius*) introduced from Europe into North America (Ownbey 1950). *Oenothera moravica* was detected soon after it arose and at that time reported as occurring only at two close-by localities in south-western Moravia (Jehlík and Rostański 1995). This newly evolved species was expected to spread, which proved to be true and additional localities of *O. moravica* have been recently recorded during a floristic inventory in the surrounding area (Danihelka in Suda and Kaplan 2018).

### *3.4.3 Revised List of Endemics*

In assessing plant taxa for inclusion on the revised list of endemics of the Czech Republic presented in this review, some subjective decisions had to be made, particularly about their taxonomic rank and delimitation. The taxonomic status mostly follows recent expert revisions of the Flora of the Czech Republic (Hejný et al. 1988, 1990, 1992; Slavík et al. 1995, 1997, 2000, 2004; Štěpánková et al. 2010), Key to the Flora of the Czech Republic (Kaplan et al. 2017a), monograph of the Czech endemics (Suda and Kaplan 2018) and the distribution records therein and in the series Distributions of Vascular Plants in the Czech Republic (Kaplan et al. 2015, 2016a, b, 2017b), which were particularly important in assessing endemic status. Only species and subspecies are considered, whereas endemic varieties and forms, such as *Salix lapponum* var. *daphneola* from the Pančavská louka peat bog in the Krkonoše Mts (Chmelař and Koblížek in Hejný et al. 1990: 458–495) or *Dianthus lumnitzeri* f. *palaviensis* (Kovanda in Hejný et al. 1990: 200–213; Weiss et al. 2002) from the Pavlov Hills, are not discussed here.

Several taxa do not appear in the current list because they have been re-evaluated taxonomically and merged with more widespread species. Doubtful taxa of very limited distribution that are not adopted and substantiated in recent revisions were generally excluded. They are often only minutely distinct from their widespread relatives. Examples include *Coronilla moravica*, which was based on a single herbarium specimen collected near Pašovice in the White Carpathians (Chrtková and Stavělová 1986) and not found in the field again. Several described species or subspecies are taxonomically uncertain and may be reclassified in future revisions. In the absence of modern monographs, these are provisionally included here among the recognized endemic taxa.

Among apomictic microspecies of *Hieracium* and *Taraxacum* only those groups that were recently revised by experts in the Czech Republic are considered. These include the majority of the subalpine groups of *Hieracium* (Chrtek 1995, 1997; Chrtek and Marhold 1998; Chrtek in Slavík et al. 2004: 504–701; Chrtek in Kaplan et al. 2017a; Chrtek in Suda and Kaplan 2018), *Taraxacum* sect. *Palustria* (Kirschner and Štěpánek 1998; Kirschner in Štěpánková et al. 2010: 56–85) and *Taraxacum* sect. *Alpestria* (Štěpánek in Štěpánková et al. 2010: 85–87; Štěpánek et al. 2011). In contrast, the subalpine *Hieracium prenanthoides* group, occurring above the timberline in the Krkonoše, Králický Sněžník and Hrubý Jeseník Mts, is likely to include local endemics (Chrtek in Slavík et al. 2004: 504–701; Krahulec 2006), but this still awaits detailed investigation (Kaplan et al. 2016a). No modern taxonomic revision is available for the *Ranunculus auricomus* complex in the Czech Republic, which may also contain endemic microspecies.

Of the facultatively apomictic *Rubus* only regional species reported exclusively from the Czech Republic in Atlas Florae Europaeae (Kurtto et al. 2010) were adopted, whereas individual plants (single bushes) and local biotypes (i.e. forms occupying an area smaller than 20 km in diameter) were not considered in accordance with the present-day specialists' approach (Weber 1977, 1995, 1996; Holub 1991, 1997).

In this chapter 82 species and subspecies are considered to be endemic to the Czech Republic and closely adjacent areas (Table 3.1), which is 2.2% of the total vascular plant diversity. Of these, 42 species and subspecies (1.1% of the total diversity) are strictly Czech endemics, i.e. they occur only within the borders of this country; the distributions of the other 40 taxa extend slightly beyond the borders of this country (mostly by less than 1 km). There is no genus endemic to the Czech Republic. For habitats and distribution of the endemics, references to selected literature on taxonomy and biology and additional notes, see Table 3.1.

The overview provided above indicates that the Czech Republic and closely adjacent areas are not particularly rich in endemic species. This results from the relatively small size of the country (78,867 km$^2$), its geographical position in Europe, absence of large mountain ranges with extensive alpine belts and the Quaternary history of the Central European landscape. The glaciations during the Pleistocene, with considerable changes of climate and vegetation repeatedly eliminated many species in the previously established floras. The climatic conditions have stabilized since the end of the Pleistocene but there was not sufficient time in the Holocene for the evolution of a highly diversified flora with many local endemics. All Czech endemics are therefore of Quaternary age (neoendemics) and often not well differentiated morphologically. There are no endemics of Tertiary age (palaeoendemics).

### *3.4.4 Distribution and Habitats of Endemics*

The endemic taxa are very unevenly distributed within the Czech Republic (Fig. 3.6). The highest numbers of endemics occur only in the higher parts of the Sudetes, which are the only mountain ranges in the Czech Republic on which there are areas above the timberline. Among the individual mountain ranges in the High Sudetes (Krkonoše, Králický Sněžník and Hrubý Jeseník), the number and size of glacial cirques and the extent of alpine and subalpine summit habitats is largest in the Krkonoše Mts, which host 30 endemics (Table 3.1). Of these, 23 taxa are confined to the Krkonoše Mts, with 4 occurring solely on the Czech side of the border (*Carex derelicta* – Fig. 3.7, *Hieracium purkynei*, *Knautia pseudolongifolia* – Fig. 3.8, *Minuartia corcontica* – Fig. 3.9) and 19 also on the Polish side (*Campanula bohemica* – Fig. 3.7, *Pedicularis sudetica* subsp. *sudetica* – Fig. 3.10, *Pilosella rubra* – Fig. 3.11, *Primula elatior* subsp. *corcontica* – Fig. 3.9, *Taraxacum alpestre* – Fig. 3.12, and 14 *Hieracium* species and subspecies – Fig. 3.8). Finally, three endemics occur also in the Hrubý Jeseník Mts (*Campanula rotundifolia* subsp. *sudetica* – Fig. 3.10, *Galium sudeticum* – Fig. 3.11, *Hieracium chlorocephalum*) and two others also in the Králický Sněžník Mts (*Hieracium schustleri*, *H. uechtritzianum*). Three *Hieracium* endemics, namely *H. chlorocephalum*, *H. pedunculare* and *H. sudetotubulosum*, were each once recorded also in the closely adjacent Jizerské hory Mts (Bräutigam 2001; Chrtek in Slavík et al. 2004: 504–701; Chrtek in Suda and Kaplan

**Table 3.1** A revised list of the species and subspecies endemic to the Czech Republic, with basic descriptions of their habitats, distributions and the most important references. Subalpine taxa occurring on the summits of the Sudetes Mts and *Sorbus* species confined to the valley of the Dyje River, whose distributions only slightly extend beyond the borders of this country (generally less than 1 km) but mostly with the great majority of their localities in the Czech Republic, are included. Endemics that occur strictly within the borders of this country are indicated as "CZ" in the column Notes. Only the most relevant references on the taxonomy and biology are cited

| Taxon | Habitat; distribution | Relevant literature | Notes |
|---|---|---|---|
| *Alchemilla obtusa* subsp. *trapezialis* | Meadow springs, alder carrs and wet meadows; about 75 localities reported at low altitudes in the western Sudetes, particularly in the foothills of the Jizerské hory and Krkonoše Mts, 330–850 m a.s.l. | Plocek (1986), Plocek in Slavík et al. (1995: 247–270), Kaplan (2012), and Trávníček in Suda and Kaplan (2018) | CZ; drastically declined, not recorded recently; taxonomically doubtful, critical revision needed |
| *Campanula bohemica* | Species-rich subalpine and montane meadows and grasslands; Krkonoše Mts,750–1580 m a.s.l., majority of localities above 1200 m | Kovanda (1975, 1977), Kovanda in Slavík et al. (2000: 719–748), Hadač (1977), Hendrych (1981a, b), Kaplan (2012), and Suda and Hanušová in Suda and Kaplan (2018) | May occasionally hybridize with *C. rotundifolia* subsp. *rotundifolia* |
| *Campanula gelida* | Summit rock outcrop and adjacent subalpine grasslands; Mt. Petrovy kameny in the Hrubý Jeseník Mts, 1438–1446 m a.s.l. | Kovanda (1968, 1977), Kovanda in Slavík et al. (2000: 719–748), Hendrych (1981a, b), Bureš (1996), Procházka and Bureš in Čeřovský et al. (1999: 69), Kaplan (2012), and Bureš and Kaplan in Suda and Kaplan (2018) | CZ |
| *Campanula rotundifolia* subsp. *sudetica* | Rocks and scree on treeless summits and in glacial cirques; 6 localities in the Krkonoše Mts and about 15 in the Hrubý Jeseník Mts, ca 950–1450 m a.s.l. | Kovanda (1977), Kovanda in Slavík et al. (2000: 719–748), Kaplan (2012), and Suda et al. in Suda and Kaplan (2018) | CZ |
| *Carex derelicta* | Subalpine spring and moist places on rocky flats; Velká Kotelní jáma cirque in the Krkonoše Mts, 1230–1330 m a.s.l. | Havlíčková (1983), Holub in Čeřovský et al. (1999: 83) (in both cases as *C. viridula* subsp. *pseudoscandinavica*), Štěpánková (2008), Kaplan (2012), Kaplan et al. (2015), and Štěpánková in Suda and Kaplan (2018) | CZ |

| | | | |
|---|---|---|---|
| *Carlina biebersteinii* subsp. *sudetica* | Grasslands on rocky slopes in glacial cirques; Velká kotlina and Malá kotlina cirques in the Hrubý Jeseník Mts, ca 1250–1320 m a.s.l. | Bureš (1996) (as *C. longifolia*), Kovanda (2002), Kovanda in Slavík et al. (2004: 356–361), Kaplan (2012), and Kaplan and Bureš in Suda and Kaplan (2018) | CZ; extirpated in the Velká kotlina cirque, recently observed only in the Malá kotlina cirque |
| *Cerastium alsinifolium* | Springs, seeps, wet margins of tracks and forest clearings in spruce and pine forests on serpentinites, less frequently in grasslands and on rock outcrops; ca 13 localities in the serpentinite area around the villages of Prameny, Sítiny and Mnichov in the Slavkovský les Mts, ca 750–880 m a.s.l. | Novák (1960), Hendrych (1981a, b), Smejkal in Hejný et al. (1990: 136–151), Klaudisová and Čeřovský in Čeřovský et al. (1999: 91), Kolář and Vít (2008), Kaplan (2012), Suda and Kaplan (2012), Vít et al. (2014), and Suda et al. in Suda and Kaplan (2018) | CZ; hybridizing with *C. arvense*, hybrids dominate on sunny and dry places, whereas "pure" plants of *C. alsinifolium* occur mainly in semi-shaded and moist habitats in forests |
| *Cortusa matthioli* subsp. *moravica* | Shaded and moist mossy places on limestone rocks and scree at the bottom of the Macocha abyss in the Moravian Karst, distr. Blansko, ca 400 m a.s.l. | Podpěra (1921, 1923), Kovanda in Hejný et al. (1992: 254–256), Jatiová and Čeřovský in Čeřovský et al. (1999: 110), Gerža (2009), Kaplan (2012), and Kaplan and Musil in Suda and Kaplan (2018) | CZ; seeds not produced, the population persists by vegetative propagation; taxonomically uncertain, requires critical re-evaluation throughout the range of the species |
| *Dactylorhiza bohemica* | Wet meadows and fens; between Jestřebí and Staré Splavy, distr. Česká Lípa, 259 m a.s.l. | Businský (1989), Průša (2005), Sczepanski and Kreutz (2007), Nordström and Hedrén (2009), Kubát in Štěpánková et al. (2010: 502–522), Kaplan (2012), and Ponert and Kaplan in Suda and Kaplan (2018) | CZ; hybridizing with *D. maculata* subsp. *maculata* and *D. majalis*, threatened by disappearance in hybrid swarms |
| *Dactylorhiza carpatica* | Calcareous fen; Březová, distr. Uherské Hradiště, in the White Carpathians Mts, 515–530 m a.s.l. | Batoušek and Kreutz (1999), Průša (2005), Kubát in Štěpánková et al. (2010: 502–522), Kaplan (2012), and Ponert and Kaplan in Suda and Kaplan (2018) | CZ; hybridizing with other *Dactylorhiza* species |
| *Dactylorhiza traunsteineri* subsp. *turfosa* | Peat bogs and transitional mires; six localities in the Šumava Mts and two in the Krušné hory Mts, 1000–1200 m a.s.l. | Procházka (1982), Průša (2005), Kubát in Štěpánková et al. (2010: 502–522); Kaplan (2012), and Ponert and Kaplan in Suda and Kaplan (2018) | CZ; hybridizing with other *Dactylorhiza* species |

(continued)

**Table 3.1** (continued)

| Taxon | Habitat; distribution | Relevant literature | Notes |
|---|---|---|---|
| *Dianthus arenarius* subsp. *bohemicus* | Open sandy grasslands and open pine forests; at Kleneč and Vražkov near Roudnice nad Labem, distr. Litoměřice, ca 200–220 m a.s.l. | Novák (1915, 1927), Toman (1970), Kovanda in Hejný et al. (1990: 200–213), Kuncová and Bělohoubek (1996), Čeřovský and Abtová in Čeřovský et al. (1999: 128), Šlechtová and Bělohoubek (2010), Kaplan (2012), Vítová et al. (2015), and Suda and Vítová in Suda and Kaplan (2018) | CZ; extinct at Vražkov, hybrids with *D. carthusianorum* occasionally occur at Kleneč |
| *Dianthus carthusianorum* subsp. *sudeticus* | Rocks and subalpine grasslands on avalanche track; Velká kotlina cirque in the Hrubý Jeseník Mts, 1160–1360 m a.s.l. | Kovanda (1980), Kovanda in Hejný et al. (1990: 200–213), Hendrych (1981a), Procházka et al. in Čeřovský et al. (1999: 129), Kaplan (2012), and Kaplan and Bureš in Suda and Kaplan (2018) | CZ |
| *Dianthus moravicus* | Rocks and rocky cliffs; six to eight localities in valleys of the Želetavka, Rokytná, Jihlava and Dyje rivers in south-western Moravia, 240–460 m a.s.l. | Kovanda (1982, 1984b), Kovanda in Hejný et al. (1990: 200–213), Čeřovský and Grulich in Čeřovský et al. (1999: 130), Kaplan (2012), and Suda et al. in Suda and Kaplan (2018) | CZ |
| *Galium sudeticum* | Open subalpine grasslands on rocks and screes in glacial cirques; six localities in the Krkonoše Mts and one in the Hrubý Jeseník Mts, ca 1000–1500 m a.s.l. | Krahulcová and Štěpánková (1998), Štursa et al. in Čeřovský et al. (1999: 163), Štěpánková and Kaplan in Slavík et al. (2000: 122–156), Kaplan (2012), Suda and Kaplan (2012), Kolář et al. (2013, 2014), and Knotek and Kolář in Suda and Kaplan (2018) | Extirpated in the Hrubý Jeseník Mts |
| *Hieracium albinum* | Subalpine grasslands and dwarf-shrub vegetation on mountain summits and in glacial cirques; about 15 sites in the Krkonoše Mts, ca 800–1300 m a.s.l. | Procházka and Chrtek in Čeřovský et al. (1999: 183), Chrtek in Slavík et al. (2004: 504–701), Kaplan (2012), Chrtek in Kaplan et al. (2017a), and Chrtek in Suda and Kaplan (2018) | |

| | | | |
|---|---|---|---|
| *Hieracium apiculatum* | Subalpine grasslands and dwarf-shrub vegetation on mountain summits and in glacial cirques; Krkonoše Mts, 920–1450 m a.s.l. | Chrtek in Slavík et al. (2004: 504–701); Kaplan (2012), Kaplan et al. (2016b), Chrtek in Kaplan et al. (2017a), and Chrtek and Kocián in Suda and Kaplan (2018) | |
| *Hieracium asperulum* | Subalpine grasslands and dwarf-shrub vegetation on mountain summits; three sites in the western part of the Krkonoše Mts, 1300–1430 m a.s.l. | Chrtek in Slavík et al. (2004: 504–701), Kaplan (2012), Chrtek in Kaplan et al. (2017a), and Chrtek in Suda and Kaplan (2018) | Extirpated on the Polish side of the mountains |
| *Hieracium chlorocephalum* | Subalpine grasslands and rocky and scree slopes on mountain summits and in glacial cirques; Krkonoše and Hrubý Jeseník Mts, also reported from one site in the Góry Izerskie Mts, 1100–1350 m a.s.l. | Chrtek in Slavík et al. (2004: 504–701), Kaplan (2012), Kaplan et al. (2016b), Chrtek in Kaplan et al. (2017a), and Chrtek and Kocián in Suda and Kaplan (2018) | |
| *Hieracium chrysostyloides* | Subalpine grasslands and dwarf-shrub vegetation on mountain summits; eight sites in the Hrubý Jeseník Mts and one in the Králický Sněžník Mts, 1430–1480 m a.s.l. | Chrtek (1995), Chrtek in Slavík et al. (2004: 504–701), Kaplan (2012), Kaplan et al. (2016b), Chrtek in Kaplan et al. (2017a), and Chrtek and Kocián in Suda and Kaplan (2018) | |
| *Hieracium corconticum* | Subalpine grasslands and dwarf-shrub vegetation; about 12 sites in the Krkonoše Mts, 1000–1400 m a.s.l. | Chrtek in Slavík et al. (2004: 504–701), Kaplan (2012), Chrtek in Kaplan et al. (2017a), and Chrtek in Suda and Kaplan (2018) | |
| *Hieracium decipiens* | Subalpine grasslands and dwarf-shrub vegetation; Krkonoše Mts, 800–1590 m a.s.l. | Chrtek in Slavík et al. (2004: 504–701), Kaplan et al. (2016b), Chrtek in Kaplan et al. (2017a), and Chrtek and Kocián in Suda and Kaplan (2018) | |
| *Hieracium fritzei* | Subalpine grasslands and dwarf-shrub vegetation on mountain summits; Krkonoše Mts, mainly in the western part, 1150–1450 m a.s.l. | Chrtek and Marhold (1998), Chrtek in Slavík et al. (2004: 504–701), Kaplan (2012), Chrtek in Kaplan et al. (2017a), and Chrtek in Suda and Kaplan (2018) | |

(continued)

**Table 3.1** (continued)

| Taxon | Habitat; distribution | Relevant literature | Notes |
|---|---|---|---|
| *Hieracium glandulosodentatum* | Subalpine grasslands and dwarf-shrub vegetation on mountain summits; Krkonoše Mts, 1000–1550 m a.s.l. | Chrtek in Slavík et al. (2004: 504–701), Kaplan (2012), Kaplan et al. (2016b), Chrtek in Kaplan et al. (2017a), and Chrtek in Suda and Kaplan (2018) | |
| *Hieracium nigrescens* | Subalpine grasslands and dwarf-shrub vegetation; Krkonoše Mts, 800–1550 m a.s.l. | Chrtek in Slavík et al. (2004: 504–701), Kaplan (2012), Kaplan et al. (2016b), Chrtek in Kaplan et al. (2017a), and Chrtek and Kocián in Suda and Kaplan (2018) | |
| *Hieracium nigrostylum* | Subalpine grasslands and dwarf-shrub vegetation; Krkonoše Mts, 800–1560 m a.s.l. | Chrtek and Marhold (1998), Chrtek in Slavík et al. (2004: 504–701), Kaplan (2012), Chrtek in Kaplan et al. (2017a), and Chrtek in Suda and Kaplan (2018) | |
| *Hieracium nivimontis* | Subalpine grasslands and dwarf-shrub vegetation on mountain summits; Králický Sněžník Mts, 1380–1424 m a.s.l. | Chrtek (1995), Chrtek in Slavík et al. (2004: 504–701), Kaplan (2012), Kaplan et al. (2016b), Chrtek in Kaplan et al. (2017a), and Chrtek and Kocián in Suda and Kaplan (2018) | Missing (probably extinct) |
| *Hieracium pedunculare* | Subalpine grasslands and dwarf-shrub vegetation; Krkonoše Mts, in the past also one site in the Góry Izerskie Mts, 800–1500 m a.s.l. | Chrtek in Slavík et al. (2004: 504–701), Kaplan (2012), Kaplan et al. (2016b), Chrtek in Kaplan et al. (2017a), and Chrtek in Suda and Kaplan (2018) | |
| *Hieracium pseudalbinum* | Subalpine grasslands and dwarf-shrub vegetation; about 10 sites in the Krkonoše Mts, 995–1300 m a.s.l. | Chrtek in Slavík et al. (2004: 504–701), Kaplan (2012), Chrtek in Kaplan et al. (2017a), and Chrtek in Suda and Kaplan (2018) | |

| | | | |
|---|---|---|---|
| *Hieracium purkynei* | Subalpine grasslands and dwarf-shrub vegetation; two localities in the western part of the Krkonoše Mts, 1300–1350 m a.s.l. | Procházka and Chrtek in Čeřovský et al. (1999: 183), Chrtek in Slavík et al. (2004: 504–701), Kaplan (2012), and Chrtek in Suda and Kaplan (2018) | CZ; extinct |
| *Hieracium riphaeum* | Subalpine grasslands and dwarf-shrub vegetation; Krkonoše Mts, 900–1500 m a.s.l. | Chrtek in Slavík et al. (2004: 504–701), Kaplan (2012), Kaplan et al. (2016b), Chrtek in Kaplan et al. (2017a), and Chrtek in Suda and Kaplan (2018) | |
| *Hieracium rohlenae* | Subalpine grasslands and dwarf-shrub vegetation; eastern part of the Krkonoše Mts, 920–1570 m a.s.l. | Chrtek and Marhold (1998), Chrtek in Slavík et al. (2004: 504–701), Kaplan (2012), Chrtek in Kaplan et al. (2017a), and Chrtek in Suda and Kaplan (2018) | |
| *Hieracium schneiderianum* | Subalpine grasslands and dwarf-shrub vegetation on mountain summits; Krkonoše Mts (mainly in the eastern part), 850–1580 m a.s.l. | Chrtek and Marhold (1998), Chrtek in Slavík et al. (2004: 504–701), Kaplan (2012), Chrtek in Kaplan et al. (2017a), and Chrtek in Suda and Kaplan (2018) | |
| *Hieracium schustleri* | Subalpine grasslands and dwarf-shrub vegetation on mountain summits; eastern part of the Krkonoše and Králický Sněžník Mts, 920–1520 m a.s.l. | Chrtek (1995, 1997), Chrtek in Slavík et al. (2004: 504–701), Kaplan (2012), Kaplan et al. (2016b), Chrtek in Kaplan et al. (2017a), and Chrtek in Suda and Kaplan (2018) | |
| *Hieracium sudetotubulosum* | Subalpine grasslands and dwarf-shrub vegetation; Krkonoše Mts, rarely also in the Góry Izerskie and Góry Stołowe Mts, 850–1580 m a.s.l. | Chrtek (1997) (as *H. tubulosum*), Chrtek in Slavík et al. (2004: 504–701) (as *H. tubulosum*), Kaplan (2012) (as *H. tubulosum*), Szeląg and Wójcik (2014), Kaplan et al. (2016b), Chrtek in Kaplan et al. (2017a), and Chrtek in Suda and Kaplan (2018) | |
| *Hieracium uechtritzianum* | Subalpine grasslands and dwarf-shrub vegetation; Krkonoše Mts (mainly in the eastern part) and Králický Sněžník Mts, 920–1520 m a.s.l. | Chrtek and Marhold (1998), Chrtek in Slavík et al. (2004: 504–701), Kaplan (2012), Chrtek in Kaplan et al. (2017a), and Chrtek in Suda and Kaplan (2018) | |

(continued)

**Table 3.1** (continued)

| Taxon | Habitat; distribution | Relevant literature | Notes |
|---|---|---|---|
| *Knautia pseudolongifolia* | Open subalpine grasslands on outcrops of carbonate rocks in a cliff above the timberline; Kotelní jámy cirque in the Krkonoše Mts, 1320–1390 m a.s.l. | Štěpánek (1989), Štěpánek in Slavík et al. (1997: 543–554), Štěpánek and Procházka in Čeřovský et al. (1999: 205), Kaplan (2012), and Kolář et al. (2015) | CZ |
| *Minuartia corcontica* | Rocky slopes and rocks in glacial cirques; 4–5 localities in the Obří důl valley in the Krkonoše Mts, ca 1100–1300 m a.s.l. | Dvořáková in Hejný et al. (1990: 101–109) (as *M. gerardii*), Dvořáková (1999, 2003), Kaplan (2012), and Chrtek et al. in Suda and Kaplan (2018) | CZ; recent occurrence confirmed only in the ravines Čertova zahrádka and Čertova rokle on eastern slopes of Mt. Studniční hora |
| *Minuartia smejkalii* | Rocky slopes and rocks in open pine forests on serpentinites; several small populations (near Mladá Vožice, Hrnčíře, Bernartice and Borek) in three areas on serpentinite outcrops in south-eastern Bohemia, 385–460 m a.s.l. | Dvořáková (1988), Dvořáková in Hejný et al. (1990: 101–109), Procházka and Klaudisová in Čeřovský et al. (1999: 245), Kolář and Vít (2008), Kaplan (2012), and Chrtek et al. in Suda and Kaplan (2018) | CZ; survives at Hrnčíře and Bernartice, extinct elsewhere |
| *Oenothera moravica* | Roadsides, railway station, ash and slag heaps and other ruderal sites; several localities in south-western Moravia, mostly in distr. Třebíč, ca 220–680 m a.s.l. | Jehlík and Rostański (1995), Jehlík in Slavík et al. (1997: 68–94), Kaplan (2012), and Danihelka in Suda and Kaplan (2018) | CZ |
| *Pedicularis sudetica* subsp. *sudetica* | Mossy springs in subalpine grasslands; Krkonoše Mts, 1210–1450 m a.s.l. | Hultén (1961), Hendrych and Hendrychová (1988), Molau and Murray (1996), Procházka et al. in Čeřovský et al. (1999: 275), Hrouda in Slavík et al. (2000: 455–461), Štursová and Kociánová (2005), Kaplan (2012), and Štech and Krahulec in Suda and Kaplan (2018) | Known from a number of localities in the past but recently observed only at about a dozen of them |
| *Pilosella rubra* | Subalpine grasslands and montane meadows; Krkonoše Mts, ca 850–1300 m a.s.l. | Chrtek in Slavík et al. (2004: 504–701), Krahulec et al. (2004, 2008, 2011), Krahulcová et al. (2004, 2013), Rosenbaumová et al. (2012), and Krahulec and Chrtek in Suda and Kaplan (2018) | Stabilized apomictic species; similar plants found in the Šumava Mts and in Westphalia, Germany, are recent hybrids with variable breeding systems |

| | | | |
|---|---|---|---|
| *Pinguicula vulgaris* subsp. *bohemica* | Mossy fens; ca 20 localities in the middle Labe River basin and a few localities around Jestřebí, distr. Česká Lípa, 180–260 m a.s.l. | Krajina (1927), Domin (1944), Hadač (1977), Bělohlávková (1989), Krahulcová and Jarolímová (1991), Studnička and Hejný (1992), Procházka and Studnička in Čeřovský et al. (1999: 282), Bělohlávková in Slavík et al. (2000: 514–517), Casper and Stimper (2009), Suda and Kaplan (2012), Kaplan (2012), and Dančák and Kaplan in Suda and Kaplan (2018) | CZ; now surviving only between Jestřebí and Staré Splavy, extinct elsewhere; hybrids with *P. vulgaris* subsp. *vulgaris* detected in two localities; claimed karyological differentiation between the subspecies not confirmed; relationships to similar forms occurring elsewhere require further study |
| *Plantago atrata* subsp. *sudetica* | Subalpine grasslands and rocks; Velká kotlina cirque in the Hrubý Jeseník Mts, 1355–1365 m a.s.l. | Holub et al. (1971), Hadač (1977), Bureš (1996), Holub in Čeřovský et al. (1999: 283), Chrtek in Slavík et al. (2000: 529–549), Kaplan (2012), Kaplan and Bureš in Suda and Kaplan (2018) | CZ |
| *Poa riphaea* | Summit rocks and rocky slopes in a glacial cirque; Mt. Petrovy kameny, Tabulové skály rock and Velká Kotlina cirque in the Hrubý Jeseník Mts, ca 1440 m a.s.l. | Jirásek and Chrtek (1963), Chrtek and Jirásek (1966), Hadač (1977), Hendrych (1981a, b), Bureš (1996), Holub in Čeřovský et al. (1999: 286), Gillespie et al. (2007), Kaplan (2012), and Hoták et al. (2013) | CZ; now surviving only on the summit rock outcrop of Mt. Petrovy kameny, extinct at the other two sites |
| *Potentilla psammophila* | Sand dunes and sandy places in open pine forests; Doksy and Bezděz, distr. Česká Lípa, ca 270–450 m a.s.l. | Soják (2009a, b), Kaplan (2012), and Kaplan in Suda and Kaplan (2018) | CZ; missing; closely related to *P. lindackeri* s. str., requires further study |
| *Primula elatior* subsp. *corcontica* | Subalpine grasslands, montane meadows and glacial cirques; several localities in the Krkonoše Mts, ca (800–)1220–1350 m a.s.l. | Domin (1930), Kovanda in Hejný et al. (1992: 246–252), Kovanda (1997), and Kaplan (2012) | Recent occurrence confirmed only in the Malá Kotelní jáma and the Velká Kotelní jáma cirques; requires further taxonomic study |

(continued)

**Table 3.1** (continued)

| Taxon | Habitat; distribution | Relevant literature | Notes |
|---|---|---|---|
| *Rubus bohemiicola* | Forest fringes, forest clearings and open forest; central, southern and eastern Bohemia, 300–685 m a.s.l. | Holub (1991), Holub in Slavík et al. (1995: 54–206), Krahulcová and Holub (1998), Trávníček and Zázvorka (2005), Kaplan (2012), and Trávníček in Suda and Kaplan (2018) | CZ |
| *Rubus brdensis* | Coniferous forests and forest fringes; south-western Bohemia, particularly Brdy Mts and their foothills; ca 300–600 m a.s.l. | Holub (1991), Holub in Slavík et al. (1995: 54–206), Krahulcová and Holub (1998), Kaplan (2012), and Trávníček in Suda and Kaplan (2018) | CZ |
| *Rubus centrobohemicus* | Mixed and coniferous forests, forest clearings and forest fringes; central Bohemia, ca 300–500 m a.s.l. | Holub (1991), Holub in Slavík et al. (1995: 54–206), Kaplan (2012), and Trávníček in Suda and Kaplan (2018) | CZ |
| *Rubus vratnensis* | Forest fringes, forest clearings and scrub; northern, central and south-western Bohemia, ca 200–500 m a.s.l. | Holub (1992), Holub in Slavík et al. (1995: 54–206), Kaplan (2012), and Trávníček in Suda and Kaplan (2018) | CZ |
| *Saxifraga rosacea* subsp. *steinmannii* | Rocks and screes in deep rocky valleys; several localities in two small areas: the valley of the Labe River south of Ústí nad Labem and that of the Jizera River at Semily, 150–385 m a.s.l. | Braun-Blanquet (1922), Hrouda and Šourková in Hejný et al. (1992: 401–422), Nepraš (2006), Nepraš et al. (2008), Kaplan (2012), and Nepraš and Kaplan in Suda and Kaplan (2018) | CZ |
| *Scilla bifolia* subsp. *rara* | Moist deciduous forest; Purkrábka forest between Suchohrdly and Těšetice, distr. Znojmo, 280–320 m a.s.l. | Trávníček (2010), Trávníček et al. (2010), Trávníček in Štěpánková et al. (2010: 613–628), Kaplan (2012), and Trávníček in Suda and Kaplan (2018) | CZ; morphologically best differentiated subspecies of *S. bifolia* |
| *Sorbus albensis* | Thermophilous woodlands and scrub on slopes and rocks; 15 localities west to north of Litoměřice in the České středohoří Mts, 180–540 m a.s.l. | Lepší et al. (2009b), Kaplan (2012), Kaplan et al. (2016b), Lepší and Lepší in Kaplan et al. (2017a), and Lepší and Lepší in Suda and Kaplan (2018) | CZ |

| | | | |
|---|---|---|---|
| *Sorbus alnifrons* | Open hornbeam forests; valley of the Jihlava River around the ruins of Templštejn castle near Moravský Krumlov, distr. Znojmo, ca 250–380 m a.s.l. | Kovanda (1996b, 1999), Kaplan (2012), Kaplan et al. (2016b), Lepší and Lepší in Kaplan et al. (2017a), and Lepší and Lepší in Suda and Kaplan (2018) | CZ |
| *Sorbus barrandienica* | Thermophilous open woodlands on limestone at the tops of hills; 10 localities in the Bohemian Karst east of Beroun, 340–450 m a.s.l. | Vít et al. (2012), Kaplan (2012), Kaplan et al. (2016b), Lepší and Lepší in Kaplan et al. (2017a), and Lepší and Lepší in Suda and Kaplan (2018) | CZ |
| *Sorbus bohemica* | Thermophilous woodlands and scrub, mostly on the tops of volcanic hills; 36 localities in the České středohoří Mts, 150–600 m a.s.l. | Kovanda (1961, 1999), Challice and Kovanda (1978), Jankun and Kovanda (1987), Kovanda in Hejný et al. (1992: 474–484), Procházka in Čeřovský et al. (1999: 349), Boublík et al. (2002), Lepší et al. (2008, 2009b), Kaplan (2012), Kaplan et al. (2016b), Lepší and Lepší in Kaplan et al. (2017a), and Lepší and Lepší in Suda and Kaplan (2018) | CZ |
| *Sorbus cucullifera* | Forest-steppes, rocky steppes, scrub, open thermophilous oak forests and their fringes; valley of the Dyje River around Hardegg along the border between Lower Austria and Moravia, 300–420 m a.s.l. | Lepší et al. (2015), Kaplan et al. (2016b), Lepší and Lepší in Kaplan et al. (2017a), and Lepší and Lepší in Suda and Kaplan (2018) | |
| *Sorbus eximia* | Open thermophilous and mesic oak forests on limestone; five localities in the Bohemian Karst south of Beroun, 350–460 m a.s.l. | Kovanda (1984a, 1999), Challice and Kovanda (1986), Jankun and Kovanda (1988), Kovanda in Hejný et al. (1992: 474–484), Lepší et al. (2008), Kaplan (2012), Vít et al. (2012), Kaplan et al. (2016b), Lepší and Lepší in Kaplan et al. (2017a), and Lepší and Lepší in Suda and Kaplan (2018) | CZ |

(continued)

**Table 3.1** (continued)

| Taxon | Habitat; distribution | Relevant literature | Notes |
|---|---|---|---|
| *Sorbus gemella* | Open thermophilous oak woodlands and pine forests often on rocky cliffs; about 12 localities (over 20 individual sites) in the area of Džbán, between the villages of Třeskonice, Břínkov and Pochvalov, distr. Louny, 350–480 m a.s.l. | Kovanda (1996a, 1999), Lepší et al. (2008, 2009a), Kaplan (2012), Kaplan et al. (2016b), Lepší and Lepší in Kaplan et al. (2017a), and Lepší and Lepší in Suda and Kaplan (2018) | CZ |
| *Sorbus milensis* | Thermophilous woodlands and scrub on basaltic rocks and screes; Milá hill near Louny in the České středohoří Mts, 380–490 m a.s.l. | Lepší et al. (2008), Kaplan (2012), Kaplan et al. (2016b); Lepší and Lepší in Kaplan et al. (2017a), and Lepší and Lepší in Suda and Kaplan (2018) | CZ |
| *Sorbus moravica* | Cliffs, screes, ravines and steep woody or shrubby slopes of a deep valley; Suchý žleb gorge near Lažánky in the Moravian Karst, 360–490 m a.s.l. | Lepší et al. (2015), Kaplan et al. (2016b), Lepší and Lepší in Kaplan et al. (2017a), and Lepší and Lepší in Suda and Kaplan (2018) | CZ |
| *Sorbus omissa* | Open oak woodlands on rocky slopes; two sections of the lower Vltava River valley between Dolany and Roztoky north of Prague, ca 200–290 m a.s.l. | Kaplan (2012), Velebil (2012), Kaplan et al. (2016b), Lepší and Lepší in Kaplan et al. (2017a), and Lepší and Lepší in Suda and Kaplan (2018) | CZ |
| *Sorbus pontis-satanae* | Cliffs, rocky slopes, open pine forests and forest clearings; Suchý žleb gorge near Lažánky in the Moravian Karst, 380–450 m a.s.l. | Lepší et al. (2015), Kaplan et al. (2016b), Lepší and Lepší in Kaplan et al. (2017a), and Lepší and Lepší in Suda and Kaplan (2018) | CZ |
| *Sorbus pauca* | Scrub and grasslands on sunny rocks and on slopes of an abandoned quarry; Malý Bezděz and Bezděz, two nearby phonolite hills near Doksy, distr. Česká Lípa, 440–580 m a.s.l. | Lepší et al. (2013a), Kaplan et al. (2016b), Lepší and Lepší in Kaplan et al. (2017a), and Lepší and Lepší in Suda and Kaplan (2018) | CZ |
| *Sorbus portae-bohemicae* | Open oak forests on slopes and rock outcrops; Porta bohemica gorge and foot of Mt. Lovoš north of Lovosice, distr. Litoměřice, 210–290 m a.s.l. | Boublík et al. (2009), Lepší et al. (2009b), Kaplan (2012), Kaplan et al. (2016b), Lepší and Lepší in Kaplan et al. (2017a), and Lepší and Lepší in Suda and Kaplan (2018) | CZ |

| | | | |
|---|---|---|---|
| *Sorbus rhodanthera* | Open oak and hornbeam woodlands on basaltic cliffs and scree; Chlumská hora hill near Manětín, distr. Plzeň-sever, 450–640 m a.s.l. | Kovanda (1996a, 1999), Lepší et al. (2008), Kaplan (2012), Kaplan et al. (2016b), Lepší and Lepší in Kaplan et al. (2017a), and Lepší and Lepší in Suda and Kaplan (2018) | CZ |
| *Sorbus sudetica* | Avalanche tracks, dwarf-shrub vegetation and subalpine grasslands; Krkonoše Mts, 10 sites in the Labský důl and Obří důl valleys on the Czech side and one site above lake Mały Staw on the Polish side, ca 1060–1350 m a.s.l. | Kárpáti (1960), Kovanda (1965, 1999), Hadač (1977), Challice and Kovanda (1978), Hendrych (1981a, b), Jankun and Kovanda (1986), Kovanda in Hejný et al. (1992: 474–484), Procházka in Čeřovský et al. (1999: 351), Nelson-Jones et al. (2002), Krahulec (2006), Kaplan (2012), and Lepší et al. in Suda and Kaplan (2018) | Extant populations only on the Czech side of the border, extirpated on the Polish side |
| *Sorbus thayensis* | Woodland fringes, forest-steppes, cliffs, open thermophilous forests, screes; valleys of the Dyje River and the Fugnitz stream around Hardegg along the border between Lower Austria and Moravia, 310–440 m a.s.l. | Lepší et al. (2015), Kaplan et al. (2016b), Lepší and Lepší in Kaplan et al. (2017a), and Lepší and Lepší in Suda and Kaplan (2018) | |
| *Taraxacum alpestre* | Slightly disturbed sites in subalpine grasslands and on rocks; summit area of the Krkonoše Mts, ca 1400–1600 m a.s.l. | Procházka and Štěpánek in Čeřovský et al. (1999: 365), Štěpánek in Štěpánková et al. (2010: 85–87), Štěpánek et al. (2011), Kaplan (2012), and Zámečník in Suda and Kaplan (2018) | |
| *Taraxacum indigenum* | Fens and wet meadows; Rovná and Lužnice in southern Bohemia, 415–425 m a.s.l. | Kirschner and Štěpánek (1998), Kirschner in Štěpánková et al. (2010: 56–85), Kaplan (2012), and Zámečník in Suda and Kaplan (2018) | CZ |

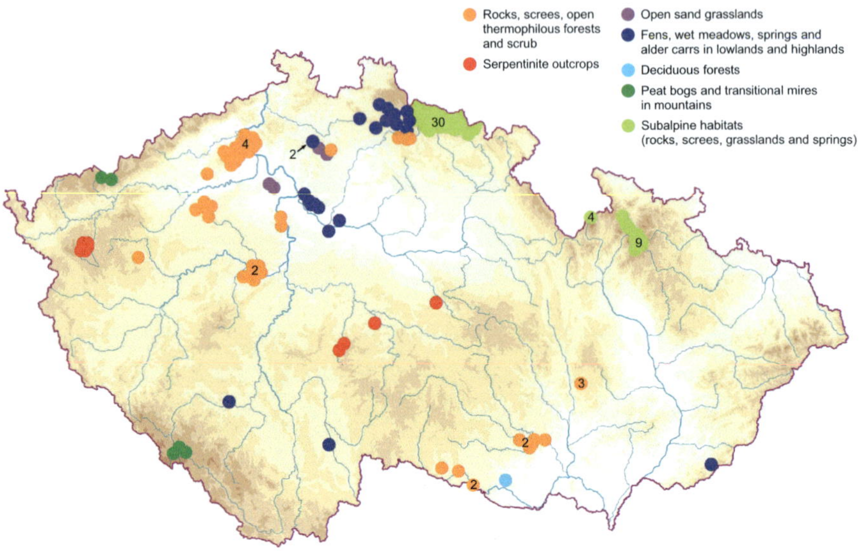


**Fig. 3.6** Areas in which Czech endemics occur and their main habitats in these areas. Habitats of endemic *Rubus* species and of *Oenothera moravica* are not mapped. Numbers of endemic taxa (if more than one) occurring in a particular habitat and area are shown

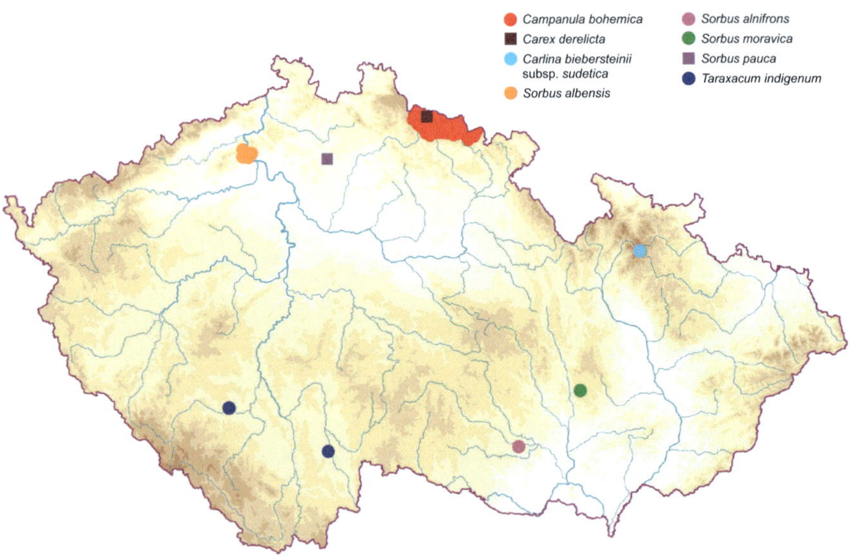


**Fig. 3.7** Distribution of selected endemic taxa in the Czech Republic

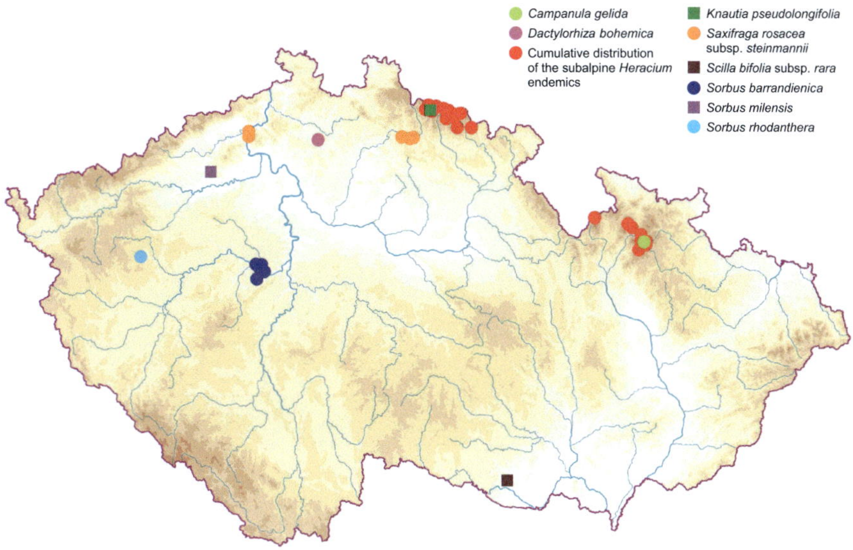


**Fig. 3.8** Distribution of selected endemic taxa in the Czech Republic

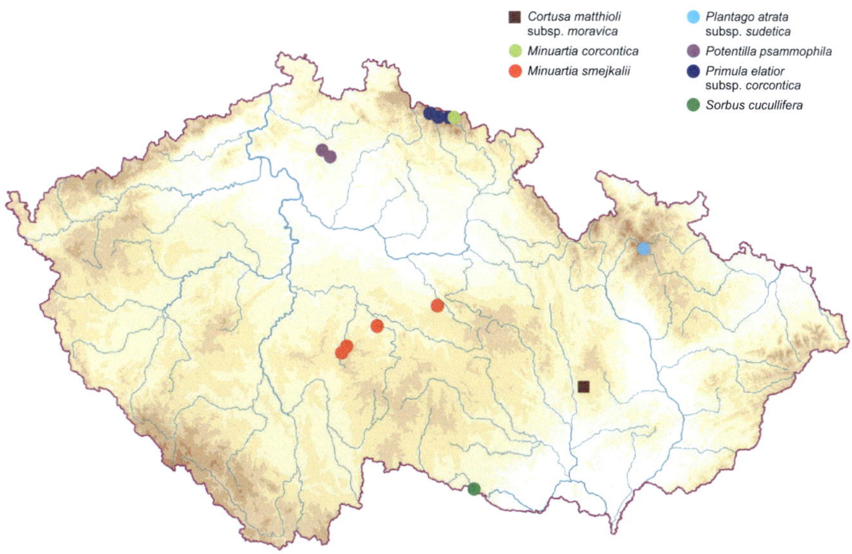


**Fig. 3.9** Distribution of selected endemic taxa in the Czech Republic

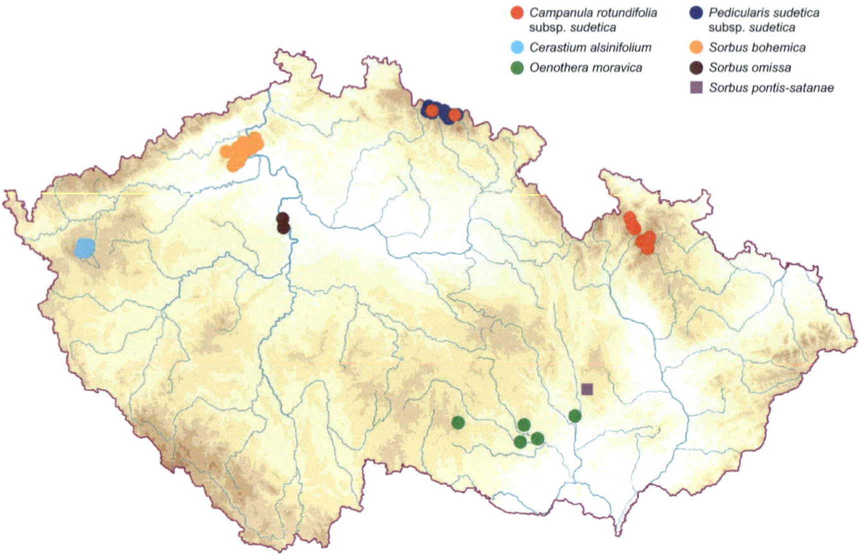


**Fig. 3.10** Distribution of selected endemic taxa in the Czech Republic

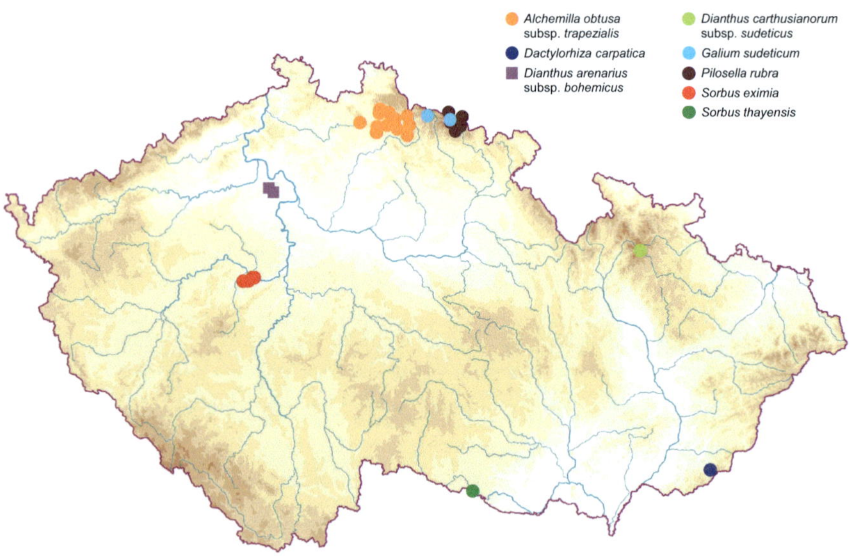


**Fig. 3.11** Distribution of selected endemic taxa in the Czech Republic

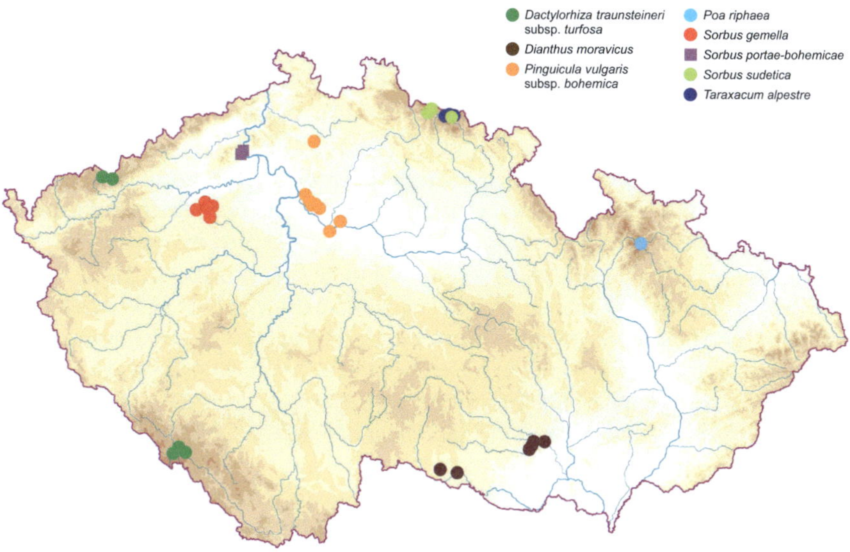


**Fig. 3.12** Distribution of selected endemic taxa in the Czech Republic

2018). However, these occurrences were associated with temporary introductions during the time of extensive land-use at high altitudes and all vanished immediately after this traditional management ceased (Krahulec 2006). All Krkonoše endemics, including those that only occur on the Polish side of the border, are discussed in detail by Krahulec (2006).

Nine Czech endemics occur in the Hrubý Jeseník Mts, mainly in the two glacial cirques there, and on adjacent mountain summits. Five of them are restricted to these mountains (*Campanula gelida* – Fig. 3.8, *Carlina biebersteinii* subsp. *sudetica* – Fig. 3.7, *Dianthus carthusianorum* subsp. *sudeticus* – Fig. 3.11, *Plantago atrata* subsp. *sudetica* – Fig. 3.9, *Poa riphaea* – Fig. 3.12), three occur also in the Krkonoše Mts (*Campanula rotundifolia* subsp. *sudetica* – Fig. 3.10, *Galium sudeticum*, *Hieracium chlorocephalum*) and one occurs also in the Králický Sněžník Mts (*Hieracium chrysostyloides*).

There is a small area of subalpine habitats and a poorly developed glacial cirque in the Králický Sněžník Mts. There is only one avalanche track, with infrequent avalanches (Krahulec 1990). Four Czech endemics occur there but only one of them (*Hieracium nivimontis*) is an endemic solely in this mountain group, while others also occur either in the Krkonoše Mts or the Hrubý Jeseník Mts.

In these mountains the endemics occur in a wide range of alpine and subalpine habitats, such as natural grasslands above the timberline, dwarf-shrub vegetation, summit rocks and rocky slopes and various sites in glacial cirques including avalanche tracks; less frequently they occur in moist places such as springs. Most of the endemics are restricted to these natural habitats but *Campanula bohemica*, *Primula elatior* subsp. *corcontica* and locally also some *Hieracium* species (such as *H. nigrescens*, *H. rohlenae* and *H. schustleri*) also occur in man-made subalpine or even montane meadows in the Krkonoše Mts. *Campanula bohemica* is the most frequent among the Sudetes endemics, while the majority of the others are rare and some of them (e.g. *Campanula gelida*, *Carex derelicta*, *Knautia pseudolongifolia* and *Poa riphaea*) only found at a single locality. One species (*Hieracium purkynei*) is extinct.

Other endemic taxa are confined to specific habitats at low altitudes (Fig. 3.6). Eighteen endemics occur on rocks, screes or in associated open thermophilous forests and grasslands. *Dianthus moravicus* (Fig. 3.12) and *Saxifraga rosacea* subsp. *steinmannii* (Fig. 3.8) occur on rocks in deep river valleys whereas *Cortusa matthioli* subsp. *moravica* (Fig. 3.9) is found only on limestone rocks and a scree at the bottom of the Macocha abyss in the Moravian Karst. The 15 endemic *Sorbus* species are most frequent in thermophilous open woodlands on rock outcrops and tops of limestone or basaltic hills (Figs. 3.7, 3.8, 3.9, 3.10, 3.11, and 3.12). *Dianthus arenarius* subsp. *bohemicus* (Fig. 3.11) and *Potentilla psammophila* (Fig. 3.9) are the only Czech endemics occurring in sandy habitats.

The serpentinite outcrops in the Bohemian Massif with their specific physical features, chemical properties (particularly the high Mg:Ca ratio, high proportion of heavy metals such as Ni, Cr and Co, and low amount of P) and specific type of vegetation (often relict open pine forests that probably persisted throughout the Holocene) provide unique habitats for two Czech endemics, *Cerastium alsinifolium* (Fig. 3.10) and *Minuartia smejkalii* (Fig. 3.9) and several other subendemic taxa, such as *Armeria elongata* subsp. *serpentini*, *Dianthus carthusianorum* subsp. *capillifrons* and *Knautia serpentinicola*.

Species-rich calcareous fens, wet meadows and mires are the habitats of *Dactylorhiza bohemica*, *D. carpatica*, *D. traunsteineri* subsp. *turfosa*, *Pinguicula vulgaris* subsp. *bohemica* and *Taraxacum indigenum*. These habitats are potentially suitable for agriculture and that is why they were drained and converted into arable land at many sites. There are currently only a small number of species-rich fens. *Dactylorhiza traunsteineri* subsp. *turfosa* still occurs at 6–7 sites (Fig. 3.12), but the other two endemic species of *Dactylorhiza* are each known only from one site (Figs. 3.8 and 3.11). *Pinguicula vulgaris* subsp. *bohemica* occurred at about 20 localities in the past (Fig. 3.12), but only one spontaneous population is now extant (it was introduced at two other sites). The distribution of *Taraxacum indigenum* (Fig. 3.7) indicates that it also was more widespread in the past but only two populations have been documented since the species was distinguished taxonomically. A somewhat wider range of wet habitats is occupied by *Alchemilla obtusa* subsp. *trapezialis*, which is reported to occur not only in wet meadows but also around springs and in alder carrs (Fig. 3.11).

The only endemic confined to moist deciduous forests is *Scilla bifolia* subsp. *rara*, which is known from a single locality (Fig. 3.8). In contrast, secondary, mostly coniferous forests and associated habitats such as forest fringes and clearings are the habitats of endemic species of *Rubus*. *Oenothera moravica*, which is still confined to a relatively limited area, occurs only in man-made habitats (Fig. 3.10).

The majority of Czech endemics are rare or strongly endangered plants and included on the Red List of the Czech flora (Grulich 2012), one is extinct (*Hieracium purkynei*) and one is missing and probably extinct (*Potentilla psammophila*).

### 3.4.5 Evolution of Endemics

The origin of the majority of Czech endemics is associated with Pleistocene glaciations. In response to climatic changes, arctic species spread from Northern Europe southwards and Central European montane and alpine species colonized sites at lower altitudes. After the retreat of the glaciers and climatic amelioration at the Pleistocene/Holocene boundary, most of these species recolonized their original distribution ranges but relict populations of some species survived at the sites where they previously found refuge. These peripheral populations became isolated from the main species range and mostly consisted of relatively few individuals, which facilitated rapid evolution by means of genetic drift. Mutations and natural selection in response to different selection pressures also played a role. Because the effect of genetic drift is stronger and faster in small populations, the isolated relict populations could potentially accumulate enough genetic differences for allopatric speciation even within the relatively short period of time since the end of the last glacial period (approximately 11,600 cal. years BP). These evolutionary processes led in some groups to substantial genetic and phenetic divergence and the respective plants are recognized as distinct endemic taxa.

Ancestors of several endemics occurred at suitable refugia on serpentinite outcrops, which are scattered in highland areas of the Bohemian Massif. These serpentinites are often covered by open pine forests, which enabled light-demanding species to persist throughout the Holocene. *Cerastium alsinifolium* is an endemic that occurs only in the serpentinite area near the village of Mnichov in the Slavkovský les Mts in western Bohemia. It apparently evolved from a relict population of an alpine species of the *C. alpinum* group that colonized Bohemia from the Alps during full-glacial periods (Novák 1960). After the expansion of forest in the Early Holocene it largely disappeared surviving only in open pine forests in serpentinite areas, where a small relict population evolved into a new species. In a similar way, *Minuartia smejkalii* appears to have evolved from populations of *M. gerardii* that were in the Alps during the glacial period and also survived in open pine forest on serpentinite outcrops in the Bohemian-Moravian Highlands (Dvořáková 1988).

Some endemics evolved from relicts of the arctic and alpine flora that survived the postglacial expansion of forest in former grasslands or similar treeless habitats above the timberline. *Campanula bohemica* and *C. gelida* (Fig. 3.13) apparently evolved from *C. scheuchzeri* (Hadač 1977; Kovanda 1977; Hendrych 1981a). This species nowadays occurs in the Alps and a few other high mountain ranges in Southern Europe but in colder periods in the past, perhaps during the last glacial period, spread to the lowlands of Central Europe and remained isolated in the Sudetes in the following postglacial period. Remnants of the isolated populations evolved into the two separate neoendemic species, *C. bohemica* in the Krkonoše Mts and *C. gelida* in the Hrubý Jeseník Mts. Similar processes of genetic and phenetic differentiation occurred in *Plantago atrata*, which diverged during the Holocene into several allopatric subspecies isolated in the high mountains in the southern half of Europe. The small population in the Hrubý Jeseník Mts is recognized as *P. atrata* subsp. *sudetica*. Another stenoendemic species in the Hrubý Jeseník Mts, *Poa riphaea*, apparently evolved from *P. glauca* s.l. populations

**Fig. 3.13** Examples of Czech endemics: (**a**) *Campanula gelida*, summit rock on Mt. Petrovy kameny, Hrubý Jeseník Mts; (**b**) *Carlina biebersteinii* subsp. *sudetica*, Malá kotlina cirque, Hrubý Jeseník Mts; (**c**) *Saxifraga rosacea* subsp. *steinmannii*, rocks in the valley of the Jizera River in Semily, northern Bohemia; (**d**) *Dactylorhiza bohemica*, fen between Jestřebí and Staré Splavy, northern Bohemia; (**e**) *Dianthus carthusianorum* subsp. *sudeticus*, Velká kotlina cirque, Hrubý Jeseník Mts; (**f**) *Sorbus sudetica*, Obří důl valley, Krkonoše Mts. Photo credits: J. Suda (a), L. Bureš (b), Z. Kaplan (c, e), J. Ponert (d) and P. Vít (f)

(Jirásek and Chrtek 1963; Hadač 1977) that spread from the Arctic to Central Europe during a glacial period and currently there are relict populations on some Central European mountains. The small relict population that survived in the Hrubý Jeseník Mts rapidly evolved into a new endemic species. *Galium sudeticum* is a tetraploid species that evolved from the *G. pumilum* complex in subalpine refugia in the Krkonoše and Hrubý Jeseník Mts. It is probably a descendant of an ancient hybridization between alpine *G. anisophyllon* and its usually mid-altitude counterpart *G. valdepilosum* (Kolář et al. 2013; Knotek and Kolář in Suda and Kaplan 2018). The populations growing on serpentinites in the Slavkovský les Mts, previously also assigned to *G. sudeticum*, are a distinct evolutionary lineage (Kolář et al. 2014). *Minuartia corcontica* is a local endemic on rocky slopes in glacial cirques in the Krkonoše Mts that apparently originated from relict populations of *M. gerardii* (Dvořáková 1999).

The *Pedicularis sudetica* complex globally consists of 7–9 taxa with arctic-alpine distributions (Hultén 1961; Molau and Murray 1996). It has the highest diversity in the arctic and northern boreal zones where there are at least six taxa, whereas only two taxa occur in southern mountain ranges. The population in the Krkonoše Mts is a relict of the arctic flora that occurred in Central Europe during the last glacial period. Now it is a remote isolated population far from the continuous range of the *P. sudetica* complex (Fig. 3.2), with nearest localities on the Kola Peninsula ~2300 km away, where *P. sudetica* subsp. *arctoeuropaea* occurs. Hultén (1961) classified the plants from the Krkonoše Mts as endemic *P. sudetica* subsp. *sudetica* (Fig. 3.5). In their recent revision, Molau and Murray (1996) point out that this population was isolated from all the arctic taxa in this complex for a very long time and argue that it is morphologically well differentiated from the others and should be treated as a separate species, *P. sudetica* s. str.

In contrast, the present-day distributions of relatives of other endemics, now restricted to refugia above the timberline, are in warmer low-altitude areas or more southern regions. Relict, mostly diploid populations of the *Knautia arvensis* complex occur in several refugia in the Bohemian Massif, within an area otherwise occupied by the widespread tetraploid *K. arvensis* subsp. *arvensis* (Štěpánek 1982, 1989; Štěpánek in Slavík et al. 1997: 543–554; Kaplan 1998; Kolář et al. 2009). These relict populations form a distinct genetic lineage (Rešetnik et al. 2014) and are morphologically, ecologically and geographically differentiated, being confined to either subalpine habitats in the Kotelní jámy cirque in the Krkonoše Mts or to open pine forest on serpentinite outcrops in Bohemia and closely adjacent Bavaria (Štěpánek 1989; Štěpánek in Slavík et al. 1997: 543–554; Kaplan 1998; Kolář et al. 2015) and are now distinguished as *K. pseudolongifolia* and *K. serpentinicola*, respectively. The origin of these relict populations is thought to be remnants of an ancestral diploid of the *K. arvensis* complex that was widespread in Central Europe in the early postglacial period before the expansion of forest (Štěpánek 1989; Kaplan 1998). Under these specific conditions, the isolated relict populations gradually diverged into new entities. The widespread tetraploid *K. arvensis* subsp. *arvensis*, now occurring in man-made habitats in Central Europe, evolved from a different lineage (Rešetnik et al. 2014) probably in Southern Europe and later colonized most of the northern half of Europe when this area was deforested by humans (Kaplan

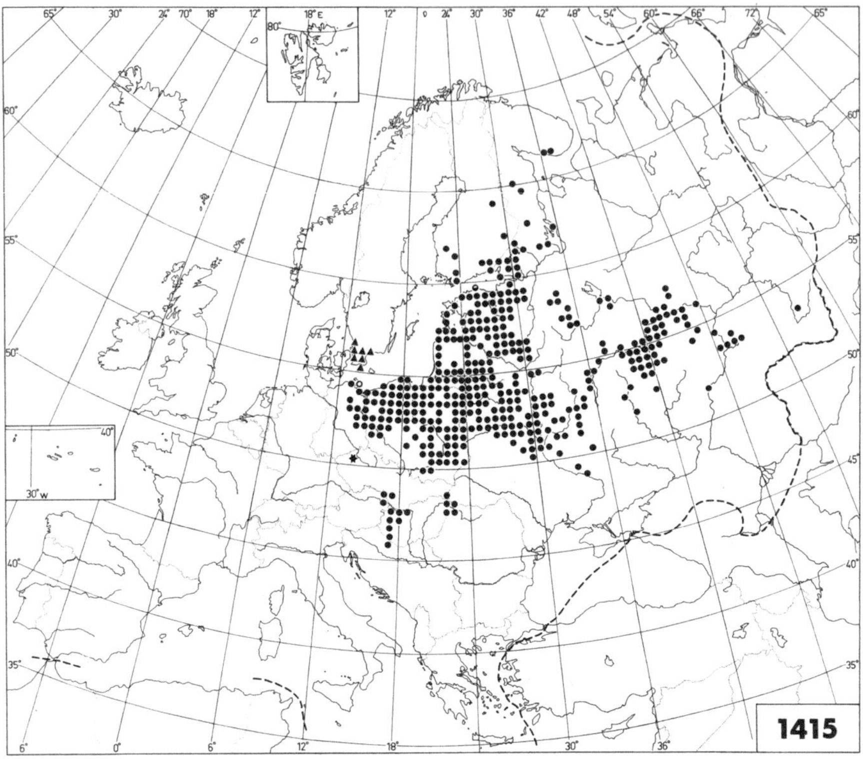


**Fig. 3.14** Total distribution of *Dianthus arenarius* s.l. The only locality of the local endemic *D. a.* subsp. *bohemicus* in central Bohemia, indicated by an asterisk, is the most south-western location for this species. The populations in eastern Austria, Slovakia and Hungary are often separated as *D. serotinus* (reproduced from Jalas and Suominen 1986 with permission of the Committee for Mapping the Flora of Europe and Societas Biologica Fennica Vanamo)

1998). A similar evolutionary history of continuous differentiation of small populations isolated in subalpine refugia may be hypothesized for *Campanula rotundifolia* subsp. *sudetica*, *Carlina biebersteinii* subsp. *sudetica* (Fig. 3.13), *Dianthus carthusianorum* subsp. *sudeticus* (Fig. 3.13) and *Primula elatior* subsp. *corcontica*, of which their closest relatives occur at low altitudes, often mainly in warm areas.

Other endemics have originated from relicts surviving in specific habitats at low altitudes. *Cortusa matthioli* subsp. *moravica* has evolved from a small population isolated in an abyss in a karst area. Both *Dianthus moravicus* and *Saxifraga rosacea* subsp. *steinmannii* (Fig. 3.13) also originated on isolated rock outcrops. *Dianthus arenarius* subsp. *bohemicus* (Fig. 3.5) is confined to open sandy habitats. The ancestral form of *D. arenarius* was apparently widespread in the northern part of Central Europe and north and east of the Carpathians (Sarmatian element) during the Early Holocene (Novák 1927). The subsequent isolation associated with the spread of forest caused it to split into 4–5 currently recognized more or less geographically vicariant subspecies. Among them, subsp. *bohemicus* is isolated in a south-westernmost outpost of this species range (Fig. 3.14). In contrast to these plants of mostly dry habitats, *Pinguicula vulgaris* subsp. *bohemica* occurs in species-rich mossy fens.

Hybridization was involved in the evolution of some other endemic species. This often occurred when two previously allopatric species temporarily came into contact during major shifts in the species' ranges triggered by climatic changes. In some groups, hybridization was followed by a change in the mode of reproduction to agamospermy.

In addition to mostly diploid sexually reproducing species, the genus *Sorbus* includes also apomictic polyploids that evolved from the hybridization of diploids (Kárpáti 1960). Apomictic *S. sudetica* (Fig. 3.13) resulted from a past hybridization of *S. chamaemespilus* and *S. aria* s.l. (Kárpáti 1960; Challice and Kovanda 1978; Jankun and Kovanda 1986; Kovanda in Hejný et al. 1992: 474–484; Nelson-Jones et al. 2002). It probably originated during a temporary co-occurrence of the parental species in the Krkonoše Mts in the Early Holocene. The later climatic and vegetation changes caused the disappearance of parental species from Krkonoše but *S. sudetica* occurs in suitable habitats in two glacial cirques.

Other apomictic *Sorbus* endemics occur in relict rocky habitats in warmer areas. Chemotaxonomic investigations indicate that *S. bohemica* originated from hybridization between *S. danubialis* and *S. torminalis* (Challice and Kovanda 1978) whereas *S. eximia* arose from backcrossing of a *S. aria* s.l. × *S. torminalis* F1 hybrid with *S. aria* s.l. (Challice and Kovanda 1986). Judging from its morphology, the triploid *S. milensis* originated from a cross between the sexual *S. aria* s.l. and *S. torminalis* (Lepší et al. 2008), while *S. albensis*, *S. portae-bohemicae* and *S. omissa* originated from crosses between *S. danubialis* and *S. torminalis* (Lepší et al. 2008; Velebil 2012), *S. pauca* is assumed to have originated from a cross between *S. aucuparia* and *S. danubialis* (Lepší et al. 2013a).

Another genus with a complicated evolutionary history and complex breeding systems is *Pilosella*. One of its endemics, *P. rubra*, originated from hybridization between *P. aurantiaca* and *P. officinarum* (formula *P. aurantiaca* > *P. officinarum*), but now it is a species stabilized by agamospermy and independent of the occurrence of its parents (Krahulec and Chrtek in Suda and Kaplan 2018).

Ancient hybridization associated with subsequent agamospermy has been the key speciation mechanisms also in *Alchemilla*, *Hieracium*, *Rubus* and *Taraxacum*. However, the parental species and the time when the currently recognized taxa originated are generally unknown because the variation and evolutionary histories within these genera are complex. Considering the fact that related sexual diploids of many of these apomictic taxa either occur in distant areas, such as the Eastern Carpathians or Southern Europe (*Hieracium* and *Taraxacum*), or even may have become extinct (*Alchemilla*), indicates they may be very old lineages. Based on the patterns of distribution of the Krkonoše endemics in these groups and of their relatives, Krahulec (2006) assumes they may be of pre-Holocene origin and occurred over a wide area at low altitudes during the glacial periods. At the end of the last (Weichselian/Würm) glaciation, some of them followed the retreat of the continental glaciers into Northern Europe while others colonized high mountain ranges in the southern half of Europe.

The recently described *Potentilla psammophila* belongs to *Potentilla* subsect. *Collinae*, which includes apomictic species that originated from hybridization between *P. argentea* and *P. verna* or *P. incana*. Among them, *P. psammophila* is a close relative of *P. lindackeri* s. str. and seven other microspecies that originated from a cross between *P. argentea* and *P. verna* (Soják 2009b).

In contrast to the relative stability of apomicts, small isolated populations of sexual plants are subject to genetic drift, which promotes speciation. It is hypothesized that the origin and evolution of *Carex derelicta* was dependent on a geographically isolated population of the North-western-European *C. scandinavica* (or a close relative of it) persisting in a refugium in the Krkonoše Mts as a glacial relict and subsequently was introgressed by another taxon of the subsect. *Serotinae*, perhaps *C. demissa* (Štěpánková 2008).

Most tetraploid European *Dactylorhiza* species, including the widespread *D. majalis* and *D. traunsteineri*, are considered to be allotetraploids. These have evolved by repeated hybridization between the diploid species *D. incarnata* s.l. and *D. fuchsii* s.l. (Hedrén 1996; Devos et al. 2006; Pillon et al. 2007; Nordström and Hedrén 2009). Morphologically, geographically or ecologically separated allotetraploids of this complex are recognized as species or subspecies. Some even hybridize with each other or backcross with parental taxa. This hybridogenous complex includes three morphologically distinct local populations in the Czech Republic, which were described as *D. bohemica* (Businský 1989; Fig. 3.13), *D. traunsteineri* subsp. *carpatica* (Batoušek and Kreutz 1999) and *D. majalis* subsp. *turfosa* (Procházka 1982).

## 3.5 Floristic Geoelements

Analysing the ranges of species provides a basis for distinguishing geographical (floristic) elements (geoelements), which are groups of species with similar distributions. The traditional concept of a geoelement is based on heuristic approaches of classification using visual inspection and comparison of distribution maps and expert judgement. There is no analytical revision of European geoelements using numerical methods mainly because there is no complete synthesis of the distribution data at the continental level. Finnie et al. (2007) attempted to provisionally identify floristic elements using a numerical analysis of the distributions of species mapped in 12 volumes of Atlas Florae Europaeae (Jalas and Suominen 1972–1994; Jalas et al. 1996, 1999), which, however, represent only about 20% of the European flora and is not based on a taxonomically representative sample. For these reasons, the floristic characteristics given here inevitably rely on the geoelements traditionally distinguished in the Central European literature.

The flora of the Czech Republic includes mainly lowland to mountain Central European plants. The Czech Republic is situated at the intersection of several important European migration routes. Consequently, the Czech flora is composed of almost all the floristic elements occurring in Central Europe. Based on similarities in the geographical distributions of species (see e.g. Hultén 1964, 1971; Meusel et al. 1965, 1978; Jalas and Suominen 1972–1994; Hultén and Fries 1986; Meusel and Jäger 1992; Jalas et al. 1996, 1999; Kurtto et al. 2004, 2007, 2010, 2013), a number of geoelements are distinguished in Central European literature (e.g. Meusel et al. 1965; Walter and Straka 1970; Hendrych 1984; Walter 1986; Slavík in Hejný et al. 1988: 65–102; Slavík in Slavík et al. 1995: 20–40). Their precise number, delimitation and hierarchical subdivision vary between authors. For example, a species that is characterized as a Sub-Arctic-alpine by one author can be viewed as Boreal by someone else. In the absence of an analytical revision based on extensive sets of complete data, only the principal geoelements traditionally distinguished in Central Europe and relevant for the Czech Republic are discussed here.

**The Central European geoelement** (in a narrow sense) includes species found only or mainly in Central Europe, in some cases extending northwards to central Scandinavia, southwards to the mountains of Southern Europe and eastwards to the Southern Ural Mts. Its species mostly occur in the zone of deciduous or mixed forests (with dominant oaks, hornbeam or beech) where the summers are mild with rather high precipitation and winters also mild and with a short period of frost. This is the most frequent geoelement in the Czech flora. Woody plants are represented by *Acer platanoides*, *A. pseudoplatanus*, *Alnus glutinosa*, *Carpinus betulus*, *Cornus sanguinea*, *Corylus avellana*, *Fagus sylvatica* (Fig. 3.15), *Fraxinus excelsior*, *Hedera helix*, *Prunus avium*, *Quercus petraea* and *Tilia cordata*, herbaceous plants include *Ajuga reptans*, *Alliaria petiolata*, *Allium ursinum*, *Aquilegia vulgaris* (Fig. 3.15), *Arum maculatum* (Fig. 3.15), *Astrantia major*, *Atropa bella-donna* (Fig. 3.15), *Briza media*, *Campanula patula*, *Carex brizoides*, *C. sylvatica*, *C. umbrosa*, *Cirsium oleraceum*, *Colchicum autumnale*, *Corydalis cava* (Fig. 3.15), *Cynosurus cristatus*, *Dentaria enneaphyllos* (Fig. 3.15), *Euphrasia officinalis* subsp. *rostkoviana*, *Euphorbia cyparissias*, *Ficaria verna*, *Galeobdolon luteum*, *G. montanum*, *Galium sylvaticum*, *Genista germanica*, *G. tinctoria*, *Geranium robertianum*, *Glyceria fluitans*, *Helianthemum grandiflorum* subsp. *obscurum*, *Hepatica nobilis*, *Hordelymus europaeus*, *Hylotelephium maximum*, *Hypericum montanum*, *Lathraea squamaria*, *Lunaria rediviva*, *Luzula luzuloides*, *Lysimachia nummularia*, *Mercurialis perennis*, *Petasites albus*, *Phyteuma spicatum*, *Pimpinella major*, *Polygonatum multiflorum*, *Pulmonaria officinalis*, *Ranunculus lanuginosus*, *Stachys sylvatica*, *Stellaria holostea* (Fig. 3.15), *Thalictrum aquilegiifolium* and *Trifolium medium*.

The distributions of some species are shifted towards Eastern Europe, often as far as the Southern Ural Mts, and these are sometimes distinguished as **the Sarmatian geoelement** (or subelement). They are more tolerant of a continental climate, which is drier and with a greater oscillation in temperature between summer and winter. In the Czech Republic they have limited distributions, often being relicts from late glacial or early postglacial ages. *Astragalus arenarius* (Fig. 3.16), *Dianthus arenarius* (Fig. 3.5), *Jurinea cyanoides* and *Thesium ebracteatum* may be given as examples.

**The Central European (sub-)alpine geoelement** includes species of mainly subalpine to alpine belts that are most frequent in the mountains in Central Europe. This

**Fig. 3.15** Examples of the Central European geoelement: (**a**) *Aquilegia vulgaris*, botanical garden, Prague; (**b**) *Dentaria enneaphyllos*, Malá Morávka, Hrubý Jeseník Mts; (**c**) *Atropa belladonna*, botanical garden, Prague; (**d**) *Arum maculatum*, Nedošínský háj wood near Litomyšl, eastern Bohemia; (**e**) *Corydalis cava*, Libice nad Cidlinou, central Bohemia; (**f**) *Stellaria holostea*, Průhonice, central Bohemia; (**g**) *Fagus sylvatica*, Nové Heřmínovy, Nízký Jeseník Mts. Photo credits: L. Hrouda (a, c), L. Bureš (b, d, g) and Z. Kaplan (e, f)

geoelement includes *Alnus alnobetula*, *Gentiana asclepiadea* (Fig. 3.17), *G. punctata* (Fig. 3.17), *Geum montanum*, *Hieracium villosum* (Fig. 3.17), *Homogyne alpina*, *Hypochaeris uniflora*, *Lilium bulbiferum* (Fig. 3.17), *Luzula sylvatica*, *Meum athamanticum*, *Pinus mugo* (Fig. 3.17), *Potentilla aurea*, *Primula minima*,

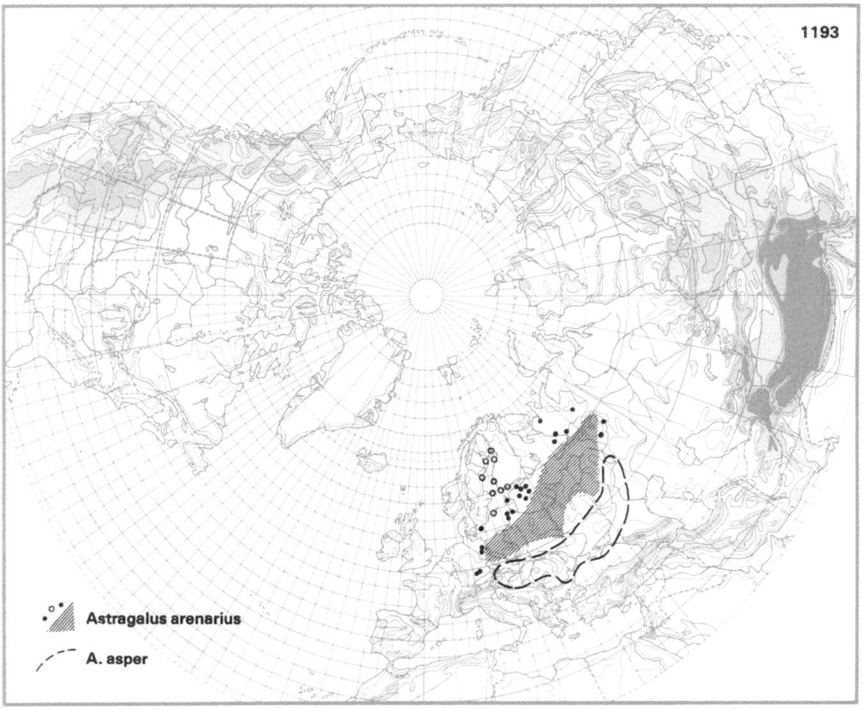


**Fig. 3.16** Distribution of *Astragalus arenarius*, a Sarmatian geoelement, and *Astragalus asper*, a Pontic geoelement. Both species reach the western limit of their distributions in Central Europe (reproduced from Hultén and Fries 1986 with permission from Koeltz Scientific Books)

*Rosa pendulina*, *Scabiosa lucida* and *Sesleria caerulea*. The species that occur mainly or exclusively at high altitudes in the Alps and Carpathians, however, are sometimes placed in the Alpine-Carpathian geoelement.

**The Carpathian geoelement** is confined to the Carpathians or in some cases extends to adjacent areas. It includes alpine, subalpine and montane species, which in the Czech Republic occur mainly in its easternmost part, particularly in the Moravskoslezské Beskydy Mts, although some of them extend westwards as far as the Krkonoše Mts. Examples include *Aconitum firmum* subsp. *moravicum*, *Cardamine amara* subsp. *opicii* (Fig. 3.18), *Centaurea oxylepis*, *Dentaria glandulosa* (Fig. 3.19), *Salix silesiaca*, *Scilla kladnii* and *Thymus pulcherrimus*.

**The Alpine geoelement** occurs mainly in the Alps and sometimes also in their foothills. Several species occur in the Czech Republic, either in the Šumava Mts or the whole of western Bohemia, but only exceptionally and only at isolated localities further east or north-east. *Erica carnea*, *Gentiana pannonica* (Fig. 3.18), *Polygala chamaebuxus*, *Salix appendiculata*, *Soldanella montana*, *Thesium rostratum* and *Willemetia stipitata* may be given as examples.

**Fig. 3.17** Examples of the Central European (sub-)alpine geoelement: (**a**) *Gentiana asclepiadea*, between Mt. Velká Jezerná and Mt. Velký Máj, Hrubý Jeseník Mts; (**b**) *Hieracium villosum*, Velká kotlina cirque, Hrubý Jeseník Mts; (**c**) *Gentiana punctata*, Mt. Vysoká hole, Hrubý Jeseník Mts; (**d**) *Lilium bulbiferum*, Pustá Rudná near Andělská Hora, Hrubý Jeseník Mts; (**e**) *Pinus mugo*, Malá kotlina cirque, Hrubý Jeseník Mts. Photo credits: L. Bureš (a, e), Z. Kaplan (b–d)

**The Arctic geoelement** mainly occurring in the Arctic tundra in areas with permafrost, is absent from the Czech Republic, but **the Arctic-alpine geoelement**, which occurs also in more southern mountains, is represented here by *Bartsia alpina*, *Carex atrata*, *C. rupestris*, *Diphasiastrum alpinum*, *Epilobium anagallidifolium*, *Gnaphalium norvegicum*, *Hieracium alpinum* (Fig. 3.18), *Luzula spicata*, *Oreojuncus trifidus*, *Pedicularis sudetica* (s.l.) (Fig. 3.2), *Pseudorchis albida*, *Rhodiola rosea*, *Salix herbacea* (Fig. 3.18) and *Saxifraga oppositifolia*. Similarly **the Sub-Arctic geoelement** extends further south than the Arctic geoelement and occurs in forest-tundra, which in the Czech Republic is represented by the Sub-Arctic-alpine geoelement with species such as *Betula nana*, *Carex capillaris*, *C. chordorrhiza*, *Eriophorum vaginatum*, *Phleum alpinum*, *Polystichum lonchitis*, *Rubus chamaemorus* (Fig. 3.5), *Salix myrtilloides*, *Selaginella selaginoides*, *Trichophorum alpinum* (Fig. 3.18) and *Viola biflora* (Fig. 3.18).

**The Boreal geoelement** includes species adapted to long winters that are distributed mainly in the boreal zone, but occur also further south mostly in the mountains,

**Fig. 3.18** Examples of different types of geoelements occurring in the Czech Republic mainly in the mountains: (**a**) Carpathian geoelement, *Cardamine amara* subsp. *opicii*, Velká kotlina cirque, Hrubý Jeseník Mts; (**b**) Alpine geoelement, *Gentiana pannonica*, Kvilda, Šumava Mts; (**c**) Arctic-alpine geoelement, *Hieracium alpinum*, Mt. Petrovy kameny, Hrubý Jeseník Mts; (**d**) Arctic-alpine geoelement, *Salix herbacea*, Mt. Petrovy kameny, Hrubý Jeseník Mts; (**e**) Sub-Arctic-alpine geoelement, *Trichophorum alpinum*, Velká kotlina cirque, Hrubý Jeseník Mts; (**f**) Sub-Arctic-alpine geoelement, *Viola biflora*, Velká kotlina cirque, Hrubý Jeseník Mts. Photo credits: L. Bureš (a, c, d, f) and Z. Kaplan (b, e)

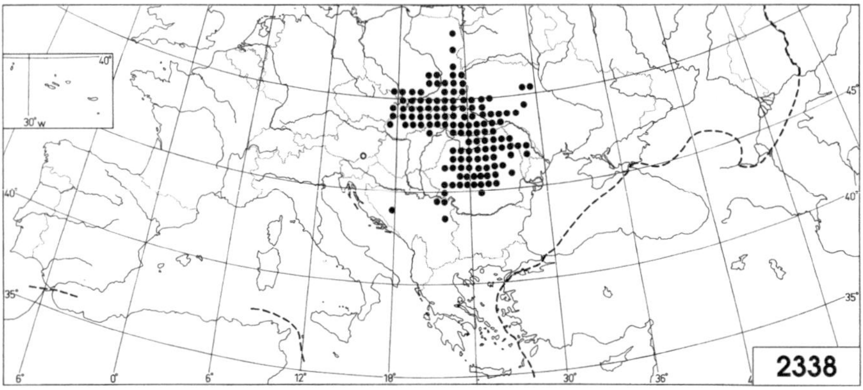


**Fig. 3.19** *Dentaria glandulosa* is an example of the Carpathian geoelement that entered the Czech Republic from the east and whose western distribution limit is in Moravia (reproduced from Jalas and Suominen 1994 with permission of the Committee for Mapping the Flora of Europe and Societas Biologica Fennica Vanamo)

usually in coniferous forests (in more oceanic areas they sometimes occur in mixed or broad-leaved deciduous forests). Their occurrence at low altitudes is often associated with wet habitats such as marshes and peat bogs. Some of these plants were probably more frequent in Central Europe during the early postglacial period. This geoelement is represented, for example, by *Calla palustris* (Fig. 3.20), *Carex canescens*, *Chimaphila umbellata* (Fig. 3.20), *Cirsium heterophyllum*, *Coeloglossum viride*, *Corallorhiza trifida*, *Epilobium angustifolium*, *Eriophorum angustifolium*, *Geranium pratense*, *Lycopodium annotinum*, *Maianthemum bifolium*, *Melampyrum sylvaticum*, *Menyanthes trifoliata*, *Persicaria amphibia*, *Rhododendron tomentosum* (Fig. 3.20), *Trientalis europaea*, *Vaccinium myrtillus* and *V. vitis-idaea* (Fig. 3.20).

**The Sub-Boreal geoelement** is similar to the Boreal geoelement but extends further south, particularly in more oceanic areas. In these areas it is more frequent in mixed and deciduous forests and at low altitudes. This geoelement is one of the most common in the Czech flora. It includes, for example, the trees *Alnus incana*, *Betula pubescens*, *Frangula alnus*, *Juniperus communis*, *Pinus sylvestris*, *Salix caprea*, *Sorbus aucuparia* and *Viburnum opulus*, and the herbaceous plants *Achillea ptarmica*, *Angelica sylvestris*, *Athyrium filix-femina*, *Bistorta officinalis*, *Cardamine pratensis*, *Carex acuta*, *Chrysosplenium alternifolium*, *Circaea alpina*, *Delphinium elatum*, *Deschampsia cespitosa*, *Dryopteris filix-mas*, *Filipendula ulmaria*, *Geum rivale*, *Glyceria maxima*, *Gnaphalium sylvaticum*, *Gymnadenia conopsea*, *Heracleum sphondylium*, *Hypericum perforatum*, *Lycopodium clavatum* (Fig. 3.20), *Melampyrum pratense*, *Melica nutans*, *Milium effusum*, *Oxalis acetosella* (Fig. 3.20), *Paris quadrifolia*, *Phalaris arundinacea*, *Poa nemoralis*, *Potamogeton natans*, *Potentilla erecta*, *Pyrola rotundifolia*, *Rumex acetosa*, *Sanguisorba officinalis*, *Scirpus sylvaticus*, *Symphytum officinale* and *Urtica dioica*.

**The South-Siberian geoelement** includes species distributed mainly in southern Siberia and adjacent areas. Those extending westwards to Europe may be recognized as **the European-South-Siberian geoelement**. They can tolerate an extremely con-

**Fig. 3.20** Examples of the Boreal and Sub-boreal geoelements: (**a**) *Calla palustris*, Malý Karlov fishpond near Čermná nad Orlicí, eastern Bohemia; (**b**) *Chimaphila umbellata*, between Plchůvky and Horní Jelení, eastern Bohemia; (**c**) *Lycopodium clavatum*, Mt. Špičák near Železná Ruda, Šumava Mts; (**d**) *Oxalis acetosella*, Velká kotlina cirque, Hrubý Jeseník Mts; (**e**) *Rhododendron tomentosum*, Rejvíz bog, Hrubý Jeseník Mts; (**f**) *Vaccinium vitis-idaea*, Mt. Vysoká hole, Hrubý Jeseník Mts. Photo credits: L. Bureš (a, d–f) and Z. Kaplan (b, c)

tinental climate with warm but short summers and long winters with severe frosts. They occur in a broad range of habitats including steppes and forests. The examples occurring in the Czech Republic are *Adenophora liliifolia*, *Artemisia campestris*, *Brachypodium pinnatum*, *Fragaria viridis*, *Filipendula vulgaris*, *Inula salicina* (Fig. 3.21), *Koeleria glauca*, *Lilium martagon* (Fig. 3.21), *Phleum phleoides*,

**Fig. 3.21** Examples of the European-South-Siberian (**a**–**c**, **f**) and Sub-Atlantic (**d**–**e**) geoelements: (**a**) *Inula salicina*, Vysoké Mýto, eastern Bohemia; (**b**) *Lilium martagon*, Merklovice near Vamberk, eastern Bohemia; (**c**) *Platanthera bifolia*, Třemešná, northern Moravia; (**d**) *Cytisus scoparius*, Dívčí Hrad near Osoblaha, northern Moravia; (**e**) *Pedicularis sylvatica*, Rejvíz bog, Hrubý Jeseník Mts; (**f**) *Veratrum album* subsp. *lobelianum*, Velká kotlina cirque, Hrubý Jeseník Mts. Photo credits: L. Bureš (a, c–f) and Z. Kaplan (b)

*Platanthera bifolia* (Fig. 3.21), *Pimpinella saxifraga*, *Rhamnus cathartica*, *Silene nutans*, *Veratrum album* subsp. *lobelianum* (Fig. 3.21) and *Vicia sylvatica*.

**The Pontic geoelement** (sometimes called the Pontic-Pannonian geoelement) is distributed mainly in the steppes and forest-steppes of southern Ukraine and southern part of European Russia. Many of these species spread westwards through the Pannonian Basin to Central Europe, or even further west. However, recent studies indicate that some may have spread along the northern foothills of the Carpathians (Cieślak 2014). The Pontic geoelement comprises species that tolerate a continental climate with warm and dry summers and extremely cool winters. Within this geoelement, Slavík (in Hejný et al. 1988: 65–102) distinguished a Pannonian subelement, which differs from the Pontic-South-Siberian element occurring along the northern coast of the Black Sea and in the south-Siberian steppes. Examples of the Pontic geoelement (in the broad sense) include *Adonis vernalis* (Fig. 3.22), *Artemisia pontica*, *Clematis integrifolia*, *C. recta*, *Cirsium pannonnicum* (Fig. 3.22), *Crepis pannonica*, *Echium maculatum*, *Galium glaucum*, *Inula germanica*, *I. hirta*, *Iris pumila* (Fig. 3.22), *Linaria genistifolia*, *Linum flavum*, *Melica transsilvanica*, *Prunus fruticosa*, *Ranunculus illyricus*, *Scabiosa ochroleuca*, *Scorzonera purpurea* and *Stipa pennata*. **The Sub-Pontic geoelement** is similarly distributed and its species have similar ecological requirements as those of the Pontic geoelement but tend to occur more frequently in open forests than on steppes. They more frequently occur in Central Europe and are more numerous there than species of the Pontic geoelement. Species occurring in the Czech Republic include *Anemone sylvestris* (Fig. 3.22), *Anthemis tinctoria*, *Astragalus glycyphyllos*, *Chamaecytisus ratisbonensis*, *Euonymus verrucosus*, *Gentiana cruciata*, *Iris aphylla*, *Ononis arvensis*, *Potentilla recta*, *Prunus spinosa*, *Securigera varia*, *Seseli annuum*, *Stachys recta*, *Tanacetum corymbosum*, *Trifolium montanum*, *Verbascum lychnitis* and *Vincetoxicum hirundinaria*.

**The Mediterranean geoelement** occurs around the Mediterranean Sea or in parts of this area. Its species are found where the winters are short, mild and wet, and are also tolerant of hot and dry summers. They extend northwards to the southern edge of Central Europe but do not reach the Czech Republic. **The Sub-Mediterranean geoelement** occurs further north and is found also in Central Europe. Those species that reach the Czech Republic mostly spread from the Balkan Peninsula and from the southern and eastern edges of the Alps and currently often occur in thermophilous oak forests. This geoelement is well represented in the Czech Republic, e.g. by *Arabis turrita*, *Bifora radians*, *Bromus erectus*, *Buglossoides purpurocaerulea*, *Carex michelii*, *Cornus mas*, *Coronilla vaginalis*, *Danthonia alpina*, *Euphorbia amygdaloides*, *Globularia bisnagarica*, *Heliotropium europaeum*, *Iris variegata* (Fig. 3.23), *Ligustrum vulgare*, *Linum tenuifolium*, *Loranthus europaeus*, *Medicago minima*, *Melampyrum arvense* (Fig. 3.4), *Melica ciliata*, *Melittis melissophyllum* (Fig. 3.23), *Microthlaspi perfoliatum*, *Muscari comosum*, *Ophrys apifera*, *Orchis purpurea* (Fig. 3.23), *Orlaya grandiflora* (Fig. 3.23), *Potentilla micrantha*, *Prunella laciniata*, *Quercus pubescens*, *Rosa gallica*, *Salvia verticillata*, *Sorbus torminalis*, *Staphylea pinnata*, *Teucrium chamaedrys*, *Trinia glauca*, *Viburnum lantana* and *Viola alba*.

**Fig. 3.22** Examples of the Pontic (**a**–**c**) and Sub-Pontic (**d**) geoelements: (**a**) *Adonis vernalis*, Boubová hill near Srbsko, Bohemian Karst, central Bohemia; (**b**) *Cirsium pannonicum*, Suchov, Bílé Karpaty Mts, south-eastern Moravia; (**c**) *Iris pumila*, Divoká Šárka valley in Prague; (**d**) *Anemone sylvestris*, Středošice, eastern Bohemia. Photo credits: Z. Kaplan (a, c), L. Hrouda (b) and L. Bureš (d)

**The Atlantic geoelement** occurs mainly along the European Atlantic seaboards and in adjacent areas from Portugal to Norway. Its species occur where the winters are mild and sometimes even frostless and the summers are mild and wet. This geoelement is in its strict sense represented in the Czech Republic only by *Erica tetralix* (questionably native), *Narthecium ossifragum* (at a single locality, now vanished) and *Trichomanes speciosum* (only as prothallium); their Czech localities are at the border of their distributions. **The Sub-Atlantic geoelement** extends further on to the European mainland, as it is slightly more tolerant of frost and drought. It is more numerous in the Czech Republic than the Atlantic geoelement, and is represented e.g. by *Aira praecox*, *Calluna vulgaris*, *Carex pseudobrizoides*, *Chrysosplenium oppositifolium* (Fig. 3.24), *Cytisus scoparius* (Fig. 3.21), *Euphrasia nemorosa* (s.l.),

**Fig. 3.23** Examples of the Sub-Mediterranean geoelement: (**a**) *Iris variegata*, Lanžhot near Břeclav, southern Moravia; (**b**) *Melittis melissophyllum*, Dománovice, central Bohemia; (**c**) *Orchis purpurea*, Halín wood near Dobruška, eastern Bohemia; (**d**) *Orlaya grandiflora*, Děvín hill, Pavlov Hills, southern Moravia. Photo credits: Z. Kaplan (a–c) and P. Veselý (d)

*Hypericum humifusum*, *H. pulchrum*, *Luronium natans*, *Lysimachia nemorum*, *Pedicularis sylvatica* (Fig. 3.21), *Polygala serpyllifolia*, *Potamogeton polygonifolius*, *Rhynchospora fusca* and *Sagina subulata*.

**The Sub-Atlantic-Sub-Mediterranean geoelement** is in many respects transitional between the two respective geoelements. It extends into Central Europe from the south or west, or both. Its species usually occur in deciduous forests where the

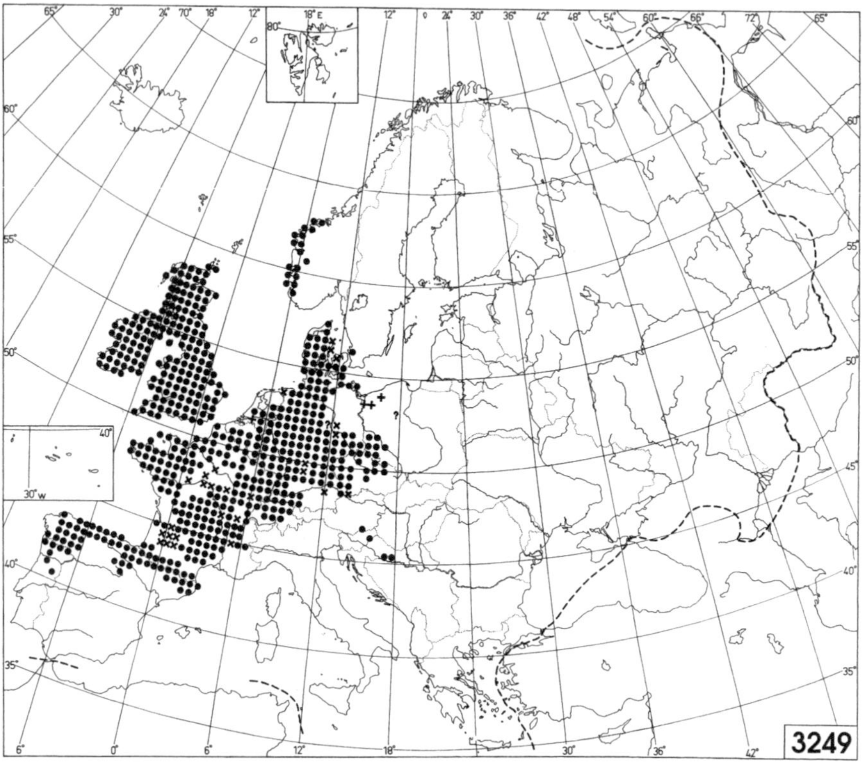


**Fig. 3.24** The Sub-Atlantic geoelement *Chrysosplenium oppositifolium* has the eastern limit of its continuous distribution in the Czech Republic (reproduced from Jalas et al. 1999 with permission of the Committee for Mapping the Flora of Europe and Societas Biologica Fennica Vanamo)

winters are mild and short and the summers not extremely dry and hot. Examples include *Asplenium scolopendrium*, *Genista pilosa*, *Hippocrepis comosa*, *Lathyrus nissolia*, *Osmunda regalis*, *Saxifraga granulata* and *Trifolium striatum*.

## 3.6 Boundary and Outlying Elements

A number of widespread species are at the limits of their distributions in the Czech Republic (Slavík in Hejný et al. 1988: 65–102). Boundary elements are at the limit of their more or less continuous distributions here, whereas outlying elements occur at a few isolated sites in the Czech Republic, which are well separated from the rest of their range. The limits of their distributions are often not the absolute limit in a particular geographical direction but rather the regional edge of the distributions.

Most frequent are thermophilous species in the Sub-Mediterranean and Pontic-South-Siberian geoelements that occur in the Pannonian Basin and have their northern or north-western (in a few cases north-eastern) boundaries in the warm and dry area in southern Moravia or even in Bohemia. These include *Allium flavum*, *Astragalus asper* (Fig. 3.16), *A. austriacus*, *Buglossoides purpurocaerulea*, *Cirsium brachycephalum*, *Clematis integrifolia*, *Cotoneaster laxiflorus*, *Crambe tataria* (Figs. 3.4 and 3.25), *Crypsis alopecuroides*, *C. schoenoides*, *Danthonia alpina*, *Draba nemorosa*, *Euphorbia epithymoides*, *Globularia bisnagarica*, *Helianthemum canum*, *Himantoglossum adriaticum*, *Iris arenaria*, *I. pumila* (Fig. 3.22), *Linum flavum*, *L. hirsutum*, *Medicago monspeliaca* (Fig. 3.26), *Minuartia glaucina*, *Notholaena marantae* (Fig. 3.27), *Onosma arenaria*, *Orlaya grandiflora* (Fig. 3.23), *Potentilla micrantha*, *P. patula*, *Prunella laciniata*, *Prunus tenella* (Fig. 3.4), *Quercus cerris* (Fig. 3.28), *Q. pubescens*, *Salvia aethiopis*, *Seseli pallasii*, *Sorbus aria*, *Thesium arvense*, *T. dollineri*, *Trinia glauca*, *Veratrum nigrum* and *Vicia pannonica* subsp. *pannonica* (Fig. 3.4). Other thermophilous species (mainly of the West-Sub-Mediterranean geoelement) came from the south-west and either reach their limits in Bohemia (such as *Anthericum liliago* and *Dianthus gratianopolitanus*; Fig. 3.4) or in Moravia (*Arenaria grandiflora* – Fig. 3.29, *Sedum reflexum*, *Teucrium botrys*). Several species spreading from the Alps also reach their northern or north-eastern limits in this country, mostly in the mountains or highlands in southern and western Bohemia, seldom in south-western Moravia. Examples include *Alnus alnobetula*, *Cyclamen purpurascens*, *Gentiana pannonica* (Fig. 3.18), *Ranunculus aconitifolius* (Fig. 3.30), *Salix appendiculata*, *Soldanella montana*, *Thesium rostratum*, *Veratrum album* subsp. *album* and *Willemetia stipitata*.

The eastern or north-eastern limits of the ranges of some Atlantic-Sub-Mediterranean species such as *Arabis pauciflora*, *Juncus subnodulosus*, *Lathyrus linifolius* and *Polygala chamaebuxus* are in western Bohemia and in a few cases in south-western Moravia. There are a relatively large number of Sub-Atlantic species that spread from the north-west, which are at the south-eastern (or eastern) limit of their continuous distributions in the Czech Republic. These include *Arnoseris minima*, *Chrysosplenium oppositifolium* (Fig. 3.24), *Corrigiola litoralis*, *Hydrocotyle vulgaris*, *Hylotelephium telephium*, *Illecebrum verticillatum*, *Juncus bulbosus*, *Lotus pedunculatus*, *Lysimachia nemorum*, *Myriophyllum alterniflorum*, *Pedicularis sylvatica* (Fig. 3.21), *Polygala serpyllifolia*, *Potentilla anglica*, *P. sterilis*, *Radiola linoides*, *Spergula morisonii*, *Stachys arvensis* and *Teesdalia nudicaulis*.

In contrast, some Sub-Arctic and Boreal species reach their regional or absolute southern distribution limits in the Czech mountains, or rarely in lowland refugia; in particular, *Betula nana*, *Calamagrostis purpurea*, *Carex vaginata*, *Pedicularis sceptrum-carolinum*, *P. sudetica* (s.l.; Fig. 3.2), *Rhododendron tomentosum* (Fig. 3.20), *Rubus chamaemorus* (Fig. 3.3), *Salix myrtilloides* and *Trichophorum cespitosum*. Another species, *Calamagrostis stricta*, occurs mainly in lowland mires. The Sarmatian element with its south-western boundary in the Czech Republic includes *Astragalus arenarius* (Fig. 3.16), *Dianthus arenarius* (Fig. 3.14), *Jurinea cyanoides*, *Pulsatilla patens* (s. str.) and *Thesium ebracteatum*.

Finally, the distributions of several species extend from the Western Carpathians or their foothills and reach their western limits in Moravia or north-eastern Bohemia. These (mostly Carpathian) taxa include *Aconitum firmum* subsp. *moravicum*, *Actaea europaea* (Fig. 3.31), *Cardamine amara* subsp. *opicii* (Fig. 3.18), *Dentaria glandulosa* (Fig. 3.19), *Hacquetia epipactis*, *Pedicularis exaltata*, *Salix silesiaca*, *Scilla*

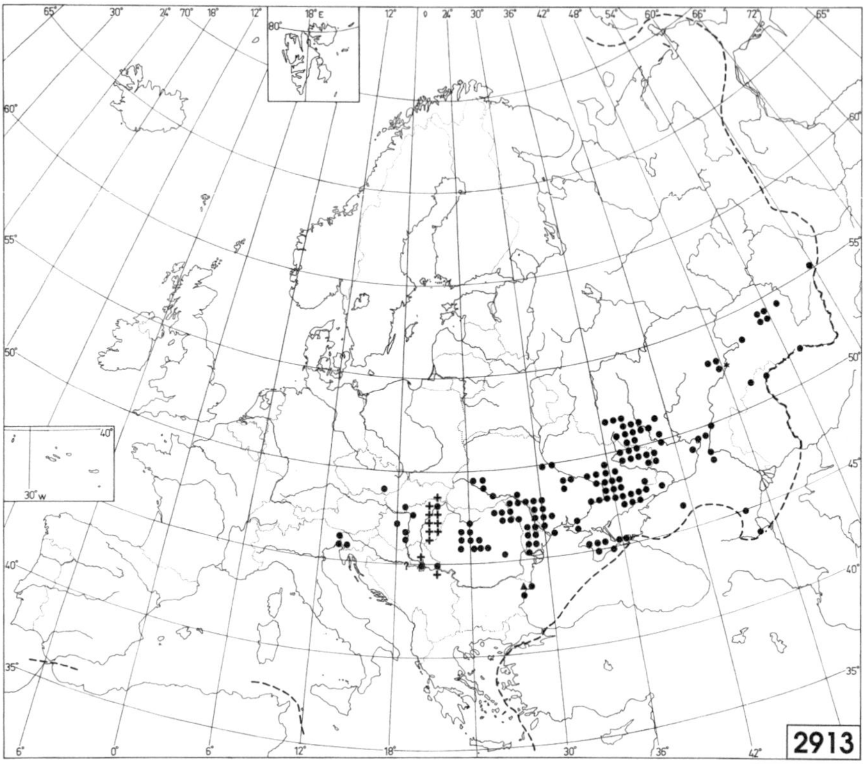


**Fig. 3.25** The European part of the extensive distribution of *Crambe tataria*. The localities in southern Moravia are at the north-western border of its distribution (reproduced from Jalas et al. 1996 with permission of the Committee for Mapping the Flora of Europe and Societas Biologica Fennica Vanamo)

*kladnii* and *Thymus pulcherrimus*. Several of the species with extensive ranges also have the western boundary of their distributions in the Czech Republic. Most of them are continental plants that occupy large areas that extend to southern Siberia. Examples include *Conioselinum tataricum*, *Crepis pannonica*, *C. sibirica* (Fig. 3.32), *Echium maculatum*, *Erigeron podolicus*, *Gypsophila paniculata*, *Helictotrichon desertorum*, *Pilosella echioides*, *P. onegensis*, *Stipa glabrata*, *Trinia ucrainica* and *Verbascum speciosum*. For many of them the Czech Republic is also at the northern limit of their distributions in Central Europe.

## 3.7 Phytochorotypes in the Czech Flora

Based on similar principles to those used to distinguish geoelements at broader scales (such as in Europe), species can be classified according to their distributions within a country. Based on a comparison of several hundred distribution maps of

**Fig. 3.26** Distribution of *Medicago monspeliaca*, the northernmost locality of which is in the České středohoří Mts (reproduced from Meusel et al. 1965 with permission from Elsevier and from Prof. Eckehart J. Jäger)

vascular plants in the Czech Republic, Slavík (1984) identified 15 basic types of regional distribution of groups of species with common ecological requirements and geographical distributions, for which he proposed the term phytochorotypes. Species are assigned to phytochorotypes based on the similarity of their distributions within the Czech Republic, which reflect not only local climate, geology and geomorphology but also migration history. Total ranges of species are not considered at this scale, which is why species of different major geoelements can be classified in the same phytochorotype. Only native species and archeophytes with relatively stable distributions were used for these comparisons.

The 15 basic phytochorotypes are divided into three groups according to their prevailing relation to one of the three principal phytogeographical regions in the Czech Republic (see below). The phytochorotypes were named after pairs of species typical of a given phytochorotype. Only a brief account of their characteristics is given here. For a description of the phytogeographical divisions in the Czech Republic and definitions of phytogeographical regions (Thermophyticum, Mesophyticum and Oreophyticum) see below. Details, characteristic species and distribution maps of the species in each phytochorotype are given by Slavík (1984). This concept was applied by Slavík in his later studies (Slavík 1998; Slavík in Kubát et al. 2002: 42–45), a slightly modified version supplemented with a group of edaphically specialized phytochorotypes is proposed in the introductory chapters of

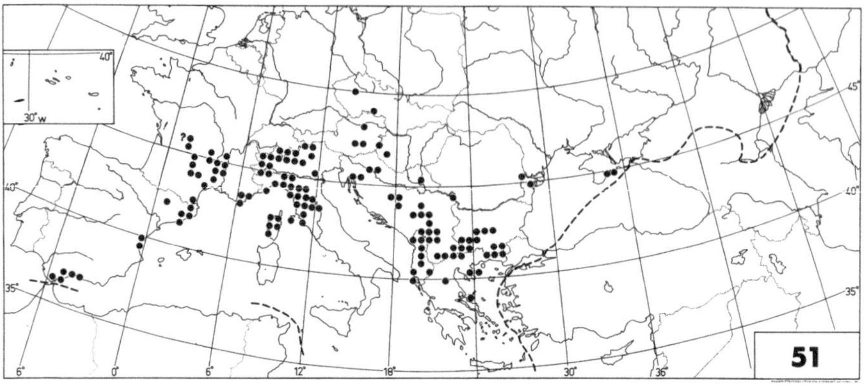


**Fig. 3.27** *Notholaena marantae* is an example of the Mediterranean geoelement that reaches the absolute northern boundary of its distribution in the Czech Republic (adopted from Jalas and Suominen 1972 with permission of the Committee for Mapping the Flora of Europe and Societas Biologica Fennica Vanamo, and supplemented with the northernmost locality recently recorded in the Bohemian Karst)

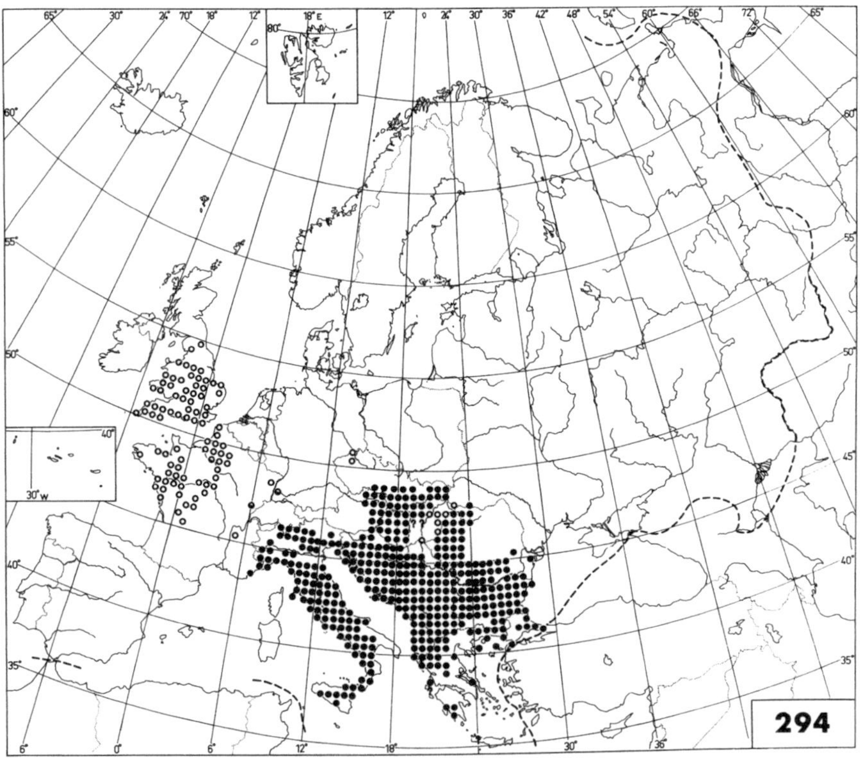


**Fig. 3.28** The European distribution of the Mediterranean geoelement *Quercus cerris*. The northern limit of its native distribution runs through southern Moravia (reproduced from Jalas and Suominen 1976 with permission of the Committee for Mapping the Flora of Europe and Societas Biologica Fennica Vanamo)

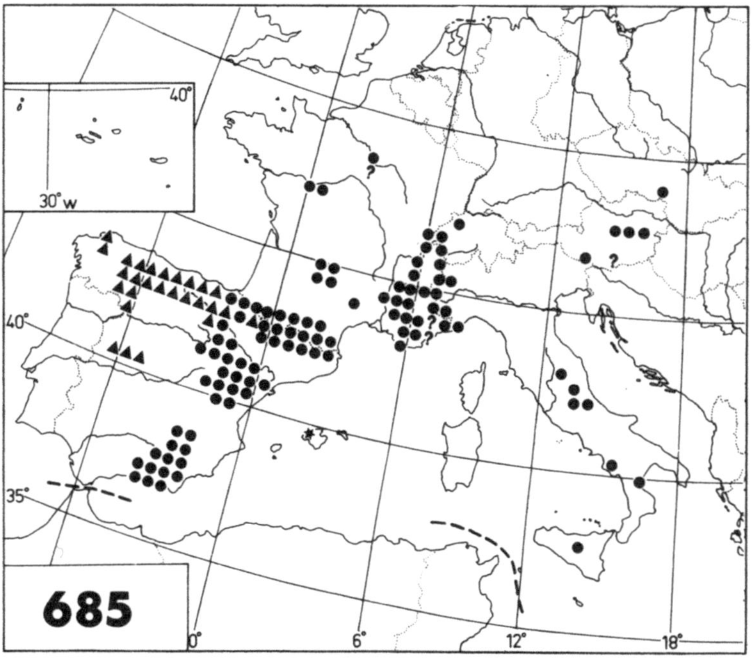


**Fig. 3.29** The European distribution of *Arenaria grandiflora*. The north-easternmost locality of this species is in the Pavlov Hills in southern Moravia (reproduced from Jalas and Suominen 1983 with permission of the Committee for Mapping the Flora of Europe and Societas Biologica Fennica Vanamo)

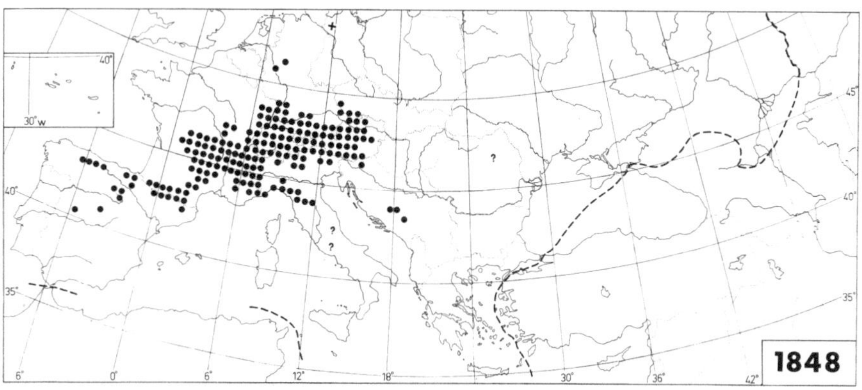


**Fig. 3.30** The core of the distribution of *Ranunculus aconitifolius* is in the Alps and its north-eastern limit in the Šumava Mts in south-western Bohemia (reproduced from Jalas and Suominen 1989 with permission of the Committee for Mapping the Flora of Europe and Societas Biologica Fennica Vanamo)

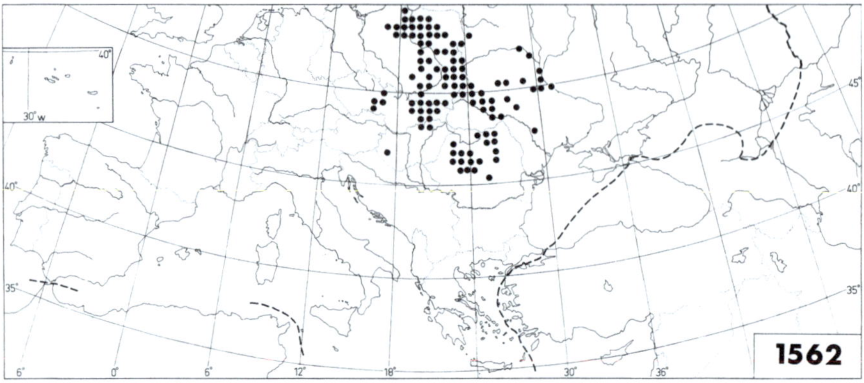


**Fig. 3.31** Distribution of *Actaea europaea*, the westernmost localities of which are in the Czech Republic (reproduced from Jalas and Suominen 1989 with permission of the Committee for Mapping the Flora of Europe and Societas Biologica Fennica Vanamo)

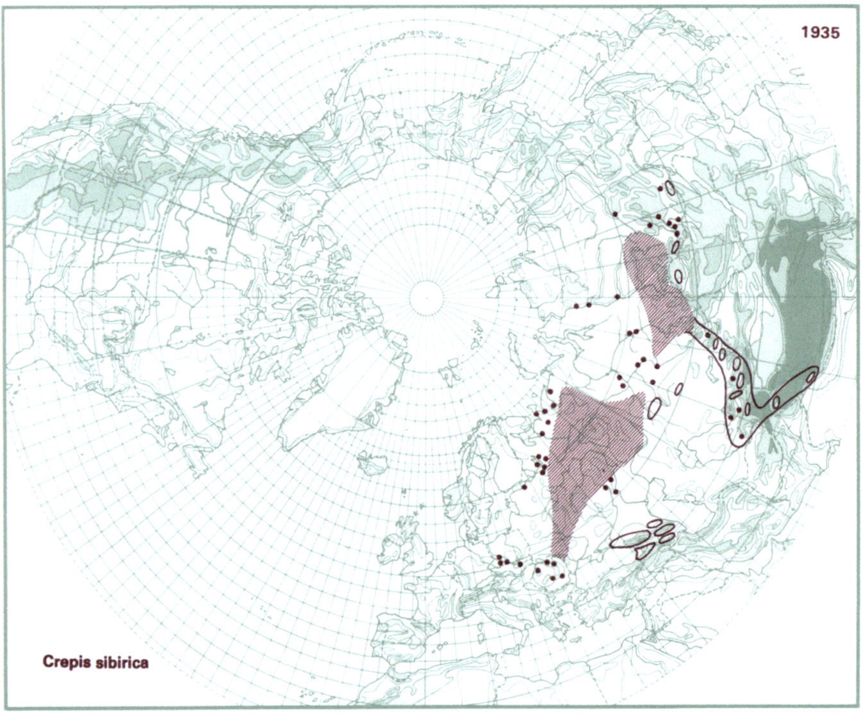


**Fig. 3.32** *Crepis sibirica* has a wide distribution in Eastern Europe and Western and Central Asia. Its westernmost locality is in the Velká kotlina cirque in the Hrubý Jeseník Mts (reproduced from Hultén and Fries 1986 with permission from Koeltz Scientific Books)

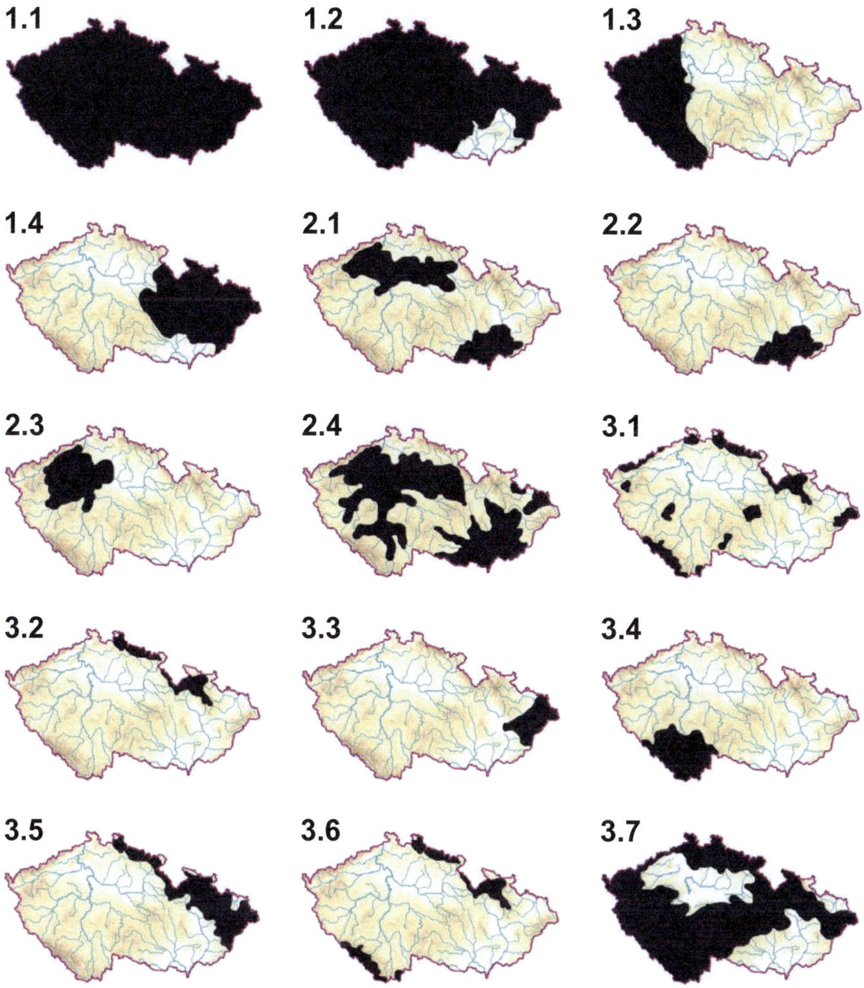


**Fig. 3.33** Schematic maps of the distributions of the basic phytochorotypes in the Czech Republic (redrawn from Slavík 1984 and slightly modified by Kaplan 2012)

the Flora of the Czech Republic (Slavík in Hejný et al. 1988: 65–102). Schematic illustrations of the 15 phytochorotypes based on Slavík's analysis and refined according to updated distribution maps are given in Fig. 3.33.

1. Species common throughout the country or distributed mainly in Mesophyticum.

   1.1 Phytochorotype: *Achillea millefolium* agg.-*Urtica dioica*

   Species distributed throughout or nearly throughout the Czech Republic.

   1.2 Phytochorotype: *Hypericum maculatum-Luzula pilosa*

   Species distributed throughout most of this country but absent or rare in the Pannonian subregion in southern Moravia.

1.3 Phytochorotype: *Chaerophyllum aureum-Polygala chamaebuxus*

Species largely confined to the western half of Bohemia, regardless of the regional phytogeographical division.

1.4 Phytochorotype: *Euphorbia amygdaloides-Galium rivale*

Species largely confined to the eastern half of the country, occurring exclusively in Moravia or reaching only the eastern part of Bohemia.

2. Species distributed mainly in Thermophyticum.

2.1 Phytochorotype: *Buglossoides purpurocaerulea-Ranunculus illyricus*

Species distributed in both Bohemian and Pannonian Thermophyticum.

2.2 Phytochorotype: *Euphorbia epithymoides-Hesperis tristis*

Species of Pannonian Thermophyticum, particularly of its warmest parts in southern Moravia, where they often reach the northern or north-western limit of their distributions.

2.3 Phytochorotype: *Erysimum crepidifolium-Lactuca perennis*

Species that occur in the Bohemian Thermophyticum, sometimes with local extensions south to the warmer parts of Mesophyticum (up to the surroundings of the city of Plzeň and the middle Vltava River valley).

2.4 Phytochorotype: *Bothriochloa ischaemum-Scabiosa ochroleuca*

Slightly thermophilous species distributed mainly in Thermophyticum and occurring also in warmer parts of Mesophyticum.

3. Species distributed mainly in Oreophyticum that occasionally occur in the cooler parts of Mesophyticum, particularly at sites with temperature inversions in valleys and ravines or peat bogs.

3.1 Phytochorotype: *Rumex arifolius-Streptopus amplexifolius*

Species that occur in all or most high mountain ranges in this country (Krušné hory, Jizerské hory, Krkonoše, Orlické hory, Králický Sněžník, Hrubý Jeseník, Moravskoslezské Beskydy and Šumava) and sometimes also in lower mountain ranges (Slavkovský les, Brdy, Novohradské hory, Bohemian-Moravian Highlands, Adršpach-Teplice Rocks and Nízký Jeseník Mts).

3.2 Phytochorotype: *Anemonastrum narcissiflorum-Delphinium elatum*

Species that occur only in the Sudetes Mts, particularly in the High Sudetes (Krkonoše, Králický Sněžník and Hrubý Jeseník) and sometimes also in other ranges of the Sudetes (Lužické hory, Jizerské hory, Adršpach-Teplice Rocks, Orlické hory and Nízký Jeseník). Some of them only occur in the Krkonoše and Hrubý Jeseník Mts, which have a well-developed subalpine belt.

3.3 Phytochorotype: *Dentaria glandulosa-Luzula luzulina*

Carpathian species that are found in the Moravskoslezské Beskydy Mts, sometimes occurring also in adjacent parts of Mesophyticum.

3.4 Phytochorotype: *Salix appendiculata-Veratrum album* subsp. *album*

Species that occur in Oreophyticum in southern Bohemia (particularly those that extend from the Alps) and in adjacent parts of Mesophyticum (mainly aquatic plants).

3.5 Phytochorotype: *Potentilla aurea-Veratrum album* subsp. *lobelianum*

Species that occur in the Sudetes and Carpathians, and occasionally in adjacent parts of Mesophyticum.

3.6 Phytochorotype: *Epilobium anagallidifolium-Trichophorum cespitosum*

Species that occur both in the Sudetes and in Oreophyticum of southern Bohemia, which is influenced by the proximity of the Alps.

3.7 Phytochorotype: *Montia fontana-Polygonatum verticillatum*

Species that are found (almost) throughout this country in Oreophyticum and cooler parts of Mesophyticum.

## 3.8 Phytogeographical Division of the Country

Another method in phytogeography, complementary to the identification of floristic elements, is dividing areas into floristic regions based on the relative internal homogeneity in their flora. In the phytogeographical division of Europe proposed by Meusel et al. (1965), the Czech Republic is situated in the temperate zone of the Holarctic floristic kingdom, and in its Middle European region. Within this region, most of the Czech Republic is situated in the Middle European subregion (and its Central European province), but the mountains of eastern Moravia and adjacent Silesia are in the Carpathian subregion (and its North-western Carpathian province). The Pannonian province in the Pontic-South-Siberian subregion does not reach the Czech Republic under this concept. There is another concept proposed by Takhtajan (1986) under which there are 20 floristic provinces in Europe, Asia Minor and the Caucasus, and in which the whole of the Czech Republic is in the Central European province.

The system of Meusel et al. (1965) was modified and refined by Slavík (in Hejný et al. 1988: 65–102) who argued that the Pannonian lowland is an isolated area of forest-steppe, separated from the Pontic-South-Siberian subregion by the Carpathian mountain range and influenced by the Sub-Mediterranean geoelement. He distin-

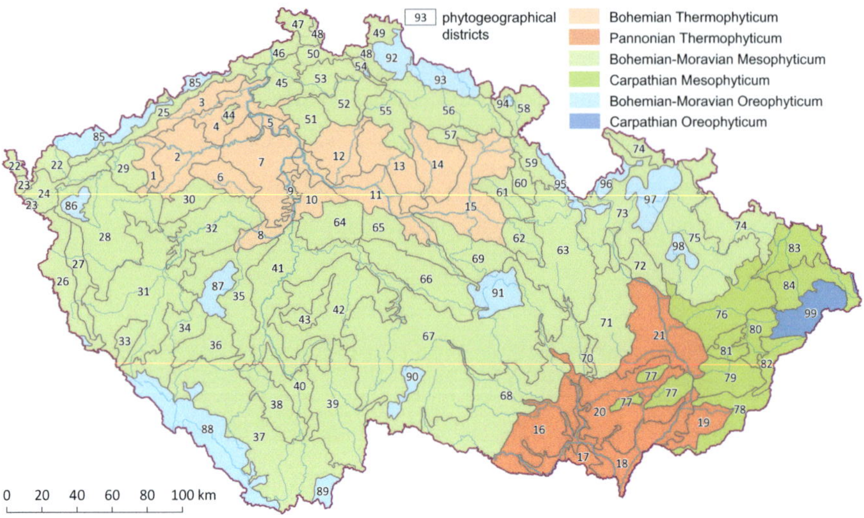


**Fig. 3.34** Phytogeographical division of the Czech Republic according to Skalický in Hejný et al. (1988: 103–121)

guishes this area as a separate Pannonian subregion and extended it (in contrast to Meusel's Pannonian province) in the north-west to southern Moravia.

Within the Czech Republic, the currently used concept of regional phytogeographical division was elaborated by phytogeographical experts of the Czech Botanical Society coordinated by V. Skalický (Skalický 1982; Skalický in Hejný et al. 1988: 103–121). This system replaced the previous phytogeographical divisions of Czechoslovakia proposed by Dostál (1957, 1960), which more closely reflected the Central European context and occurrence of geoelements. The final version of his division (Dostál 1960) included three regions in the Czech Republic, namely Hercynicum, with Central European temperate forest vegetation; Pannonicum, with thermophilous vegetation; and Carpaticum occidentale, with vegetation typical of the Western Carpathians. The concept published by Skalický is the basic system used for recording the distributions of species and subspecies in the Flora of the Czech Republic (Hejný et al. 1988, 1990, 1992; Slavík et al. 1995, 1997, 2000, 2004; Štěpánková et al. 2010) as well as in many floristic and taxonomic studies. Species and plant communities diagnostic and characteristic of each region are listed by Skalický (in Hejný et al. 1988: 103–121).

A four-level hierarchical system of phytogeographical units (phytochoria) is used in the current phytogeographical division of the Czech Republic (for details, see Kaplan 2012 and Electronic Appendices 1–3 therein). Three principal phytogeographical regions are recognized, based on the dominant flora and vegetation that reflects specific regional geomorphological and climatic conditions: Thermophyticum, Mesophyticum and Oreophyticum (Fig. 3.34). Mesophyticum is the basic region with flora and vegetation typical of the Central European temperate zone. In terms of the altitudinal zonation of vegetation (Skalický in Hejný et al. 1988: 103–121), Mesophyticum is situated in the supracolline and submontane belts. The other two regions are partly extrazonal. Thermophyticum includes warm areas with thermophilous flora and vegetation, often characteristic of the submeridional floristic zone. It includes lowland and colline belts. In contrast, Oreophyticum is a cold region with mountain flora and vegetation corresponding to that of forests in the boreal zone, with smaller areas above the timberline similar to habitats in the arctic zone; thermophilous species generally do not occur there except at some protected sites in glacial cirques. It is situated in the montane, supramontane, subalpine and alpine belts.

Each of the phytogeographical regions has a characteristic flora and vegetation. Thermophyticum is characterized by the occurrence of basiphilous thermophilous oak forests of the *Quercion pubescenti-petraeae*, *Aceri tatarici-Quercion* and *Quercion petraeae* alliances and oak-hornbeam forests of the *Carpinion betuli* alliance, low steppic scrub of the *Prunion fruticosae* alliance and particularly dry grasslands of the *Festuco-Brometea* class, while beech forests are nearly absent. There are some remains of softwood floodplain forests of the *Salicion albae* alliance and hardwood floodplain forests of the *Alnion incanae* alliance in lowland basins. Other communities differentiating Thermophyticum from Mesophyticum include the alliances *Geranion sanguinei*, *Festucion valesiacae*, *Onopordion acanthii*, *Spergulo arvensis-Erodion cicutariae*, *Eragrostion cilianensi-minoris*, *Corynephorion canescentis* and *Hydrocharition morsus-ranae*. This region largely overlaps the distribution of loess deposits. Peat bogs are absent, but remnants of calcareous fens are present in some areas (*Junco subnodulosi-Schoenetum nigricantis*, *Cladietum marisci*) and locally there are remnants of saltmarshes and saline grasslands (*Meliloto dentati-Bolboschoenion maritimi*, *Salicornion prostratae*, *Puccinellion limosae*, *Juncion gerardii*). Differential species of Thermophyticum include *Adonis vernalis* (Fig. 3.22), *Astragalus exscapus* (Fig. 3.4), *Buglossoides purpurocaerulea*, *Cytisus procumbens*, *Loranthus europaeus*, *Lotus maritimus*, *Quercus pubescens*, *Rumex hydrolapathum*, *Sesleria uliginosa*, *Sium latifolium*, *Viburnum lantana* and *Viola ambigua*.

The potential natural vegetation in Mesophyticum comprises mainly various types of mesic beech (*Fagion sylvaticae*, *Luzulo-Fagion sylvaticae*) or hornbeam (*Carpinion betuli*) forests. Communities differentiating Mesophyticum from Oreophyticum include grasslands of the alliances *Arrhenatherion elatioris*, *Molinion caeruleae* and *Bromion erecti*, mesic herbaceous forest fringes of the *Trifolion medii* alliance and some specific communities such as the annual vegetation on the wet bottoms of drained fishponds of the *Eleocharition ovatae* alliance. Communities differentiating Mesophyticum from Thermophyticum include the alliances *Fagion sylvaticae*, *Luzulo-Fagion sylvaticae* and *Trifolion medii*.

Oreophyticum is dominated by coniferous forests (mainly acidophilous spruce forests) or by mixed forests with a high abundance of conifers. The area of open vegetation is relatively small. Natural subalpine and alpine grasslands occur only at the highest altitudes of the Czech Republic while most of the present-day subalpine and montane meadows are a consequence of human activity. The following montane, subalpine and alpine communities are characteristic of this region: *Luzulo-Fagion sylvaticae*, *Piceion abietis*, *Alnetum incanae*, *Pinion mugo*, *Salicion silesiacae*, *Loiseleurio procumbentis-Vaccinion*, *Juncion trifidi*, *Nardo strictae-Caricion bigelowii*, *Agrostion alpinae*, *Calamagrostion villosae*, *Calamagrostion arundinaceae*, *Adenostylion alliariae*, *Dryopterido filicis-maris-Athyrion distentifolii*, *Rumicion alpini*, *Nardion strictae*, *Androsacion alpinae*, *Littorellion uniflorae* and *Oxycocco microcarpi-Empetrion hermaphroditi*. Differential species include montane and sub-boreal species, such as *Aconitum firmum*, *A. plicatum* (Fig. 3.5), *Carex pauciflora*, *Hieracium alpinum* (Fig. 3.18), *Ligusticum mutellina*, *Listera cordata* (Fig. 3.1), *Luzula sylvatica*, *Meum athamanticum*, *Oreojuncus trifidus*, *Pinus mugo* (Fig. 3.17), *Swertia perennis*, *Trichophorum cespitosum* and *Viola biflora* (Fig. 3.18), other species mainly occurring in the Czech Republic in Oreophyticum are for example *Anemonastrum narcissiflorum* (Fig. 3.1), *Betula nana*, *Delphinium elatum* (Fig. 3.1), *Doronicum austriacum* (Fig. 3.1), *Gentiana asclepiadea* (Fig. 3.17), *Homogyne alpina*, *Huperzia selago*, *Potentilla aurea* and *Streptopus amplexifolius*.

Each of these three regions is subdivided into two provinces (Fig. 3.34). Thermophyticum consists of two separate areas, one being Bohemian Thermophyticum, which is an isolated area of thermophilous vegetation in the northern half of Bohemia, the other is Pannonian Thermophyticum, which is located in southern Moravia and connected to the forest-steppe area in the Pannonian Basin. Mesophyticum and Oreophyticum are each subdivided based on their assignment to major mountain systems and floristic differences reflecting the proportion of Hercynian vs. Carpathian flora and the gradient of oceanity vs. continentality (the latter increasing from the west to the east). Accordingly, most of the Mesophyticum associated with the Bohemian Massif is recognized as Bohemian-Moravian Mesophyticum whereas the smaller part in eastern Moravia and that associated with the Western Carpathians belongs to Carpathian Mesophyticum. In parallel, Oreophyticum is divided into Bohemian-Moravian Oreophyticum and Carpathian Oreophyticum. Based on the specific and relatively uniform composition of the local flora and vegetation and particularly the contrast with that in neighbouring areas, these provinces are further subdivided into 99 phytogeographical districts (for details, see Kaplan 2012), of which 21 are in Thermophyticum, 63 in Mesophyticum and 15 in Oreophyticum. Some of the districts, which are more heterogeneous than others, are subdivided into subdistricts. Many of these phytochoria do not have sharp boundaries and are separated from the others by transitional zones. Exceptions are the boundaries that follow sharp changes in natural conditions, such as the geological boundary between calcium-rich rocks and acidic sands.

**Acknowledgements** I thank Jiří Danihelka, Milan Chytrý and Marinus Werger for reading and commenting on the manuscript. Leo Bureš, Martin Čertner, Lubomír Hrouda, Josef Kučera, Jan Lepš, Marek Mejstřík, Jan Ponert, Miroslav Pospíšil, Jan Suda, Milan Štech, Pavel Veselý, Petr Vít and Jana Vítová kindly provided photographs of plants. Ondřej Hájek provided a basic map of the Czech Republic for mapping the distribution of endemics and produced the map of the phytogeographical division. Tony Dixon kindly improved my English. Financial support was provided by the Centre of Excellence Pladias (project no. 14-36079G) from the Czech Science Foundation.

# References


Batoušek P, Kreutz CAJ (1999) *Dactylorhiza traunsteineri* (Sauter) Soó subsp. *carpatica* – eine neue *Dactylorhiza*-Unterart aus den Weißen Karpaten (Bílé Karpaty) in der Tschechischen Republik. Journal Europäischer Orchideen 31:678–702

Bělohlávková R (1989) Tučnice česká – *Pinguicula bohemica* Krajina. In: Slavík B (ed) Vybrané ohrožené druhy flóry ČSR [Selected endangered species of the flora of the Czech Socialist Republic]. Studie ČSAV 1989(10):81–23

Binney H, Edwards M, Macias-Fauria M, Lozhkin A, Anderson P, Kaplan JO, Andreev A, Bezrukova E, Blyakharchuk T, ... Zernitskaya V (2017) Vegetation of Eurasia from the last glacial maximum to present: key biogeographic patterns. Quaternary Science Reviews 157:80–97

Birks HJB, Willis KJ (2008) Alpines, trees, and refugia in Europe. Plant Ecology and Diversity 1:147–160

Boublík K, Lepší M, Lepší P (2002) Jeřáb český (*Sorbus bohemica*) v Českém středohoří [*Sorbus bohemica* in the České středohoří Mts]. Severočeskou přírodou 33–34:55–72

Boublík K, Lepší M, Lepší P (2009) *Sorbus portae-bohemicae* M. Lepší, P. Lepší, P. Vít & K. Boublík. In: Hadinec J, Lustyk P (eds) Additamenta ad floram Reipublicae Bohemicae VIII [Additions to the flora of the Czech Republic VIII]. Zprávy České botanické společnosti 44, pp 303–304

Braun-Blanquet J (1922) Fam. *Saxifragaceae*, Steinbrechgewächse. In: Hegi G (ed) Illustrierte Flora von Mitteleuropa 4(2). J. F. Lehmann, München, pp 562–657

Bräutigam S (2001) Ungewöhnliche Arten der Gattung *Hieracium* subgen. *Hieracium* aus dem Isergebirge im Görlitzer Herbarium. Przyroda Sudetów Zachodnich 4:15–18

Bureš L (1996) Studium populací ohrožených druhů rostlin v CHKO Jeseníky [Study of populations of threatened plant species in the Jeseníky protected landscape area]. Příroda 6:7–38

Businský R (1989) *Dactylorhiza bohemica* – nový druh objevený v severních Čechách [*Dactylorhiza bohemica* – a new species discovered in North Bohemia]. Preslia 61:289–314

Casper SJ, Stimper R (2009) Chromosome numbers in *Pinguicula* (*Lentibulariaceae*): survey, atlas, and taxonomic conclusions. Plant Systematics and Evolution 277:21–60

Čeřovský J, Feráková V, Holub J, Maglocký Š, Procházka F (eds) (1999) Červená kniha ohrožených a vzácných druhů rostlin a živočichů ČR a SR 5. Vyšší rostliny [Red data book of threatened species of plants and animals in the Czech Republic and Slovak Republic 5. Higher plants]. Príroda, Bratislava

Challice J, Kovanda M (1978) Flavonoids as markers of taxonomic relationships in the genus *Sorbus* in Europe. Preslia 50:305–320

Challice J, Kovanda M (1986) Flavonoids of *Sorbus eximia*. Preslia 58:165–167

Chrtek J Jr (1995) Notes on *Hieracium alpinum* and *Hieracium nigrescens* groups (section *Alpina* Fries) in the Eastern Sudeten (Mt. Králický Sněžník, the Hrubý Jeseník Mts). Preslia 67:97–106

Chrtek J Jr (1997) Taxonomy of the *Hieracium alpinum* group in the Sudeten Mts, the West and the Ukrainian East Carpathians. Folia Geobotanica et Phytotaxonomica 32:67–91

Chrtek J, Jirásek V (1966) Zur Systematik von *Poa laxa* Haenke. Novitates Botanicae ex Instituto et Horto Botanico Universitatis Carolinae 1966:11–17

Chrtek J Jr, Marhold K (1998) Taxonomy of the *Hieracium fritzei* group (*Asteraceae*) in the Sudeten Mts and the West Carpathians. (Studies in *Hieracium* sect. *Alpina* II.) Phyton (Horn) 37:181–217

Chrtková A, Stavělová E (1986) *Coronilla moravica* Chrtková et Stavělová spec. nova from Czechoslovakia. Folia Geobotanica et Phytotaxonomica 21:325–328

Chytrý M, Horsák M, Syrovátka V, Danihelka J, Ermakov N, German DA, Hájek M, Hájek O, Hájková P, ... Valachovič M (2017) Refugial ecosystems in central Asia as indicators of biodiversity change during the Pleistocene–Holocene transition. Ecological Indicators 77:357–367

Cieślak E (2014) Phylogeography of Pontic-Pannonian species in Central Europe. Polish Botanical Studies 30:1–53

Danihelka J (1995) O druhu *Artemisia pancicii* v České republice a příbuzných taxonech [On the species *Artemisia pancicii* in the Czech Republic and on related taxa]. Zprávy České botanické společnosti 30:81–90

Danihelka J (2013) Botanické součty, rozdíly a podíly [Botanical counts]. Živa 61:69–72

Danihelka J, Marhold K (2003) Validation of the name *Artemisia pancicii* (*Asteraceae*). Willdenowia 33:251–254

Danihelka J, Chrtek J Jr, Kaplan Z (2012) Checklist of vascular plants of the Czech Republic. Preslia 84:647–811

Deffontaine V, Libois R, Kotlik P, Sommer R, Nieberding C, Paradis E, Searle JB, Michaux J (2005) Beyond the Mediterranean peninsulas: evidence of central European glacial refugia for a temperate forest mammal species, the bank vole (*Clethrionomys glareolus*). Molecular Ecology 14:1727–1739

Delforge P (2006) Orchids of Europe, North Africa and the Middle East, 3rd edn. A. & C. Black, London

Devos N, Raspé O, S-H O, Tyteca D, Jacquemart A-L (2006) The evolution of *Dactylorhiza* (*Orchidaceae*) allotetraploid complex: insights from nrDNA sequences and cpDNA PCR-RFLP data. Molecular Phylogenetics and Evolution 38:767–778

Domin K (1904) Dritter Beitrag zur Kenntnis der Phanerogamenflora von Böhmen. Sitzungsberichte der Königlichen Böhmischen Gesellschaft der Wissenschaften in Prag, Mathematisch-naturwissenschaftliche Classe 1904(18):1–81

Domin K (1930) O variabilitě prvosenky bledožluté (*Primula elatior* Hill.) v Československu [On the variation of *Primula elatior* in Czechoslovakia]. Věda přírodní 11:236–242

Domin K (1944) Poznámky o proměnlivosti tučnic v Střední Evropě [Notes on variation of butterworts in Central Europe]. Věda přírodní 23:21–23

Dostál J (1957) Fytogeografické členění ČSR [Phytogeographical division of the Czechoslovak Republic]. Sborník Československé společnosti zeměpisné 62:1–18

Dostál J (1960) The phytogeographical regional distribution of the Czechoslovak flora. Sborník Československé společnosti zeměpisné 65:193–202

Dostál J (1989) Nová květena ČSSR [New flora of the Czechoslovak Socialist Republic]. Academia, Praha

Dvořáková M (1988) *Minuartia smejkalii*, eine neue Art aus der *Minuartia gerardii*-Gruppe (*Caryophyllaceae*). Preslia 60:1–9

Dvořáková M (1999) Zwei neue zur Sektion *Polymechana* gehörende *Minuartia*-Arten (*Caryophyllaceae*). Preslia 70(1998):335–338

Dvořáková M (2003) *Minuartia pauciflora*, das karpatische Endemit aus der *M. verna*-Gruppe. Preslia 75:349–356

Ehrich D, Alsos IG, Brochmann C (2008) Where did the northern peatland species survive the dry glacials: cloudberry (*Rubus chamaemorus*) as an example. Journal of Biogeography 35:801–814

Finnie TJR, Preston CD, Hill MO, Uotila P, Crawley MJ (2007) Floristic elements in European vascular plants: an analysis based on Atlas Florae Europaeae. Journal of Biogeography 34:1848–1872

Gerstberger P (2003) *Potentilla*. In: Weber HE (ed) Gustav Hegi, Illustrierte Flora von Mitteleuropa, Ed. 3, 4(2C). Parey, Berlin, pp 109–205

Gerža M (2009) Endemismus v České republice. Rostliny – 1. část [Endemism in the Czech Republic. Plant endemism, Part I]. Ochrana přírody 64(2):12–15

Gillespie LJ, Archambault A, Soreng RJ (2007) Phylogeny of *Poa* (*Poaceae*) based on trnT-trnF sequence data: major clades and basal relationships. Aliso 23:420–434

Granoszewski W (2003) Late Pleistocene vegetation history and climatic changes at Horoszki Duże, E Poland. Acta Palaeobotanica, Suppl 4:3–95

Gregor T, Müller F (2005) Verbreitung und Ökologie von *Potentilla lindackeri* Tausch in Sachsen. Sächsische Floristische Mitteilungen 9:68–81

Grulich V (2012) Red List of vascular plants of the Czech Republic: 3rd edition. Preslia 84:631–645

Hadač E (1966) Rozšíření černýše českého (*Melampyrum bohemicum* Kerner) v Československu [Distribution of *Melampyrum bohemicum* Kerner in Czechoslovakia]. Preslia 38:403–412

Hadač E (1977) Poznámky o endemických rostlinách České socialistické republiky [Notes on endemic plants of the Czech Socialist Republic]. Zprávy Československé botanické společnosti 12:1–15

Havlíčková J (1983) *Carex flava*-complex in the Czech lands II. Notes on the geographical distribution. Preslia 55:245–263

Hedrén M (1996) Genetic differentiation, polyploidization and hybridization in northern European *Dactylorhiza* (*Orchidaceae*): evidence from allozyme markers. Plant Systematics and Evolution 201:31–55

Hejný S, Slavík B, Chrtek J, Tomšovic P, Kovanda M (eds) (1988) Květena České socialistické republiky 1 [Flora of the Czech Socialist Republic 1]. Academia, Praha

Hejný S, Slavík B, Hrouda L, Skalický V (eds) (1990) Květena České republiky 2 [Flora of the Czech Republic 2]. Academia, Praha

Hejný S, Slavík B, Kirschner J, Křísa B (eds) (1992) Květena České republiky 3 [Flora of the Czech Republic 3]. Academia, Praha

Hendrych R (1981a) Bemerkungen zum Endemismus in der Flora der Tschechoslowakei. Preslia 53:97–120

Hendrych R (1981b) Rostlinné endemity a jejich zastoupení na území Československa [Plant endemics and their representation in the territory of Czechoslovakia]. Živa 29:7–9, 45–46, 123–126

Hendrych R (1984) Fytogeografie [Phytogeography]. Státní pedagogické nakladatelství, Praha

Hendrych R, Hendrychová H (1988) Die *Pedicularis* Arten der Tschechoslowakei, früher und jetzt. Acta Universitatis Carolinae, Biologica 32:403–465

Hensen I, Kilian C, Wagner V, Durka W, Pusch J, Wesche K (2010) Low genetic variability and strong differentiation among isolated populations of the rare steppe grass *Stipa capillata* L. in Central Europe. Plant Biology 12:526–536

Hewitt GM (1999) Post-glacial re-colonization of European biota. Biological Journal of the Linnean Society 68:87–112

Holub J (1962) *Helictotrichon desertorum* (Less.) Nevskij – ein bedeutsames Relikt der tschechoslowakischen Flora. Acta Universitatis Carolinae, Biologica 1962:153–188

Holub J (1979) Some novelties of the Czechoslovak flora. Preslia 51:281–282

Holub J (1991) Eight new *Rubus* species described from Czech Republic. Folia Geobotanica et Phytotaxonomica 26:331–340

Holub J (1992) A preliminary checklist of *Rubus* species occurring in the Czech Republic. Preslia 64:97–132

Holub J (1997) Some considerations and thoughts on the pragmatic classification of apomictic *Rubus* taxa. Osnabrücker Naturwissenschaftliche Mitteilungen 23:147–155

Holub J, Měsíček J, Javůrková V (1970) Annotated chromosome counts of Czechoslovak plants (1–15). (Materials for "Flóra ČSSR" – 1). Folia Geobotanica et Phytotaxonomica 5:339–368

Holub J, Měsíček J, Javůrková V (1971) Annotated chromosome counts of Czechoslovak plants (16–30). (Materials for "Flóra ČSSR" – 2). Folia Geobotanica et Phytotaxonomica 6:179–214

Holub J, Procházka F, Čeřovský J (1979) Seznam vyhynulých, endemických a ohrožených taxonů vyšších rostlin květeny ČSR (1. verze) [The list of extinct, endemic and threatened taxa of vascular plants of the Czech Socialistic Republic (1st version)]. Preslia 51:213–237

Horsák M, Chytrý M, Hájková P, Hájek M, Danihelka J, Horsáková V, Ermakov N, German DA, Kočí M, … Valachovič M (2015) European glacial relict snails and plants: environmental context of their modern refugial occurrence in southern Siberia. Boreas 44:638–657

Hošek J, Pokorný P, Kubovčík V, Horáček I, Žáčková P, Kadlec J, Rojik F, Lisá L, Bučkuliaková S (2014) Late glacial climatic and environmental changes in eastern-central Europe: correlation of multiple biotic and abiotic proxies from the Lake Švarcenberk, Czech Republic. Palaeogeography, Palaeoclimatology, Palaeoecology 396:155–172

Hoták Z, Štěpánek J, Plačková I, Jarolímová V (2013) *Poa riphaea* – an endangered stenoendemic species of the Hrubý Jeseník Mts (Eastern Sudetes). Preslia 85:81–96

Hubberten HW, Andreev A, Astakhov VI, Demidov I, Dowdeswell JA, Henriksen M, Hjort C, Houmark-Nielsen M, Jakobsson M, ... Svendsen JI (2004) The periglacial climate and environment in northern Eurasia during the last glaciation. Quaternary Science Reviews 23:1333–1357

Hultén E (1961) Two *Pedicularis* species from NW America, *P. albertae* n. sp. and *P. sudetica* sensu lat. Svensk Botanisk Tidskrift 55:193–204
Hultén E (1964, 1971) The circumpolar plants, vols 1, 2. Almqvist & Wiksell, Stockholm
Hultén E, Fries M (1986) Atlas of North European vascular plants north of the Tropic of Cancer 1–3. Koeltz Scientific Books, Königstein
Jalas J, Suominen J (eds) (1972–1994) Atlas Florae Europaeae 1–10. Committee for Mapping the Flora of Europe & Societas Biologica Fennica Vanamo, Helsinki
Jalas J, Suominen J, Lampinen R (eds) (1996) Atlas Florae Europaeae 11. Committee for Mapping the Flora of Europe & Societas Biologica Fennica Vanamo, Helsinki
Jalas J, Suominen J, Lampinen R, Kurtto A (eds) (1999) Atlas Florae Europaeae 12. Committee for Mapping the Flora of Europe & Societas Biologica Fennica Vanamo, Helsinki
Jankovská V (2007) Giant Mountains and pollenanalytical research: new results and interesting palaeobotanical findings. Opera Corcontica 44:227–242
Jankovská V, Pokorný P (2008) Forest vegetation of the last full-glacial period in the Western Carpathians (Slovakia and Czech Republic). Preslia 80:307–324
Jankun A, Kovanda M (1986) Apomixis in *Sorbus sudetica* (Embryological studies in *Sorbus* 1). Preslia 58:7–19
Jankun A, Kovanda M (1987) Apomixis and origin of *Sorbus bohemica* (Embryological studies in *Sorbus* 2). Preslia 72:187–230
Jankun A, Kovanda M (1988) Apomixis at the diploid level in *Sorbus eximia* (Embryological studies in *Sorbus* 3). Preslia 60:193–213
Janská V, Jiménez-Alfaro B, Chytrý M, Divíšek J, Anenkhonov O, Korolyuk A, Lashchinskyi N, Culek M (2017) Palaeodistribution modelling of European vegetation types at the last glacial maximum using modern analogues from Siberia: prospects and limitations. Quaternary Science Reviews 159:103–115
Jehlík V, Rostański K (1995) Notes on the genus *Oenothera* subsect. *Oenothera* (*Onagraceae*) in the Czech Republic. Folia Geobotanica et Phytotaxonomica 30:435–444
Jirásek V, Chrtek J (1963) *Poa riphaea* (A. et Gr.) Fritsch, ein Endemit des Gesenkes. Novitates Botanicae 1963:20–27
Kajtoch Ł, Cieślak E, Varga Z, Paul W, Mazur MA, Sramkó G, Kubisz D (2016) Phylogeographic patterns of steppe species in Eastern Central Europe: a review and the implications for conservation. Biodiversity and Conservation 25:2309–2339
Kaplan Z (1998) Relict serpentine populations of *Knautia arvensis* s.l. (*Dipsacaceae*) in the Czech Republic and an adjacent area of Germany. Preslia 70:21–31
Kaplan Z (2012) Flora and phytogeography of the Czech Republic. Preslia 84:505–573
Kaplan Z, Danihelka J, Štěpánková J, Bureš P, Zázvorka J, Hroudová Z, Ducháček M, Grulich V, Řepka R, ... Trávníček B (2015) Distributions of vascular plants in the Czech Republic. Part 1. Preslia 87:417–500
Kaplan Z, Danihelka J, Lepší M, Lepší P, Ekrt L, Chrtek J Jr, Kocián J, Prančl J, Kobrlová L, Hroneš M, Šulc V (2016a) Distributions of vascular plants in the Czech Republic. Part 3. Preslia 88:459–544
Kaplan Z, Danihelka J, Štěpánková J, Ekrt L, Chrtek J Jr, Zázvorka J, Grulich V, Řepka R, Prančl J, ... Brůna J (2016b) Distributions of vascular plants in the Czech Republic. Part 2. Preslia 88:229–322
Kaplan Z, Danihelka J, Chrtek J Jr, Kirschner J, Kubát K, Štech M, Štěpánek J (eds) (2017a) Klíč ke květeně České republiky [Key to the Flora of the Czech Republic], 2nd edn. Praha, Academia
Kaplan Z, Danihelka J, Koutecký P, Šumberová K, Ekrt L, Grulich V, Řepka R, Hroudová Z, Štěpánková J, ... Wild J (2017b) Distributions of vascular plants in the Czech Republic. Part 4. Preslia 89:115–201
Kárpáti Z (1960) Die *Sorbus*-Arten Ungarns und der angrenzenden Gebiete. Feddes Repertorium 62:71–331

Kirschner J (1989) Bika klasnatá – *Luzula spicata* (L.) DC. In: Slavík B (ed) Vybrané ohrožené druhy flóry ČSR [Selected endangered species of the flora of the Czech Socialist Republic]. Studie ČSAV 1989(10):11–23

Kirschner J, Štěpánek J (1986) Towards a monograph of *Taraxacum* sect. *Palustria* (Studies in *Taraxacum* 5). Preslia 58:97–116

Kirschner J, Štěpánek J (1994) *Taraxacum* sect. *Palustria* (*Compositae*) in Bohemia. A contribution to the RAMSAR sites. Thaiszia 4:125–170

Kirschner J, Štěpánek J (1998) A monograph of *Taraxacum* sect. *Palustria*. Institute of Botany, Academy of Sciences of the Czech Republic, Průhonice

Kochjarová J (1997) Náčrt taxonomickej problematiky rodu *Tephroseris* v Západných Karpatoch [Outline of taxonomic problems in *Tephroseris* in the Western Carpathians]. Preslia 69:71–93

Kochjarová J (1998) Poznámky k rozšíreniu, cenológii a ohrozenosti populácií zástupcov rodu *Tephroseris* (Rchb.) Rchb. na Slovensku. II: *T. longifolia* subsp. *moravica* v Západných Karpatoch [Notes on distribution, cenology and threat of populations of *Tephroseris* (Rchb.) Rchb. in Slovakia. II: *T. longifolia* subsp. *moravica* in the Western Carpathians]. Bulletin Slovenskej botanickej spoločnosti 20:69–77

Kolář F, Vít P (2008) Endemické rostliny českých hadců. 3. Rožec kuřičkolistý a kuřička Smejkalova [Endemic species of Czech serpentinites 3. *Cerastium alsinifolium* and *Minuartia smejkalii*]. Živa 56:111–113

Kolář F, Štech M, Trávníček P, Rauchová J, Urfus T, Vít P, Kubešová M, Suda J (2009) Towards resolving the *Knautia arvensis* agg. (*Dipsacaceae*) puzzle: primary and secondary contact zones and ploidy segregation at landscape and microgeographic scales. Annals of Botany 103:963–974

Kolář F, Lučanová M, Vít P, Urfus T, Chrtek J, Fér T, Ehrendorfer F, Suda J (2013) Diversity and endemism in deglaciated areas: ploidy, relative genome size and niche differentiation in the *Galium pusillum* complex (*Rubiaceae*) in Northern and Central Europe. Annals of Botany 111:1095–1108

Kolář F, Lučanová M, Koutecký P, Dortová M, Knotek A, Suda J (2014) Spatio-ecological segregation of diploid and tetraploid cytotypes of *Galium valdepilosum* in central Europe. Preslia 86:155–178

Kolář F, Kaplan Z, Suda J, Štech M (2015) Populations of *Knautia* in ecologically distinct refugia on the Hercynian massif belong to two endemic species. Preslia 87:363–386

Kovanda M (1961) Spontaneous hybrids of *Sorbus* in Czechoslovakia. Acta Universitatis Carolinae, Biologica 1:41–83

Kovanda M (1965) Taxonomie jeřábu sudetského [*Sorbus sudetic*a (Tausch) Hedl.] a otázka jeho endemismu [Taxonomy of *Sorbus sudetica* and the question on its endemism]. Československá ochrana prírody 2:47–62

Kovanda M (1968) New taxa and combinations in the subsection *Heterophyll*a (Witas.) Fed. of the genus *Campanula* L. Folia Geobotanica et Phytotaxonomica 3:407–411

Kovanda M (1975) *Campanula corcontica*, a superfluous name. Preslia 47:262–266

Kovanda M (1977) Polyploidy and variation in the *Campanula rotundifolia* complex. Part II. (Taxonomic). 2. Revision of the groups *Vulgares* and *Scheuchzerianae* in Czechoslovakia and adjacent regions. Folia Geobotanica et Phytotaxonomica 12:23–89

Kovanda M (1980) A neglected endemic of the Eastern Sudeten. Preslia 52:117–126

Kovanda M (1982) *Dianthus gratianopolitanus*: variability, differentiation and relationships. Preslia 54:223–242

Kovanda M (1984a) A new hybridogeneous *Sorbus*. Preslia 56:169–172

Kovanda M (1984b) Chromosome numbers in selected angiosperms (2). Preslia 56:289–301

Kovanda M (1996a) New taxa of *Sorbus* from Bohemia (Czech Republic). Verhandlungen der Zoologisch-Botanischen Gesellschaft in Österreich 133:319–345

Kovanda M (1996b) Observations on *Sorbus* in Southwest Moravia (Czech Republic) and adjacent Austria I. Verhandlungen der Zoologisch-Botanischen Gesellschaft in Österreich 133:347–369

Kovanda M (1997) Observations on *Primula elatior* in the High Sudeten Mts. Thaiszia 7:17–28

Kovanda M (1999) Jeřáby (*Sorbus*) České republiky a jejich ochrana [*Sorbus* species in the Czech Republic and their conservation status]. Příroda 15:31–47

Kovanda M (2002) Observations on *Carlina biebersteinii*. Thaiszia 12:75–82

Krahulcová A, Holub J (1998) Chromosome number variation in the genus *Rubus* in the Czech Republic. III. Preslia 70:33–50

Krahulcová A, Jarolímová V (1991) Relationship between *Pinguicula bohemica* Krajina and *Pinguicula vulgaris* L. (*Lentibulariaceae*) from the karyological point of view. Preslia 53:323–328

Krahulcová A, Štěpánková J (1998) Serpentine and polyploid differentiation within *Galium pumilum* agg. (*Rubiaceae*) in eastern C. Europe. Folia Geobotanica 33:87–102

Krahulcová A, Papoušková S, Krahulec F (2004) Reproduction mode in the allopolyploid facultatively apomictic hawkweed *Hieracium rubrum* (*Asteraceae*, *H.* subgen. *Pilosella*). Hereditas 141:19–30

Krahulcová A, Krahulec F, Bräutigam S, Chrtek J Jr (2013) Chromosome numbers and reproductive systems of selected representatives of *Pilosella* from the Krkonoše Mts (the Sudetes Mts). Part 3. Preslia 85:179–192

Krahulec F (1990) Alpine vegetation of the Králický Sněžník Mts (E. Sudeten Mts). Preslia 62:307–322

Krahulec F (2006) Species of vascular plants endemic to the Krkonoše Mts (Western Sudetes). Preslia 78:503–516

Krahulec F, Krahulcová A, Fehrer J, Bräutigam S, Plačková I, Chrtek J Jr (2004) The Sudetic group of *Hieracium* subgen. *Pilosella* in the Krkonoše Mts: a synthetic view. Preslia 76:223–243

Krahulec F, Krahulcová A, Fehrer J, Bräutigam S, Schuhwerk F (2008) The structure of the agamic complex of *Hieracium* subgen. *Pilosella* in the Šumava Mts and its comparison with other regions in Central Europe. Preslia 80:1–26

Krahulec F, Krahulcová A, Rosenbaumová R, Plačková I (2011) Production of polyhaploids by facultatively apomictic *Pilosella* can result in formation of new genotypes via genome doubling. Preslia 83:471–490

Krajina V (1927) *Pinguicula bohemica*, species nova e sectione *Pionophyllum* DC. Věstník Královské české společnosti nauk, Třída mathematicko-přírodovědecká 1926(15):1–13

Kramp K, Huck S, Niketić M, Tomović G, Schmitt T (2009) Multiple glacial refugia and complex postglacial range shifts of the obligatory woodland plant species *Polygonatum verticillatum* (*Convallariaceae*). Plant Biology 11:392–404

Kubát K, Hrouda L, Chrtek J Jr, Kaplan Z, Kirschner J, Štěpánek J (eds) (2002) Klíč ke květeně České republiky [Key to flora of the Czech Republic]. Academia, Praha

Kuncová J, Bělohoubek J (1996) Sledování a posilování populace hvozdíku písečného českého (*Dianthus arenarius* subsp. *bohemicus*) v národní přírodní památce Kleneč [Monitoring and enhancement of the *Dianthus arenarius* subsp. *bohemicus* population in the National Nature Monument Kleneč]. Příroda 6:163–168

Kuneš P, Pelánková B, Chytrý M, Jankovská V, Pokorný P, Petr L (2008) Interpretation of the last-glacial vegetation of eastern-central Europe using modern analogues from Siberia. Journal of Biogeography 35:2223–2236

Kúr P, Košnar J, Koutecký P, Tremetsberger K, Štech M (2016) Origin of *Spergularia* × *kurkae*, a hybrid between the rare endemic *S. echinosperma* and its widespread congener *S. rubra*. Preslia 88:391–407

Kurtto A, Lampinen R, Junikka L (eds) (2004) Atlas Florae Europaeae 13. Committee for Mapping the Flora of Europe & Societas Biologica Fennica Vanamo, Helsinki

Kurtto A, Fröhner SE, Lampinen R (eds) (2007) Atlas Florae Europaeae 14. Committee for Mapping the Flora of Europe & Societas Biologica Fennica Vanamo, Helsinki

Kurtto A, Weber HE, Lampinen R, Sennikov AN (eds) (2010) Atlas Florae Europaeae 15. Committee for Mapping the Flora of Europe & Societas Biologica Fennica Vanamo, Helsinki

Kurtto A, Sennikov AN, Lampinen R (eds) (2013) Atlas Florae Europaeae 16. Committee for Mapping the Flora of Europe & Societas Biologica Fennica Vanamo, Helsinki

Lang G (1994) Quartäre Vegetationsgeschichte Europas. Gustav Fischer, Jena

Lepší M, Vít P, Lepší P, Boublík K, Suda J (2008) *Sorbus milensis*, a new hybridogenous species from northwestern Bohemia. Preslia 80:229–244

Lepší M, Kaplan Z, Lepší P (2009a) *Sorbus gemella* Kovanda. Severočeskou přírodou 40:122–126
Lepší M, Vít P, Lepší P, Boublík K, Kolář F (2009b) *Sorbus portae-bohemicae* and *Sorbus albensis*, two new endemic apomictic species recognized based on a revision of *Sorbus bohemica*. Preslia 81:63–89
Lepší M, Lepší P, Sádlo J, Koutecký P, Vít P, Petřík P (2013a) *Sorbus pauca* species nova, the first endemic species of the *Sorbus hybrida* group for the Czech Republic. Preslia 85:63–80
Lepší M, Lepší P, Vít P (2013b) *Sorbus quernea*: taxonomic confusion caused by the naturalization of an alien species, *Sorbus mougeotii*. Preslia 85:159–178
Lepší M, Lepší P, Koutecký P, Bílá J, Vít P (2015) Taxonomic revision of *Sorbus* subgenus *Aria* occurring in the Czech Republic. Preslia 87:109–162
Ložek V (1973) Příroda ve čtvrtohorách [Nature in the quaternary]. Academia, Praha
Ložek V (2007) Zrcadlo minulosti. Česká a slovenská krajina v kvartéru [Mirror of the past. Czech and Slovak landscape in the quaternary]. Dokořán, Praha
Ložek V (2009a) Refugia, migrace a brány. I. Ohlédnutí za starými problémy [Refugia, migration and gateways. I. Review of old problems]. Živa 57:146–149
Ložek V (2009b) Refugia, migrace a brány. II. Ve světle dnešních poznatků [Refugia, migration and gateways. II. In the light of recent knowledge]. Živa 57:194–198
Ložek V (2011) Po stopách pravěkých dějů. O silách, které vytvářely naši krajinu [Tracing the prehistoric events. On the forces that formed our landscape]. Dokořán, Praha
Magyari EK, Chapman JC, Passmore DG, Allen JRM, Huntley JP, Huntle B (2010) Holocene persistence of wooded steppe in the Great Hungarian Plain. Journal of Biogeography 37:915–935
Martinovský JO (1965) *Stipa rubens* P. Smirn. – das Rötliche Federgras – eine für die Tschechoslowakei und für Mitteleuropa neue Art. Preslia 37:156–169
Martinovský JO (1969) Česnek tuhý (*Allium strictum* Schrad.) památná rostlina naší květeny [*Allium strictum* – a memorial plant of the Czech flora]. Severočeskou přírodou 1:1–35
Martinovský JO (1971) Srovnávací fytogeografická studie k problematice středoevropské stepi [Comparative phytogeographical study on Central European steppes]. Severočeskou přírodou 2:43–107
Martinovský JO (1975) Taxonomische Studie über die *Stipa*-Serie *Dasyphyllae*. XXV. Beitrag zur Kenntnis der Gattung *Stipa*. Preslia 47:249–261
Meusel H, Jäger EJ (eds) (1992) Vergleichende Chorologie der zentraleuropäischen Flora 3. Gustav Fischer, Jena
Meusel H, Jäger E, Weinert E (1965) Vergleichende Chorologie der zentraleuropäischen Flora 1. Gustav Fischer, Jena
Meusel H, Jäger E, Rauschert S, Weinert E (1978) Vergleichende Chorologie der zentraleuropäischen Flora 2. Gustav Fischer, Jena
Mitka J (2003) The genus *Aconitum* L. (*Ranunculaceae*) in Poland and adjacent countries. Institute of Botany of the Jagiellonian University, Kraków
Molau U, Murray DF (1996) Taxonomic revision of the *Pedicularis sudetica* complex (*Scrophulariaceae*): the Arctic species. Symbolae Botanicae Upsalienses 31(3):33–46
Müller UC, Pross J, Bibus E (2003) Vegetation response to rapid climate change in Central Europe during the past 140,000 yr based on evidence from the Füramoos pollen record. Quaternary Research 59:235–245
Nelson-Jones EB, Briggs D, Smith AG (2002) The origin of intermediate species of genus *Sorbus*. Theoretical and Applied Genetics 105:953–963
Nepraš K (2006) *Saxifraga rosacea* subsp. *steinmannii* (Tausch) Holub u Ústí nad Labem [*Saxifraga rosacea* subsp. *steinmannii* (Tausch) Holub near Ústí nad Labem]. Severočeskou přírodou 38:153–154
Nepraš K, Kroufek R, Bultas P (2008) Příspěvek ke květeně Českého středohoří [Contribution to the flora of the České středohoří Mts]. Severočeskou přírodou 39:59–72
Nordström S, Hedrén M (2009) Genetic diversity and differentiation of allopolyploid *Dactylorhiza* (*Orchidaceae*) with particular focus on the *Dactylorhiza majalis* ssp. *traunsteineri*/*lapponica* complex. Biological Journal of the Linnean Society 97:52–67
Novák FA (1915) Kritická studie o *Dianthus arenarius* L. a jemu blízce příbuzných druzích a o jeho stanovišti v Čechách [Critical study on *Dianthus arenarius* L. and its closely related spe-

cies and habitat in Bohemia]. Věstník Královské české společnosti nauk, Třída mathematicko-přírodovědecká 1915(8):1–27
Novák FA (1927) Monografická studie o postglaciálních druzích rodu *Dianthus* ze sekce *Plumaria* (Opiz) [A monographic study of postglacial species of *Dianthus* sect. *Plumaria* (Opiz)]. Věstník Královské české společnosti nauk, Třída mathematicko-přírodovědecká 1927(9):1–100
Novák FA (1960) Fylogenese serpentinových typů [Phylogeny of serpentinite types]. Preslia 32:1–8
Nováková H, Rydlo J (1978) *Epipactis albensis* – nový autogamický druh z okruhu *Epipactis helleborine* agg. (*Orchidaceae*) [*Epipactis albensis* – a new autogamous species of the *Epipactis helleborine* complex (*Orchidaceae*)]. Preslia 50:161–171
Ownbey M (1950) Natural hybridization and amphiploidy in the genus *Tragopogon*. American Journal of Botany 37:487–499
Parisod C (2008) Postglacial recolonisation of plants in the western Alps of Switzerland. Botanica Helvetica 118:1–12
Petit RJ, Brewer S, Bordács S, Burg K, Cheddadi R, Coart E, Cottrell J, Csaikl UM, van Dam B, ... Kremer A (2002) Identification of refugia and post-glacial colonisation routes of European white oaks based on chloroplast DNA and fossil pollen evidence. Forest Ecology and Management 156:49–74
Petr L, Novák J (2014) High vegetation and environmental diversity during the Late Glacial and Early Holocene on the example of lowlands in the Czech Republic. Biologia 69:847–862
Petr L, Sádlo J, Žáčková P, Lisá L, Novák J, Rohovec J, Pokorný P (2014) Late-Glacial and Holocene environmental history of an oxbow wetland in the Polabí lowland (river Elbe, Czech Republic); a context-dependent interpretation of a multi-proxy analysis. Folia Geobotanica 49:137–162
Pillon Y, Fay MF, Hedrén M, Bateman RM, Devey DS, Shipunov AB, van der Bank M, Chase MW (2007) Evolution and temporal diversification of western European polyploid species complexes in *Dactylorhiza* (*Orchidaceae*). Taxon 56:1185–1208
Plocek A (1986) New taxa and combinations in *Alchemilla* for the flora of Slovakia and the flora of Czechoslovakia (1). Folia Geobotanica et Phytotaxonomica 21:423–427
Podpěra J (1921) Fytogeografický rozbor nálezu *Cortusa Matthioli* L. na Moravě [Phytogeographical analysis of the find of *Cortusa Matthioli* in Moravia]. Sborník Klubu přírodovědeckého v Brně 3(1920):51–70
Podpěra J (1923) Studien über den Formenkreis der *Cortusa Matthioli* L. Beihefte zum Botanischen Centralblatt, Zweite Abteilung 39:276–287
Pokorný P (2002) A high-resolution record of Late-Glacial and Early-Holocene climatic and environmental change in the Czech Republic. Quaternary International 91:101–122
Pokorný P (2011) Neklidné časy. Kapitoly ze společných dějin přírody a lidí [Unstable times. Chapters from the common history of nature and people]. Dokořán, Praha
Pokorný P, Klimešová J, Klimeš L (2000) Late Holocene history and vegetation dynamics of a floodplain alder carr: a case study from eastern Bohemia, Czech Republic. Folia Geobotanica et Phytotaxonomica 35:43–58
Pokorný P, Sádlo J, Bernardová A (2010) Holocene history of *Cladium mariscus* (L.) Pohl in the Czech Republic. Implications for species population dynamics and palaeoecology. Acta Palaeobotanica 50:65–76
Procházka F (1982) *Dactylorhiza majalis* subsp. *turfosa*, nové plemeno prstnatce májového [*Dactylorhiza majalis* subsp. *turfosa*, a new subspecies of the broad-leaved marsh orchid]. Preslia 54:289–295
Průša D (2005) Orchideje České republiky [Orchids of the Czech Republic]. Computer Press, Brno
Reisch C, Poschlod P, Wingender R (2003) Genetic variation of *Saxifraga paniculata* Mill. (*Saxifragaceae*): molecular evidence for glacial relict endemism in central Europe. Biological Journal of the Linnean Society 80:11–21
Rešetnik I, Frajman B, Bogdanović S, Ehrendorfer F, Schönswetter P (2014) Disentangling relationships among the diploid members of the intricate genus *Knautia* (*Caprifoliaceae*, *Dipsacoideae*). Molecular Phylogenetics and Evolution 74:97–110

Rosenbaumová R, Krahulcová R, Krahulec F (2012) The intriguing complexity of parthenogenesis inheritance in *Pilosella rubra* (*Asteraceae*, *Lactuceae*). Sexual Plant Reproduction 25:185–196
Rotreklová O, Krahulcová A, Mráz P, Mrázová V, Mártonfiová L, Peckert T, Šingliarová B (2005) Chromosome numbers and breeding systems of some European species of *Hieracium* subgen. *Pilosella*. Preslia 77:177–195
Rull V (2008) Microrefugia. Journal of Biogeography 36:481–484
Rull V (2010) On microrefugia and cryptic refugia. Journal of Biogeography 37:1623–1625
Rybníčková E, Rybníček K (2014) Palaeovegetation in the Pavlovské vrchy hills region (South Moravia, Czech Republic) around 25,000 BP: the Bulhary core. Vegetation History and Archaeobotany 23:719–728
Sádlo J (2000) Původ travinné vegetace slatin v Čechách: sukcese kontra cenogeneze [The origin of grassland vegetation of fen peats in the Czech Republic: succession versus coenogenesis]. Preslia 81:173–185
Sádlo J (2007) Diverzita vegetace České republiky, její příčiny a historický vývoj [Diversity of vegetation of the Czech Republic, its determinants and history]. In: Chytrý M (ed) Vegetace České republiky 1. Travinná a keříčková vegetace [Vegetation of the Czech Republic 1. Grassland and heathland vegetation]. Academia, Praha, pp 53–64
Sádlo J, Bufková I (2002) Vegetace Vltavského luhu na Šumavě a problém reliktních praluk [Vegetation of the Vltava river alluvial plain in the Šumava Mts (Czech Republic) and the problem of relict primary meadows]. Preslia 74:67–83
Sczepanski S, Kreutz K (2007) Beiträge zur Kenntnis der europäischen und vorderasiatischen *Dactylorhiza*-Arten. Berichte aus den Arbeitskreisen Heimische Orchideen 24:91–106
Šefl J (2007) Jeřáby na Podyjí [Rowans in the Podyjí National Park]. Thayensia (Znojmo) 7:121–151
Skalický V (1982) Regionálně-fytogeografické členění ČSR [Regional phytogeographical division of the Czech Socialist Republic]. In: Chrtek J, Slavík B, Tomšovic P (eds) Směrnice pro zpracování Květeny ČSR [Guidelines to elaboration of the Flora of the Czech Socialist Republic], pp 106–120 & supplementary map. Botanický ústav ČSAV, Průhonice
Slavík B (1984) Grundlegende Phytochorotypen der Tschechischen Sozialistischen Republik. Preslia 56:241–265
Slavík B (1998) Phytocartographical syntheses of the Czech Republic 3. Academia, Praha
Slavík B, Smejkal M, Dvořáková M, Grulich V (eds) (1995) Květena České republiky 4 [Flora of the Czech Republic 4]. Academia, Praha
Slavík B, Chrtek J Jr, Tomšovic P (eds) (1997) Květena České republiky 5 [Flora of the Czech Republic 5]. Academia, Praha
Slavík B, Chrtek J Jr, Štěpánková J (eds) (2000) Květena České republiky 6 [Flora of the Czech Republic 6]. Academia, Praha
Slavík B, Štěpánková J, Štěpánek J (eds) (2004) Květena České republiky 7 [Flora of the Czech Republic 7]. Academia, Praha
Šlechtová A, Bělohoubek J (2010) Hvozdík písečný český – záchranný program běží [The Bohemian sand pink (*Dianthus arenarius* subsp. *bohemicus*) – action plan is running]. Ochrana přírody 65(3):18–23
Smejkal M (1963) Taxonomické studie československých druhů rodu *Euphrasia* L. [Taxonomic study of the Czechoslovak species of the genus *Euphrasia* L.]. Biologické práce Slovenskej akadémie vied 9(9):1–83
Soják J (2009a) *Potentilla lindackeri* a *P. psammophila* (podsekce *Collinae*) v Čechách [*Potentilla lindackeri* and *P. psammophila* (subsection *Collinae*) in Bohemia (Notes on *Potentilla* XXVII.)]. Zprávy České botanické společnosti 44:11–22
Soják J (2009b) *Potentilla psammophila* (*Rosaceae*), a new species from N Bohemia. Willdenowia 39:59–61
Štech M (2006) Was sind *Melampyrum subalpinum*, *M. angustissimum* und *M. bohemicum*? Neilreichia 4:221–234
Štěpánek J (1982) Die Chromosomenzahlen von tschechoslowakischen Arten der Gattung *Knautia* L. (*Dipsacaceae*). Folia Geobotanica et Phytotaxonomica 17:359–386

Štěpánek J (1989) Chrastavec rolní krkonošský – *Knautia arvensis* (L.) Coulter subsp. *pseudolongifolia* (Szabó) O. Schwarz. In: Slavík B (ed) Vybrané ohrožené druhy flóry ČSR [Selected endangered species of the flora of the Czech Socialist Republic]. Studie ČSAV 10:25–36

Štěpánek J, Kirschner J, Jarolímová V, Kirschnerová L (2011) *Taraxacum nigricans*, *T. alpestre* and allies in the *Taraxacum* sect. *Alpestria*: taxonomy, geography and conservation status. Preslia 83:537–564

Štěpánková J (2008) *Carex derelicta*, a new species from the Krkonoše Mountains (Czech Republic). Preslia 80:389–397

Štěpánková J, Chrtek J Jr, Kaplan Z (eds) (2010) Květena České republiky 8 [Flora of the Czech Republic 8]. Academia, Praha

Stewart JR, Lister AM (2001) Cryptic northern refugia and the origins of the modern biota. Trends in Ecology and Evolution 16:608–613

Stewart JR, Lister AM, Barnes I, Dalén L (2010) Refugia revisited: individualistic responses of species in space and time. Proceedings of the Royal Society B: Biological Sciences 277:661–671

Studnička M, Hejný S (1992) Does *Pinguicula bohemica* exist? Carnivorous Plant Newsletter 21(3):64–67

Štursová H, Kociánová M (2005) Poznámky k rozšíření, biologii a ekologii všivce *Pedicularis sudetica* subsp. *sudetica* [Notes on distribution, biology and ecology of *Pedicularis sudetica* subsp. *sudetica*]. Opera Corcontica 43:157–178

Suda J, Kaplan Z (2012) Rostlinný endemismus a endemity české květeny [Plant endemism and endemics of the Czech flora]. Živa 60:168–174

Suda J, Kaplan Z (eds) (2018) Endemické rostliny České republiky [Endemic plants of the Czech Republic]. Academia, Praha

Svobodová H, Soukupová L, Reille M (2002) Diversified development of mountain mires, Bohemian Forest, Central Europe, in the last 13,000 years. Quaternary International 91:123–135

Szafer W (1912) Eine Dryas-Flora bei Krystynopol in Galicien. Bulletin International de l'Academie des Sciences de Cracovie, Classe des Sciences Mathematiques et Naturelles, Série B, Sciences Naturelles 1912:1103–1123

Szeląg Z, Wójcik G (2014) *Hieracium sudetotubulosum* (*Asteraceae*) rediscovered outside the Karkonosze Mts. Polish Botanical Journal 59:117–119

Taberlet P, Fumagalli L, Wust-Saucy A-G, Cosson J-F (1998) Comparative phylogeography and postglacial colonization routes in Europe. Molecular Ecology 7:453–464

Takhtajan A (1986) Floristic regions of the world. University of California Press, Berkeley

Tarasov PE, Volkova VS, Webb T, Guiot J, Andreev AA, Bezusko LG, Bezusko TV, Bykova GV, Dorofeyuk N, ... Sevastyanov DV (2000) Last glacial maximum biomes reconstructed from pollen and plant macrofossil data from northern Eurasia. Journal of Biogeography 27:609–620

Toman M (1970) Lokalita hvozdíku písečného (*Dianthus arenarius* L.) u Klenče a její ochrana [The site of *Dianthus arenarius* L. near Kleneč and its conservation]. Studie ČSAV 1970(7):163–187

Tomšovic P (1989) Bytel položený – *Kochia prostrata* (L.) Schrad. In: Slavík B (ed) Vybrané ohrožené druhy flóry ČSR [Selected endangered species of the flora of the Czech Socialist Republic]. Studie ČSAV 1989(10):151–163

Tralau H (1963) The recent and fossil distribution of some boreal and arctic montane plants in Europe. Arkiv før Botanik, Series 2(5):533–582

Trávníček B (2010) Ladoňky (rod *Scilla* s. lat.) v České republice. I. Původní taxony [Squills (*Scilla* s. lat.) in the Czech Republic. I. Native taxa]. Zprávy České botanické společnosti 45:1–23

Trávníček B, Zázvorka J (2005) Taxonomy of *Rubus* ser. *Discolores* in the Czech Republic and adjacent regions. Preslia 77:1–88

Trávníček B, Duchoslav M, Šarhanová P, Šafářová L (2010) Squills (*Scilla* s. lat., *Hyacinthaceae*) in the flora of the Czech Republic, with taxonomical notes on Central-European squill populations. Acta Musei Moraviae, Scientiae biologicae 94(2009):157–205

Tutin TG, Heywood VH, Burges NA, Moore DM, Valentine DH, Walters SM, Webb DA (eds) (1964–1980) Flora Europaea 1–5. Cambridge University Press, Cambridge

Velebil J (2012) *Sorbus omissa*, a new endemic hybridogenous apomictic species from the valley of the lower Vltava river valley. Preslia 84:375–390

Vít P, Lepší M, Lepší P (2012) There is no diploid apomict among Czech *Sorbus* species: a biosystematic revision of *S. eximia* and discovery of *S. barrandienica*. Preslia 84:71–96

Vít P, Wolfová K, Urfus T, Tájek P, Suda J (2014) Interspecific hybridization between rare and common plant congeners inferred from genome size data: assessing the threat to the Czech serpentine endemic *Cerastium alsinifolium*. Preslia 86:95–117

Vítová J, Vít P, Suda J (2015) Rare occurrence of reciprocal hybridization in a sympatric population of the Czech stenoendemic *Dianthus arenarius* subsp. *bohemicus* and widespread *D. carthusianorum*. Preslia 87:329–345

von Nägeli C, Peter A (1885) Die Hieracien Mittel-Europas. Monographische Bearbeitung der Piloselloiden mit besonderer Berücksichtigung der mitteleuropäischen Sippen. R. Oldenbourg, München

Walter H (1986) Allgemeine Geobotanik, 2nd edn. Eugen Ulmer, Stuttgart

Walter H, Straka H (1970) Arealkunde. Floristisch-historische Geobotanik, 2nd edn. Eugen Ulmer, Stuttgart

Weber HE (1977) Die ehemalige und jetzige Brombeerflora von Mennighüffen, Kreis Herford, Ausgangsgebiet der europäischen *Rubus*-Forschung durch K. E. A. Weihe (1779–1834). Bericht des Naturwissenschaftlichen Vereins für Bielefeld 23:161–193

Weber HE (1995) *Rubus*. In: Weber HE (ed) Gustav Hegi, Illustrierte Flora von Mitteleuropa, Ed. 3, 4(2A). Blackwell, Berlin, pp 284–595

Weber HE (1996) Former and modern taxonomic treatment of the apomictic *Rubus* complex. Folia Geobotanica et Phytotaxonomica 31:373–380

Weber HE (2000) Zur Erforschung der Gattung *Rubus* in der Tschechischen Republik [On the research of the genus *Rubus* in the Czech Republic]. Preslia 72:231–239

Weiss H, Dobeš C, Schneeweiss GM, Greimler J (2002) Occurrence of tetraploid and hexaploid cytotypes between and within populations in *Dianthus* sect. *Plumari*a (*Caryophyllaceae*). New Phytologist 156:85–94

West RG (2000) Plant life of the quaternary cold stages. Cambridge University Press, Cambridge

Willis KJ, van Andel TH (2004) Trees or no trees? The environments of central and eastern Europe during the last glaciation. Quaternary Science Reviews 23:2369–2387

Windmaißer T, Reisch C, Kattari S, Heubl G (2016) Glacial refugia and postglacial expansion of the alpine–prealpine plant species *Polygala chamaebuxus*. Ecology and Evolution 6:7809–7819

Zahn KH (1921–1923) *Compositae – Hieracium*. In: Engler A (ed) Das Pflanzenreich 75–77, 79, 82 (IV/280). Wilhelm Engelmann, Leipzig, pp 1–1705

Zázvorka J (1984) *Orobanche coerulescens* v Československu [*Orobanche coerulescens* in Czechoslovakia]. Severočeskou přírodou 16:1–23

# Chapter 4
# Bryoflora of the Czech Republic

**Jan Kučera**

**Abstract** Bryophytes are an intrinsic part of various vegetation types in Central European landscapes. The Czech bryoflora is made up of 866 species (207 liverworts, 655 mosses and 4 hornworts) of which 59 are considered to be extinct or have vanished. The great majority are native species, while only four species are of demonstrably alien origin. Two of the alien mosses are invasive and about ten native species or species earlier reported from neighbouring countries are regarded as expanding. The majority of Czech bryophytes belong to temperate montane elements, with a significant proportion of boreal and suboceanic taxa. Arctic-alpine and boreal taxa are at the highest risk of extinction as they are confined to relict habitats. Probably no species of bryophyte is endemic to the Czech Republic. As in the other industrial countries with intensive agriculture and forestry, a significant proportion of the bryoflora is at risk; nearly a third of the species in this country are evaluated as threatened according to IUCN criteria.

## 4.1 Introduction

Bryophytes typically occur in most Central European vegetation types as a minor but intrinsic component of an ecosystem, except in *Sphagnum* peatbogs, in which they dominate. Despite their low biomass, the species diversity of bryophytes in the Czech Republic is surprisingly high. The last census of bryophytes (Kučera et al. 2012) credibly recorded a total of 863 species in the Czech Republic, including 207 liverworts, 652 mosses and 4 hornworts, plus 5 subspecies and 23 generally recognized varieties. Since then, one other species has been recorded and two species added as a result of taxonomic revisions. The exact numbers are, however, significantly affected by the taxonomic concepts adopted and level of acceptance of earlier records unsupported by herbarium specimens. Moreover, a large number of the reported taxa is believed to be regionally extinct or has been missing for many

J. Kučera (✉)
Department of Botany, University of South Bohemia,
Branišovská 1760, 370 05 České Budějovice, Czech Republic
e-mail: kucera@prf.jcu.cz



M. Chytrý et al. (eds.), *Flora and Vegetation of the Czech Republic*, Plant and Vegetation 14, DOI 10.1007/978-3-319-63181-3_4

decades (38 species were, at the end of 2016, considered extinct and 21 missing for more than 30 years). Still, the number of extant bryophytes in the Czech Republic exceeds 800 species.

## 4.2 Species Composition and Diversity of Bryophytes in the Czech Republic

The liverworts are assigned to 76 genera, the mosses to 194 and the hornworts to 3 genera (Kučera et al. 2012). However, the generic treatments are experiencing a period of turmoil because of the ongoing molecular-phylogenetic revisions. For instance, the recently published World Checklist of Hornworts and Liverworts (Söderström et al. 2016) accepts four additional genera not recognized by Kučera et al. (2012) and merges two of the recognized genera with others. For the mosses, a comparable recent synopsis is absent, which makes the generic assignments more difficult. Even less stable is the situation in family treatments. For instance, while the largest liverwort families of the last pre-molecular Czech checklist (Váňa 1997) were *Lophoziaceae* (36 species), followed by *Scapaniaceae* (25), according to the current checklist (Söderström et al. 2016) these are *Scapaniaceae* (29 species), followed by *Anastrophyllaceae* (17), while the *Lophoziaceae* now have only 11 species. In mosses, the three richest families in the pre-molecular treatment were *Pottiaceae* (82), followed by *Bryaceae* (61), *Dicranaceae* (58) and *Grimmiaceae* (52). Currently, based on various recent treatments, the largest family is still *Pottiaceae* (88), followed by *Grimmiaceae* (55) and *Brachytheciaceae* (40), while the number of species in *Bryaceae* has decreased to 38 and in *Dicranaceae* to 18. The largest genera are still *Sphagnum* and *Bryum* (each with 34 currently recognized species), although in the latter genus, narrower concepts are becoming increasingly popular.

The number of species of bryophytes recorded in the Czech Republic reflects multiple factors, the most significant being the area, ecosystem diversity and the level of bryofloristic exploration. Therefore the comparison of species numbers with neighbouring countries is difficult. For example, the number of species of bryophytes recorded in Germany (ca 1150, www.moose-deutschland.de) significantly exceeds that in the Czech Republic mainly because of this country's larger area (357,168 $km^2$ as compared to 78,866 $km^2$) and greater diversity of habitats. In the neighbouring Poland, also much larger than the Czech Republic (312,679 $km^2$), only approximately 910 species of bryophytes are recorded, i.e. nearly the same number as in the Czech Republic despite the narrower species concepts commonly used by Polish authors (Ochyra et al. 2003; Szweykowski 2006). Nevertheless, Poland might be generally less intensively explored compared to Germany and the Czech Republic, as evidenced by the large number of additions to the Polish bryoflora in recent years. On the other hand, the bryoflora of the Austrian federal state of Carinthia (9536 $km^2$) exceeds that in the Czech Republic (893 species, Köckinger et al. 2008), while there are only 659 species of bryophytes in the somewhat larger Hungary (93,030 km$^2$) (Papp et al. 2010). The latter number, nevertheless, is a consequence of not only the smaller diversity of ecosystems but also the historically lower intensity of floristic research on bryophytes.

The vast majority of the taxa occurring in the Czech Republic are native. Alien bryophytes are generally extremely rare; the only two naturalized alien species (*Campylopus introflexus* and *Orthodontium lineare*) are generally regarded as invasive in Central Europe (Essl and Lambdon 2009) and two species (the liverwort *Lunularia cruciata* and the moss *Didymodon umbrosus*) are considered to be casual aliens. About ten other species native to the Czech Republic or neighbouring countries are regarded as spreading, with a few additional uncertain cases (Kučera et al. 2012). This situation strongly contrasts with the Czech vascular flora, in which about 1300 out of more than 3550 species are considered naturalized or casual aliens (Danihelka et al. 2012). Low numbers of alien bryophyte taxa are probably most attributable to virtually no deliberate introductions as a consequence of their low horticultural attractiveness, although ephemeral escapes of tropical and subtropical species from greenhouses are rarely recorded even in the Czech Republic. Among the spreading species, most numerous are the epiphytes, some of them with still only a rare occurrence, mostly in north-western Bohemia (*Orthotrichum pulchellum* and *O. rogeri*), some of them however are already widespread (*Dicranoweisia cirrata* and *Dicranum tauricum*). The spread of *Dicranum tauricum* has been mapped in detail by Stebel et al. (2012).

## 4.3 Threats

A significant percentage of the Czech bryoflora is thought to be threatened. The level of threat has been assessed using IUCN criteria since 1993 and the last evaluation (Kučera et al. 2012) listed 32.6% of the evaluated taxa as threatened or extinct. An additional 13.5% of the taxa qualified for Near Threatened and Data Deficient categories, which left only 54% of the taxa categorized as of Least Concern. The level of threat posed to the Czech bryoflora appears to be similar to the situation in most Central European countries, such as Switzerland (37.0% threatened, 1.4% extinct; Schnyder et al. 2004), Slovakia (34.2% threatened, 2.9% extinct; Kubinská et al. 2001) and Hungary (26.3% threatened, 0.5% extinct; Papp et al. 2010), and the differences are in the reported percentage of Data Deficient taxa (e.g., 6.1% in the Czech Republic compared to 21.1% in Hungary), which might reflect the different ways of evaluating taxa in different countries. Notable representatives of strongly endangered bryophytes are the species occurring in lowland rich fens (*Bryum longisetum*, *Drepanocladus lycopodioides*, *Meesia longiseta* and *M. uliginosa*; all of them probably went extinct in the Czech fens). These fens started to be converted to agricultural land before the first bryologists of the early nineteenth century could report the occurrence of boreal fen species. Perhaps less pronounced in comparison to lichens are the negative changes in the epiphytic flora. Still epiphytes nearly completely disappeared in northern Bohemia as a consequence of industrial pollution, which peaked in the early 1980s. However, the epiphytic bryoflora, in particular in the Krušné hory Mts, has experienced a remarkable recovery, which to-date has only been intensively studied on the Saxon side of this mountain range (cf. Seifert 2004). This re-establishment of bryophytes has also occurred on the Czech side, although it is less well documented because fewer surveys of epiphytes have been carried out there. Only four epiphytic mosses (*Neckera pumila, Nyholmiella gymnostoma, Paraleucobryum sauteri and Ulota drummondii*) are regarded extinct in the Czech Republic.

## 4.4 Phytogeographic Aspects

Currently there is no phytogeographic analysis of the Czech bryoflora and it is unlikely the situation will change substantially in the near future, as the distributions of many taxa are likely to be revised as a result of molecular studies, revealing morphologically cryptic diversity and at the same time merging putatively endemic taxa with those known in other areas. A brief synthetic review of the European phytogeography of bryophytes has recently been published by Frahm (2012). In agreement with other recent treatments, he refutes the pre-Pleistocene origin of any elements in the Central European bryoflora, which parallels the situation in the flora of vascular plants (see Chap. 3, this book). The oldest elements of our bryoflora might be relics from glacial and early postglacial periods, although some of these arctic-alpine and boreo-montane species could have migrated to the Czech territory later in the Holocene. These elements, which can be found both in high-altitude ecosystems and in relict lowland habitats (various types of fens, steppes, and open sites in deep river valleys), are generally at greatest risk of extinction. There has been a strong decline in the abundance of arctic-alpine elements as a direct consequence of successional changes connected with the warming of the climate. These elements are the largest group among Critically Endangered taxa (*Anthelia juratzkana, Dicranum elongatum*, *Grimmia elatior*, *Gymnomitrion corallioides* (Fig. 4.1a), *Kiaeria falcata* and *Lophozia wenzelii*), despite occurring in relatively stable habitats, such as those of the epilithic species. Many of these even became extinct during the past few decades (*Arctoa fulvella*, *Grimmia unicolor*, *Gymnomitrion adustum* and *Ochyraea smithii*). The bryophytes bound to snow beds probably became extinct with the disappearance of snow beds from high-altitude areas in the Krkonoše Mts (*Gymnomitrion brevissimum*, *Pohlia obtusifolia* and *Polytrichastrum sexangulare*). Another large group of extinct and critically endangered bryophytes are the boreo-montane species living in rich fens (see above).

The basic arrangement of zonal phytogeographical elements reflects the temperature-related (north-south and altitudinal) and oceanity gradients. Most members of the Czech bryoflora are species distributed in the temperate and boreal zones, and with no markedly pronounced affinity to oceanity, although the Czech bryoflora definitely has many representatives of elements that are considered to be

**Fig. 4.1** Examples of remarkable Czech bryophytes: (**a**) *Gymnomitrion corallioides*, Malá Studniční jáma cirque in the Krkonoše Mts; (**b**) *Anastrepta orcadensis*, Mt. Medvědí vrch in the Hrubý Jeseník Mts; (**c**) *Campylopus subulatus*, Píšťala Nature Monument near Černá Voda, Silesia; (**d**) *Orthotrichum pulchellum*, Mt. Studenec in the Lužické hory Mts, northern Bohemia; (**e**) *Sphagnum lindbergii*, Hančova louka site in the Krkonoše Mts; (**f**) *Acaulon triquetrum*, Zimarky hill near Velké Bílovice, southern Moravia. Photo credits: J. Kučera (a, e), Š. Koval (b, c, f), V. Plášek (d)

suboceanic. Such species more or less reach the eastern limit of their European distribution in the Czech Republic or at least markedly decrease in abundance towards the east. *Anastrepta orcadensis* (Fig. 4.1b), *Cephalozia macrostachya*, *Kurzia* spp., *Microlejeunea ulicina*, *Nardia compressa*, *Odontoschisma sphagni* and *Scapania compacta* are examples of suboceanic liverworts. In mosses, the percentage of suboceanic elements is smaller because of the general preference of liverworts for humid conditions and greater tolerance of mosses to drought. Examples of suboceanic mosses include *Campylopus* spp. (Fig. 4.1c), *Dicranodontium uncinatum*, *Kindbergia praelonga*, *Fissidens rufulus*, *Hookeria lucens*, *Hypnum imponens*, *Isothecium myosuroides*, *Mnium hornum*, *Plagiothecium undulatum*, *Rhabdoweisia crenulata*, *Thamnobryum alopecurum* and *Zygodon dentatus*. Some suboceanic elements cannot tolerate the high level of continentality in the Czech Republic, although many of them are recorded close to the Czech border in Germany, such as the liverworts *Frullania microphylla*, *Metzgeria temperata* and *Solenostoma paroicum*, or the mosses *Isothecium holtii*, *Leptodontium flexifolium*, *Nogopterium gracile*, *Pseudohygrohypnum eugyrium*, *Racomitrium obtusum*, *Syntrichia pagorum* and *Zygodon conoideus*. Suboceanic taxa make up the major part of those species that are spreading (*Orthotrichum pulchellum* (Fig. 4.1d), *Microlejeunea ulicina*, *Campylopus introflexus*, *C. pyriformis*), although other suboceanic bryophytes are now extinct in the Czech Republic (*Gymnomitrion obtusum*, *Neckera pumila*, *Pallavicinia lyellii*, *Ptychomitrium polyphyllum*, *Sphagnum austinii* and *Ulota drummondii*). Species with a subcontinental distribution are much rarer in the Czech Republic. Most of them are Pontic elements that reached what is now the Czech Republic probably along the Danube-Pannonian migration route (the liverworts *Asterella saccata* and *Mannia fragrans*, the mosses *Hilpertia velenovskyi* and *Syntrichia caninervis*) but there are also rare examples of eastern, boreal elements (*Callicladium haldanianum*, *Polytrichum pallidisetum* and also the common *Eurhynchium angustirete*, which becomes increasingly rare west of the Czech border). Circumarctic or circumboreal taxa at their southern distribution limit in the Czech Republic are also very rare. Well-known examples currently occur mostly in the Krkonoše Mts. These are the mosses *Sphagnum lindbergii* (Fig. 4.1e), occurring also in bogs in the Krušné hory Mts, or *Sphagnum jensenii* on the Polish side of the Krkonoše Mts, just a few dozen metres from the Czech border. Further examples include *Discelium nudum* occurring sporadically throughout the country and the aquatic moss *Dichelyma falcatum*, recorded from the Šumava Mts and another locality in southern Bohemia, in which its occurrence was probably casual. Still rarer is the representation of alpine species not occurring in northern parts of Europe. Perhaps the only examples are *Hypnum fertile*, known from a single locality in southern Bohemia, *Plagiothecium neckeroideum*, occurring sporadically in the Šumava Mts, and *Streblotrichum enderesii* with a single historical record from the Krkonoše Mts. While the species of the arctic element are generally considered glacial relics, the species of the Mediterranean element are probably later colonists, which reached Central Europe via several migration routes around the Alps. Typical representatives of the Mediterranean element are many species of the moss family *Pottiaceae*, such as *Acaulon triquetrum* (Fig. 4.1f), *Crossidium squamiferum*,

*Pottiopsis caespitosa, Tortella squarrosa, Tortula atrovirens, T. inermis, Funariaceae,* such as *Entosthodon fascicularis* and *E. pulchellus,* and *Bryaceae,* such as *Bryum torquescens*. Mediterranean liverworts are very rare in the Czech Republic. An interesting example is the single occurrence of *Targionia hypophylla* (Fig. 4.2a) around warm air efflux sites (ventaroles) on Mt. Boreč in the České středohoří Mts in northern Bohemia, which was earlier believed to be a relict population that dates back to the Tertiary (Pilous 1959). In addition to typically Mediterranean species, thermophilous bryophytes in the Czech Republic have often migrated to Central Europe from the Pannonian Basin (the Pontic species mentioned above plus, e.g., the mosses *Didymodon acutus, D. cordatus* (Fig. 4.2b) and *Pterygoneurum lamellatum*). A rare element of the thermophilous flora is *Frullania inflata*, a species with an Illyric-Insubric distribution range, known from several locations in southern Moravia.

The dispersal capacity of spore-producing bryophytes is generally considerably greater than that of seed plants. Consequently, small distribution ranges are rare or at least do not exist for a long time in bryophytes. There are very few examples of convincingly stenoendemic species of bryophytes in Europe and except for one uncertain case none is presently considered to occur only in the Czech Republic. Several dozens of widely distributed species were originally described from the Czech Republic, including *Racomitrium sudeticum* (Fig. 4.2c), described from the Krkonoše Mts, and *Fossombronia wondraczekii* and *Hilpertia velenovskyi* (Fig. 4.2d), described from localities in Prague. The same applies to *Bryum moravicum*, described by Podpěra as a southern-Moravian endemic, which was recently shown to be the oldest name for a widely distributed species (Kučera and Holyoak 2005). Even very recently, three taxa were described from localities restricted to the Czech Republic: *Platyhypnidium grolleanum*, which is probably only a rheophytic modification of the widespread *Rhynchostegium riparioides*, and two *Orthotrichum* taxa, *O. moravicum* (Fig. 4.2e) and *O. affine* var. *bohemicum*. *Orthotrichum moravicum* was recently recorded from Slovakia and Tajikistan (Plášek and Číhal 2012) and *O. affine* var. *bohemicum* from Slovakia, Poland, Ukraine, Belgium and the USA (Plášek et al. 2016). An interesting example of a putatively stenoendemic species described from the Czech Republic and not so far recorded from outside Central Europe, is the hornwort *Anthoceros neesii* (Fig. 4.2f). Although it occurs in the common and widely distributed habitat of stubble fields in submontane regions on non-calcareous substrates, it seems to be surprisingly rare and was long regarded as having vanished from the Czech Republic.

## 4.5 Bryophytes in Vegetation

The composition of the bryoflora reflects both general macroclimatic conditions, which affect the local species pool, and specific site conditions, determined by the substrate and geomorphology. In contrast to the vascular plant flora, the role of the micro-relief and substrate conditions is generally much greater for bryophytes as a

**Fig. 4.2** Examples of remarkable Czech bryophytes: (**a**) *Targionia hypophylla*, Mt. Boreč in the České středohoří Mts, northern Bohemia; (**b**) *Didymodon cordatus* near Mělčany, southern Moravia; (**c**) *Racomitrium sudeticum*, Malá Studniční jáma cirque in the Krkonoše Mts; (**d**) *Hilpertia velenovskyi*, Červený kopec National Nature Monument in Brno, southern Moravia; (**e**) *Orthotrichum moravicum* near Chladná voda creek in the Moravskoslezské Beskydy Mts; (**f**) *Anthoceros neesii* near Rýmařov, northern Moravia. Photo credits: Š. Koval (a, d, f), J. Kučera (b, c), V. Plášek (e)

consequence of both their smaller size and greater sensitivity to environmental conditions in the absence of barriers to evaporation from the surface of moss gametophytes. This might result in vastly different bryofloras occurring in one type of vegetation under similar climatic conditions. The following paragraphs briefly mention the bryofloristically conspicuous features of the major natural vegetation types occurring in four zonal and extrazonal biomes in the Czech Republic (see Fig. 7.1 in Chap. 7, this book).

Within the forest-steppe biome of the Czech Republic, thermophilous oak forests generally harbour a very poor bryoflora with almost no notable species. One exception is the extremely rare *Dicranum muehlenbeckii*. However, where local conditions in the warmest and driest areas do not allow for the existence of continuous forest or where man has deforested an area, interesting thermophilous bryophytes occur. The richest communities with mostly Mediterranean elements can be found on calcareous rocks (e.g. *Grimmia anodon*, *G. teretinervis*, *G. tergestina*, *Schistidium brunnescens* and *Syntrichia montana*; Fig. 4.3a), on base-rich soil (*Tortella squarrosa*) and notably also on the loess cliffs, which support the existence of subcontinental elements of cold steppes (*Didymodon cordatus, Hilpertia velenovskyi* and *Pterygoneurum lamellatum*).

Mesic broad-leaved forests in the broad-leaved deciduous forest biome may be both relatively poor in bryophytes, as in oak-hornbeam or beech forests without rock outcrops, and rich, as in moist (sub)montane beech and beech-fir forests on rugged topography. The bryoflora composition is also significantly affected by management. Rapid turnover of trees and removal of coarse woody debris has an adverse effect on epiphytic and epixylic bryoflora, whereas old-growth forests, such as the famous Žofínský prales in southern Bohemia, are important biodiversity hot-spots (Kučera 2009; Táborská et al. 2015; an example of a rare epiphytic moss is *Anacamptodon splachnoides*; Fig. 4.3b). As in the case of the thermophilous forests, locally richest or bryofloristically most interesting sites in the landscapes with mesic broad-leaved forests are often the rocky habitats. The most interesting examples are the boulder screes in the České středohoří Mts in northern Bohemia, with many prominent glacial relicts such as the liverworts *Gymnomitrion corallioides* and *Sphenolobus saxicola* (Němcová 2000), or the sandstone pseudokarst areas in northern Bohemia, also with a large numbers of relict species (e.g. *Dicranum elongatum* and *Hygrobiella laxifolia*). Another bryologically important type of azonal vegetation in the deciduous forest biome is the local occurrence of fens and bogs. The best preserved rich fens with the bryologically most interesting flora are in the Doksy area in northern Bohemia, the Bohemian-Moravian Highlands and the Třeboň Basin in southern Bohemia.

Montane coniferous forests (the spruce taiga), despite long-lasting human management, are home to a significant number of bryophytes often occurring abundantly on the floor of forests even in the absence of diversified micro-relief. This is particularly the case in water-logged forests and forest on boulder screes in topographic situations with a high air humidity. Large pleurocarpous mosses (*Eurhynchium angustirete, Hylocomium splendens* and *Thuidium tamariscinum*) together with similarly robust species of *Dicranum*, often dominate the forest floor

**Fig. 4.3** Examples of remarkable Czech bryophytes: (**a**) *Syntrichia montana* near Řeznovice, southern Moravia; (**b**) *Anacamptodon splachnoides*, Venušiny misky National Nature Monument near Černá Voda, Silesia; (**c**) *Dicranum spurium* near Lniště, southern Bohemia; (**d**) *Andreaea frigida*, Pod Koulemi valley, Krkonoše Mts. Photo credits: J. Kučera (a, c, d) and Š. Koval (b)

and prevent smaller species from establishing. This results in a relatively low species diversity except in topographically heterogeneous areas. Spruce forests richest in bryophyte species are typically those occurring in lower parts of glacial cirques in the Krkonoše, Hrubý Jeseník and Šumava Mts. The glacial cirques are possibly the bryologically richest sites in the Czech Republic, because of the extremely diversified microsite conditions, substrate types and high humidity (Kučera et al. 2004). Lowland pine taiga, which occurs locally on sandy substrates, also mostly has a well-developed moss layer, dominated typically by large species of *Dicranum* (*D. polysetum*, *D. scoparium*, *D. spurium*; Fig. 4.3c), *Leucobryum* and *Campylopus* (notably *C. flexuosus*). Bryologically important in the coniferous forest biome (both in the montane and lowland types) is also the local occurrence of acidic bogs, dominated by species of *Sphagnum*, which constitute the major part of their biomass.

Subalpine tundra is generally poor in bryophyte species. This particularly applies to the subalpine grasslands dominated by *Nardus stricta*, where niches for mosses only occur locally on elevated landforms. On the other hand, rock outcrops and screes are a refuge for arctic-alpine relict species such as the liverwort genera *Gymnomitrion, Marsupella* and *Tetralophozia*, and moss species *Andreaea rothii, A. frigida* (Fig. 4.3d) and *Grimmia elongata*. They are strongly endangered by successional changes, as the altitudinal and spatial extent of the tundra biome in the Czech Republic is extremely limited. The reasons underlying these changes are multiple; Treml et al. (2016) recently argued for the major importance of the absence of cattle grazing for the advance of the treeline, in addition to the increasing temperatures over the last century (Migała et al. 2016).

**Acknowledgements** I am grateful to Štěpán Koval and Vítězslav Plášek for providing photographs, Milan Chytrý and Zdeněk Kaplan for helpful comments on earlier versions of the text and Tony Dixon for the English proofreading.

## References

Danihelka J, Chrtek J Jr, Kaplan Z (2012) Checklist of vascular plants of the Czech Republic. Preslia 84:647–811

Essl F, Lambdon PW (2009) Alien bryophytes and lichens of Europe. In: DAISIE, Handbook of alien species in Europe. Springer, Dordrecht, pp 29–41

Frahm J-P (2012) The phytogeography of European bryophytes. Botanica Serbica 36:23–36

Köckinger H, Suanjak M, Schriebl A, Schröck C (2008) Die Moose Kärntens. Naturwissenschaftlicher Verein für Kärnten, Klagenfurt

Kubinská A, Janovicová K, Šoltés R (2001) Červený zoznam machorastov Slovenska (december 2001) [Red List of bryophytes in Slovakia (December 2001)]. Ochrana prírody, Suppl 20:31–43

Kučera J (2009) Bryoflora of the Žofínský Prales nature reserve (Novohradské hory Mts., South Bohemia). Silva Gabreta 15:97–120

Kučera J, Holyoak DT (2005) Lectotypification of *Bryum moravicum* Podp. (*Bryopsida*: *Bryaceae*). Journal of Bryology 27:161–162

Kučera J, Buryová B, Plášek V, Váňa J, Zmrhalová M (2004) Bryophytes of the glacial cirques in the Giant Mountains and Hrubý Jeseník Mts (the Czech Republic). Opera Corcontica 41:170–184

Kučera J, Váňa J, Hradílek Z (2012) Bryophyte flora of the Czech Republic: updated checklist and Red List and a brief analysis. Preslia 84:813–850

Migała K, Urban G, Tomczyński K (2016) Long-term air temperature variation in the Karkonosze mountains according to atmospheric circulation. Theoretical and Applied Climatology 125:337–351

Němcová L (2000) Die Moose auf den Blockhalden des Böhmischen Mittelgebirges und der benachbarten Gebiete – eine erste Skizze. Acta Universitatis Purkynianae, Studia Biologica 4:97–111

Ochyra R, Żarnowiec J, Bednarek-Ochyra H (2003) Census catalogue of Polish mosses. Biodiversity of Poland 3:1–372

Papp B, Erzberger P, Ódor P, Hock Z, Szövényi P, Szurdoki E, Tóth Z (2010) Updated checklist and red list of Hungarian bryophytes. Studia Botanica Hungarica 41:31–59

Pilous Z (1959) Mechorosty Státní přírodní rezervace Borečský vrch v Českém Středohoří [Bryophytes of the Borečský vrch Nature Reserve in the České středohoří Mts]. Ochrana přírody 14:97–99

Plášek V, Číhal L (2012) New national and regional bryophyte records, 32. 22. *Orthotrichum moravicum* Plášek & Sawicki (Tajikistan). Journal of Bryology 34:239

Plášek V, Fialová L, Skoupá Z, Sawicki J (2016) New national and regional bryophyte records, 46. 16. *Orthotrichum affine* var. *bohemicum* (Slovakia). Journal of Bryology 38:52

Schnyder N, Bergamini A, Hofmann H, Müller N, Schubiger-Bossard C, Urmi E (2004) Rote Liste der gefährdeten Moose der Schweiz. BUWAL, Bern

Seifert E (2004) Epiphyten im Wandel: Zur Vorkommen epiphytischer Moose im Erzgebirge. Beiträge zum Naturschutz im Mittleren Erzgebirgskreis 3:77–86

Söderström L, Hagborg A, von Konrat MJ, Bartholomew-Began SE, Bell D, Briscoe LRE, Brown EA, Cargill DC, da Costa DP, ... Zhu R-L (2016) World checklist of hornworts and liverworts. PhytoKeys 59: 1–828

Stebel A, Virchenko VM, Plášek V, Ochyra R, Bednarek-Ochyra H (2012) Range extension of *Orthodicranum tauricum* (*Bryophyta, Dicranaceae*) in Central-East Europe. Polish Botanical Journal 57:119–128

Szweykowski J (2006) An annotated checklist of Polish liverworts and hornworts. Biodiversity of Poland 4:1–114

Táborská M, Přívětivý T, Vrška T, Ódor P (2015) Bryophytes associated with two tree species and different stages of decay in a natural fir-beech mixed forest in the Czech Republic. Preslia 87:387–401

Treml V, Šenfeldr M, Chuman T, Ponocná T, Demková K (2016) Twentieth century treeline ecotone advance in the Sudetes Mountains (Central Europe) was induced by agricultural land abandonment rather than climate change. Journal of Vegetation Science 27:1209–1221

Váňa J (1997) Bryophytes of the Czech Republic: an annotated check-list of species (1). Novitates Botanicae Universitatis Carolinae 11:39–89

# Chapter 5
# Lichen Biota of the Czech Republic

**Zdeněk Palice**

**Abstract** A basic overview of the current state of knowledge on the lichen biota in the Czech Republic is provided. Lichens of particular interest, e.g. those described from this country, globally rare species and endangered bioindicators are dealt with in more detail. Important lichenological localities in this country are briefly mentioned. Biogeographical aspects of Czech lichen biota are discussed and examples of various biogeographical elements given. Changes in abundance, especially of epiphytic lichens, both on a long time scale and in recent years, are evaluated, especially in relation to the latest national Red List of lichens. The phenomenon of recent recolonization and gradual spread of nitrophilous and previously rare or new acidophilous lichens is outlined.

## 5.1 Introduction

The Czech Republic is a lichenologically rather well known area, although bryophytes and vascular plants are much better explored in this country. This is mainly because of the still unresolved taxonomy of many groups of lichens. In addition, many lichens are small, easily overlooked and impossible to identify in the field, and many new records result from accidental ex situ discoveries in herbarium collections.

In the latest Czech lichen checklist (Liška and Palice 2010) 1526 species are listed. The landlocked position of the Czech Republic, the absence of high mountains and the rarity of limestone outcrops at high altitudes preclude the occurrence of a potentially more diverse lichen biota. In spite of that the number of lichens is comparable to that recorded in Slovakia and Poland, both countries with recently updated checklists (Guttová et al. 2013; Fałtynowicz and Kossowska 2016). In Central Europe the richest lichen biotas by far are those of Austria (Hafellner and Türk 2016) and Germany (Wirth et al. 2013), where the number of species exceeds 2300 and 2100, respectively. This is apparently due to the significant number of species in the Alps and larger area of Germany.

Z. Palice (✉)
Institute of Botany, The Czech Academy of Sciences, 252 43 Průhonice, Czech Republic

Department of Botany, Faculty of Science Charles University, Benátská 2, 128 01 Praha 2, Czech Republic
e-mail: zdenek.palice@ibot.cas.cz



M. Chytrý et al. (eds.), *Flora and Vegetation of the Czech Republic*, Plant and Vegetation 14, DOI 10.1007/978-3-319-63181-3_5

According to Liška and Palice (2010) a notable part (~37%) of the Czech lichen biota is threatened (IUCN Red List categories CR, EN and VU) and 9% of previously recorded species are now extinct (RE category). Not currently threatened lichens (NT and LC categories) comprise ~24% of the total. The very high percentage of species classified as DD (data deficient, ~22%) and taxonomically unresolved and dubious NE taxa (not evaluated, ~7%) clearly indicate that the knowledge of this country's lichen biota is still insufficient. In particular many crustose species are poorly known. Interestingly, more than 100 lichen taxa are recorded only from one locality in the Czech Republic, even when dubious taxa (NE) are excluded. Crustose microlichens constitute about three quarters of the total number of the lichens in this country, while only each fourth lichen taxon is a macrolichen. Making up almost half of the total number, saxicolous lichens are the largest ecological group and the majority of the questionable taxa (NE) come from rocky substrates. The percentage of epiphytes (including lignicoles) slightly exceeds one third of the lichens and approximately every sixth lichen may be ranked among terricolous or epibryophytic species.

## 5.2 Pecularities of the Czech Lichen Biota in the European Context

Diverse geology, geomorphology and preservation of endangered ecosystems make the Czech Republic suitable for various ecological groups of lichens. Some of them are narrow niche-specialists adapted to specific environmental and microclimatic conditions, and show specific substrate requirements (ultramafic rocks, lime-enriched silicate rocks in open high-mountain habitats), being in some cases vicariants of more widespread species (see e.g. Guttová et al. 2014). Although these rare lichens may be relatively distinct and easily identifiable, some of them seem to be very local, being recorded only a few times globally. Such rarities in the Czech lichen biota include *Diploschistella athalloides*, *Gyalecta sudetica*, *Harpidium rutilans*, *Porpidia nadvornikiana* (see Vězda and Liška 1999), *Psorotichia taurica* (Czeika et al. 2004) and *Solenopsora liparina* (Fig. 5.1e; Guttová et al. 2014).

A good example of a specialist lichen is the crustose *Porpidia nadvornikiana* (Fig. 5.1a) confined to serpentinite rocks (see Fig. 1.7 in Chap. 1, this book, for distribution of serpentinite). It is a globally rare species, recorded so far only in

**Fig. 5.1** (continued) sandstone outcrops above the Křinice rivulet in the Bohemian Switzerland National Park; (**c**) *Lobaria amplissima*, a critically endangered species in the Czech Republic surviving on the last veteran trees in the humid forest area west of Modrava, Šumava Mts; (**d**) *Arctoparmelia centrifuga*, a glacial relict occurring on boulder fields on Mt. Luč in the Šumava Mts, which is probably the largest locality for this species in Central Europe outside the Alps; (**e**) *Solenopsora liparina*, a lichen confined to serpentinite outcrops with its northenmost known European locality near Raškov in northern Moravia; (**f**) *Lobaria pulmonaria*, a rapidly declining species of lichen typical of old-growth forests with the richest Czech population near Modrava in the Šumava Mts; (**g**) *Cladonia stellaris*, a rare boreal species in Central Europe occurring in relictual habitats; serpentinites near the Želivka water reservoir in central Bohemia. Photo credits: P. Uhlík (a, b, g), F. Bouda (c, d, f) and J. P. Halda (e)

**Fig. 5.1** Examples of rare lichens occurring in the Czech Republic (**a**) *Porpidia nadvornikiana*, a globally rare lichen confined to serpentinite outcrops, described from the Czech Republic and not known outside Europe; Dukovanský mlýn near Mohelno in south-western Moravia; (**b**) *Cladonia subcervicornis*, primarily a coastal lichen with only a few inland localities in Central Europe;

Europe and with most records from the Czech Republic, from where it was described by Vězda (1972). Another peculiarity is the rare *Cladonia krogiana* found on ultramafic outcrops (peridotite-gabbro) at the Ranský Babylon site (Žďárské vrchy Mts; Palice, unpubl.). This single species of *Cladonia* producing xanthones was described from Norway as growing on mossy rocks mainly near lakes (Løfall and Timdal 2002), while at the Czech locality it was collected in a relatively dry habitat, a low ultramafic boulder on a hill plateau. This dual ecology is not fully understood but may be attributable to specific properties of ultramafic rocks that sometimes support lichens otherwise known mainly as epiphytes in more humid woodland areas, e.g. *Normandina pulchella*, which is often associated with liverworts of the genus *Frullania*.

Worth mentioning are also dealpine occurrences of predominantly arctic-alpine crustose species like *Dimelaena oreina* and *Pleopsidium flavum* on outcrops of hard proterozoic silicite at low altitudes of the Bohemian Massif, including in the capital city of Prague.

Endemism is very rare among European lichens. Probably there is no lichen taxon endemic to the Czech Republic, although some species are documented only from their type localities in this country. These include, except several dubious taxa, *Aspicilia serpentinicola* from the Mohelno serpentinite steppe in south-western Moravia, *Polyblastia suzae* from serpentinites near the village of Raškov in northern Moravia (see Vězda and Liška 1999), the epiphytic *Lecanographa aggregata* collected by J. Suza in 1920 on road-side trees near the town of Blansko in central Moravia (Egea and Torrente 1994) or the recently described saxicole *Bacidina flavoleprosa* discovered near the town of Sedlčany in central Bohemia (Czarnota and Guzow-Krzemińska 2012). Further information may be found in the Czech Lichen Type Database (Liška 2016).

Rare epiphytic and epixylic lichens are mainly niche specialists in old-growth forests, which require long forest continuity and are susceptible to forest management. Their rarity may be also partly linked to the migration history of their preferred phorophyte, e.g. *Abies alba*, although recent finds also come from other trees. These include some species of *Biatora*, in Central Europe known only from a few woodland sites, such as the south-western Bohemian occurrences of *B. ligni-mollis* and *B. mendax* (Printzen and Palice 1999; Malíček and Palice 2013).

### 5.2.1 Biogeographical Patterns

The distribution of lichens depends on a complex set of evolutionary, historical, environmental and ecological factors. Since lichens produce small, easily spreading diaspores, many species have large distribution ranges, hence (sub)cosmopolitan distributions are not rare. Widespread, supposedly Holarctic species often have outlying localities also in high mountains in the tropics or in the Southern Hemisphere. Therefore, applying a strict phytogeographic framework does not appear to be fully appropriate in lichenology. Since many groups of lichens are undercollected, their records may reflect intense research activity in particular regions, mainly in some parts of Europe and North America, although these species are actually more widespread.

Lichens with **boreal affinities** include mainly epiphytes typical of the boreal forest zone, which are recorded mainly in montane spruce forests or pine mire woodlands in the Czech Republic. Some of these lichens are rare, preferably occurring in old-growth forests (e.g. *Alectoria sarmentosa*, *Lecanactis abietina* and *Mycoblastus sanguinarius*) and some others are widespread across the country since they can grow in spruce plantations (e.g. *Chaenotheca ferruginea* and *Hypocenomyce scalaris*). Examples of relict boreal epiphytic species are *Melanohalea olivacea* and *M. septentrionalis*, recorded mainly on raised bogs in the Šumava Mts before World War II (see Vězda and Liška 1999). While these lichens are occasionally dominant on *Betula* trunks in Fennoscandia, declining relict populations in the Czech Republic were apparently unable to cope with the changing environment, especially with acidic pollution. Indicators of old-growth boreal forest in temperate mountain forests include *Chaenotheca sphaerocephala* and other rarities recorded only recently in the Šumava Mts, such as *Lecidea betulicola* (Palice 1999), *L. coriacea* (Holien et al. 2016) and *Myochroidea porphyrospoda* (Printzen et al. 2008). Also some terricolous boreal species are declining, for example the decorative *Cladonia stellaris* (Fig. 5.1g), no longer present at many of its former sites and currently known only from about two dozen dispersed localities in Bohemia (Fig. 5.2a). A majority of the Czech epiphytic lichens are **temperate broad-leaved forest species**, occurring mainly on the bark of the two dominant trees of these forests, oak (*Quercus* spp.) and beech (*Fagus sylvatica*). Although many generalist epiphytes occur on both oak and beech trees, different microclimates in warmer and drier oak-dominated forests and colder and wetter beech forests account for different assemblages of lichens at least in their understoreys, with the most characteristic lichen species being *Flavoparmelia caperata* in oak forests and *Pyrenula nitida* in beech forests. Notably old-growth forests dominated by beech have the highest diversity of epiphytic lichens (Hofmeister et al. 2016), with beech being the most preferred phorophyte. Forest continuity over several centuries and the range of microhabitats available in the best preserved fir-beech primeval forests account for the uniqueness and extraordinarily high diversity of epiphytic lichens in the Central European context. In these woodlands there are both lichens with oceanic and boreal affinities. Well-preserved old-growth forests in the Boubínský prales (Šumava Mts) and Žofínský prales (Novohradské hory Mts) National Nature Reserves harbour several sensitive macrolichens (e.g. *Lobaria pulmonaria* and *Menegazzia terebrata*) and also rare crustose specialists confined to specific niches such as exposed roots, weathered bark of old trees or decaying wood of standing trees. These crustose species are otherwise only exceptionally found in Central Europe (*Arthonia incarnata*, *Biatora ligni-mollis*, *Cliostomum leprosum* and *Pachyphiale carneola*; Svoboda 2009; Malíček and Palice 2013; Malíček et al. 2014). Unlike some mountain beech and spruce forests, oak forests have been subject to strong human interventions for millenia and subject to marked changes after the cessation of traditional management in the twentieth century. Numerous species of lichen characteristic of oak woodlands have recently

(a) *Cladonia stellaris*

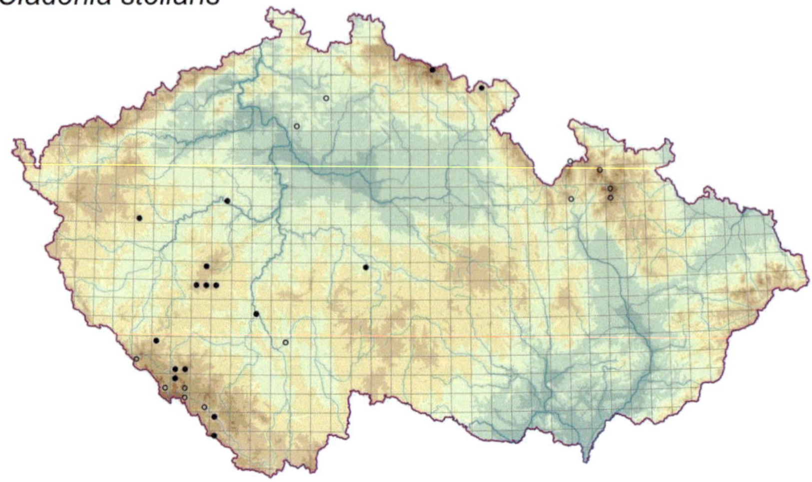

(b) *Lobaria pulmonaria*

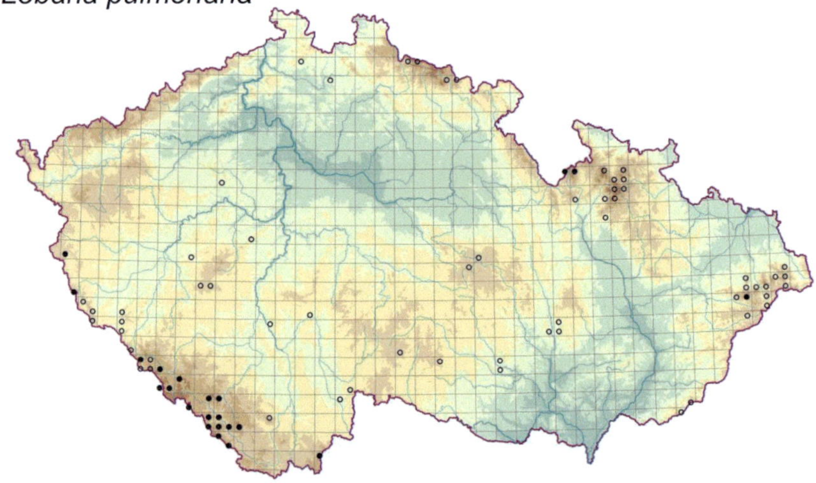

**Fig. 5.2** Distribution of two rare macrolichen species in the Czech Republic: (**a**) *Cladonia stellaris* and (**b**) *Lobaria pulmonaria. Empty circles* refer to the records from the period 1900–1989 only, *full circles* to the records since 1990 (data were extracted from the NDOP database of the Nature Conservation Agency of the Czech Republic and reliable literature sources; the maps were prepared by O. Hájek)

declined in abundance or become extinct (Svoboda et al. 2011). The lichens favouring oaks are also disadvantaged by the low buffering capacity of their bark (Frahm et al. 2009). Despite this, less fragmented lowland floodplain forests (e.g. upstream of the confluence of the Dyje and Morava rivers in southern Moravia) and wooded middle and lower parts of the valleys in the southern part of the Bohemian Massif (e.g. Vltava and Oslava rivers) harbour a surprisingly high diversity of crustose lichens. They have been underexplored for a long time and recent surveys yielded unexpected discoveries such as *Arthonia pruinata*, *Bactrospora dryina*, *Bacidia auerswaldii*, *Buellia violaceofusca* and *Dendrographa decolorans*, all species that were either considered extinct or were newly recorded for this country (Malíček et al. 2014; Šoun et al. 2015; Vondrák et al. 2016). Lichen biota in lowland forests suffers from overall eutrophication. What is more, several species are confined to the rough bark of old oak trees; some of them (e.g. *Calicium quercinum*, *Lecanographa amylacea* and *L. lyncea*) have not been observed for almost a century and are considered extinct in many Central European countries.

**Lichens with oceanic affinities** generally have large discontinuous distribution ranges and tend to occur patchily, being confined only to sites with relatively stable humidity and temperature; in addition, more southerly distributed taxa occur only at sites with mild winters. Permanent high humidity may locally compensate for an otherwise unsuitable macroclimate. Molecular markers helped to identify a sterile specimen of the predominantly coastal epiphytic species *Caloplaca ulcerosa* living on shaded limestone rocks in the foothills of the Šumava Mts (Vondrák et al. 2009b). *Pilophorus strumaticus*, an unmistakable macrolichen occurring on wet siliceous rocks, was traditionally considered a typical example of an oceanic Northern European endemic (Dahl 1998). However, it was unexpectedly discovered on hardly accessible rock-walls in the Labský důl valley in the Krkonoše Mts (Halda et al. 2011). For both species these are the only Central European and at the same time most continental records. Similarly, glacial cirques in the Šumava Mts with lakes and stable and permanently wet microclimate support predominantly coastal species that only exceptionally occur inland, such as *Rinodina interpolata*, which was found on a rock-wall at Černé Lake (Palice 1999).

Most lichens with oceanic affinities in the Czech Republic occur in northern Bohemian sandstone landscapes. The rare *Bunodophoron melanocarpum* is known only from this area, now surviving only in the Adršpach-Teplice Rocks. There are isolated populations of the epiphytic microlichens *Micarea pycnidiophora* and *Phaeographis inusta*, and the mainly coastal macrolichen *Cladonia subcervicornis* (Fig. 5.1b) in Saxon-Bohemian Switzerland (Palice et al. 2007).

**Lichens with Mediterranean affinities** occur mainly in dry areas of central and northern Bohemia and southern Moravia, e.g. in the Bohemian Karst and the Pavlov Hills, the valleys of the lower stretches of the Berounka and Vltava rivers, and on the volcanic hills in the České středohoří Mts. Terricolous and saxicolous lichens with Mediterranean affinities were surveyed by Suza (1937, 1943), who mentions several characteristic taxa, like the terricolous *Fulgensia fulgens*. Interestingly, some calcicolous taxa have a restricted distribution within the Czech Republic, being known only from its Carpathian part. The best examples are *Lecanora pruinosa*, *Psora vallesiaca* and *Toninia toninioides*, all recorded only on the Jurassic limestone rocks on Kotouč hill near Štramberk in north-eastern Moravia (Suza 1943), which is the westernmost outpost of their Carpathian distribution. The remnants of their habitat on the south-facing slopes of Kotouč hill were destroyed by limestone quarrying after World War II and all three species became extinct there. Also occurrences of some calcicolous lichens in the Moravian Karst that are absent from Bohemia, e.g. *Toninia taurica* (Peksa 2008a), may be interpreted as outposts of their Carpathian range (Peksa 2008a). Several species of the families *Teloschistaceae* and *Candelariaceae* with mainly Mediterranean distributions probably reach their northern distribution limits in the Czech Republic, Germany and Slovakia; this applies to e.g. *Caloplaca inconnexa* (Vondrák et al. 2007), *C. austrocitrina* (Vondrák et al. 2009a), *C. emilii*, *C. interfulgens* (Vondrák et al. 2013), *C. xerica*, *Candelariella plumbea* and *C. viae-lacteae* (Malíček et al. 2014).

**Continental steppe elements** are few, including mainly terricolous lichens with scattered records of them occurring in dry and warm areas in Central Europe that are the westernmost limits of their distribution ranges. The most favourable conditions for these species are in south-eastern Moravia and in the České středohoří Mts in northern Bohemia. Many of them are associated with loess, e.g. *Gyalidea asteriscus* (Suza 1936) and *Rinodina terrestris* (Vězda 1970). Jurassic limestone rocks in the Pavlov Hills in southern Moravia are the only known site in the Czech Republic for *Xanthoria papillifera* (Malíček et al. 2014).

Most **arctic-alpine lichens** are saxicolous and terricolous species, which occur in the Czech Republic mainly in the Sudetes Mts. Some high-mountain species are also found in other mountain ranges. *Arctoparmelia centrifuga* (Fig. 5.1d), for instance, is known from a few sites on the summits of the Krkonoše Mts and there is a large population (perhaps one of the richest in Central Europe) on Mt. Luč in the Šumava Mts (Černohorský 1961). This lichen is considered to be a glacial relict in Central Europe (Wirth et al. 2013). Some high-mountain elements occur also in the Brdy Mts in central Bohemia, especially on cold boulder fields, where *Brodoa intestinifomis*, *B. atrofusca* (Peksa 2008b) and *Miriquidica deusta* (Malíček et al. 2015) are recorded, the latter being known in this country only from this area. The most prominent arctic-alpine lichens are mentioned in the next subchapter.

## 5.3 Sites of Particular Lichenological Interest in the Czech Republic

The majority of the lichenologically unique localities are in the Sudetes Mts. A particularly remarkable site is the **Obří důl** valley (Riesengrund) in the Krkonoše Mts, which encompasses glacial cirques and the smaller famous sites **Čertova zahrádka** (Teufelsgärtchen) and **Rudník** (Kiesberg). Obří důl is *locus classicus* for several distinct species discovered and described by Julius Flotow and Gustav Körber in the mid-nineteenth century and Eugen Eitner at the beginning of the twentieth century. *Gyalecta friesii* (Koerber 1855), a rare boreal species, was apparently extirpated at its type locality because the humid old-growth forest, its typical habitat, was largely destroyed. There are no recent records of this species in the Alps, and viable populations are known only in northern Eurasia. *Sporodictyon cruentum* is an example of a more widespread amphibious lichen described from this locality (Körber 1853). The

Obří důl valley is rich geologically: it includes crystalline limestones, on which *Collema undulatum* was found and described (Flotow 1850), as well as metamorphic metal-rich rocks providing niches for specific assemblages of ferrophilous lichens. This also applies to two species described by Eitner (1911) from the abandoned mine area at Rudník, namely *Rhizocarpon pycnocarpoides*, recently once again recognized as taxonomically distinct (Westberg et al. 2015), and *Lecanora subaurea*, which was originally described by Eitner as *L. aurea*. Mt. **Sněžka** (Schneekoppe, Śnieżka), the highest mountain in the country (1603 m), hosts the richest populations of several arctic-alpine species in the Czech Republic, such as *Alectoria nigricans*, *Allantoparmelia alpicola*, *Flavocetraria cucullata* and *F. nivalis*. The rare arctic-alpine terricolous calciphilous species *Caloplaca nivalis* was described from the Sněžka slopes facing the Obří důl valley (Körber 1853), where patches of base-rich rocks occur. Five specialized lichenicolous lichens with reduced thalli that are usually confined to a single crustose host lichen are so far known in the Czech Republic only from Mt. Sněžka; they are *Caloplaca epithallina*, *C. magni-filii*, *Lecanora latro*, *Protoparmelia phaeonesos* and *Rhizocarpon pusillum* (see Vězda and Liška 1999; Vondrák et al. 2007). The wind-exposed, nutrient-enriched schist outcrop **Petrovy kameny** (Peterstein) in the Hrubý Jeseník Mts harbours several lichens that are not recorded elsewhere in the country (*Anaptychia bryorum*, *Fulgensia schistidii*, *Melanohalea infumata* and *Phaeophyscia kairamoi*) or only rarely in the nearby Velká kotlina cirque (*Acarospora badiofusca* s. str., *Caloplaca ammiospila*; Suza 1933; Vondrák and Malíček 2015). Likewise the sheltered lime-enriched schist outcrops on Mt. **Studénková hole** (Brünnelheide) are unique, being the only site in the country for *Gyalecta sudetica* (type locality), *Porina mammillosa* and *Leucocarpia biatorella* (see Vězda and Liška 1999; Peksa 2008a).

Although the serpentinite area near **Mohelno** in south-western Moravia is not extraordinarily rich in lichens, it is one of the most unique lichenological localities in Central Europe. Three accepted species were described from this site: *Aspicilia serpentinicola* (known only from this site), *Acarospora suzae* and *Psorotichia moravica*. This locality hosts also other globally rare species such as *Gyalidea asteriscus*, *Harpidium rutilans*, *Lecanora laatokkaensis* and *Toninia cinereovirens*.

Remnants of old-growth deciduous forest surrounding the raised-bog area **Modravské slatě** (Weitfäller Filze) near Modrava village in the Šumava Mts harbour luxuriant suboceanic epiphytic assemblages with high proportions of lichens with a cyanobacterial photobiont. Already in the early twentieth century this locality was known to be unique (Hilitzer 1925). Thanks to high precipitation (close to the highest values in the Czech Republic), high air humidity with numerous foggy days, its topography and long forest continuity, this area hosts very rich epiphytic assemblages similar to those in lichen-rich oceanic Western European woodlands. This site was the last known locality of the sensitive native woodland indicator *Usnea longissima* (Hilitzer 1924) and it is also the country's last refuge for some critically endangered macrolichens, namely *Lobaria amplissima* (Fig. 5.1c) (Liška et al. 1996). The most viable population of *Lobaria pulmonaria* (Fig. 5.1f), an indicator of old-growth forests, produces apothecia at this site.

## 5.4 Effects of Environmental Changes on Lichen Biota

The decline in rare sensitive lichens was already notified by Anders (1935), who recorded their disappearance and extirpation in northern Bohemia. Suza (1944) and Kuťák (1952) reported the decline and local extinctions of some old-growth forest lichens in the Žďárské vrchy Mts and the Krkonoše Mts, respectively. The gradual disappearance of the sensitive forest species *Lobaria pulmonaria* is documented by Liška and Pišút (1990). Although this species declined dramatically in abundance during the twentieth century, eventually being known only from the Šumava Mts, Novohradské hory Mts and Králický Sněžník Mts, recent intensified field research resulted in its rediscovery in the Český les Mts in western Bohemia and on Mt. Kněhyně in the Moravskoslezské Beskydy Mts (Fig. 5.2b).

Several sensitive species may have been declining in abundance already during the eighteenth and nineteenth centuries. This applies to the most sensitive old-growth forest species and to some Mediterranean or high-mountain lichens that are at their distributional and ecological limits in the Czech Republic. For instance, *Solorina crocea*, a species of arctic-alpine snow-patches is known only from nineteenth century records from Mt. Sněžka in the Krkonoše Mts and Mt. Studénková hole in the Hrubý Jeseník Mts (Koerber 1855; Suza 1933).

Acidic air pollution by sulphur dioxide peaked in this country in the 1970s and 1980s. It was especially detrimental to epiphytes (Anděl and Černohorský 1978; Liška 1994). In the most polluted areas in northern Bohemia and Silesia and in industrial centres lichens have almost disappeared, leaving behind “lichen deserts”. Acidification also caused changes in the pH values of bark of phorophytes and even in less affected areas a shift in substrate preference was noted e.g. for the rare old-growth forest species *Biatora fallax* (Printzen and Palice 1999) or the common epiphyte *Lecanora pulicaris* (Malíček 2014). Among the terricolous and epibryophytic lichens the most striking decline was that of the cephalodiate macrolichens capable of fixing nitrogen. *Stereocaulon tomentosum*, a conspicuous lichen on acidic sandy soils, namely in heathlands (see Suza 1946) used to be “locally frequent” (Černohorský et al. 1956), but is now considered extinct in this country. Cephalodiate members of *Peltigera* (*P. aphthosa*, *P. leucophlebia* and *P. venosa*) are close to extinction in the Czech Republic and currently only known from a few sites in the Šumava Mts, Krkonoše Mts, Králický Sněžník Mts and Hrubý Jeseník Mts.

Acidifying emissions in Europe have been declining substantially since 1990 (European Environment Agency 2015) but in forests close to some industrial centres the effects of acidification are still discernible in their impoverished lichen biota in which the “weedy” *Lecanora conizaeoides* is dominant. This is also supported by traditional forestry practices preferring spruce (*Picea abies*) plantations. Since the decline in sulphur dioxide emissions, nitrogen has become the predominant acidifying agent (Moldanová et al. 2011) and dry nitrogen deposits may also support the increase of nitrophytic species (Frahm et al. 2009). A persisting issue is the non-declining ammonium emissions and dust from traffic, agriculture and other local sources. This has resulted in striking changes in the epiphytic lichen biota during the past decades, mainly in a continuous increase of nitrophytes in open landscape, such as in parks and on roadside trees, which are often covered by *Xanthoria parietina*. Also nitrophytic crustose species such as *Bacidina neosquamulosa* or *Candelariella efflorescens* agg. have increased remarkably in these habitats.

An interesting phenomenon is the **(re)colonization** of former "lichen deserts" and lichen-impoverished landscapes after desulphurization of coal power plants in the early 1990s, especially in western Bohemia. Lichen recolonizers were noted by lichenologists soon after 2000 but the observations were only published quite recently (e.g. Steinová et al. 2013; Šoun et al. 2015). The empty niches are re-occupied by assemblages of ecologically and biogeographically rather diverse species growing mainly on well-lit sites on branchlets of shrubs (e.g. *Prunus spinosa* and *Crataegus* spp.) in pastureland or on twigs of larch trees in plantations. In addition to generalists, such as *Evernia prunastri*, *Hypogymnia physodes* and *Parmelia sulcata*, both lichens with suboceanic (e.g. *Hypotrachyna revoluta*) and boreal distributions (e.g. *Evernia divaricata*, *E. mesomorpha* and *Hypogymnia bitteri*) may spread side by side, and these assemblages include even numerous nitrophilous species (e.g. *Melanelixia subaurifera* and *Xanthoria polycarpa*). Some of the colonizers with suboceanic distributions (e.g. *Flavoparmelia soredians* and *Punctelia borreri*) are assumed to have spread from Western Europe (Steinová et al. 2013; Šoun et al. 2015). (Re)expansion is documented also for predominantly foliicolous lichens that were until recently known mainly from the Šumava Mts or were even missing (Palice 1999; see also Vězda and Liška 1999). An example is a lichen known also from the tropics, *Fellhanera bouteillei*, which is native to humid lowland sites. This historically sparsely recorded species was last reported in 1947 and rediscovered at the beginning of the twenty-first century. Recently it spread into various parts of the Czech Republic including Prague, formerly heavily polluted areas in central and northern Bohemia, and at high altitudes in the Šumava Mts (Malíček and Palice 2013) and the Krkonoše Mts. It is thus tempting to attribute this expansion to global warming. Similar gradual spread is also recorded for mosses of the family *Orthotrichaceae* (Kučera et al. 2012). However, some changes in the epiphytic lichen biota are too fast and recent for a full understanding and some even may turn out to be episodic.

Probably no lichens are unequivocally known to be alien in Europe (Essl and Lambdon 2009), and even the widespread toxitolerant *Lecanora conizaeoides* might have been native to Central Europe regardless of its extensive spread over much of the continent since the 1950s. Dozens of presumably native calciphilous lichens have adapted to man-made substrates like mortar, concrete, bricks and roof-tiles. Some species in these habitats belong to the most common ubiquitous lichens, which usually are recorded on calcareous rock outcrops. They may switch also to secondarily base-enriched siliceous rocks (e.g. gravestones) and to dust-impregnated wood. Examples of widespread ubiquitous lichens are *Acarospora moenium*, *Caloplaca decipiens*, *Lecanora dispersa* and *L. muralis*. The high dispersal potential of lichens and their ability to occupy transient habitats is documented by unexpected finds in antropogenic habitats of the rare high-mountain species *Rinodina castanomelodes* (concrete at high altitudes on Mt. Sněžka, Krkonoše Mts; Vondrák et al. 2006) and *Placidiopsis oreades* (sedimentation basin at the foothills of the Adršpach-Teplice Rocks at an altitude of 500 m; Peksa 2009). In Central Europe both species are otherwise known only from the Alps and Tatra Mts.

## 5.5 Changes in Our Knowledge of the National Lichen Biota

Lichenofloristic research has experienced a renaissance in the Czech Republic during the last few years, which is mirrored e.g. by additions of more than 100 new species records for the country since issuing the latest version of the Czech lichen checklist (e.g. Malíček and Palice 2013; Malíček et al. 2014; Vondrák et al. 2016). Irrespective of the fact that several taxa were also merged or excluded (e.g. Malíček 2014; Svoboda et al. 2014), to date the number of lichen species recorded in the Czech Republic has exceeded 1630. Surprisingly, most new records are represented by non-cryptic morphospecies identifiable also by non-molecular means. Some DD and NE taxa are currently studied by specialists and new results are to be published in the next few years, therefore the percentage of DD and NE species is likely to be reduced substantially. Intensive field research and targeted search for RE taxa at their former localities since 2010 has yielded records of approximately 30 species of the total of 138 listed by Liška and Palice (2010); hence the percentage of RE species has decreased to 6–7% of the total number of lichens. Also the status of several recolonizers ranked as CR in the Red List (Liška and Palice 2010), including *Evernia divaricata*, *E. mesomorpha*, *Hypogymnia bitteri*, *Nephromopsis laureri* and *Usnea scabrata*, will need to be reconsidered. These species, most of them primarily related to old-growth boreal forests, seem to be adaptive recolonizers in other habitats (e.g. Steinová et al. 2013).

During the last twelve years ten new species were described (based on holotypes) from the Czech Republic: *Agonimia flabelliformis*, *Aspicilia serpentinicola*, *Bacidia pycnidiata*, *Bacidina flavoleprosa*, *Biatora radicicola*, *Caloplaca emilii*, *C. microstepposa*, *C. soralifera*, *C. substerilis* and *Lecania leprosa* (Czarnota and Coppins 2006; Vondrák and Hrouzek 2006; Reese Næsborg 2008; Czarnota and Guzow-Krzemińska 2012; Guzow-Krzemińska et al. 2012; Nordin 2013; Vondrák et al. 2013; Frolov et al. 2016; Printzen et al. 2016). Given the recent boost in lichenological research in this country, new discoveries and significant modifications of the national list of lichens are to be expected in the near future.

**Acknowledgements** I thank František Bouda, Josef Halda and Petr Uhlík for providing photographs, Ondřej Hájek for preparing maps based on data extracted from the NDOP database of the Nature Conservation Agency of the Czech Republic, Milan Chytrý, Jiří Danihelka and Marinus Werger for improving a draft version of the text, and Tony Dixon for English proofreading. This study was supported by the project no. RVO67985939 (Academy of Sciences of the Czech Republic).

## References

Anděl P, Černohorský Z (1978) Lišejníky a znečištění ovzduší na Liberecku [Lichens and air pollution in the Liberec District]. Preslia 50:341–359

Anders J (1935) Im Verschwinden begriffene und verschwundene Flechtenarten in Nordböhmen. Berichte der Deutschen Botanischen Gesellschaft 53:319–330

Černohorský Z (1961) Die Flechte *Parmelia centrifuga* (L.) Ach. im Böhmerwald. Preslia 33:359–364

Černohorský Z, Nádvorník J, Servít M (1956) Klíč k určování lišejníků ČSR [Key to identification of lichens of Czechoslovakia], vol 1. Nakladatelství Československé akademie věd, Praha

Czarnota P, Coppins BJ (2006) A new *Bacidia* with long-necked pycnidia from Central Europe. Lichenologist 38:407–410

Czarnota P, Guzow-Krzemińska B (2012) ITS rDNA data confirm a delimitation of *Bacidina arnoldiana* and *B. sulphurella* and support a description of a new species within the genus *Bacidina*. Lichenologist 44:743–755

Czeika H, Czeika G, Guttová A, Farkas E, Lőkös L, Halda J (2004) Phytogeographical and taxonomic remarks on eleven species of cyanophilic lichens from Central Europe. Preslia 76:183–192

Dahl E (1998) The phytogeography of northern Europe (British Isles, Fennoscandia and adjacent areas). Cambridge University Press, Cambridge

Egea JM, Torrente P (1994) El género de hongos liquenizados *Lecanactis* (*Ascomycotina*). Bibliotheca Lichenologica 54:1–205

Eitner E (1911) Dritter Nachtrag zur schlesischen Flechtenflora. Jahresbericht der Schlesischen Gesellschaft für vaterländische Cultur, Abt II, 88:20–60

Essl F, Lambdon PW (2009) The alien bryophytes and lichens of Europe. In: DAISIE, Handbook of alien species in Europe. Springer, Dordrecht, pp 29–41

European Environment Agency (2015) Air pollution fact sheet 2014. Czech Republic. European Environment Agency, Copenhagen

Fałtynowicz W, Kossowska M (2016) The lichens of Poland. A fourth checklist. Acta Botanica Silesiaca Monographiae 8:3–122

Flotow J (1850) Ueber Collemaceen. Epistel an Schaerer in Bern, geschrieben im März 1841. Linnaea 23:147–172

Frahm J-P, Thönnes D, Hensel S (2009) Ist der Anstieg nitrophiler Flechten an Bäumen auf eine Erhöhung des Borken-pHs zurückzuführen? Archive for Lichenology 1:1–10

Frolov I, Vondrák J, Fernández-Mendoza F, Wilk K, Khodosovtsev A, Halıcı MG (2016) Three new, seemingly-cryptic species in the lichen genus *Caloplaca* (*Teloschistaceae*) distinguished in two-phase phenotype evaluation. Annales Botanici Fennici 53:243–262

Guttová A, Lackovičová A, Pišút I (2013) Revised and updated checklist of lichens of Slovakia (May 2013). Biologia 68:845–850

Guttová A, Zozomová-Lihová J, Timdal E, Kučera J, Slovák M, Piknová K, Paoli L (2014) First insights into genetic diversity and relationships of European taxa of *Solenopsora* (*Catillariaceae, Ascomycota*) with implications for their delimitation. Botanical Journal of the Linnean Society 176:203–223

Guzow-Krzemińska B, Halda JP, Czarnota P (2012) A new *Agonimia* from Europe with a flabelliform thallus. Lichenologist 44:55–66

Hafellner J, Türk R (2016) Die lichenisierten Pilze Österreichs – eine neue Checkliste der bisher nachgewiesenen Taxa mit Angaben zu Verbreitung und Substratökologie. Stapfia 104:1–216

Halda J, Hauer T, Kociánová M, Mühlsteinová R, Řeháková K, Šťastná P (2011) Biodiverzita cévnatých rostlin, lišejníků, sinic a řas na skalách s ledopády v Labském dole [Biodiversity of vascular plants, lichens, cyanophytes and algae on rocks with icefalls in the Labský důl valley]. Opera Corcontica 48:45–68

Hilitzer A (1924) Addenda ad lichenographiam Bohemiae. Acta Botanica Bohemica 3:3–15

Hilitzer A (1925) Étude sur la végétation épiphyte de la Bohême. Spisy vydávané přírodovědeckou fakultou Karlovy university 41:1–202

Hofmeister J, Hošek J, Malíček J, Palice Z, Syrovátková L, Steinová J, Černajová I (2016) Large beech (*Fagus sylvatica*) trees as 'lifeboats' for lichen diversity in central European forests. Biodiversity and Conservation 25:1073–1090

Holien H, Palice Z, Björk CR, Goward T, Spribille T (2016) *Lecidea coriacea* sp. nov., a lichen species from oldgrowth boreal and montane forests in Europe and North America. Herzogia 29:412–420

Koerber [Körber] GW (1855) Systema Lichenum Germaniae. Trewendt & Granier, Breslau

Körber GG [recte GW] (1853) Sertum Sudeticum, continens novas Lichenum species. In: Denkschrift der Schlesischen Gesellschaft für vaterländische Cultur. Josef Max & Komp., Breslau, pp 231–238

Kučera J, Váňa J, Hradílek Z (2012) Bryophyte flora of the Czech Republic: updated checklist and Red List and a brief analysis. Preslia 84:813–850

Kuťák V (1952) Lišejníky v Krkonoších [Lichens in the Krkonoše Mts]. Časopis Národního musea, Oddíl přírodovědný 121:106–116

Liška J (1994) Bioindikace znečištění ovzduší v České republice pomocí lišejníků [The use of epiphytic lichens for biondication of air pollution in the Czech Republic]. Příroda 1:7–21

Liška J (2016) Czech database of lichen type material. Herzogia 29:814–818

Liška J, Palice Z (2010) Červený seznam lišejníků České republiky (verze 1.1) [Red List of lichens of the Czech Republic (version 1.1)]. Příroda 29:3–66

Liška J, Pišút I (1990) Verbreitung der Flechte *Lobaria pulmonaria* (L.) Hoffm. in der Tschechoslowakei. Biologia 45:23–30

Liška J, Dětinský R, Palice Z (1996) Importance of the Šumava Mts. for the biodiversity of lichens in the Czech Republic. Silva Gabreta 1:71–81

Løfall BP, Timdal E (2002) *Cladonia krogiana*, a new xanthone-containing species from Norway. Lichenologist 34:277–281

Malíček J (2014) A revision of the epiphytic species of the *Lecanora subfusca* group (*Lecanoraceae*, *Ascomycota*) in the Czech Republic. Lichenologist 46:489–513

Malíček J, Palice Z (2013) Lichens of the virgin forest reserve Žofínský prales (Czech Republic) and surrounding woodlands. Herzogia 26:253–292

Malíček J, Palice Z, Vsondrák J (2014) New lichen records and rediscoveries from the Czech Republic and Slovakia. Herzogia 27:257–284

Malíček J, Peksa O, Steinová J (2015) Lišejníky sutí v jižních Brdech [Lichens of boulder screes in the southern Brdy Hills]. Bryonora 56:24–44

Moldanová J, Grennfelt P, Jonsson Å, Simpson D, Spranger T, Aas W, Munthe J, Rabl A (2011) Nitrogen as a threat to European air quality. In: Sutton MA, Howard C, Erisman JW, Billen G, Bleeker A, Grennfelt P, van Grinsven H, Grizzetti B (eds) The European nitrogen assessment. Cambridge University Press, Cambridge, pp 405–433

Nordin A (2013) On *Aspicilia serpentinicola* and some other *Aspicilia* names. Graphis Scripta 25:18–20

Palice Z (1999) New and noteworthy records of lichens in the Czech Republic. Preslia 71:289–336

Palice Z, Slavíková-Bayerová Š, Peksa O, Svoboda D, Kučerová L (2007) The lichen flora of the Bohemian Switzerland National Park (Czech Republic). In: Härtel H, Cílek V, Herben T, Jackson A, Williams RB (eds) Sandstone landscapes. Academia, Praha, pp 200–204

Peksa O (ed) (2008a) Zajímavé lichenologické nálezy [Interesting records of lichens] III. Bryonora 41:21–24

Peksa O (ed) (2008b) Zajímavé lichenologické nálezy [Interesting records of lichens] IV. (*Parmeliaceae*). Bryonora 42:30–37

Peksa O (2009) Lichens – species composition and diversity. In: Neustupa J, Škaloud P, Peksa O, Kubátová A, Soldán Z, Černá K, Prášil K, Bukovská P, Vojta J, ... Škaloudová M, The biological soil crusts in Central European ecosystems, with special reference to taxonomic structure and ecology of the surface crusts at Czech ore-waste and ash-slag sedimentation industrial basins. Novitates Botanicae Universitatis Carolinae 19(2008): 38–40

Printzen C, Palice Z (1999) The distribution, ecology and conservational status of the lichen genus *Biatora* in central Europe. Lichenologist 31:319–335

Printzen C, Spribille T, Tønsberg T (2008) *Myochroidea*, a new genus of corticolous, crustose lichens to accommodate the *Lecidea leprosula* group. Lichenologist 40:195–207

Printzen C, Halda JP, McCarthy JW, Palice Z, Rodriguez-Flakus P, Thor G, Tønsberg T, Vondrák J (2016) Five new species of *Biatora* from four continents. Herzogia 29:566–585

Reese Næsborg R (2008) Taxonomic revision of the *Lecania cyrtella* group based on molecular and morphological evidence. Mycologia 100:397–416

Šoun J, Vondrák J, Bouda F (2015) Vzácné a málo známé druhy lišejníků Třebíčska a okolí [Rare and little known species in the Třebíč Region and its surroundings]. Bryonora 56:1–23

Steinová J, Bouda F, Halda JP, Kukwa M, Malíček J, Müller A, Palice Z, Peksa O, Schiefelbein U, … Vondrák J (2013) Lichens recorded during the 16th meeting of the bryological and lichenological section CBS in the Slavkovský les mountains, April 2009. Bryonora 51:1–14

Suza J (1933) Der Peterstein in den Ostsudeten im Lichte der lichenologischen Durchforschung. Časopis Moravského zemského musea v Brně 28–29:507–532

Suza J (1936) *Solorinella asteriscus* Anzi in der Flechtenflora der Lößsteppe Mitteleuropas. Věstník Královské české společnosti nauk. Třída mathematicko-přírodovědecká 1935/1:1–35

Suza J (1937) Pozoruhodné lišejníky československé xerothermní oblasti [Noteworthy lichens of Czechoslovak xerothermic area]. I. Časopis Národního musea 111:136–156

Suza J (1943) Meridionální vlivy v lišejníkové flóře Západních Karpat [Meridional influences in the lichen flora of western Carpathians]. Věstník Královské české společnosti nauk, Třída mathematicko-přírodovědecká 1942/16:1–47

Suza J (1944) K lichenologickému svérázu Žďárských hor [On the lichenological peculiarity of the Žďárské hory Mts]. Práce Moravské přírodovědecké společnosti 16:1–15

Suza J (1946) K lichenologickému rázu středoevropských vřesovin, především xerothermních obvodů [On the lichenological nature of central European heathlands, especially in xerothermic areas]. Věstník Královské české společnosti nauk, Třída mathematicko přírodovědecká 1944/18:1–35

Svoboda D (2009) Zajímavé lichenologické nálezy [Interesting records of lichens] V. Bryonora 44:39–40

Svoboda D, Peksa O, Veselá J (2011) Analysis of the species composition of epiphytic lichens in central European oak forests. Preslia 83:129–144

Svoboda D, Halda JP, Malíček J, Palice Z, Šoun J, Vondrák J (2014) Lišejníky Českého krasu: shrnutí výzkumů a soupis druhů [Lichens of the Bohemian Karst (Central Bohemia, Czech Republic): a summary of previous surveys and a checklist of the area]. Bohemia Centralis 32:213–265

Vězda A (1970) Neue oder wenig bekannte Flechten in der Tschechoslowakei. I. Folia Geobotanica et Phytotaxonomica 5:307–337

Vězda A (1972) *Haplocarpon nadvornikianum* Vězda sp. n. Preslia 44:208–212

Vězda A, Liška J (1999) Katalog lišejníků České republiky [A catalogue of lichens of the Czech Republic]. Institute of Botany, Academy of Sciences of the Czech Republic, Průhonice

Vondrák J, Hrouzek P (2006) *Caloplaca soralifera*, a new species from Europe. Graphis Scripta 17:6–15

Vondrák J, Malíček J (2015) *Teloschistaceae* Velké kotliny a Petrových kamenů v Hrubém Jeseníku [*Teloschistaceae* of the localities Velká kotlina and Petrovy kameny in the Hrubý Jeseník Mts]. Bryonora 56:45–55

Vondrák J, Kocourková J, Palice Z, Liška J (2006) New and noteworthy lichens in the Czech Republic – genus *Rinodina*. In: Lackovičová A, Guttová A, Lisická E, Lizoň P (eds) Central European lichens – diversity and threat. Mycotaxon Ltd., Ithaca, pp 109–117

Vondrák J, Kocourková J, Palice Z, Liška J (2007) New and noteworthy lichens in the Czech Republic – genus *Caloplaca*. Preslia 39:163–184

Vondrák J, Říha P, Arup U, Søchting U (2009a) The taxonomy of the *Caloplaca citrina* group (*Teloschistaceae*) in the Black Sea region; with contributions to the cryptic species concept in lichenology. Lichenologist 41:571–604

Vondrák J, Šoun J, Arup U, Aptroot A, Redchenko O (2009b) *Caloplaca ulcerosa*, a maritime species in Europe with a remarkable occurrence in the Czech Republic. Bryonora 44:1–7

Vondrák J, Frolov I, Říha P, Hrouzek P, Palice Z, Nadyeina O, Halıcı G, Khodosovtsev A, Roux C (2013) New crustose *Teloschistaceae* in Central Europe. Lichenologist 45:701–722

Vondrák J, Malíček J, Palice Z, Coppins B, Kukwa M, Czarnota P, Sanderson N, Acton A (2016) Methods for obtaining more complete species lists in surveys of lichen biodiversity. Nordic Journal of Botany 34:619–626

Westberg M, Timdal E, Asplund J, Bendiksby M, Haugan R, Jonsson F, Larsson P, Odelvik G, Wedin M, Millanes AM (2015) New records of lichenized and lichenicolous fungi in Scandinavia. MycoKeys 11:33–61

Wirth V, Hauck M, Schultz M (eds) (2013) Die Flechten Deutschlands. Band 1 & 2. Ulmer, Stuttgart

# Chapter 6
# History of Czech Vegetation Since the Late Pleistocene

**Petr Kuneš and Vojtěch Abraham**

**Abstract** A long-term perspective is a crucial dimension for understanding the present-day composition and structure of the Czech flora and vegetation. We outline processes that were important for the development of the present-day diversity of flora and vegetation including extinctions of taxa and ecological mechanisms operating within glacial-interglacial cycles. Further, we present the history of vegetation during the key stages in the glacial and postglacial periods. First, we outline the pattern in the vegetation during the last glacial, including a discussion of the existence of refugia for trees. We further describe the changes in vegetation during the Late Glacial, which were mostly the results of abrupt climatic events. We also present a new synthesis of the Holocene regional development in vegetation based on a Landscape Reconstruction Algorithm, which results in different regional vegetation trajectories and three main phases in the development of vegetation. Finally, we give some examples of the histories of local vegetation at several sites mainly based on plant macrofossils.

## 6.1 Introduction

Palaeoecological research has a long tradition in the Czech Republic. It has produced an outstanding quantity of records of Late-Quaternary flora and changes in vegetation (Kuneš et al. 2009), which is still growing. Data collected mostly as sedimentary pollen, plant macrofossils or woody charcoal allow us to understand mainly the dynamics of vegetation in this area during the postglacial period. The Czech Republic is situated in the temperate zone where the present-day diversity and structure of vegetation is a legacy of the glacial-interglacial cycles during the Quaternary. However, to understand the effects of periods before the Late Quaternary we have to look outside the borders of this country because of a general lack of primary data for the Czech territory.

P. Kuneš (✉) • V. Abraham
Department of Botany, Faculty of Science, Charles University,
Benátská 2, 128 01 Praha 2, Czech Republic
e-mail: petr.kunes@natur.cuni.cz; vojtech.abraham@gmail.com



M. Chytrý et al. (eds.), *Flora and Vegetation of the Czech Republic*, Plant and Vegetation 14, DOI 10.1007/978-3-319-63181-3_6

## 6.2 Quaternary Impoverishment of Temperate Flora

The Quaternary climatic cycles had a profound effect on the diversity of the temperate flora in Europe. Extinctions of numerous temperate woody taxa were caused by the geographical constraints to migration (Huntley 1993) and ecological limitations of these taxa in cold and dry environments, coupled with their rather low evolutionary adaptability (Svenning 2003). This resulted in the present-day impoverishment mainly in the number of genera in Central Europe compared to similar ecosystems in North America or East Asia. At the same time, survival of temperate trees in geographically restricted and isolated refugia led to evolutionary differentiation of vicariant species (Huntley 1993). Impoverishment of the flora is recorded in many long records spanning a major part of the Quaternary (van der Hammen et al. 1971; Tzedakis et al. 2006) and even in more fragmented records identifying significant boundaries in ecosystem development since the Tertiary (Zagwijn 1960). The most important genera widely occurring in the European Pliocene that became extinct include *Carya*, *Castanea*, *Cedrus*, *Cryptomeria*, *Eucommia*, *Liquidambar*, *Liriodendron*, *Magnolia*, *Morus*, *Nyssa*, *Ostrya*, *Pterocarya*, *Sciadopitys*, *Sequoia*, *Thuja*, *Tsuga* and *Zelkova* (for more complete list see electronic appendix in Svenning 2003). Some of these taxa, e.g. *Carya*, *Castanea*, *Celtis*, *Eucommia* and *Tsuga*, were still important in Early to Middle Pleistocene interglacials (Lang 1994). Some taxa even reappeared or regained dominance after regional decline or extinction in the Middle to Late Pleistocene, such as *Buxus*, *Fagus*, *Hedera*, *Ilex* and *Pterocarya*. These taxa were recorded occurring in the Czech territory in the Holsteinian interglacial (Břízová 1994). Their occurrences are indicative of warm periods in the Quaternary.

## 6.3 Past Interglacials

Human-induced changes in the vegetation cannot be quantified without understanding long-term vegetation dynamics under natural conditions. The current warm period (interglacial), the Holocene (the last ~11,700 years[1]), is characterized by the continuous effect of humans on ecosystems for many thousands of years resulting in quite distinct climate dynamics (Ruddiman et al. 2011). Prior to this period there were at least ten interglacial periods during the last 800 ka (Past Interglacials Working Group of PAGES 2016; Fig. 6.1), when the glacial/interglacial cycles switched with a periodicity of ~100 ka. During this period long extensive glaciations occurred together with a clear differentiation and alternation between glacial and interglacial floras (Birks and Willis 2008). Vegetation during the interglacials evolved under predominantly natural forces, but exceptions show local patterns in the hominin exploitation of forests (Pop and Bakels 2015).

Iversen's glacial/interglacial cycle (Iversen 1960) describes the main patterns in the changes in ecosystem under alternating long periods of cold climate and short

[1] Henceforth we indicate age as 'ka', which means in thousands of calendar years before the year AD 2000 (B2K).

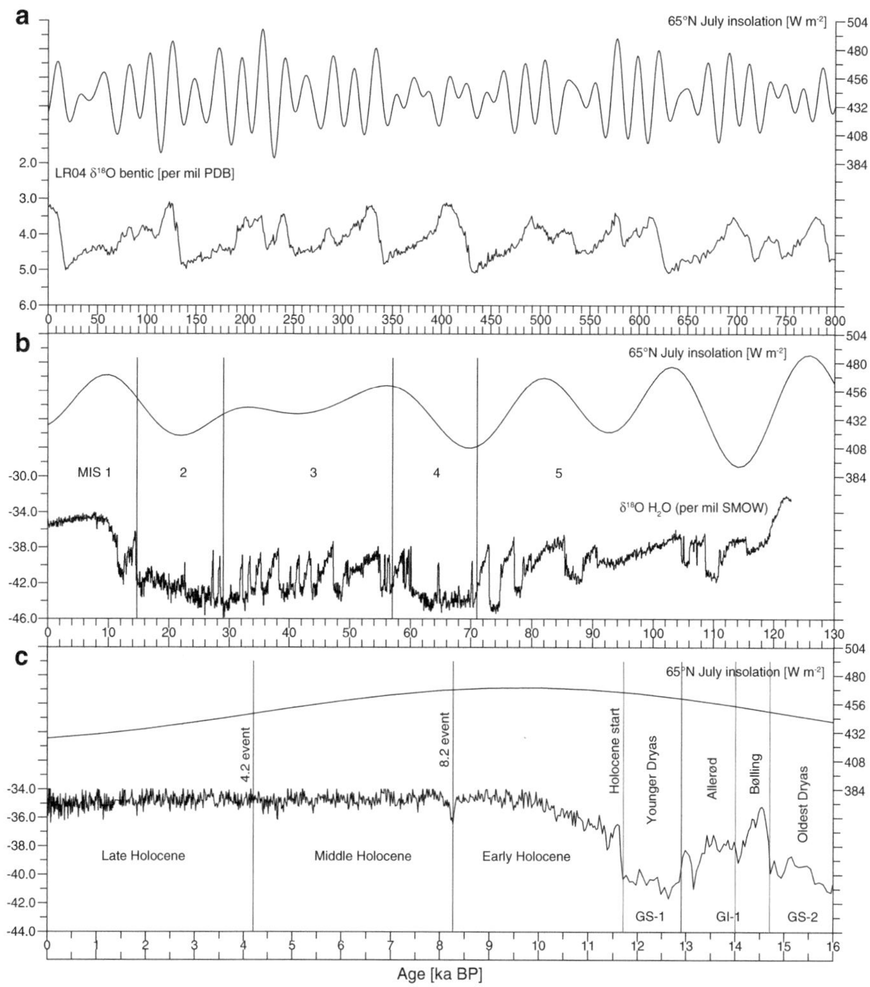


**Fig. 6.1** Astronomic parameters and climatic records during the Middle and Late Quaternary. The *upper smoothed line* in each part (**a**–**c**) shows July insolation calculated for 65°N (Berger and Loutre 1991) and in (**a**) the *lower curve* shows the proxy for temperature represented by the $\delta^{18}O$ record from the LR04 stack of 57 globally distributed sites (Lisiecki and Raymo 2005), in (**b**) the lower curve represents $\delta^{18}O$ record from the Greenland ice core NGRIP (Andersen et al. 2004) for the last climatic cycle, and in (**c**) the lower curve represents $\delta^{18}O$ record from the Greenland ice core GISP2 (Stuiver et al. 1995) for the Late Glacial and Holocene. Particular periods and events discussed in the text are indicated for the postglacial period; *orange vertical lines* indicate warming, *blue lines* cooling

warm stages such as the development of fertile soils under interglacial climates. When the interglacial periods are looked at in more detail, slow changes in ecosystems can be observed that are more due to soil nutrient dynamics than climatic forcing. This interglacial model for changes in ecosystems and nutrients predicts an initial rise in plant biomass along with high nutrient availability, followed by a slow biomass and nutrient decline. These changes are accompanied by compositional changes in vegetation with increasing abundance of oligocratic species (Birks 1986). Four phases in the development of vegetation can be identified within the interglacial cycle: protocratic (pioneer phase with rising temperatures, light demanding vegetation and unleached soils), mesocratic (high temperatures, closed forest and fertile soils), oligocratic (decreasing forest cover and deteriorating soils) and telocratic (decreasing temperatures, open vegetation and infertile soils; Andersen 1994). Present-day undisturbed soils of a very old age show similarly low levels of accessible phosphorus as late-interglacial ecosystems (Wardle et al. 2004). Although the vegetation successions during each interglacial follow different trajectories, they exhibit similar patterns, which can be explained by long-term development of the soil characterized by initially high ecosystem productivity followed by a slow and long decline (Kuneš et al. 2011).

Studies of the dynamics of past interglacial vegetation can enhance our understanding of a long discussed question of vegetation openness during the Holocene (Svenning 2002; Birks 2005; Mitchell 2005), in providing two contradictory hypotheses of 'high forest' (Iversen 1960; Bradshaw et al. 2003) and 'wood pasture' (Vera 2000). Quantitative reconstruction of the vegetation for the interglacial periods clearly shows that the landscape in the oceanic part of Northern Europe became almost completely forested during the mesocratic stages (Kuneš et al. 2011, 2013). These findings contradict Vera's grazing theory, despite the fact that there were sufficient herbivores present during past interglacials (Bradshaw et al. 2003). A recent study of wood-pasture theory based on fossil dung-beetle assemblages indicates that vegetation in past interglacials (particularly during the Eemian) was much more open and diverse due to a significant presence of large herbivores (Sandom et al. 2014). This discrepancy between palaeovegetation data and beetle remains could be explained by Sandom's et al. (2014) interpretation of the 'too long' last interglacial (132–110 ka), which includes the early stage of the last glacial with climatically-driven landscape openness. Yet future systematic multi-proxy research of past interglacials should test the hypothesis that the past interglacials allowed the development of dense forests in temperate Europe and large deforestation happened only during the Holocene as a result of human activity rather than herbivory.

## 6.4 Last Glacial

Vegetation underwent significant changes during the last glacial cycle, responding to mildly cooling conditions during the early phase and more extreme cold conditions in the later phase. At the same time it experienced frequent, millennial-scale changes in climate (Helmens 2014; Fig. 6.1b). The Last Glacial Maximum (LGM; 21 ka) or the Marine Isotope Stage (MIS) 2 (29–14 ka) is considered to have been a bottleneck for present-day temperate taxa in Europe. Many went extinct over vast areas during the cold, dry and unstable climatic conditions in this glaciation and survived only in limited refugia (Birks and Willis 2008). A heated ongoing debate on the possibility of northerly refugia of many temperate taxa brings a lot of confusion (Tzedakis et al. 2013), which can only be resolved based on robust palaeoecological evidence, which is still missing (de Lafontaine et al. 2014).

Last-glacial palaeobotanical data in the Czech Republic come mostly from MIS 3 (57–29 ka; Lisiecki and Raymo 2005; Fig. 6.1b), a period for which climatic simulations indicated presence of relatively warm intervals (Barron and Pollard 2002). In addition, a high-resolution analysis of charcoal tree rings from southern Moravia indicates rapid changes in climate at that time (Beresford-Jones et al. 2011). Charcoal data collected mostly at sites of human habitation suggest presence of boreal trees (*Larix*, *Picea*, *Pinus cembra*) and locally several temperate trees in Moravia in the period 40–25 ka, such as *Abies*, *Corylus*, *Fagus*, *Fraxinus*, *Quercus*, *Taxus* or *Ulmus* (Willis and van Andel 2004). Presence of these taxa in charcoal cannot be interpreted in terms of the quantitative extent of woodland cover. However, their presence in vegetation is also supported by pollen records at Jablůnka and Týn nad Bečvou in Carpathian valleys and foothills in eastern Moravia (Jankovská 2008; Jankovská and Pokorný 2008) and Bulhary in a lowland landscape in southern Moravia (Rybníčková and Rybníček 2014). These pollen data, which also span the period of approximately 40–25 ka BP, show repeated occurrence of pollen of mesophilous trees, indicating that this area could have been a microrefugium for these trees. Survival of forest vegetation is indicated by the earlier classical interpretation of Frenzel (1968) and vegetation models (Huntley et al. 2003).

A growing demand for a more accurate interpretation of pollen data resulted in comparative studies of hypothesized modern analogues of the last glacial vegetation in southern Siberia (Pelánková and Chytrý 2009; Chytrý et al. 2010; Horsák et al. 2010; Fig. 6.2). A quantitative comparison of modern pollen spectra from Siberia with fossil pollen spectra from the Czech Republic and Slovakia supports the hypothesis of the occurrence of forest stands during MIS 3 in the Western Carpathians; these stands were mainly interpreted as taiga and hemiboreal forest (Kuneš et al. 2008a). In many places the woodlands could have been dominated by *Larix*, not only in the Carpathians, for example, but also in central Bohemia (Jankovská and Pokorný 2008). Larch can thrive under very severe climatic conditions (Chytrý et al. 2008; Fig. 6.2a) while producing no or only very little pollen. This makes it difficult to record this species in pollen spectra and the interpretation of these spectra is then biased towards open land. Fossil records, however, did not confirm the LGM survival of forest tree taxa in the territory of the current Czech Republic, because they either pre-date or post-date this event. The closest records of LGM vegetation originate from the Pannonian Plain and Eastern Carpathians, with continuous curves of mesophilous tree pollen recorded only in the Carpathians (Magyari et al. 2014). Modern pollen analogues of the three Hungarian and Romanian LGM sites indicate a highly diverse pattern of vegetation dominated by various types of grassland with possible occurrence of limited patches of forest dominated by coniferous and deciduous boreal trees (*Larix*, *Pinus sylvestris*, *P. cembra*, *Picea*, *Juniperus*, *Alnus*). Moreover, LGM snail assemblages also indirectly support the hypothesis of the survival of mesophilous trees in the Carpathians (Juřičková et al. 2014a). A unique pollen record most probably running through the LGM situated near the alpine timberline in the Krkonoše Mts indicates presence of tree taxa including *Pinus*, *Betula*, *Picea* and *Alnus* (Engel et al. 2010). However, due to erosion during the glacial and high ratio of redeposited pollen the authors mostly infer treeless, sparse vegetation until the end of the Younger Dryas.

An early occurrence and spread of some mesophilous trees during the Late-Glacial phase (16–11.7 ka; Fig. 6.1c) indicate the proximity of their refugia

**Fig. 6.2** Examples of current vegetation analogous to vegetation of different periods during the late Pleistocene: (**a**) Dry and cold steppe with patches of *Larix* forest on north-facing slopes and sheltered sites in valleys in northern-central Mongolia and (**b**) the continental steppe-tundra in the southern part of the Russian Altai Mts, a mosaic of dry grassland on drier landforms and patches of dwarf birches of *Betula nana* group (*B. rotundifolia*) on topographically wetter sites, are potentially analogous to the vegetation that existed in Central Europe during full glacials; (**c**) Open woodland of *Pinus sylvestris* and *Betula pendula* with a species-rich herbaceous plant layer in the lowland forest-steppe of southern Siberia near Novosibirsk is a potentially analogous vegetation to the zonal forests in Central Europe in the Late Glacial and Early Holocene; (**d**) Closed-canopy broad-leaved deciduous forest with *Tilia cordata* and *Acer platanoides* in the foothills of the Southern Ural Mts in Russia is potentially analogous to the zonal forests in Central European lowlands in the Middle Holocene. Photo credits: V. Abraham (a) and M. Chytrý (b–d)

(Tzedakis et al. 2013). For example, existence of spruce refugia in the Western Carpathians is supported by its early spread documented by both pollen (Labe lowland in central Bohemia; Petr et al. 2014) and plant macrofossils (northern Bohemia; Nováková 2000; Pokorný 2003). Temperate deciduous trees (*Corylus*, *Quercus*, *Ulmus* and *Tilia*) are recorded by continuous pollen curves only 80 km from the Czech border in south-western Slovakia (Petr et al. 2013). The development of vegetation during the Late Glacial phase is recorded at many different sites, which indicates high diversity in landscape-scale vegetation (e.g. Svobodová 1997; Pokorný 2002; Jankovská 2006; Břízová 2009; Petr and Novák 2014). According to combined Greenland ice core and volcanic stratigraphies (Blockley et al. 2012) the Late Glacial period starts with a slow warming but still cold stadial phase (Oldest Dryas, GS-2a, 16–14.7 ka) followed by a warm spell indicating the start of an interstadial (often called Bølling-Allerød according to terrestrial stratigraphies;

14.7–12.9 ka). The interstadial was shortly interlaid by one cold event, the Older Dryas (GI-1d), and concludes by one longer cold oscillation called Younger Dryas (glacial stadial GS-1; Blockley et al. 2012). Temperatures during the initial Bølling warming rose by up ~7 °C per century (Lowe et al. 1994), which might be comparable with the rate of warming at the start of the Holocene.

According to the modern pollen analogues of late-glacial vegetation (Kuneš et al. 2008a) the landscape during the Bølling-Allerød interstadial was covered by a diverse assemblage of patchy vegetation. Mostly treeless shrubby tundra (dominated by *Betula nana*) with patches of dry grassland prevailed in the Šumava and Krkonoše Mts (Jankovská 2006; Engel et al. 2010; Fig. 6.2b), whereas lowland sites seem to have been at least partly covered by hemiboreal forest or taiga (*Betula*, *Larix*, *Pinus sylvestris* and *P. cembra*; Petr and Novák 2014; Fig. 6.2c). A unique high-resolution record of the late-glacial interstadial and stadial comes from the Švarcenberk site in the Třeboň Basin (Pokorný 2002; Hošek et al. 2014). This former lake probably originated from the thawing of permafrost, which started during the Bølling period. Landscape during the Allerød (14–12.9 ka) was gradually afforested by pine-dominated woodland but the proportion of steppe and tundra elements remained high. A short climatic oscillation within this interstadial is reflected in the increase in *Artemisia* and *Juniperus* and can be interpreted as the so-called Gerzensee oscillation (13.2–13 ka; Ammann 2000; van Raden et al. 2013) or GI-1b (Blockley et al. 2012). The Younger Dryas stadial (12.9–11.7 ka) started with rapid cooling accompanied by an increase in grasses, *Artemisia* and *Juniperus*, indicating an opening up of the landscape. Drier climate in the later part of this stadial is indicated by different proxies, which is in concordance with other records from Western and Northern Europe (Walker 1995). It seems that taxa present in this area responded quite quickly to rapid warming during the interstadial (expanding *Pinus*, aquatic plants) in contrast to their slow response with a time lag to rapid cooling (Ammann et al. 2000).

A different pattern in the vegetation from that in the rest of the country can be observed in the lowlands of Bohemia and Moravia. Although poorly dated, a general development can be observed at the Hrabanovská černava site in central Bohemia (Petr and Novák 2014) and Vracov site in southern Moravia (Rybníčková and Rybníček 1972; Svobodová 1997). The Bohemian site seems to reflect a part of the interstadial when the landscape was generally covered by pine woodlands, whereas the Younger Dryas was responsible for a decrease in trees accompanied first by an abrupt increase in wetland shrubs (*Salix*, *Alnus*) and then an increase of drier steppe elements in the later part (*Helianthemum*, *Artemisia*, *Thalictrum*). The Moravian site is more difficult to interpret as it probably covers, according to the dating, a longer period of the interstadial (Kuneš et al. 2015). During the Allerød there was an increase in *Betula* pollen, whereas during the Younger Dryas *Pinus* reached its maximum with only a small amount of herbaceous plant pollen. Landscape prior to these periods was generally more open indicated by higher proportion of herbaceous plant pollen, but it is important to have a more precise dating of these sediments in order to interpret these periods in finer detail. The close proximity of this site to possible glacial refugia of temperate trees is reflected in the immediate increase in *Quercus*, *Corylus*, *Alnus* and *Picea* with climate warming.

Typical late-glacial interstadial plants include the woody taxa *Pinus cembra* and *Larix decidua*, which are recorded at many sites. Shrubs of *Betula nana*, *Alnus*

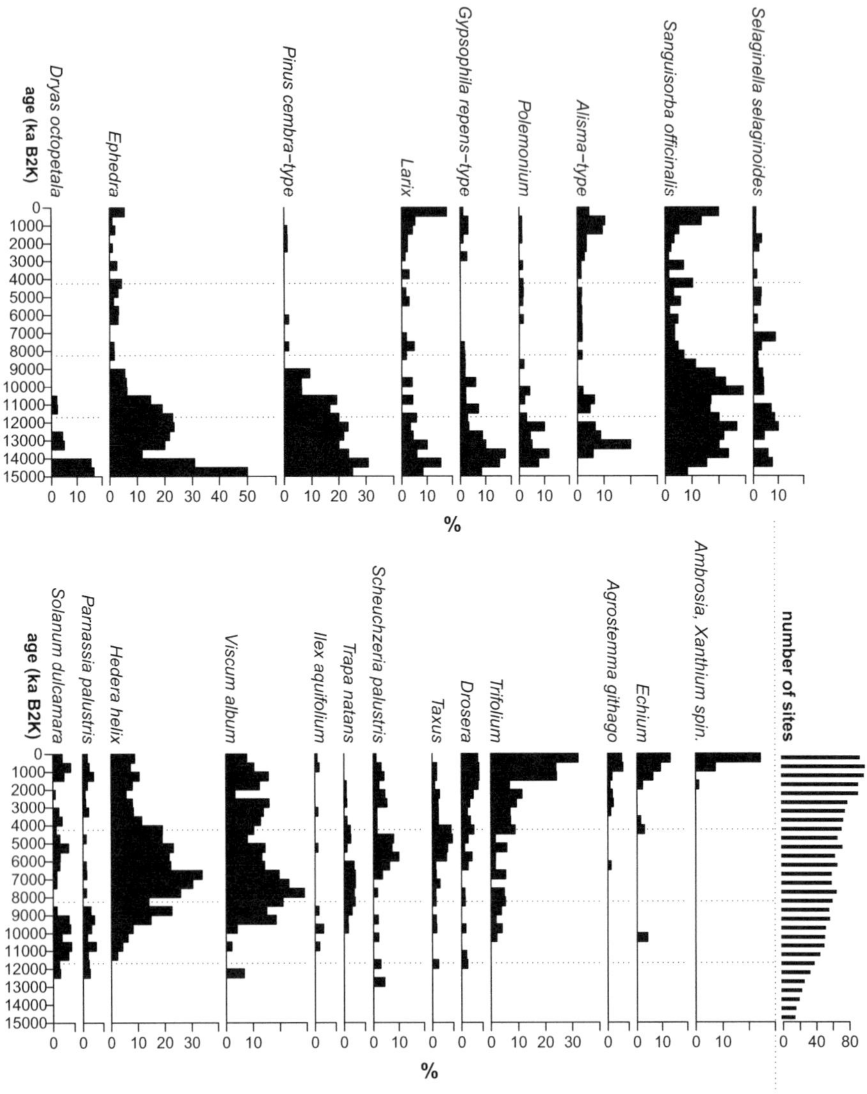


**Fig. 6.3** Summary stratigraphic diagram of occurrence of selected rare pollen taxa in the Czech Republic. *Horizontal axis* shows percentage of sites where the taxon is present, considering all available sites per time window

*alnobetula* and *Juniperus* were common during the stadials and in the mountains. *Ephedra*, occurring at warmer and drier sites, indicates steppic vegetation. There was a remarkable increase in indicators of cooling, such as *Amaranthaceae*, *Artemisia*, *Helianthemum* and *Thalictrum*. *Dryas octopetala* is also a significant indicator of cooling, but it is rarely found in Czech deposits (Fig. 6.3).

## 6.5 Holocene

From a present-day perspective the Younger Dryas/Holocene transition (~11.7 ka; Fig. 6.1c) was a major climatic event. Initial rapid warming was accompanied by vegetation change that resulted in long-term forest succession lasting until now. Despite minor climatic and environmental variations over the last 11 ka the Holocene is a geological period representing the present time, which is a warm and forested phase (interglacial) of the Quaternary. Therefore, we apply the uniformitarianism principle assuming the stability of several factors in order to reconstruct quantitatively the vegetation. The methods we use here have the advantage of accounting for some biases associated with pollen analysis. The representation of pollen taxa in an assemblage is influenced by differential pollen production, pollen dispersal, taphonomy and the spatial structure of the vegetation around the sedimentation site (Prentice 1985; Sugita 1994). Using vegetation estimates instead of pollen percentages allows to relate the results for a defined space. Its extent is given by the size of sedimentation basin. Recent studies provide past vegetation estimates for ten circular regions in the Czech Republic, each with a 60 km radius (Abraham et al. 2016, 2017; Szabó et al. 2017; Figs. 6.4 and 6.5) calculated using the REVEALS model (Sugita 2007a). The taxa that are dominant in the vegetation are determined by availability of species-specific estimates of pollen productivity validated using information on present-day vegetation (Abraham et al. 2014). Rare taxa also constitute an important component for studying the vegetation pattern in the Holocene, therefore we extract additional sequences available in the Czech Quaternary Palynological Database (PALYCZ; Kuneš et al. 2009; Table 6.1 and Fig. 6.5) and transform the pollen abundances of important indicative taxa to a presence/absence scale in 500-year time windows.

Stratigraphy of the Holocene before the discovery of radiocarbon dating in the 1940s was based on the sedimentology and degree of peat decomposition, presence of tree trunks and diagnostic macrofossils (Birks and Seppä 2010). Changes from dark peat with trunks to sedge or *Sphagnum* peat was interpreted as a change from dry to moist climate. The two alternations recorded split the Holocene into four (or five) periods: (Preboreal), Boreal, Atlanticum, Subboreal and Subatlanticum. This Blytt-Sernander system was developed in Scandinavia (Blytt 1882; Sernander 1908) and adopted in many parts of the world. The pioneer palynological synthesis for Central Europe (Firbas 1949) divided the history of the Holocene using biostratigraphic pollen zones linked to this periodization: pine, hazel, mixed-oak-woodland, beech, hornbeam or a similar succession according to the region. Another periodization of the Holocene relevant for the Czech Republic is based on palaeomalacological data and accumulations of calcareous tufa (Ložek 1964).

The long-term succession described by the REVEALS model provides a more precise insight in terms of the abundances of taxa in the vegetation but with a coarser temporal resolution based on 500-year intervals. Six groups of vegetation are distinguished by hierarchical clustering of vegetation estimates in 10 regions in the Czech Republic. Figure 6.6A shows that the first group of clusters on the left side (a–c) is

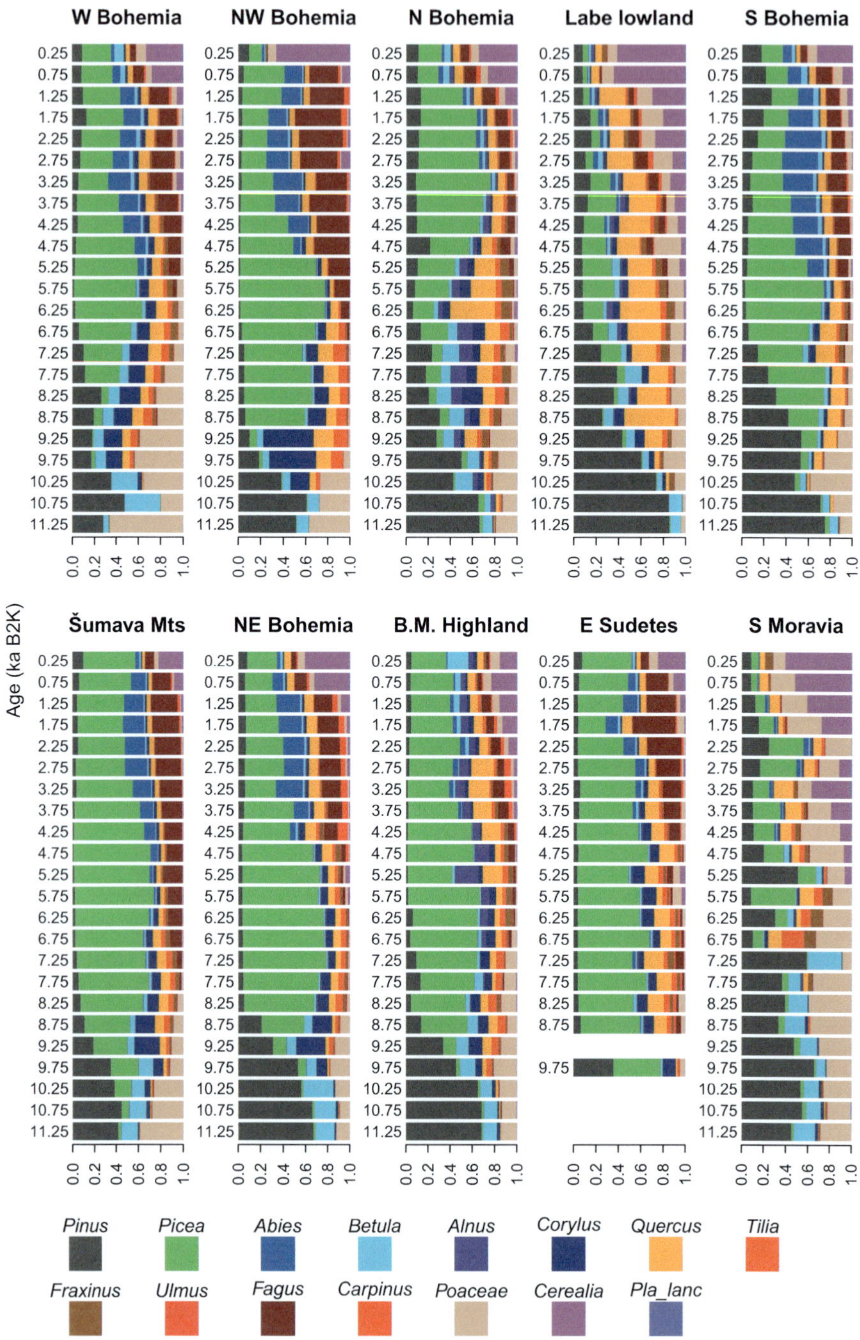


**Fig. 6.4** Regional vegetation estimates (REVEALS): proportions of 15 taxa for ten regions of the Czech Republic and 23 time windows of 500 years displayed on the scale before the year AD 2000. The order of the taxa in the graph follows the order in the legend. Pla_lanc = *Plantago lanceolata*

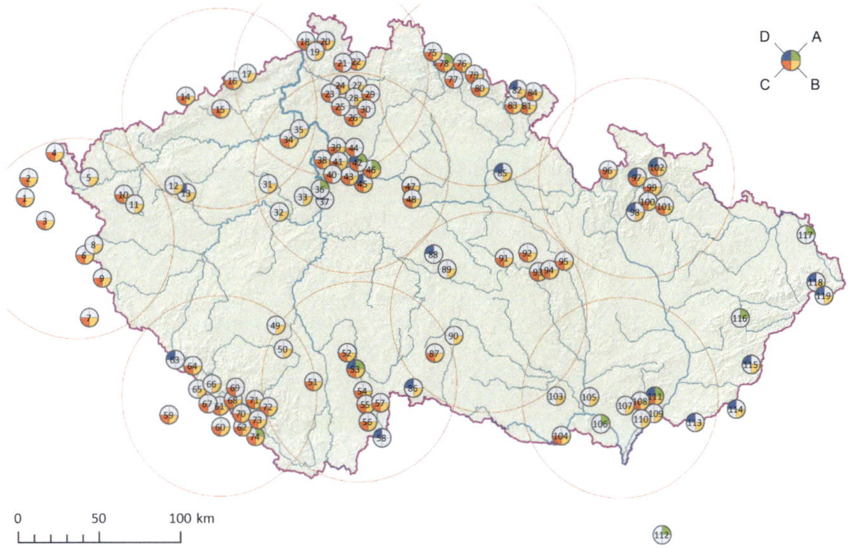


**Fig. 6.5** Map of the Czech Republic showing position of all sites used in the palaeoecological reconstructions. (**A**) Pleistocene record, (**B**) pollen sequences used for extraction of indicative taxa (Fig. 6.3), (**C**) pollen sequences used for REVEALS reconstruction (Fig. 6.4 and 6.6), (**D**) record of macrofossils. Numbers refer to sites listed in Table 6.1 (the map was prepared by O. Hájek)

recorded in the Early Holocene in all regions and in the Middle Holocene in lowlands. The second group of clusters on the right side (d–f) is recorded in the Middle Holocene in the highlands and in the Late Holocene in all regions. The first group comprises (a) semi-open pine woodlands, (b) mixed oak woodlands and (c) hazel woodlands. The second group includes (d) vegetation subject to human activity, (e) fir-beech forests and (f) spruce forests. The boundaries between groups in different regions tend to split the whole dataset into three general phases: Early (11.7–8.5 ka), Middle (8.5–4.5 ka) and Late Holocene (4.5–0 ka; Fig. 6.6B). Interestingly, these REVEALS-based vegetation phases fit the recently proposed subdivision of the Holocene into three phases, which depends on two important breakpoints detectable in the whole Northern Hemisphere. The Early/Middle Holocene boundary is the 8.2 ka event, a short period of ~160 years with a drier and colder climate (Alley et al. 1997). The Middle/Late Holocene boundary is the 4.2 ka event, a period of ~375 years of widespread period of aridification followed by a shift to cooler and wetter conditions (Walker et al. 2012). Both climatic events were connected with the releases of freshwater into the North Atlantic by ice rafting, known as Bond event 5 and 3, however forcing mechanisms behind the 4.2 ka event are less obvious than for the 8.2 ka event (Bond et al. 1997). Here we adopt this three-phase subdivision of the Holocene.

**Table 6.1** List of sites shown in the map (Fig. 6.5). co – number of pollen sequences from the site; number of dated pollen samples in the PALYCZ database in four periods: PG Pleniglacial, LG Lateglacial, EH Early Holocene, MH Middle Holocene, LH Late Holocene (number of samples is indicated for each period along with colouring – red = high, green = low, rows with blank cells refer to macrofossil sites or older periods discussed in the text); reference – the latest or the most relevant reference to the site. All unpublished sequences reported here ("unpubl") were used only for the REVEALS estimates, see details in Abraham et al. (2016)

| | Site name | co | LH | MH | EH | LG | PG | Reference |
|---|---|---|---|---|---|---|---|---|
| 1 | Seelohe | 1 | 30 | 21 | 0 | 0 | 0 | Hahne (1992) |
| 2 | Weissenstadter Forst | 1 | 13 | 21 | 0 | 0 | 0 | Hahne (1992) |
| 3 | Wolfslohe | 1 | 9 | 5 | 0 | 0 | 0 | Hahne (1992) |
| 4 | Sauborst | 1 | 23 | 13 | 1 | 0 | 0 | Hahne (1992) |
| 5 | SOOS | 1 | 0 | 15 | 23 | 2 | 0 | Suda (2012) |
| 6 | Weiherlohe | 1 | 4 | 6 | 18 | 0 | 0 | Knipping (1997) |
| 7 | Kulzer Moos | 2 | 47 | 74 | 17 | 18 | 10 | Knipping (1997) |
| 8 | Brentenlohe | 1 | 14 | 5 | 4 | 3 | 6 | Knipping (1997) |
| 9 | Windbruch | 1 | 33 | 32 | 0 | 0 | 0 | Knipping (1997) |
| 10 | Vlček | 1 | 16 | 18 | 3 | 0 | 0 | Švarcová (2012) |
| 11 | Číhaná | 1 | 0 | 17 | 29 | 10 | 0 | Švarcová (2012) |
| 12 | Veselov | 1 | 63 | 2 | 8 | 2 | 0 | Kozáková et al. (2015) |
| 13 | Vladař | 1 | 87 | 0 | 0 | 0 | 0 | Pokorný et al. (2006) |
| 14 | Mothäuser Heide | 1 | 87 | 47 | 21 | 0 | 0 | Lange et al. (2005) |
| 15 | Komořanské jezero | 2 | 23 | 16 | 16 | 0 | 0 | Jankovská and Pokorný (2013) |
| 16 | Fláje - Kiefern | 1 | 12 | 11 | 19 | 1 | 0 | Jankovská et al. (2007) |
| 17 | Georgenfelder Hochmoor | 1 | 31 | 0 | 0 | 0 | 0 | Stebich and Litt (1997) |
| 18 | Jelení louže | 1 | 19 | 20 | 0 | 0 | 0 | Pokorný and Kuneš (2005) |
| 19 | Nad Dolským mlýnem | 1 | 18 | 7 | 4 | 0 | 0 | Abraham (2006) |
| 20 | Pryskyřičný důl | 1 | 52 | 0 | 0 | 0 | 0 | Abraham and Pokorný (2008) |
| 21 | Mařeničky | 1 | 9 | 2 | 6 | 20 | 0 | Peša and Kozáková (2012) |
| 22 | Rozmoklá Žába | 1 | 10 | 9 | 4 | 8 | 1 | Kozáková et al. (2015) |
| 23 | Milčanský rybník | 1 | 27 | 9 | 0 | 0 | 0 | L. Petr, unpubl |
| 24 | Česká Lípa | 1 | 11 | 3 | 0 | 0 | 0 | P. Kuneš, unpubl |
| 25 | Jestřebské blato | 1 | 27 | 7 | 7 | 4 | 0 | L. Petr, unpubl |
| 26 | Okna | 1 | 27 | 17 | 12 | 4 | 0 | Pokorný et al. (2017) |
| 27 | Ploučnice-bary | 1 | 18 | 0 | 5 | 1 | 0 | Štor et al. (2016) |
| 28 | Skřítkův hrnec | 1 | 15 | 13 | 0 | 0 | 0 | Novák et al. (2015) |
| 29 | Držník | 1 | 8 | 7 | 7 | 5 | 0 | H. Svitavská-Svobodová, unpubl |
| 30 | Břehynský rybník | 2 | 16 | 23 | 15 | 1 | 0 | Novák et al. (2012); H. Svitavská-Svobodová, unpubl |
| 31 | Rynholec | 1 | 14 | 19 | 21 | 6 | 0 | Pokorný (2005) |
| 32 | Hýskov | 1 | 13 | 0 | 0 | 0 | 0 | Žák et al. (2010) |
| 33 | Břve | 1 | 40 | 0 | 0 | 0 | 0 | Pokorný and van der Knaap (2011) |
| 34 | Zahájí | 2 | 78 | 42 | 25 | 0 | 0 | Pokorný et al. (2015) |
| 35 | Vrbka | 1 | 50 | 0 | 0 | 0 | 0 | Pokorný (2016) |
| 36 | Praha-Podbaba | 1 | 0 | 0 | 0 | 0 | 1 | Jankovská and Pokorný (2008) |
| 37 | Praha Nové Město | 1 | | | | | | Pokorná et al. (2014) |
| 38 | Tišice | 2 | 52 | 87 | 0 | 0 | 0 | Dreslerová and Pokorný (2004) |
| 39 | Mělnický úval | 1 | 0 | 8 | 8 | 10 | 4 | Petr and Novák (2014) |
| 40 | Kozly | 2 | 55 | 20 | 12 | 0 | 0 | Petr and Pokorný (2008); L. Petr, unpubl |
| 41 | Chrást | 2 | 27 | 26 | 44 | 1 | 0 | Břízová (1999); L. Petr, unpubl |
| 42 | Chrást u přejezdu | 1 | 0 | 10 | 8 | 42 | 0 | Petr et al. (2014) |
| 43 | Borek | 1 | 4 | 5 | 15 | 0 | 0 | P. Kuneš, unpubl |
| 44 | Košátky | 1 | 80 | 1 | 0 | 0 | 0 | R. Kozáková, unpubl |
| 45 | Stará Boleslav | 2 | 92 | 0 | 0 | 0 | 0 | Kozáková et al. (2014) |
| 46 | Hrabanovská černava | 1 | 12 | 12 | 8 | 22 | 17 | Petr and Novák (2014) |
| 47 | Libice nad Cidlinou | 1 | 17 | 0 | 0 | 0 | 0 | Kozáková et al. (2014) |
| 48 | Hradištko | 1 | 70 | 0 | 0 | 0 | 0 | Kozáková et al. (2014) |
| 49 | Kožlí | 1 | 14 | 5 | 1 | 2 | 0 | Pokorný and Kuneš (2009) |
| 50 | Řežabinec | 1 | 23 | 3 | 8 | 0 | 0 | Rybníčková and Rybníček (1985) |

(continued)

**Table 6.1** (continued)

| | | | | | | | | |
|---|---|---|---|---|---|---|---|---|
| 51 | Zbudovská blata | 2 | 20 | 14 | 12 | 5 | 0 | Rybníčková et al. (1975) |
| 52 | Borkovická blata | 2 | 23 | 17 | 16 | 38 | 14 | Jankovská (1980) |
| 53 | Švarcenberk | 2 | 4 | 55 | 34 | 102 | 0 | Hošek et al. (2014) |
| 54 | Mokré louky (South) | 1 | 9 | 35 | 38 | 0 | 0 | Jankovská (1987) |
| 55 | Branná | 1 | 40 | 36 | 0 | 0 | 0 | Jankovská (1980) |
| 56 | Červené blato | 1 | 9 | 11 | 11 | 19 | 8 | Jankovská (1980) |
| 57 | Barbora | 1 | 22 | 37 | 44 | 0 | 0 | Jankovská (1980) |
| 58 | Velanská cesta | 1 | | | | | | Jankovská (1970) |
| 59 | Dösingerried | 1 | 22 | 24 | 20 | 1 | 0 | Stalling (1987) |
| 60 | Sonndorf | 1 | 14 | 26 | 10 | 0 | 0 | Stalling (1987) |
| 61 | Finsterauer Filz | 1 | 6 | 16 | 19 | 0 | 0 | Stalling (1987) |
| 62 | Heidemühle (Beerenfilz) | 1 | 27 | 22 | 6 | 0 | 0 | Stalling (1987) |
| 63 | Černé jezero | 1 | | | | | | Vočadlová et al. (2015) |
| 64 | Hůrecká slať | 1 | 44 | 37 | 19 | 9 | 0 | Svobodová et al. (2002) |
| 65 | Rokytecká slať | 1 | 23 | 23 | 21 | 0 | 0 | Svobodová et al. (2002) |
| 66 | Rybárenská slať | 1 | 22 | 23 | 13 | 0 | 0 | Svobodová et al. (2002) |
| 67 | Březník | 1 | 21 | 21 | 1 | 0 | 0 | H. Svitavská-Svobodová, unpubl |
| 68 | Knížecí pláně - Buková slať | 1 | 13 | 11 | 14 | 2 | 0 | Svobodová et al. (2001) |
| 69 | Chalupská slať | 1 | 27 | 23 | 17 | 0 | 0 | H. Svitavská-Svobodová, unpubl |
| 70 | Strážená slať | 1 | 27 | 25 | 15 | 1 | 0 | Svobodová et al. (2001) |
| 71 | Malá niva | 1 | 34 | 19 | 0 | 0 | 0 | Svobodová et al. (2002) |
| 72 | Soumarské rašeliniště | 1 | 19 | 16 | 10 | 5 | 0 | Svobodová et al. (2001) |
| 73 | Mrtvý luh | 2 | 114 | 75 | 5 | 0 | 0 | Svobodová et al. (2001); H. Svitavská-Svobodová, unpubl |
| 74 | Plešné jezero | 1 | 22 | 4 | 7 | 18 | 16 | Jankovská (2006) |
| 75 | Hala Izerska | 1 | 26 | 25 | 4 | 0 | 0 | Skrzypek et al. (2009) |
| 76 | Pančavská louka | 1 | 130 | 0 | 0 | 0 | 0 | Speranza et al. (2000) |
| 77 | Labská louka | 1 | 45 | 4 | 0 | 0 | 0 | H. Svitavská-Svobodová, unpubl |
| 78 | Labský důl | 1 | 11 | 20 | 23 | 4 | 21 | Engel et al. (2010) |
| 79 | Úpské rašeliniště | 6 | 155 | 6 | 0 | 0 | 0 | Svobodová (2002); Speranza (2000) |
| 80 | Černá hora | 2 | 51 | 0 | 0 | 0 | 0 | Speranza et al. (2000) |
| 81 | Anenské údolí | 1 | 13 | 10 | 3 | 0 | 0 | Pokorný and Kuneš (2005) |
| 82 | Vlčí rokle | 1 | 11 | 15 | 11 | 10 | 0 | Kuneš and Jankovská (2000) |
| 83 | Teplické údolí | 1 | 30 | 23 | 1 | 0 | 0 | Kuneš and Jankovská (2000) |
| 84 | Vernéřovice | 1 | 11 | 6 | 15 | 6 | 0 | Peichlová (1979) |
| 85 | Na bahně | 1 | 25 | 0 | 0 | 0 | 0 | Pokorný et al. (2000) |
| 86 | Bláto | 1 | 17 | 16 | 15 | 12 | 0 | Rybníček and Rybníčková (1968) |
| 87 | Řásná | 2 | 43 | 37 | 50 | 5 | 0 | Szabó et al. (2017) |
| 88 | Chraňbož | 1 | | | | | | Jankovská (1971) |
| 89 | Palašiny | 1 | 8 | 8 | 5 | 0 | 0 | Jankovská (1990) |
| 90 | Loučky | 1 | 14 | 13 | 12 | 3 | 0 | Rybníčková (1974) |
| 91 | Kameničky | 1 | 7 | 5 | 9 | 8 | 0 | Rybníčková and Rybníček (1988) |
| 92 | Lubenský les | 1 | 9 | 13 | 8 | 8 | 0 | Szabó et al. (2017) |
| 93 | Jedlová (B-M Highland) | 1 | 4 | 2 | 13 | 9 | 0 | Szabó et al. (2017) |
| 94 | Šestidomí | 1 | 0 | 0 | 21 | 8 | 0 | Szabó et al. (2017) |
| 95 | Hradec nad Svitavou | 1 | 16 | 3 | 2 | 3 | 0 | Szabó et al. (2017) |
| 96 | Kralický Sněžník | 1 | 35 | 0 | 0 | 0 | 0 | Novák et al. (2010) |
| 97 | Vozka | 1 | 30 | 4 | 0 | 0 | 0 | Dudová et al. (2013) |
| 98 | Skřítek | 1 | 8 | 21 | 25 | 0 | 0 | Dudová et al. (2014) |
| 99 | Velký Děd | 1 | 12 | 3 | 0 | 0 | 0 | Rybníček and Rybníčková (2004) |
| 100 | Velký Máj | 1 | 15 | 0 | 0 | 0 | 0 | Rybníček and Rybníčková (2004) |

(continued)

**Table 6.1** (continued)

| | | | | | | | |
|---|---|---|---|---|---|---|---|
| 101 Mezikotlí | 1 | 27 | 0 | 0 | 0 | 0 | Treml et al. (2008) |
| 102 Rejvíz | 1 | 36 | 19 | 7 | 0 | 0 | Dudová et al. (2010) |
| 103 Olbramovice | 1 | 25 | 1 | 0 | 0 | 0 | Svobodová (1997) |
| 104 Dvůr Anšov | 1 | 18 | 17 | 22 | 0 | 0 | Svobodová (1997) |
| 105 Velké Němčice | 1 | 8 | 0 | 0 | 0 | 0 | Svobodová (1990) |
| 106 Bulhary | 1 | 0 | 0 | 0 | 0 | 70 | Rybníčková and Rybníček (2014) |
| 107 Čejčské jezero | 1 | 15 | 14 | 11 | 0 | 0 | Břízová (2009) |
| 108 Svatobořice-Mistřín | 1 | 33 | 12 | 3 | 0 | 0 | Svobodová (1989) |
| 109 Dúbrava MS4 | 1 | 38 | 3 | 0 | 0 | 0 | Jamrichová et al. (2013) |
| 110 Dúbrava MS3 | 1 | 38 | 0 | 0 | 0 | 0 | Jamrichová et al. (2013) |
| 111 Vracov | 1 | 33 | 11 | 17 | 33 | 8 | Kuneš et al. (2015) |
| 112 Šúr | 1 | | | | | | Petr et al. (2013) |
| 113 Machová | 1 | 21 | 0 | 0 | 0 | 0 | Rybníček and Rybníčková (2008) |
| 114 Tlstá hora | 1 | 43 | 7 | 0 | 0 | 0 | Rybníček and Rybníčková (2008) |
| 115 Královec | 1 | 26 | 0 | 0 | 0 | 0 | Rybníček and Rybníčková (2008) |
| 116 Jablůnka | 1 | 0 | 0 | 0 | 0 | 29 | Jankovská and Pokorný (2008) |
| 117 Stonava | 1 | | | | | | Břízová (1994) |
| 118 Horní Lomná | 1 | 16 | 7 | 0 | 0 | 0 | Rybníček and Rybníčková (2008) |
| 119 Kubriková | 1 | 27 | 0 | 0 | 0 | 0 | Rybníček and Rybníčková (2008) |

### 6.5.1 Early Holocene

The initial stage of the Holocene vegetation development, characterized by semi-open pine woodlands (Fig. 6.6; cluster a), is the protocratic phase (see Sect. 6.3). The REVEALS estimates quantify the landscape openness as 13–28% (which is the estimated land-cover of *Poaceae*) along with 8–15% of the landscape covered by *Betula* and 38–65% by *Pinus*. Both trees were undoubtedly present in the current Czech territory during the Younger Dryas, so with Early Holocene climate warming they could rapidly spread as during other warm oscillations of the glacial period. The landscape was characterized by good light availability and generally neutral to basic soils even in areas with acidic bedrock. Carbonates were formed in a layer of loess, which covered the whole landscape after the ice age except for exposed sites in the mountains. Soils were still poorly developed and surface was loose. Numerous habitats with these environmental characteristics and their combinations maintained the glacial flora at detectable frequencies in the landscape, including *Ephedra*, *Gypsophila repens*-type, *Selaginella selaginoides* and *Alnus alnobetula*. Their gradual disappearance during the Holocene resulted in the restricted areas of their

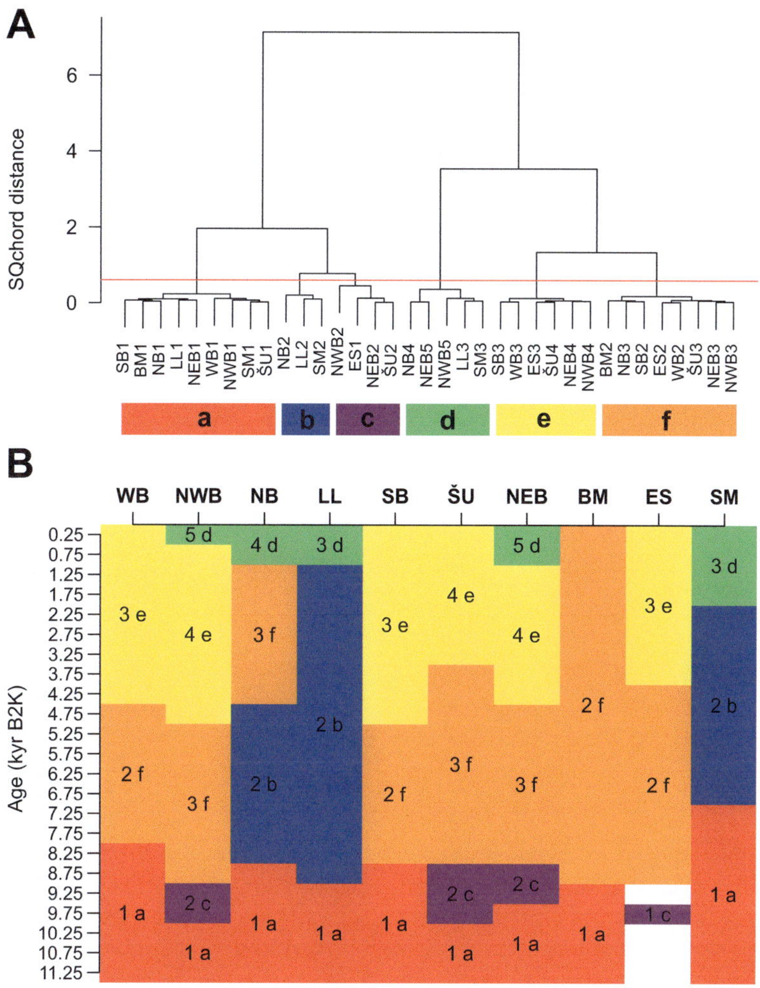


**Fig. 6.6** (**A**) Ward's hierarchical clustering of squared chord distances based on the average composition of each vegetation phase. Clusters indicated by different colours represent: (*a*) semi-open pine woodlands; (*b*) mixed oak woodlands; (*c*) hazel woodlands; (*d*) vegetation subject to human activity; (*e*) fir-beech forests; and (*f*) spruce forests. (**B**) Temporal distribution of the eight vegetation clusters in particular regions. Numbers represent significant pollen zones. The regions are abbreviated as follows: *WB* Western Bohemia, *NWB* North-western Bohemia, *NB* Northern Bohemia, *LL* Labe lowland, *SB* Southern Bohemia, *ŠU* Šumava Mts, *NEB* North-eastern Bohemia, *BM* Bohemian-Moravian Highlands, *ES* Eastern Sudetes Mts, *SM* Southern Moravia

current distribution or absence in the Czech Republic. Some of the common pollen types from the Younger Dryas and Early Holocene, e.g. *Helianthemum*, *Sanguisorba officinalis* and *Alisma*-type decreased before the Middle Holocene while currently they are relatively common again due to the presence of anthropogenic habitats (Fig. 6.3).

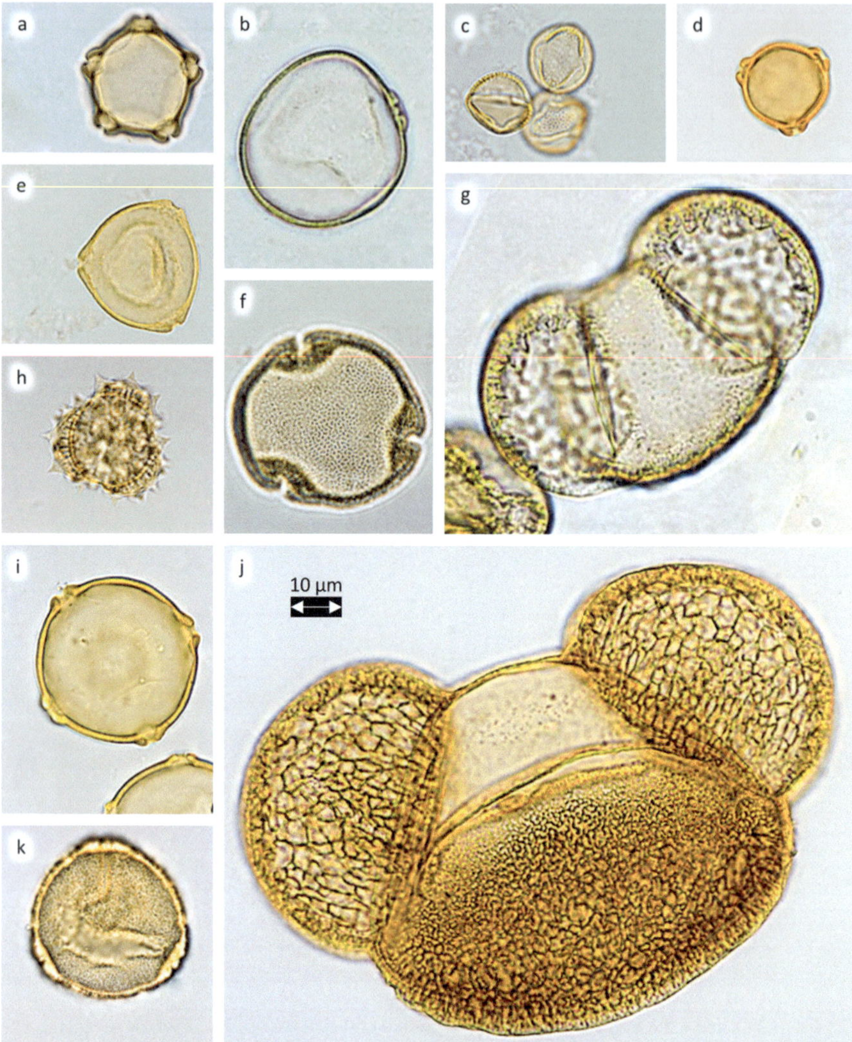


**Fig. 6.7** Examples of abundant pollen types from postglacial sediments in the Czech Republic: (**a**) *Alnus glutinosa*-type; (**b**) *Poaceae* excluding cultivated cereals; (**c**) *Salix* (homobrochate form); (**d**) *Betula*; (**e**) *Corylus*; (**f**) *Tilia*; (**g**) *Pinus* haploxylon type (includes *Pinus cembra*); (**h**) *Anthemis arvensis*-type; (**i**) *Carpinus*; (**j**) *Abies*; (**k**) *Fagus*. The same magnification is used for all photographs. Photo credits: V. Abraham (a, c–f, h–k) and P. Kuneš (b, g)

Both of the glacial tree dominants, *Pinus cembra* and *Larix decidua*, have similar patterns to those of herbaceous and shrub taxa. The closest areas of the current natural distribution of *Pinus cembra* are in the Alps and Western Carpathians. Pollen of *Pinus cembra* is difficult to distinguish from *Pinus sylvestris*-type (Fig. 6.7g) and numerous recordings of its presence during the Holocene in the Czech Republic were considered to be problematic. However, the recent discovery of macrofossils of *Pinus cembra* from the first Holocene millennia in the Bohemian Paradise supports these pollen findings (Pokorný et al. 2017). *Larix* is also native to the Alps and Western Carpathians, but its native range extends to the Hrubý Jeseník Mts in northern Moravia and Silesia (Dudová et al. 2013, 2014), where morphologically and genetically distinct populations survive (Lewandowski 1997; Wagner et al. 2015). In the rest of the Czech Republic, pollen (Fig. 6.3) and macrofossils (Opravil 1972) indicate that *Larix* was present here until historical times. Subsequently, man could have participated on its extinction.

The second group of vegetation typical of the Early Holocene (Fig. 6.6; cluster c) is still represented by *Pinus* (5–23%), but is dominated by *Picea* (9–40%) and more thermophilous *Corylus* (14–34%). This cluster firstly appears in the 10 ka window, however there was a higher percentage of *Corylus* (17%) in the previous time window of 10.5–10 ka in north-western Bohemia (Fig. 6.4). *Corylus* expanded to the whole of the northern part of Central Europe around 10.65 ka. Such a synchronous expansion on a large subcontinental scale could only have been triggered by a climatic event (Giesecke et al. 2011), perhaps connected with abrupt warming of the North Atlantic Ocean around 10.3 ka, known as the end of Bond event 7 (Bond et al. 1997). The geographical proximity of north-western Bohemia to the Atlantic explains why the expansion of *Corylus* in the rest of the regions is slightly delayed. In addition, the percentages of *Corylus* decrease eastwards, associated with an increase in *Picea*, which was spreading westwards. Millennial differences in the occurrence of records of spruce in the easternmost and the westernmost regions (Fig. 6.6B) agree with the timing observed at the continental scale (Latałowa and van der Knaap 2006), at which spruce spread in three millennial steps: (1) area spanning from Slovakia (Western Carpathians) to the south-eastern Czech Republic, (2) the rest of Czech Republic except the extreme west and (3) westernmost Bohemia. For the spruce expansion there is no evidence of a typical response to a climatic trigger that would be mirrored by a synchronous increase across a large geographic area. The pattern in the expansion visible in the pollen data could indicate a front moving from east to west; however, phylogeography indicates possible refugia on the southern Bohemian Massif and its foreland (Tollefsrud et al. 2008). The later spread in the Šumava Mts than in the Western Carpathians could be explained by the smaller size of the initial population. The hazel group of vegetation occurs in mountainous regions, which may be due to the mesoclimatic conditions in areas with steep topographical gradients providing a suitable level of humidity for both of the dominants, *Corylus* and *Picea*. However, the increase in *Corylus* is noticeable also in lowland areas, especially those close to sites that were inhabited in the Mesolithic. In the light of these findings *Corylus* is a native species whose early spread could have been supported by humans. Moreover, a synthesis of pollen data from the Early Holocene/Mesolithic period revealed indicators of human habitation, such as *Calluna vulgaris*, *Plantago lanceolata*, *Solanum* or *Pteridium* (Kuneš et al. 2008b). Site-specific finds connected to the Mesolithic period are discussed below.

### 6.5.2 Middle Holocene

During the Middle Holocene the vegetation clearly differed in the lowlands on the one hand, and at middle altitudes and in the mountains on the other hand. Mixed oak woodland (Fig. 6.6; cluster b) is a regional group of vegetation recorded in three lowland areas characterized by broad-leaved trees (*Quercus* 12–24%, *Fraxinus* 2–7%, *Tilia* 2–5%), coniferous trees (*Pinus* 9–21%, *Picea* 9–25%) and landscape openness indicated by *Poaceae* (8–18%). Spruce forest (Fig. 6.6; cluster f; *Picea* 49–64%) was the prevailing regional group of vegetation at middle altitudes and in the mountains.

The dominance of spruce at middle and high altitudes of the Bohemian Massif during the Middle Holocene is in marked contrast with the vegetation reconstructions for the rest of Central Europe and some previous reconstructions for the Czech Republic. This discrepancy can be explained, firstly, by differences in the climatic and edaphic conditions in the Bohemian Massif compared to the other parts of Central Europe at similar altitudes. Secondly, interpretations of pollen percentages may vary among authors. Local presence of trees on bogs creates a putative source of a pollen signal for the presence of *Picea* and may affect interpretations. This objection is justified, but individual spruce trees also currently grow on peat bogs. However, the REVEALS reconstruction is based on the same sites that were included in the validation set and REVEALS estimates from the shallowest depths fit present-day vegetation (Abraham et al. 2014). Moreover, occurrence of spruce is documented in historical records for sites other than bogs at middle altitudes (Rybníček and Rybníčková 1978; Szabó et al. 2017). Higher precipitation together with the chemical composition of spruce litter might lead to soil acidification and subsequent podzolization (Emmer et al. 1998). However, most of the soils in the Czech Republic are currently cambisols, which have supposedly developed under deciduous forests. Effects of extensive spruce forest at middle and high altitudes on soil development needs further investigation.

Regional estimates of vegetation, which can be viewed as an average of all the sites included, are general trends. Variations within a region can only be revealed by focusing on well-studied areas. The vegetation history in sandstone landscapes in the northern part of Bohemia is a good example. Steep topographical gradients provide spatial heterogeneity different from the rest of the landscape (see Chap. 7, Sect. 7.4, this book). Pedoanthracological evidence from the Bohemian Paradise shows codominance of *Quercus* and *Pinus* in the Middle Holocene period (Novák et al. 2015), while this area is assigned to the broad region of north-western Bohemia with dominance of spruce in the REVEALS reconstructions. Detailed comparison of pollen curves from one region also reveals large temporal variability in the time of spread of individual taxa (see Appendix 1 in Abraham et al. 2016). *Fagus* occurred at the Jelení louže site in Bohemian Switzerland ~5.5 ka (Pokorný and Kuneš 2005) and 1000 years later at the Okna site in the Doksy area, approximately 50 km to the south-east from Jelení louže (Pokorný et al. 2017).

After the cold and dry 8.2 ka event, the climate became moister and warmer and remained so throughout the first half of the Middle Holocene. The accumulation rate of calcareous tufa in the village of Svatý Jan pod Skalou in the Bohemian Karst in central Bohemia is a proxy record of humidity in a low-altitude region. The accumulation started around 9.55 ka and maximal growth was reached between 8.45 and 6.55 ka (Žák et al. 2002). The warm and humid conditions accelerated biogenic processes of pedogenesis, soils reached their maximum fertility and ecosystems their highest productivity. Mixed oak woodlands dominating in this period are characteristic of the mesocratic phase in interglacial dynamics (see Sect. 6.3; Fig. 6.2d). Humid conditions also accelerated the growth and accumulation of peat. After 6.55 ka the climate became more variable in terms of humidity (Jäger 1969; Žák et al. 2002; Dreslerová 2012) and already around 7 ka pollen of *Scheuchzeria palustris*, which is an indicator of dystrophic peat hollows (Hájek and Hájková 2011), increased especially at mountain sites (Fig. 6.3). Short periods (<1000 years) of dry climate alternating with humid conditions could have caused the creation of bog hollows as a consequence of peat decomposition followed by flooding. During this period high pollen frequencies of *Taxus* are recorded (at ~10% of all the sites, Fig. 6.3) in mid-altitude areas with steep topographic gradients.

The footprint of the Holocene climatic optimum in the pollen spectra are high frequencies of *Hedera*, *Viscum* and *Ilex* (Fig. 6.3). Their present-day distribution in Scandinavia is sharply limited by temperature, for example *Hedera* today cannot persist at sites with a mean daily temperature of less than −2 °C in the coldest month (Iversen 1944). Therefore, these species are used as pollen indicators of past climate. All of them occurred in the current Czech territory already before the end of the Early Holocene and by the end of the Middle Holocene *Hedera* and *Viscum* occurred at ~30% of the sites studied, which is approximately three times more than their present-day frequency. *Ilex* requires an even higher minimum temperature and a more oceanic climate, thus we interpret the occurrence of the very scarce pollen grains of this species as an indication of its past distribution in the Czech territory (see Lang 1994), rather than contamination of pollen record, as interpreted by Rybníčková (1974).

The Holocene climatic optimum was very favourable for forest expansion, which raises the biogeographical question of the survival of the vegetation associated with open areas during this period (Gradmann 1933; Firbas 1949; Vera 2000). REVEALS estimates of *Poaceae* (>9% during the whole Holocene) for southern Moravia and the Labe Lowland indicate the continuous existence of open areas in the landscape (Fig. 6.4). Our finding is supported by molecular data on the disjunctive floristic and faunistic elements of continental steppes (Kajtoch et al. 2016), fossil assemblages of snails that dwell exclusively in open habitats (Juřičková et al. 2013, 2014b; Pokorný et al. 2015) or the widespread presence of chernozem soils whose genesis is restricted to areas not covered by forest (Antoine et al. 2013).

The first farmers settled in the lowlands of both Bohemia and Moravia around 7.5 ka, approximately 600 years after the onset of the Middle Holocene. In addition to the cultivation of cereals (see complete history in Dreslerová and Kočár 2012) the introduction of agriculture boosted the genesis of an anthropogenic flora. Human activity triggered a decrease in beta diversity (Šizling et al. 2016) as can be seen in the quantities of particular taxa. For example, *Plantago lanceolata* increased around 5.5 ka (see Appendix in Abraham et al. 2016). Its occurrence at this time coincides with the occupation of areas by humans during the Late Neolithic. Similarly, the occurrence of *P. lanceolata* around 4.5 ka in the Šumava Mts and southern Bohemia coincides with the colonization of southern Bohemia in the Early Bronze Age.

The mixed oak woodland group of vegetation started to dominate around the Early/Middle Holocene transition in two lowland regions in Bohemia, whereas in southern Moravia it occurred one millennium later around 7 ka (Fig. 6.6B). The later transition from semi-open pine to oak woodland might have been delayed because of the more continental climate in southern Moravia. However, another significant driver of the spread of oak could be the intensity of human occupation (Kuneš et al. 2015) because prehistoric forest management may have favoured oak woodland (Jamrichová et al. 2013). Human occupation has always been associated with the occurrence of oaks. For example, all archaeo-anthracological assemblages from the whole prehistory from sites at altitudes of 180–230 m in Moravia are dominated by oak (Novák et al. 2017). The association between human occupation and oaks may be due to the fact that areas suitable for settlement overlapped with areas potentially suitable for oak. However, there is also the causal effect of a long-term human land use on the species composition of woodland, which favours trees resistant to coppicing, including oak.

### *6.5.3 Late Holocene*

The Late Holocene vegetation consisted of four groups: mixed oak woodland, spruce forest, fir-beech forest and vegetation affected by humans. Diverse timing and trajectories in the vegetation history in the Late Holocene result from a combination of soil development, disturbance, inter-specific competition and local environmental conditions. Depending on local conditions the changes in regional vegetation occurred over a long period of 5–3.5 ka.

Mixed oak forest in northern Bohemia had already changed to spruce forest by 4.5 ka (Fig. 6.6B). The Late Holocene spread of spruce in this lowland region is not anomalous from a wider Central European perspective. It occurred in south-eastern Poland, the Harz Mts and other small areas on the periphery of the main distribution and altitudinal range of spruce (Latałowa and van der Knaap 2006). We attribute the late spread of *Picea* in northern Bohemia to the acidification of soils in the Holocene. Only the lowland regions along the Labe River and in southern Moravia were too dry and warm for spruce. The gradual decrease in soil fertility started the oligocratic phase of the interglacial cycle (see Sect. 6.3). Long-term effect of precipitation led to mechanical and chemical outwash of calcium, and phosphorus also became less accessible to plants at this stage.

At the beginning of the Late Holocene most of the sites suitable for forests were already occupied by trees, so natural disturbance or human activity played a more important role in creating and maintaining areas of open landscape than during the earlier phases. Competitive abilities and other traits of both established and incoming species determined the success of the taxa that were spreading. Successful spread of *Fagus* in the mountains of Central Europe (e.g. the Šumava Mts) was attributed to natural disturbance, whereas its spread in the lowlands (Germany and the Czech Republic) was favoured by human activity. Natural disturbances tend to occur more frequently in forests in the mountains than in the lowlands due to the steeper slopes, higher wind speeds and more severe winters (Küster 1997). Human activity as a factor favouring the spread of *Fagus* is also reported in southern Scandinavia (Bradshaw and Lindbladh 2005), but not in continental Hungary, where the spread was triggered by a shift in climate resulting in greater availability of moisture during the growing-season (Magyari et al. 2010). Considering *Picea* and *Fagus* as established and incoming species, respectively, *Fagus* is ecologically stronger, because it can regenerate under the dense canopy of spruce better than spruce itself (Dobrovolny 2016) and being a tall tree it can effectively shed the surface under its canopy.

*Tilia*, *Fraxinus*, *Acer* and *Ulmus* can slow down the acidification process by physiologically recycling calcium, as there is a high concentration of calcium in their leaves, which is easily released from decaying leaf litter (Wäreborn 1969). These trees can maintain the calcium-rich characteristics of their sites by soil-vegetation feedback and could have affected the long-term succession during the Late Holocene. Disturbances can help in the establishment of new species (e.g. *Fagus* instead of *Tilia*), which cause a reduction in calcium retention accelerating acidification. Classical example of such an event was discovered in the Kokořín sandstone area (see Chap. 1, Fig. 1.4, this book). The trigger for the acidification process was human occupation of its sandstone plateaus in the Late Bronze Age (Lusatian culture). The palaeomalacological and sedimentary record from sandstone gorges reveals a decrease in snail diversity and increase in erosion of loess from the plateaus (Ložek 2007). Unfortunately, this site did not contain any palaeobotanical material, but obviously the environmental change was strong and very rapid. We assume that mixed oak woodland and partly spruce forest changed to beech forest. Other sandstone areas in Bohemia that were not affected by human activity show a gradual change in tree composition (e.g. Jelení louže in northern Bohemia; Novák et al. 2015) and gradual increase in $Ca^{2+}$ and $Mg^{2+}$ concentrations (due to increased leaching from the soils) in the peat in the Anenské údolí gorge in north-eastern Bohemia (Pokorný and Kuneš 2005).

*Carpinus* reached its maximal representation in the landscape (around 5%) in north-eastern Bohemia from 4.25 to 1.75 ka. The lowest percentage cover of *Carpinus* in vegetation (<2%) was in the Šumava Mts, southern Moravia, and western and southern Bohemia (Fig. 6.4). The contrast between its greater abundance and more rapid spread in north-eastern regions and low abundance and gradual spread in the south-western regions confirms the old hypothesis that *Carpinus* spread into the Czech territory from the north-east (Rybníček and Rybníčková 1996).

However, the spread was probably also influenced by human activity. Charcoal of *Fagus*, *Carpinus* and *Abies* is recorded for the Neolithic. Occurrence of *Carpinus* remained stable during the whole prehistory, whereas records of *Abies* and *Fagus* were occasional in the Middle Holocene and increased in the Late Holocene (Novák et al. 2017). This suggests that these three trees that spread in the Late Holocene were present in the Czech territory much earlier, especially close to human settlements. *Carpinus* regrows easily from the stumps of felled trees, which means it was able to survive in forests exploited by humans. *Abies* spread into the Czech territory together with *Fagus*, only in northern and north-western Bohemia it spread slightly later. The spread and decrease in abundance of *Abies* is connected with forest management, such as livestock grazing, litter raking, selective logging and changes in game population size (Kozáková et al. 2011).

Regional vegetation dependent on human activity (Fig. 6.6; cluster d) appeared only in the last two millennia. The percentage of dominant *Cerealia* varies from 33 to 61%. However, quantifying *Cerealia* is still difficult, because their pollen productivity used in the REVEALS models is probably underestimated. Present-day varieties of autogamous cereals have a much lower productivity than cereals that were cultivated during earlier periods, especially the anemogamous *Secale*. Intensity of human activity recorded in archaeological data indicates periods with extensive human occupation (Late Bronze Age and Late Iron Age) and periods when the incidence of archaeological findings falls below that recorded for the period of Neolithic occupation, especially the Eneolithic (= Chalcolitic) and Migration periods (Dreslerová 2011). Succession at Vladař hill fort after its complete abandonment 2000 years ago lasted for 500 years, starting with pine-birch and continuing via hazel scrub and oak woodland to fir-beech forest with spruce (Pokorný et al. 2006). However, nature in an area does not always return to the close-to-natural state recorded before the onset human activity. New settlements in previously non-colonized areas cause irreversible changes in the soil, erosion and genesis of alluvial habitats. This process culminated in the High and Late Middle Ages (thirteenth to fifteenth century), but earlier cases are not rare. The Late Iron Age occupation at Okna (Doksy area, northern Bohemia) initiated the accumulation of clay after several millennia of peat growth. A quick change from *Sphagnum* peatbog with sedges and occasional pools (*Utricularia*) to monospecific *Phragmites* beds occurred (Dreslerová et al. 2013). Such incidental changes can be contrasted with the intentional creation of new habitats. The human environment during prehistory was divided between managed woodland and production areas including fields. Part of the production area was also fallow land, which was alternated with arable land every 10–15 years (Sádlo et al. 2005). From the Early Bronze Age to the Older Iron Age this environment was colonized by on average 3.6 newly arriving species/100 years (Kočár et al. 2015). During the Younger Iron Age the spread of the use of scythes enabled the development of hay meadows. The rate of introduction of archaeophytes immediately increased to 5 species/100 years. Human activity at the end of the Late Iron Age also started to enlarge the open areas above the timberline on the Hrubý Jeseník summits (Novák et al. 2010). Nevertheless, the summits of Czech mountains were also affected in the Modern period by grassland management.

### 6.5.4 The Middle Ages and Modern Period

The key points in the development of the Czech flora and landscape occurred during the last two millennia along with the strong effect of human activity. The agricultural system during the Early Middle Ages still followed the farming practices used earlier in prehistory. Increasing population growth during the High Middle Ages required higher crop yields. The arable land was divided between summer (*Triticum*) and winter cereals (*Hordeum*, *Avena*, *Secale*). Both winter and summer cultures together with fallow land formed the three-field system. A high and extensive record of *Secale* and the weed *Centaurea cyanus* is a characteristic pollen signal of the High Middle Ages (Kozáková et al. 2009). Moreover, there are plant macrofossils of the whole group of weeds occurring in winter cereals (e.g. *Adonis aestivalis*, *Agrostemma githago*, *Bromus secalinus* and *Vicia hirsuta*). Unique archaeobotanical material from Žatec shows an increase in weeds already in the eleventh century, dating the first evidence of the three-field system to the Early Middle Ages (Kočár et al. 2010).

The high-medieval transition was a large-scale deforestation event. Human colonization reached the highlands and other peripheral areas, which were previously only subjected to sporadic visits. Social and economic transformation clearly defined the ownership of the land. Thus, in addition to the magnitude, the nature of the colonization changed. For the first time in history, it was a well-organized intervention with central planning and assignment of landscape functions (Sádlo et al. 2005). This change is even visible in the regional vegetation estimates. A comparison of the estimates of Holocene vegetation produced by the REVEALS model and the map of potential natural vegetation (Neuhäuslová et al. 1998) indicates that the highest similarity was recorded during the period 1.5–1 ka, i.e. the time window preceding the High Middle Ages. It means that the Holocene forest succession at this time is sufficiently young to fulfil the present perspective of climax vegetation (all tree taxa are present), but at the same time it is not so young as to be influenced by intense human activity (Abraham et al. 2016).

The most important change in forest vegetation was brought about by new practices in forest management during the Modern Period. The slow development can be illustrated by an example from Bohemian Switzerland, for which archival sources and pollen data are available. The development of the glass industry during the Baroque period (seventeenth to eighteenth centuries) required increasingly more timber. The annual forest exploitation at the end of the seventeenth century was three times greater than earlier in the seventeenth century, but forest management still relied on natural forest regeneration. Pollen signals indicate simultaneous decrease in *Fagus* and *Abies* and increase in *Betula* and *Calluna*. In 100 years the overall stock of timber was reduced to one half, thus reforestation was introduced in order to exploit the timber more intensively. The most favourite species planted was spruce, because it grows relatively fast. In Bohemian Switzerland, many exotic trees were also used, the most important being *Pinus strobus* for resin production. The time of maturation of the introduced spruce is recorded in pollen diagrams by an increase in *Picea* pollen (Abraham 2006).

## 6.6 Local Flora and Vegetation History

Apart from pollen analysis, plant macrofossils (or macroremains; Fig. 6.8) provide a source of valuable information for studying the development of the flora and vegetation. Plant macrofossils include any part of a plant visible to the naked eye and manipulable by hand (Birks 2013). Such characteristics indicate certain limitations on movement, therefore macrofossils are more likely to reflect the local vegetation at or around a particular site than pollen. Recent developments in the Landscape Reconstruction Algorithm have enabled the determination of the source of the pollen and by application of pollen production and dispersal models the reconstruction of local patterns in vegetation within few hundred m of a particular site (Sugita 2007b). Consequently, pollen can provide valuable information about the local history of a site in addition to that provided by macrofossils, but application of such approaches is still under development in the Czech Republic (but see Abraham et al. 2017). It has, however, resulted in robust results in other areas in Europe (Overballe-Petersen et al. 2013; Cui et al. 2014; Hultberg et al. 2015).

Plant macrofossils were originally used to infer the development of plant communities preserved in autochthonous peat. A pioneer study in the Czech Republic analyzed plant macrofossil assemblages in the Bláto mire near Nová Bystřice in the Bohemian-Moravian Highlands (Rybníček and Rybníčková 1968) in order to determine long-term changes in local plant communities. The authors were able to distinguish eight phases of mire development in the Holocene from aquatic plant communities through marsh and *Phragmites* fen and finally to bog vegetation. In a later synthesis of multiple peat profiles K. Rybníček generalized succession in aquatic/mire plant communities at larger spatial scale of Central Europe (Rybníček 1973; Birks and Birks 1980). The main conclusion of this synthesis was that mire plant communities were not stable or continuous in time and that they respond to changing environmental conditions.

In many cases plant macrofossils complemented other proxies (mainly pollen) for producing a more complex and accurate picture of the vegetation and environmental reconstruction. Studies focused on the reconstruction of local vegetation and explanation of mire origin in the Bohemian-Moravian Highlands (Jankovská 1971),

**Fig. 6.8** Examples of plant macrofossils from postglacial sediments in the Czech Republic: (**a**) *Cladium mariscus* seed; (**b**) *Trapa natans* leaf; (**c**) *Rubus idaeus* seed. Photo credits: A. Potůčková (a, b) and P. Žáčková (c)

processes linked with infilling of former lakes and terrestrialization of aquatic ecosystems (see *Trapa natans* in Fig. 6.3) at Velanská cesta and Švarcenberk in southern Bohemia (Jankovská 1970; Pokorný and Jankovská 2000; Bešta et al. 2009) and at Vracov in southern Moravia (Rybníček 1983). Another study concentrated on an oxbow lake infilling linked with medieval settlement at Stará Boleslav in central Bohemia (Kozáková et al. 2014). A study at Chrást in the Labe lowland recorded a high diversity of oxbow lake and mire vegetation and the rapid changes it underwent during the period of Late-Glacial climatic instability (Petr et al. 2014). Another study in the Šumava Mts near Černé Lake provides a good example of how plant macrofossils contribute to the understanding of local establishment and expansion of trees around a site, in this case namely spruce and fir (Vočadlová et al. 2015).

The complex topography of the Bohemian sandstone pseudokarst areas provides conditions for peat growth at the bottom of gorges (Kuneš et al. 2007). Not only pollen data from these deposits were used to infer the vegetation growing in the broader sandstone areas but also plant macroremains were studied to explain the origin of the peat. For example, plant macrofossil assemblages found at the Vlčí rokle site (Adršpach-Teplice Rocks in eastern Bohemia) provided an explanation of the origin and local development of peat deposits (Nováková 2000). The entire postglacial period was recorded there as an autogenic succession leading from sedge fen through *Eriophorum* peat bog to *Sphagnum* peat bog and eventually turning into forested mire. The mire/peatbog development was recurrently interrupted by fire and influxes of eroded sand.

Plant macrofossils are also one of the main sources of data used to account for the origin and development of spring fens in the Carpathian flysch borderland in the Czech Republic and Slovakia (Rybníčková et al. 2005), where two main phases were determined at several sites: the older initial forest spring and the younger treeless fen vegetation, which survive to the present day. The origin of different spring fens varies from the Middle to Late Holocene and spans over ~5000 years. Their species composition is influenced mainly by the calcium content of the water and bedrock quality.

A study of plant macrofossil succession contributed to our understanding of the local development of mountain bogs in the Hrubý Jeseník Mts (Dudová et al. 2010, 2013, 2014). As in previously mentioned studies, several developmental stages were found leading from the origin of the site (spring fen or mire) through reed swamp into ombrotrophic bog, and often to human-managed forest plantations. These studies emphasize the importance of local plants and their contribution to autogenic successional changes in local environmental conditions.

Macrofossils can further be used to derive quantitative environmental reconstructions. At the Skřítek site in the Hrubý Jeseník Mts (Dudová et al. 2014) transfer functions were used to link plant macrofossils to plant ecological indicators for pH and moisture derived from an extensive modern dataset of mire assemblages of vascular plants and bryophytes. The results confirmed the autogenic mire succession leading to gradual pH and moisture decline starting with the Middle Holocene (~8–7 ka), which was also supported by other proxies.

Dynamics of alder carr communities were also explored with the help of palaeoecological data, mainly pollen and plant macroremains, in a lowland in eastern Bohemia (Pokorný et al. 2000). Alder carr (forested swamp) is a transitional ecosystem in the terrestrialization of freshwater bodies, often leading to oligotrophic mires. Therefore, its long-term persistence as recorded at the Na bahně site needs to be driven by specific dynamics. A cyclic model proposed by Jeník (1980) hypothesized that autogenic succession starts as an open mire dominated by sedges and other light-demanding taxa. These plants accumulate peat that grows above the groundwater table and enables the colonization by alder seedlings and the development of alder carr. Mature alder trees decrease the groundwater table, which leads to sediment mineralization. Eventually alder stands die back and decreased evapotranspiration then increases the level of the groundwater table until the cycle is completed.

The local history of particular genera or species could be better determined by using plant macrofossils because pollen analysis offers only limited taxonomic resolution. The genus *Carex* has been explored at around 30 sites throughout the postglacial with the aim of explaining historical legacies in the present-day distribution of species of this genus and their value as environmental indicators (Jankovská and Rybníček 1988). Occurrence of the rare sedge *Cladium mariscus* in both the present and the past is reviewed by Pokorný et al. (2010a), who conclude that the relict status of this taxon is supported by plant macrofossils (Fig. 6.8a). These authors suggest that its metapopulation survived in central Bohemia throughout the entire Holocene.

Macrofossils can also provide valuable information related to the history of human settlements even if derived from natural or semi-natural deposits. Plant remains may indicate the subsistence strategies and diet of the Mesolithic human population (Divišová and Šída 2015; Pokorný et al. 2017), which is illustrated by the study of the former lake Švarcenberk deposits (Pokorný et al. 2010b). This study suggested that Mesolithic people possibly included *Rubus idaeus* or *Trapa natans* in their diet. Local vegetation succession after abandonment of the Iron Age hillfort at Vladař in western Bohemia was also traced with help of plant macrofossils. This study shows the abandonment of local agriculture and slow process of afforestation (Pokorný et al. 2006). A unique study of a medieval fishpond in Prague provides a description of the vegetation in the suburbs and the environmental changes in the High Middle Ages (Pokorná et al. 2014). Plant macrofossils revealed a high rate of eutrophication of the aquatic environment accompanied by gradual changes in the vegetation in the surroundings indicating deforestation and increase in ruderal plants.

Local studies are also valuable for informing nature conservation about the conservation value of ecosystems (Vegas-Vilarrúbia et al. 2011). An interdisciplinary study that focused on the history of the vegetation at a southern Moravian deciduous oak woodland site (Dúbrava near Hodonín), which is under natural protection due to its high biodiversity and possible relict status, asked whether this site was historically stable (Jamrichová et al. 2013). This study revealed a dramatic change in the fourteenth century from shrubby vegetation into an oak woodland accompanied by a change in human management at this site. These finding indicate that the current status of the oak woodland is a result of medieval oak promotion followed by its reduction by modern plantation forestry that favours pine. The species that today are linked to oak woodlands had to survive in other habitat types before the fourteenth century. Local palaeoecological studies also help develop management plans for other protected areas, e.g. in the northern Bohemian sandstone landscapes. Here, the long-term perspective informed the nature conservation managers about the historical natural vegetation (Kuneš et al. 2007; Abraham and Pokorný 2008; Pokorný et al. 2008) and the role of a natural fires as an important factor maintaining certain vegetation types (Adámek et al. 2015; see Chap. 9, Sect. 9.7.3, this book).

**Acknowledgements** We thank all the contributors to the Czech Quaternary Pollen Database, which made this synthesis possible, Jiří Sádlo for fruitful ideas for this study, Milan Chytrý, Anna Potůčková and Pavla Žáčková for providing photos, Milan Chytrý for valuable comments on an earlier version of the manuscript and Tony Dixon for language editing. We also thank the Czech Science Foundation for financial support (projects no. 16-06915S and 16-10100S).

## References

Abraham V (2006) Přirozená vegetace a její změny v důsledku kolonizace a lesnického hospodaření v Českém Švýcarsku [The natural vegetation of Bohemian Switzerland and its changes as an impact of habitation and forest management]. MSc thesis, Charles University, Praha

Abraham V, Pokorný P (2008) Vegetační změny v Českém Švýcarsku jako důsledek lesnického hospodaření – pokus o kvantitativní rekonstrukci [Vegetation changes in Czech Switzerland as a result of forestry management – an attempt at quantitative reconstruction on the basis of pollen analyses and historical sources]. In: Beneš J, Pokorný P (eds) Bioarcheologie v České republice [Bioarchaeology in the Czech Republic]. Jihočeská Univerzita, PřF; Archeologický ústav AV ČR, Praha, pp 443–470

Abraham V, Oušková V, Kuneš P (2014) Present-day vegetation helps quantifying past land cover in selected regions of the Czech Republic. PLoS One 9:e100117

Abraham V, Kuneš P, Petr L, Svitavská-Svobodová H, Kozáková R, Jamrichová E, Švarcová MG, Pokorný P (2016) A pollen-based quantitative reconstruction of the Holocene vegetation updates a perspective on the natural vegetation in the Czech Republic and Slovakia. Preslia 88:409–434

Abraham V, Novák J, Houfková P, Petr L, Dudová L (2017) A Landscape Reconstruction Algorithm and pedoanthracological data reveal late holocene woodland history in the lowlands of the NE Czech Republic. Review of Palaeobotany and Palynology 244:54–64

Adámek M, Bobek P, Hadincová V, Wild J, Kopecký M (2015) Forest fires within a temperate landscape: a decadal and millennial perspective from a sandstone region in Central Europe. Forest Ecology and Management 336:81–90

Alley RB, Mayewski PA, Sowers T, Stuiver M, Taylor KC, Clark PU (1997) Holocene climatic instability: a prominent, widespread event 8200 yr ago. Geology 25:483–486

Ammann B (2000) Biotic responses to rapid climatic changes: introduction to a multidisciplinary study of the Younger Dryas and minor oscillations on an altitudinal transect in the Swiss Alps. Palaeogeography, Palaeoclimatology, Palaeoecology 159:191–201

Ammann B, Birks HJB, Brooks SJ, Eicher U, von Grafenstein U, Hofmann W, Lemdahl G, Schwander J, Tobolski K, Wick L (2000) Quantification of biotic responses to rapid climatic changes around the Younger Dryas – a synthesis. Palaeogeography, Palaeoclimatology, Palaeoecology 159:313–347

Andersen ST (1994) History of the terrestrial environment in the quaternary of Denmark. Bulletin of the Geological Society of Denmark 41:219–228
Andersen KK, Azuma N, Barnola J-M, Bigler M, Biscaye P, Caillon N, Chappellaz J, Clausen HB, Dahl-Jensen D, … White JWC (2004) High-resolution record of northern hemisphere climate extending into the last interglacial period. Nature 431:147–151
Antoine P, Rousseau D-D, Degeai J-P, Moine O, Lagroix F, Kreutzer S, Fuchs M, Hatté C, Gauthier C, ... Lisá L (2013) High-resolution record of the environmental response to climatic variations during the last interglacial–glacial cycle in Central Europe: the loess-palaeosol sequence of Dolní Věstonice (Czech Republic). Quaternary Science Reviews 67:17–38
Barron E, Pollard D (2002) High-resolution climate simulations of oxygen isotope stage 3 in Europe. Quaternary Research 58:296–309
Beresford-Jones D, Taylor S, Paine C, Pryor A, Svoboda J, Jones M (2011) Rapid climate change in the upper Palaeolithic: the record of charcoal conifer rings from the Gravettian site of Dolní Věstonice, Czech Republic. Quaternary Science Reviews 30:1948–1964
Berger A, Loutre MF (1991) Insolation values for the climate of the last 10 million years. Quaternary Science Reviews 10:297–317
Bešta T, Šafránková J, Pouzar M, Novák J, Nováková K (2009) Late Pleistocene–early Holocene transition recorded in the sediments of a former shallow lake in the Czech Republic. Hydrobiologia 631:107–120
Birks HJB (1986) Late-Quaternary biotic changes in terrestrial and lacustrine environments, with particular reference to north-west Europe. In: Berglund BE (ed) Handbook of Holocene palaeoecology and palaeohydrology. Wiley, Chichester, pp 3–65
Birks HJB (2005) Mind the gap: how open were European primeval forests? Trends in Ecology and Evolution 20:154–156
Birks HH (2013) Plant macrofossil introduction. In: Elias SA, Mock CJ (eds) Encyclopedia of Quaternary science, 2nd edn. Elsevier, Amsterdam, pp 593–612
Birks HJB, Birks HH (1980) Quaternary palaeoecology. Blackburn Press, New Jersey
Birks HJB, Seppä H (2010) Late-Quaternary palaeoclimatic research in Fennoscandia – a historical review. Boreas 39:655–673
Birks HJ, Willis K (2008) Alpines, trees, and refugia in Europe. Plant Ecology & Diversity 1:147–160
Blockley SPE, Lane CS, Hardiman M, Rasmussen SO, Seierstad IK, Steffensen JP, Svensson A, Lotter AF, Turney CSM, Bronk Ramsey C (2012) Synchronisation of palaeoenvironmental records over the last 60,000 years, and an extended INTIMATE event stratigraphy to 48,000 b2k. Quaternary Science Reviews 36:2–10
Blytt A (1882) Die Theorie der wechselnden kontinentalen und insularen Klimate. Botanische Jahrbücher 2:1–50
Bond G, Showers W, Cheseby M, Lotti R, Almasi P, deMenocal P, Priore P, Cullen H, Hajdas I, Bonani G (1997) A pervasive millennial-scale cycle in North Atlantic Holocene and glacial climates. Science 278:1257–1266
Bradshaw RHW, Lindbladh M (2005) Regional spread and stand-scale establishment of *Fagus sylvatica* and *Picea abies* in Scandinavia. Ecology 86:1679–1686
Bradshaw RHW, Hannon GE, Lister AM (2003) A long-term perspective on ungulate-vegetation interactions. Forest Ecology and Management 181:267–280
Břízová E (1994) Vegetation of Holsteinian interglacial in Stonava-Horní Suchá (Ostrava region). Anthropozoikum 21:29–56
Břízová E (2009) Quaternary environmental history of the Čejčské Lake (S. Moravia, Czech Republic). Bulletin of Geosciences 84:637–652
Chytrý M, Danihelka J, Kubešová S, Lustyk P, Ermakov N, Hájek M, Hájková P, Kočí M, Otýpková Z, ... Pišút I (2008) Diversity of forest vegetation across a strong gradient of climatic continentality: Western Sayan Mountains, southern Siberia. Plant Ecology 196:61–83
Chytrý M, Danihelka J, Horsák M, Kočí M, Kubešová S, Lososová Z, Otýpková Z, Tichý L, Martynenko VB, Baisheva EZ (2010) Modern analogues from the Southern Urals provide

insights into biodiversity change in the early Holocene forests of Central Europe. Journal of Biogeography 37:767–780

Cui Q-Y, Gaillard M-J, Lemdahl G, Stenberg L, Sugita S, Zernova G (2014) Historical land-use and landscape change in southern Sweden and implications for present and future biodiversity. Ecology and Evolution 4:3555–3570

de Lafontaine G, Amasifuen Guerra CA, Ducousso A, Petit RJ (2014) Cryptic no more: soil macrofossils uncover Pleistocene forest microrefugia within a periglacial desert. New Phytologist 204:715–729

Divišová M, Šída P (2015) Plant use in the mesolithic period. Archaeobotanical data from the Czech Republic in a European context – a review. Interdisciplinaria Archaeologica 6:95–106

Dobrovolny L (2016) Density and spatial distribution of beech (*Fagus sylvatica* L.) regeneration in Norway spruce (*Picea abies* (L.) Karsten) stands in the central part of the Czech Republic. iForest – Biogeosciences and Forestry 9:666–672

Dreslerová D (2011) Přírodní prostředí a pravěké zemědělské společnosti (na území Čech) [Natural environment and prehistoric farmer societies in Bohemia]. PhD thesis, Charles University, Praha

Dreslerová D (2012) Human response to potential robust climate change around 5500 cal BP in the territory of Bohemia (Czech Republic). Interdisciplinaria Archaeologica 3:43–56

Dreslerová D, Kočár P (2012) Trends in cereal cultivation in the Czech Republic from the Neolithic to the Migration period (5500 B.C.–A.D. 580). Vegetation History and Archaeobotany 22:257–268

Dreslerová D, Waldhauser J, Abraham V, Kočár P, Meduna P, Sádlo J (2013) Bezdězsko–Dokesko v pravěku a laténské sídliště v Oknech [The Bezděz–Doksy region (Northern Bohemia) in prehistory and the La Tène settlement at Okna]. Archeologické rozhledy 65:535–573

Dudová L, Hájek M, Hájková P (2010) The origin and vegetation development of the Rejvíz pine bog and the history of the surrounding landscape during the Holocene. Preslia 82:223–246

Dudová L, Hájková P, Buchtová H, Opravilová V (2013) Formation, succession and landscape history of central-European summit raised bogs: a multiproxy study from the Hrubý Jeseník Mountains. The Holocene 23:230–242

Dudová L, Hájková P, Opravilová V, Hájek M (2014) Holocene history and environmental reconstruction of a Hercynian mire and surrounding mountain landscape based on multiple proxies. Quaternary Research 82:107–120

Emmer IM, Fanta J, Kobus AT, Kooijman A, Sevink J (1998) Reversing borealization as a means to restore biodiversity in central-European mountain forests – an example from the Krkonoše Mountains, Czech Republic. Biodiversity and Conservation 7:229–247

Engel Z, Nývlt D, Křížek M, Treml V, Jankovská V, Lisá L (2010) Sedimentary evidence of landscape and climate history since the end of MIS 3 in the Krkonoše Mountains, Czech Republic. Quaternary Science Reviews 29:913–927

Firbas F (1949) Spät- und nacheiszeitliche Waldgeschichte Mitteleuropas nördlich der Alpen. Allgemeine Waldgeschichte. Gustav Fischer, Jena

Frenzel B (1968) Pleistocene vegetation of Northern Eurasia – recent vegetation of Northern Eurasia resulted from a relentless contest between steppe and forest. Science 161:637–649

Giesecke T, Bennett KD, Birks HJB, Bjune AE, Bozilova E, Feurdean A, Finsinger W, Froyd C, Pokorný P, Rösch M (2011) The pace of Holocene vegetation change – testing for synchronous developments. Quaternary Science Reviews 30:2805–2814

Gradmann R (1933) Die Steppenheidetheorie. Geographische Zeitschrift 39:265–278

Hahne J (1992) Untersuchungen zur spät- und postglazialen Vegetationsgeschichte im nordöstlichen Bayern (Bayerisches Vogtland, Fichtelgebirge, Steinwald). Flora 187:169–200

Hájek M, Hájková P (2011) Vegetace slatinišť, přechodových rašelinišť a vrchovištních šlenků (*Scheuchzerio palustris-Caricetea nigrae*) [Vegetation of fens, transitional mires and bog hollows (*Scheuchzerio palustris-Caricetea nigrae*)]. In: Chytrý M (ed) Vegetace České republiky 3. Vodní a mokřadní vegetace [Vegetation of the Czech Republic 3. Aquatic and wetland vegetation]. Academia, Praha, pp 614–704

Helmens KF (2014) The last interglacial–glacial cycle (MIS 5–2) re-examined based on long proxy records from central and northern Europe. Quaternary Science Reviews 86:115–143

Horsák M, Chytrý M, Pokryszko BM, Danihelka J, Ermakov N, Hájek M, Hájková P, Kintrová K, Kočí M, ... Valachovič M (2010) Habitats of relict terrestrial snails in southern Siberia: lessons for the reconstruction of palaeoenvironments of full-glacial Europe. Journal of Biogeography 37:1450–1462

Hošek J, Pokorný P, Kubovčík V, Horáček I, Žáčková P, Kadlec J, Rojik F, Lisá L, Bučkuliaková S (2014) Late glacial climatic and environmental changes in eastern-central Europe: correlation of multiple biotic and abiotic proxies from the Lake Švarcenberk, Czech Republic. Palaeogeography, Palaeoclimatology, Palaeoecology 396:155–172

Hultberg T, Gaillard M-J, Grundmann B, Lindbladh M (2015) Reconstruction of past landscape openness using the Landscape Reconstruction Algorithm (LRA) applied on three local pollen sites in a southern Swedish biodiversity hotspot. Vegetation History and Archaeobotany 24:253–266

Huntley B (1993) Species-richness in north-temperate zone forests. Journal of Biogeography 20:163–180

Huntley B, Alfano MJ, Allen JRM, Pollard D, Tzedakis PC, de Beaulieu JL, Gruger E, Watts B (2003) European vegetation during marine oxygen isotope stage-3. Quaternary Research 59:195–212

Iversen J (1944) *Viscum, Hedera* and *Ilex* as climate indicators. Geologiska Föreningen i Stockholm Förhandlingar 66:463–483

Iversen J (1960) Problems of the early post-glacial forest development in Denmark. Danmarks Geologiske Undersøgelse IV række 4:1–32

Jäger KD (1969) Climatic character and oscillations of the subboreal period in the dry regions of the central European highlands. In: Wright HE Jr (ed) Quarternary geology and climate. National Academy of Sciences, Washington, DC, pp 38–42

Jamrichová E, Szabó P, Hédl R, Kuneš P, Bobek P, Pelánková B (2013) Continuity and change in the vegetation of a Central European oakwood. The Holocene 23:46–56

Jankovská V (1970) Ergebnisse der Pollen- und Grossrestanalyse des Moors "Velanská cesta" in Südböhmen. Folia Geobotanica et Phytotaxonomica 5:43–60

Jankovská V (1971) The development of vegetation on the western slopes of the Bohemian-Moravian uplands during the late Holocene period: a study based on pollen and macroscopic analyses. Folia Geobotanica et Phytotaxonomica 6:281–302

Jankovská V (2006) Late Glacial and Holocene history of Plešné Lake and its surrounding landscape based on pollen and palaeoalgological analyses. Biologia 61:371–385

Jankovská V (2008) Slovak and Moravian Carpathians in the last glacial period – an island of "Siberian taiga" in Europe. Phytopedon 7:122–130

Jankovská V, Pokorný P (2008) Forest vegetation of the last full-glacial period in the Western Carpathians (Slovakia and Czech Republic). Preslia 80:307–324

Jankovská V, Pokorný P (2013) Reevaluation of the palaeoenvironmental record of the former Komořanské jezero lake: late-glacial and Holocene palaeolimnology and vegetation development in north-western Bohemia, Czech Republic. Preslia 85:265–287

Jankovská V, Rybníček K (1988) The genus *Carex* in the late glacial and Holocene of Czechoslovakia. Aquatic Botany 30:23–37

Jeník J (1980) Struktura slatinné olšiny (*Carici elongatae-Alnetum*) v regresivní fázi [Structure of an alder carr (*Carici elongatae-Alnetum*) in a regression phase]. In: Zborník referátov z 3. zjazdu Slovenskej botanickej spoločnosti při SAV. Slovenská botanická spoločnosť, Zvolen, pp 53–57

Juřičková L, Horáčková J, Jansová A, Ložek V (2013) Mollusc succession of a prehistoric settlement area during the Holocene: a case study of the České středohoří Mountains (Czech Republic). The Holocene 23:1811–1823

Juřičková L, Horáčková J, Ložek V (2014a) Direct evidence of central European forest refugia during the last glacial period based on mollusc fossils. Quaternary Research 82:222–228

Juřičková L, Horsák M, Horáčková J, Abraham V, Ložek V (2014b) Patterns of land-snail succession in Central Europe over the last 15,000 years: main changes along environmental, spatial and temporal gradients. Quaternary Science Reviews 93:155–166

Kajtoch Ł, Cieślak E, Varga Z, Paul W, Mazur MA, Sramkó G, Kubisz D (2016) Phylogeographic patterns of steppe species in eastern Central Europe: a review and the implications for conservation. Biodiversity and Conservation 25:2309–2339

Knipping M (1997) Pollenanalytische Untersuchungen zur Siedlungsgeschichte des Oberpfälzer Waldes. Telma 27:61–74

Kočár P, Čech P, Kozáková R, Kočárová R (2010) Environment and economy of the early medieval settlement in Žatec. Interdisciplinaria Archaeologica 1:45–60

Kočár P, Pokorná A, Komárková V (2015) Synantropní flóra pravěkých sídlišť ve světle makrozbytkové analýzy [Anthropogenic flora of prehistoric settlements according to macro-remain analysis]. Zprávy České botanické společnosti 50:301–314

Kozáková R, Pokorný P, Havrda J, Jankovská V (2009) The potential of pollen analyses from urban deposits: multivariate statistical analysis of a data set from the medieval city of Prague, Czech Republic. Vegetation History and Archaeobotany 18:477–488

Kozáková R, Šamonil P, Kuneš P, Novák J, Kočár P, Kočárová R (2011) Contrasting local and regional Holocene histories of *Abies alba* in the Czech Republic in relation to human impact: evidence from forestry, pollen and anthracological data. The Holocene 21:431–444

Kozáková R, Pokorný P, Mařík J, Čulíková V, Boháčová I, Pokorná A (2014) Early to high medieval colonization and alluvial landscape transformation of the Labe valley (Czech Republic): evaluation of archaeological, pollen and macrofossil evidence. Vegetation History and Archaeobotany 23:701–718

Kozáková R, Pokorný P, Peša V, Danielisová A, Čuláková K, Svitavská-Svobodová H (2015) Prehistoric human impact in the mountains of Bohemia. Do pollen and archaeological data support the traditional scenario of a prehistoric "wilderness"? Review of Palaeobotany and Palynology 220:29–43

Kuneš P, Pokorný P, Jankovská V (2007) Post-glacial vegetation development in sandstone areas of the Czech Republic. In: Härtel H, Cílek V, Herben T, Jackson A, Williams R (eds) Sandstone landscapes. Academia, Praha, pp 244–257

Kuneš P, Pelánková B, Chytrý M, Jankovská V, Pokorný P, Petr L (2008a) Interpretation of the last-glacial vegetation of eastern-central Europe using modern analogues from southern Siberia. Journal of Biogeography 35:2223–2236

Kuneš P, Pokorný P, Šída P (2008b) Detection of the impact of early Holocene hunter-gatherers on vegetation in the Czech Republic, using multivariate analysis of pollen data. Vegetation History and Archaeobotany 17:269–287

Kuneš P, Abraham V, Kovářík O, Kopecký M, PALYCZ contributors (2009) Czech Quaternary Palynological Database – PALYCZ: review and basic statistics of the data. Preslia 81:209–238

Kuneš P, Odgaard BV, Gaillard M (2011) Soil phosphorus as a control of productivity and openness in temperate interglacial forest ecosystems. Journal of Biogeography 38:2150–2164

Kuneš P, Kjærsgaard Sørensen M, Buylaert J-P, Murray AS, Houmark-Nielsen M, Odgaard BV (2013) A new middle Pleistocene interglacial record from Denmark: chronostratigraphic correlation, palaeovegetation and fire dynamics. Boreas 42:596–612

Kuneš P, Svobodová-Svitavská H, Kolář J, Hajnalová M, Abraham V, Macek M, Tkáč P, Szabó P (2015) The origin of grasslands in the temperate forest zone of east-central Europe: long-term legacy of climate and human impact. Quaternary Science Reviews 116:15–27

Küster H (1997) The role of farming in the postglacial expansion of beech and hornbeam in the oak woodlands of central Europe. The Holocene 7:239–242

Lang G (1994) Quartäre Vegetationsgeschichte Europas: Methoden und Ergebnisse. G. Fischer, Jena

Lange E, Christl A, Joosten H (2005) Ein Pollendiagramm aus der Mothäuser Heide im oberen Erzgebirge unweit des Grenzüberganges Reitzenhain. In: Sachenbacher P, Einicke R, Beier H-J (eds) Kirche und geistiges Leben im Prozess des mittelalterlichen Landesausbaus in Ostthüringen/Westsachsen. Beier & Beran, Langenweissbach, pp 153–169

Latałowa M, van der Knaap WO (2006) Late Quaternary expansion of Norway spruce *Picea abies* (L.) Karst. in Europe according to pollen data. Quaternary Science Reviews 25:2780–2805

Lewandowski A (1997) Genetic relationships between European and Siberian larch, *Larix* spp. (*Pinaceae*), studied by allozymes. Is the Polish larch a hybrid between these two species? Plant Systematics and Evolution 204:65–73

Lisiecki LE, Raymo ME (2005) A Pliocene-Pleistocene stack of 57 globally distributed benthic $\delta^{18}O$ records. Paleoceanography 20:PA1003

Lowe JJ, Ammann B, Birks HH, Björck S, Coope GR, Cwynar L, de Beaulieu JL, Mott RJ, Peteet DM, Walker MJC (1994) Climatic changes in areas adjacent to the North Atlantic during the last glacial-interglacial transition (14–9 ka BP): a contribution to IGCP-253. Quaternary Science Reviews 9:185–198

Ložek V (1964) Quartärmollusken der Tschechoslowakei. Nakladatelství Československé akademie věd, Praha

Ložek V (2007) Zrcadlo minulosti. Česká a slovenská krajina v kvartéru [Mirror of the past. Czech and Slovak landscape in the Quaternary]. Dokořán, Praha

Magyari EK, Chapman JC, Passmore DG, Allen JRM, Huntley JP, Huntley B (2010) Holocene persistence of wooded steppe in the Great Hungarian Plain. Journal of Biogeography 37:915–935

Magyari EK, Kuneš P, Jakab G, Sümegi P, Pelánková B, Schäbitz F, Braun M, Chytrý M (2014) Late Pleniglacial vegetation in eastern-central Europe: are there modern analogues in Siberia? Quaternary Science Reviews 95:60–79

Mitchell FJG (2005) How open were European primeval forests? Hypothesis testing using palaeoecological data. Journal of Ecology 93:168–177

Neuhäuslová Z, Blažková D, Grulich V, Husová M, Chytrý M, Jeník J, Jirásek J, Kolbek J, Kropáč Z, … Sádlo J (1998) Mapa potenciální přirozené vegetace České republiky [Map of potential natural vegetation of the Czech Republic]. Academia, Praha

Novák J, Petr L, Treml V (2010) Late-Holocene human-induced changes to the extent of alpine areas in the East Sudetes, Central Europe. The Holocene 20:895–905

Novák J, Sádlo J, Svobodová-Svitavská H (2012) Unusual vegetation stability in a lowland pine forest area (Doksy region, Czech Republic). The Holocene 22:947–955

Novák J, Svoboda J, Šída P, Prostředník J, Pokorný P (2015) A charcoal record of Holocene woodland succession from sandstone rock shelters of North Bohemia (Czech Republic). Quaternary International 366:25–36

Novák J, Abraham V, Kočár P, Petr L, Kočárová R, Nováková K, Houfková P, Jankovská V, Vaněček Z (2017) Middle- and upper-Holocene woodland history in central Moravia (Czech Republic) reveals biases of pollen and anthracological analysis. The Holocene 27:349–360

Nováková D (2000) Rekonstrukce paleoekologických poměrů rašelinišť NPR Adršpašsko-teplické skály metodou analýzy makrozbytků [Reconstruction of palaeoecological conditions of peat bogs in the protected area of Adršpach-Teplice rocks based on plant macrofossil analysis]. MSc thesis, Charles University, Praha

Opravil E (1972) Výsledky analýz makrozbytků z rašeliny ve Františkových Lázních [Results of plant macrofossil analysis from peat in Františkovy Lázně]. Památky archeologické 63:429–431

Overballe-Petersen MV, Nielsen AB, Bradshaw RHW (2013) Quantitative vegetation reconstruction from pollen analysis and historical inventory data around a Danish small forest hollow. Journal of Vegetation Science 24:755–771

Past Interglacials Working Group of PAGES (2016) Interglacials of the last 800,000 years. Reviews of Geophysics 54:162–219

Pelánková B, Chytrý M (2009) Surface pollen-vegetation relationships in the forest-steppe, taiga and tundra landscapes of the Russian Altai Mountains. Review of Palaeobotany and Palynology 157:253–265

Peša V, Kozáková R (2012) Die nacheiszeitliche Landschaftsentwicklung des Lausitzer Gebirges. In: Puttkammer T (ed) Auf den Spuren der Germanen. Begleitband zur Wanderausstellung. Museum der Westlausitz, Kamenz, pp 128–143

Petr L, Novák J (2014) High vegetation and environmental diversity during the late glacial and early Holocene on the example of lowlands in the Czech Republic. Biologia 69:847–862

Petr L, Žáčková P, Grygar TM, Píšková A, Křížek M, Treml V (2013) Šúr, a former late-glacial and Holocene lake at the westernmost margin of the Carpathians. Preslia 85:239–263

Petr L, Sádlo J, Žáčková P, Lisá L, Novák J, Rohovec J, Pokorný P (2014) Late-Glacial and Holocene environmental history of an oxbow wetland in the Polabí lowland (river Elbe, Czech Republic); a context-dependent interpretation of a multi-proxy analysis. Folia Geobotanica 49:137–162

Pokorná A, Houfková P, Novák J, Bešta T, Kovačiková L, Nováková K, Zavřel J, Starec P (2014) The oldest Czech fishpond discovered? An interdisciplinary approach to reconstruction of local vegetation in mediaeval Prague suburbs. Hydrobiologia 730:191–213

Pokorný P (2002) A high-resolution record of Late-Glacial and Early-Holocene climatic and environmental change in the Czech Republic. Quaternary International 91:101–122

Pokorný P (2003) Rostlinné makrozbytky [Plant macrofossils]. In: Svoboda J (ed) Mezolit severních Čech [Mesolithic of northern Bohemia]. Archeologický ústav AV ČR, Brno, pp 272–273

Pokorný P, Jankovská V (2000) Long-term vegetation dynamics and the infilling process of a former lake (Švarcenberk, Czech Republic). Folia Geobotanica 35:433–457

Pokorný P, Kuneš P (2005) Holocene acidification process recorded in three pollen profiles from Czech sandstone and river terrace environments. Ferrantia 44:101–107

Pokorný P, Klimešová J, Klimeš L (2000) Late Holocene history and vegetation dynamics of a floodplain alder carr: a case study from eastern Bohemia, Czech Republic. Folia Geobotanica 35:43–58

Pokorný P, Boenke N, Chytráček M, Nováková K, Sádlo J, Veselý J, Kuneš P, Jankovská V (2006) Insight into the environment of a pre-Roman Iron Age hillfort at Vladař, Czech Republic, using a multi-proxy approach. Vegetation History and Archaeobotany 15:419–433

Pokorný P, Kuneš P, Abraham V (2008) Holocenní vývoj vegetace v Českém Švýcarsku [The Holocene vegetation development in Bohemian Switzerland]. In: Bauer P, Kopecký V, Šmucar J (eds) Labské pískovce – historie, příroda a ochrana území [Elbe Sandstones – history, nature and conservation of the area]. AOPK, Správa CHKO Labské pískovce, Děčín, pp 35–49

Pokorný P, Sádlo J, Bernardová A (2010a) Holocene history of *Cladium mariscus* (L.) Pohl in the Czech Republic. Implications for species population dynamics and palaeoecology. Acta Palaeobotanica 50:65–76

Pokorný P, Šída P, Chvojka O, Žáčková P, Kuneš P, Světlík I, Veselý J (2010b) Palaeoenvironmental research of the Schwarzenberg Lake, southern Bohemia, and exploratory excavations of this key Mesolithic archaeological area. Památky archeologické 101:5–38

Pokorný P, Chytrý M, Juřičková L, Sádlo J, Novák J, Ložek V (2015) Mid-Holocene bottleneck for central European dry grasslands: did steppe survive the forest optimum in northern Bohemia, Czech Republic? The Holocene 25:716–726

Pokorný P, Novák J, Šída P, Divišová M, Kozáková R, Abraham V (2017) I. Vývoj vegetace severočeských pískovcových území od pozdního glaciálu po střední holocén [Vegetation development of northern-Bohemian sandstone areas since the Late Glacial to the Middle Holocene]. In: Svoboda J (ed) Mezolit severních Čech 2 [Mesolithic of northern Bohemia 2]. Archeologický ústav AV ČR, Brno, pp 11–37

Pop E, Bakels C (2015) Semi-open environmental conditions during phases of hominin occupation at the Eemian interglacial basin site Neumark-Nord 2 and its wider environment. Quaternary Science Reviews 117:72–81

Prentice IC (1985) Pollen representation, source area, and basin size: toward a unified theory of pollen analysis. Quaternary Research 23:76–86

Ruddiman WF, Kutzbach JE, Vavrus SJ (2011) Can natural or anthropogenic explanations of late-Holocene $CO_2$ and $CH_4$ increases be falsified? The Holocene 21:865–879

Rybníček K (1973) Comparison of the present and past mire communities of Central Europe. In: Birks HJB, West RG (eds) Quaternary plant ecology. Blackwell, Oxford, pp 237–261

Rybníček K (1983) The environmental evolution and infilling process of a former lake near Vracov (Czechoslovakia). Hydrobiologia 103:247–250

Rybníček K, Rybníčková E (1968) The history of flora and vegetation on the Bláto mire in south-east Bohemia (palaeoecological study). Folia Geobotanica et Phytotaxonomica 3:117–142
Rybníček K, Rybníčková E (1978) Palynological and historical evidence of virgin coniferous forests at middle altitudes in Czechoslovakia. Vegetatio 36:95–103
Rybníček K, Rybníčková E (1996) Czech and Slovak Republics. In: Berglund BE, Birks HJB, Ralska-Jasiewiczowa M, Wright HE (eds) Paleoecological events during the last 15000 years. Wiley, Chichester, pp 488–490
Rybníček K, Rybníčková E (2008) Upper Holocene dry land vegetation in the Moravian–Slovakian borderland (Czech and Slovak Republics). Vegetation History and Archaeobotany 17:701–711
Rybníčková E (1974) Die Entwicklung der Vegetation und Flora im südlichen Teil der Böhmisch-Mährischen Höhe während des Spätglazials und Holozäns. Academia, Praha
Rybníčková E, Rybníček K (1972) Erste Ergebnisse paläogeobotanischer Untersuchungen des Moores bei Vracov, Südmähren. Folia Geobotanica et Phytotaxonomica 7:285–308
Rybníčková E, Rybníček K (2014) Palaeovegetation in the Pavlovské vrchy hills region (South Moravia, Czech Republic) around 25,000 BP: the Bulhary core. Vegetation History and Archaeobotany 23:719–728
Rybníčková E, Hájková P, Rybníček K (2005) The origin and development of spring fen vegetation and ecosystems–palaeogeobotanical results. In: Poulíčková A, Hájek M, Rybníček K (eds) Ecology and palaeoecology of spring fens of the West Carpathians. Palacký University, Olomouc & Academy of Sciences of the Czech Republic, Masaryk University Brno, Olomouc, pp 29–57
Sádlo J, Pokorný P, Hájek P, Dreslerová D, Cílek V (2005) Krajina a revoluce: významné přelomy ve vývoji kulturní krajiny českých zemí [Landscape and revolution: significant shifts in the development of cultural landscape in the Bohemian lands]. Malá Skála, Praha
Sandom CJ, Ejrnæs R, Hansen MDD, Svenning J-C (2014) High herbivore density associated with vegetation diversity in interglacial ecosystems. Proceedings of the National Academy of Sciences of the USA 111:4162–4167
Sernander R (1908) On the evidences of postglacial changes of climate furnished by the peat-mosses of northern Europe. Geologiska Föreningen i Stockholm Förhandlingar 30:465–473
Šizling AL, Pokorný P, Juřičková L, Horáčková J, Abraham V, Šizlingová E, Ložek V, Tjørve E, Tjørve KMC, Kunin W (2016) Can people change the ecological rules that appear general across space? Global Ecology and Biogeography 25:1072–1084
Stebich M, Litt T (1997) Das Georgenfelder Hochmoor ein Archiv für Vegetations-, Siedlungs- und Bergbaugeschichte. Leipziger Geowissenschaften 5:209–216
Štor T, Sádlo J, Abraham V, Martínek K (2016) Změny fluviálního stylu během svrchního pleistocénu a holocénu na příkladu nivy řeky Ploučnice v severních Čechách [The fluvial style changes during the Upper Pleistocene and Holocene: case study of Ploučnice River, Northern Bohemia]. Zprávy o geologických výzkumech 49:123–127
Stuiver M, Grootes PM, Braziunas TF (1995) The GISP2 $\delta^{18}O$ climate record of the past 16,500 years and the role of the sun, ocean, and volcanoes. Quaternary Research 44:341–354
Suda T (2012) Historie vegetace Chebské pánve ze sedimentárního záznamu lokality SOOS [History of vegetational development in the Cheb Basin from sedimentary record of the SOOS national nature reserve]. MSc thesis, Charles University, Praha
Sugita S (1994) Pollen representation of vegetation in Quaternary sediments: theory and method in patchy vegetation. Journal of Ecology 82:881–897
Sugita S (2007a) Theory of quantitative reconstruction of vegetation I: pollen from large sites REVEALS regional vegetation composition. The Holocene 17:229–241
Sugita S (2007b) Theory of quantitative reconstruction of vegetation II: all you need is LOVE. The Holocene 17:243–257
Švarcová MG (2012) Postglaciální historie lokálních fenoménů horské vegetace západních Čech [Postglacial vegetation history of local phenomena in western Bohemia]. MSc thesis, Charles University, Praha
Svenning J-C (2002) A review of natural vegetation openness in north-western Europe. Biological Conservation 104:133–148

Svenning J-C (2003) Deterministic Plio-Pleistocene extinctions in the European cool-temperate tree flora. Ecology Letters 6:646–653

Svobodová H (1989) Rekonstrukce přírodního prostředí a osídlení v okolí Mistřína. Palynologická studie [Reconstruction of natural environment and human settlement round about Mistřín. A palynological study]. Památky archeologické 80:188–206

Svobodová H (1997) Die Entwicklung der Vegetation in Südmähren (Tschechien) während des Spätglazials und Holozäns – eine palynologische Studie. Verhandlungen der Zoologisch-Botanischen Gesellschaft in Österreich 134:317–356

Szabó P, Kuneš P, Svobodová-Svitavská H, Švarcová MG, Křížová L, Suchánková S, Müllerová J, Hédl R (2017) Using historical ecology to reassess the conservation status of coniferous forests in Central Europe. Conservation Biology 31:150–160

Tollefsrud MM, Kissling R, Gugerli F, Johnsen O, Skrøppa T, Cheddadi R, van der Knaap W, Latałowa M, ... Sperisen C (2008) Genetic consequences of glacial survival and postglacial colonization in Norway spruce: combined analysis of mitochondrial DNA and fossil pollen. Molecular Ecology 17:4134–4150

Tzedakis PC, Hooghiemstra H, Pälike H (2006) The last 1.35 million years at Tenaghi Philippon: revised chronostratigraphy and long-term vegetation trends. Quaternary Science Reviews 25:3416–3430

Tzedakis PC, Emerson BC, Hewitt GM (2013) Cryptic or mystic? Glacial tree refugia in northern Europe. Trends in Ecology and Evolution 28:696–704

van der Hammen T, Wijmstra TA, Zagwijn WH (1971) The flora record of the late Cenozoic of Europe. In: Turekian KK (ed) The late Cenozoic glacial ages. Yale University Press, New Haven, pp 391–424

van Raden UJ, Colombaroli D, Gilli A, Schwander J, Bernasconi SM, van Leeuwen J, Leuenberger M, Eicher U (2013) High-resolution late-glacial chronology for the Gerzensee lake record (Switzerland): $\delta^{18}O$ correlation between a Gerzensee-stack and NGRIP. Palaeogeography, Palaeoclimatology, Palaeoecology 391, Part B:13–24

Vegas-Vilarrúbia T, Rull V, Montoya E, Safont E (2011) Quaternary palaeoecology and nature conservation: a general review with examples from the neotropics. Quaternary Science Reviews 30:2361–2388

Vera FWM (2000) Grazing ecology and forest history. CABI Publishing, New York

Vočadlová K, Petr L, Žáčková P, Křížek M, Křížová L, Hutchinson SM, Šobr M (2015) The Lateglacial and Holocene in Central Europe: a multi-proxy environmental record from the Bohemian Forest, Czech Republic. Boreas 44:769–784

Wagner S, Litt T, Sánchez-Goñi M-F, Petit RJ (2015) History of *Larix decidua* Mill. (European larch) since 130 ka. Quaternary Science Reviews 124:224–247

Walker MJC (1995) Climatic changes in Europe during the last glacial/interglacial transition. Quarternary International 28:63–76

Walker MJC, Berkelhammer M, Björck S, Cwynar LC, Fisher DA, Long AJ, Lowe JJ, Newnham RM, Rasmussen SO, Weiss H (2012) Formal subdivision of the Holocene series/epoch: a discussion paper by a working group of INTIMATE (Integration of ice-core, marine and terrestrial records) and the Subcommission on Quaternary Stratigraphy (International Commission on Stratigraphy). Journal of Quaternary Science 27:649–659

Wardle DA, Walker LR, Bardgett RD (2004) Ecosystem properties and forest decline in contrasting long-term chronosequences. Science 305:509–513

Wäreborn I (1969) Land molluscs and their environments in an oligotrophic area in southern Sweden. Oikos 20:461–479

Willis KJ, van Andel TH (2004) Trees or no trees? the environments of central and eastern Europe during the Last Glaciation. Quaternary Science Reviews 23:2369–2387

Zagwijn WH (1960) Aspects of the Pliocene and early Pleistocene vegetation in the Netherlands. Uitgevers-mij. "Ernest van Aelst", Maastricht

Žák K, Ložek V, Kadlec J, Hladíková J, Cílek V (2002) Climate-induced changes in Holocene calcareous tufa formations, Bohemian Karst, Czech Republic. Quarternary International 91:137–152

# Chapter 7
# Current Vegetation of the Czech Republic

**Milan Chytrý**

**Abstract** The Czech Republic belongs to the temperate broad-leaved deciduous forest biome for its largest part, but two of its dry lowland areas belong to the forest-steppe biome. Landscapes corresponding to the coniferous forest biome and alpine tundra occur mainly as temperate orobiomes in small areas. Eight altitudinal vegetation belts from lowland to alpine are distinguished. There are high diversities of different vegetation types mainly in deep river valleys in the Bohemian Massif, karst areas, sandstone pseudokarst areas, on solitary volcanic hills, in glacial cirques, lowland riverine landscapes and serpentinite areas. Potential natural vegetation across most of the country is deciduous and mixed forests of beech, oak, hornbeam and noble hardwoods, and coniferous forests of spruce and fir. However, large areas of these forests have been cleared or converted into forest plantations. Open landscape is covered mainly by arable land and perennial grassland. Diversity, ecology, distribution, history and dynamics of the different vegetation types defined in the national vegetation classification are described here in detail.

## 7.1 Introduction

Temperate zone of Central Europe, which includes the whole of the Czech Republic, is characterized by potential dominance of broad-leaved deciduous forest (Bohn et al. 2000–2003; Ellenberg and Leuschner 2010). However, it is a densely populated area with a long history of human impact, which results in a heterogeneous landscape mosaic of forests, arable land, perennial grasslands and human settlements. This corresponds to a relatively high diversity of vegetation types, ranging from natural to anthropogenic. In a European context, the diversity of vegetation types in the Czech Republic is slighly below the average, lagging behind countries with high mountains or a sea coast and those that include both temperate and Mediterranean areas. Nevertheless it is higher than in North-western and Northern Europe (Jiménez-Alfaro et al. 2014). Detailed phytosociological classification and description of the different vegetation types in this country is available in the

M. Chytrý (✉)
Department of Botany and Zoology, Masaryk University,
Kotlářská 2, 611 37 Brno, Czech Republic
e-mail: chytry@sci.muni.cz



M. Chytrý et al. (eds.), *Flora and Vegetation of the Czech Republic*, Plant and Vegetation 14, DOI 10.1007/978-3-319-63181-3_7

monograph Vegetation of the Czech Republic (Chytrý 2007, 2009, 2011, 2013), simplified classification in the Habitat Catalogue of the Czech Republic (Chytrý et al. 2010) and the hypothetical distribution of natural vegetation is provided in the Map of Potential Natural Vegetation of the Czech Republic (Neuhäuslová et al. 1997, 1998, 2001). This chapter summarizes general vegetation patterns and provides a brief overview of the main vegetation types with remarks on their ecology, distribution, history and dynamics.

## 7.2 Biomes

In global biome classifications, the whole of the Czech Republic is assigned to the temperate broad-leaved and mixed forest biome (e.g. Olson et al. 2001), also called the nemoral zonobiome in Walter's biome classification (Breckle 2002). This biome is predominant in most European mid-latitude areas, from the Atlantic seaboard to the Southern Ural Mts. Although the broad-leaved deciduous forest biome covers the largest part of the Czech Republic, there are smaller areas within the country with extrazonal and borderline occurrences of other biomes, namely forest-steppe, coniferous forest (taiga) and mountain tundra (Fig. 7.1).

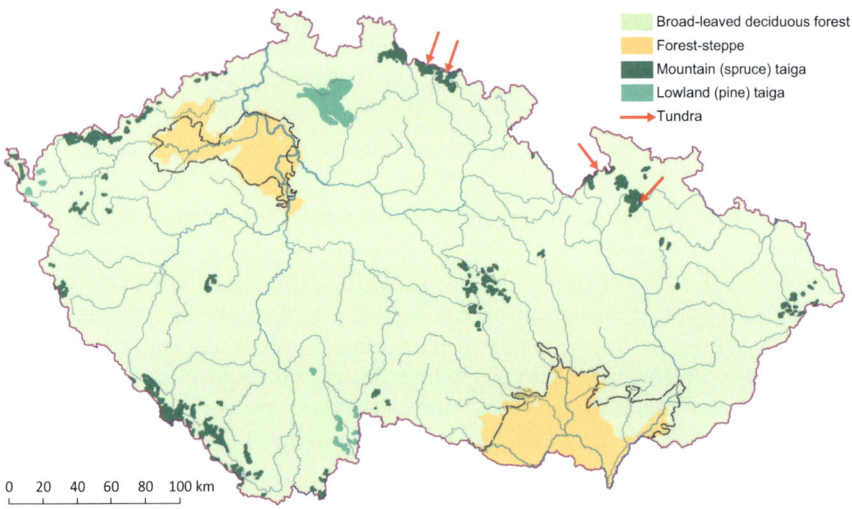


**Fig. 7.1** Approximate distribution of Czech biomes. Forest-steppe biome is delimited in two ways: (1) as areas with annual precipitation lower than 525 mm and mean annual temperature higher than 8.25 °C (*yellow areas*; climate data from Tolasz et al. 2007) or (2) as grouped phytogeographical districts of Skalický (1988) with predominance of forest-steppe vegetation (*black broken line*). The taiga areas are partly delimited based on the interpretation of mapping units corresponding to this biome on the Map of Potential Natural Vegetation of the Czech Republic (Neuhäuslová et al. 1997). The tundra biome corresponds to areas above the natural timberline; because of their small size, they are indicated by *red arrows* (all the maps in this chapter were prepared by O. Hájek)

**The broad-leaved deciduous forest biome** occurs from the lowlands up to 1000–1200 m a.s.l., but is absent from the driest lowlands, where it is replaced by forest-steppe. Dominant zonal types of natural vegetation are forests of *Fagus sylvatica* with an admixture of *Abies alba* and *Picea abies* at middle and high altitudes and of *Carpinus betulus* at low altitudes. Other broad-leaved trees (*Acer platanoides*, *A. pseudoplatanus*, *Alnus glutinosa*, *A. incana*, *Fraxinus excelsior*, *Quercus petraea*, *Q. robur*, *Tilia cordata*, *T. platyphyllos*, *Ulmus glabra* and *U. laevis*) occur especially at sites where the composition of tree species has changed as a result of past forest management or natural disturbances, or at azonal sites such as floodplains, scree slopes, ravines, dry slopes with shallow soil and rock outcrops. Natural treeless vegetation is very rare and spatially restricted, e.g. on cliffs. Large areas of potential broad-leaved forest are now covered by coniferous plantations, arable land and different types of semi-dry, mesic and wet grasslands.

**The forest-steppe biome** is a zonoecotone between the nemoral and continental zonobiome in Walter's classification. It is present in two lowland areas in the Czech Republic that are characterized by a dry and warm climate. Forest-steppe in northern and central Bohemia, which occurs in the rain shadow of the Krušné hory Mts, is an isolated island of this biome. In contrast, forest-steppe in southern Moravia, which occurs in the less pronounced rain shadow of the Bohemian-Moravian Highlands, is a part of the continuous forest-steppe area that extends from the central part of the Pannonian Basin in Hungary through eastern Austria and south-western Slovakia to southern Moravia. Both the Bohemian and Moravian forest-steppe areas are generally characterized by an annual precipitation lower than ~525 mm and a mean annual temperature higher than ~8.25 °C. The temperature regime of the former area is less continental, with warmer winters and cooler springs, summers and autumns, but these differences between regions do not exceed 1 °C in the mean monthly temperature (Tolasz et al. 2007). Palaeoecological evidence based on both pollen and mollusc records (Ložek 2011; Kuneš et al. 2015; Pokorný et al. 2015) indicates continuous existence of forest-steppe in these areas throughout the Holocene, including the Middle Holocene period with a higher precipitation (see Chap. 6, this book).

Forest-steppe is characterized by a mosaic of forests and open land. The former include *Quercus pubescens* and *Q. petraea* forests on dry soils, *Carpinus betulus* forests on mesic soils and floodplain forests along rivers. Treeless vegetation includes various types of dry grasslands, ranging from rock-outcrop steppe through short-grass steppe with *Festuca* and *Stipa* (on shallow rendzina/ranker or deep chernozem over loess) to semi-dry tall-grass steppe with *Brachypodium pinnatum*. Characteristic patterns of the exposure-related forest-steppe (Hais et al. 2016) develop in hilly landscapes, with forests confined to north-facing and dry grasslands to south-facing slopes. There are intermittently wet meadows on the floodplains and different types of dry to mesic scrub at various sites. The forest-steppe biome also includes inland saline vegetation around mineral springs, which was nearly destroyed by draining except for a few remaining fragments. Vegetation in forest-steppe landscapes has been continuously managed by humans since the onset of the Neolithic. Large areas of chernozem steppes were converted into arable land

and the area of forest reduced. The relative extent of forest and non-forest areas that would occur under natural conditions is unknown.

**The coniferous forest (taiga) biome** is represented by two distinct types in the Czech Republic, the spruce taiga developed in the mountains and pine taiga occurring in some lowland basins. Mountain taiga, dominated by *Picea abies* (dark taiga), occurs on high mountain ranges in the belt between the mid-altitude broad-leaved deciduous forests and the timberline. It is a temperate orobiome according to Walter's biome classification, but it shares many features with boreal taiga. In flat or gently undulating landscapes, especially on summit plateaus of the Šumava Mts, spruce forests are associated with bogs, either open or covered with shrubby *Pinus mugo* or arboreal *Pinus uncinata* subsp. *uliginosa*. The replacement vegetation of the mountain taiga includes low-productive grasslands.

The areas of lowland (pine) taiga in the Czech Republic bear features of the hemiboreal forest zone rather than of true boreal taiga. These areas include basins near Doksy in northern Bohemia, Cheb Basin in western Bohemia and Třeboň Basin in southern Bohemia. The mean altitude of the bottoms of these basins is around 290 m in the former and around 450 m in the latter two. All of these basins are characterized by nutrient-poor acidic bedrock (siliceous sand, sandstone, but also less acidic clay), frequent accumulations of cold air resulting in temperature inversions and a high groundwater table, which is currently maintained by the systems of fishponds built in the Middle Ages and early Modern Period. Dominant vegetation in the lowland taiga is acidophilous forest of *Pinus sylvestris* (light taiga) with dwarf shrubs, especially *Vaccinium* spp. and local admixtures of *Picea abies* and broad-leaved deciduous trees, especially *Quercus petraea* or *Q. robur.* The areas of lowland taiga also contain bogs, most of which are naturally forested with *Pinus sylvestris* or *P. uncinata* subsp. *uliginosa*, and minerotrophic mires. Although the persistence of lowland taiga may be partly due to human activity (nutrient depletion by litter raking and forest grazing, replacement of broad-leaved trees by pine), it could be natural vegetation especially in the Doksy area, which has been poorly inhabited because of its infertile soils (predominantly lowland podzols). Analyses of fossil charcoal and pollen records from this region indicate that pine forests dominated this area throughout the Holocene, possibly maintained by recurrent wildfires (Jankovská 1992; Novák et al. 2012).

**The alpine tundra biome** is also a part of the temperate orobiome, although with some similarities and historical links to Arctic tundra. It occurs above the timberline on the highest summits of the Sudetes (Krkonoše, Králický Sněžník and Hrubý Jeseník Mts), being a remnant of a presumably extensive area occupied by tundra at middle and high altitudes in the full glacial. The hypothesis of continuous occurrence of the tundra biome on the Krkonoše summits throughout the Holocene is supported by the occurrence of patterned grounds (Treml et al. 2010), which would probably be destroyed if overgrown by forest. Further support for this hypothesis is the occurrence of presumably relict tundra species (e.g. *Carex bigelowii*, *Pedicularis sudetica* and *Rubus chamaemorus*) and of neoendemics of the genus *Hieracium* confined to open tundra habitats (Soukupová et al. 1995; see also Table 3.1 in Chap. 3, this book). The alpine tundra includes mainly grasslands consisting of *Avenella flexuosa*, *Festuca supina* and *Nardus stricta*, and heathlands of *Calluna vulgaris*. The surface microtopography of the mires on the summit plateaus of the Krkonoše Mts is similar to the string-and-flark pattern found in Scandinavian aapa mires (Jeník and Soukupová 1992).

## 7.3 Altitudinal Vegetation Belts

In general, altitude is correlated with both temperature (negatively) and precipitation (positively) in the Czech Republic: low altitudes are usually warm and dry, whereas high altitudes are cool and wet. However, there are some anomalies, most notably in the low-altitude areas of north-eastern Moravia and Silesia. Therefore, altitudes of individual vegetation belts can vary across the country. Consequently, there are different classifications of vegetation belts in Czech botanical and geographical literature (Holub and Jirásek 1967; Kučera 2000). Currently the widely accepted classification is the one proposed by Skalický (1988) with eight altitudinal vegetation belts:

**Lowland (planar) belt** includes areas adjacent to large rivers at altitudes below 210 m, locally below 240 m, with deciduous broad-leaved floodplain forests, wetlands, inundated meadows, sandy grasslands and saline habitats. Main crops include wheat, oil-seed rape, maize, sugar beet, vegetables, barley, grapes, hop and thermophilous fruit trees such as apricots and peaches.

**Colline belt** includes upland areas at altitudes below 500 m, although on some hills in dry and warm areas this belt locally reaches to higher altitudes. Typical vegetation is thermophilous oak forests, oak-hornbeam forests, steppe grasslands and dry and mesic scrub. The lowland and colline belts together include the entire area of the forest-steppe biome and also some low-altitude areas of the broad-leaved deciduous forest biome. The spectrum of planted crops is similar to that in the lowland belt.

**Upper-colline (supracolline) belt** overlaps the colline belt in its altitudinal range, but occurs in cooler and wetter areas, especially in basins and on upland plateaus. It is characterized by oak-hornbeam, acidophilous oak, fir and beech forests, and in basins also by mires. Deforested areas were converted not only into arable land but also different types of wet, mesic or dry grasslands. Main crops include oil-seed rape, wheat, barley, rye and potatoes. Fruit trees are mainly apples, pears, cherries and plums.

**Submontane belt** occurs mainly between altitudes of 450 and 800 m, although it can locally occur at higher or lower altitudes. It is dominated by beech or fir-beech forests, in deforested areas also by mesic and wet meadows and pastures. Main crops are wheat, barley, oats, rye, potatoes and oil-seed rape; fruit trees are the same as in the upper-colline belt.

**Montane belt** includes mainly areas at altitudes of 750–1100 m. It is characterized by fir-spruce-beech forests and spruce forests at water-logged sites in association with minerotrophic mires and bogs. Replacement vegetation includes

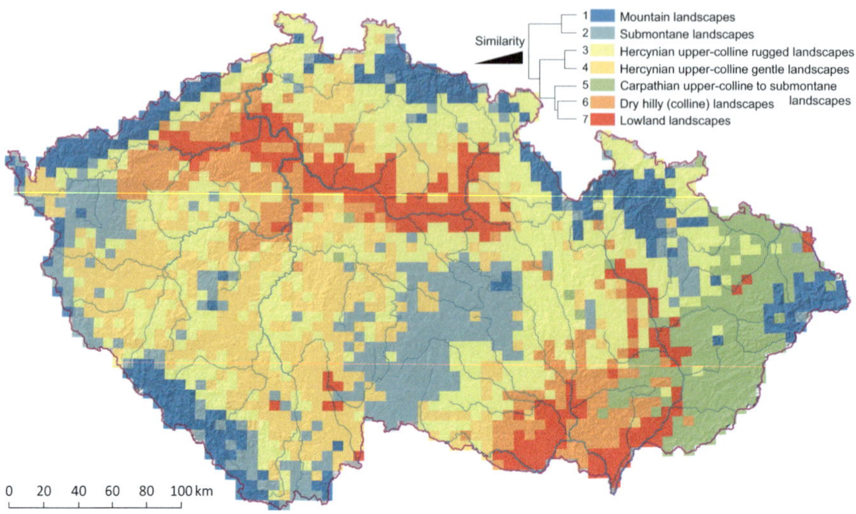


**Fig. 7.2** Landscape types defined by cluster analysis of habitat types (= broad vegetation types) recorded in grid cells (from Divíšek et al. 2014, slightly modified)

mesotrophic and oligotropic grasslands used for grazing or hay making. In the past, potato fields occurred in this belt mainly at lower altitudes, but they were converted to grasslands in the 1990s.

**Upper-montane (supramontane) belt** at altitudes 1000–1370 m is dominated by natural spruce forests. In the northern mountain ranges the altitude of this belt increases with climate continentality from the west (Krušné hory Mts) to the east (Moravskoslezské Beskydy Mts), and the belt also occurs at higher altitudes in the Šumava Mts. Deforested areas are used as low-productive meadows and pastures.

**Subalpine belt** ranges from 1200 to 1500 m a.s.l. In the Krkonoše Mts it is characterized by *Pinus mugo* krummholz, but it also includes birch-willow scrub, tall-forb vegetation and grasslands in the cirques, open park-like woodland at the timberline and closed grasslands of *Avenella flexuosa*, *Calamagrostis villosa*, *Deschampsia cespitosa*, *Molinia caerulea* and *Nardus stricta* above the timberline. In the Králický Sněžník and Hrubý Jeseník Mts, it comprises a treeline ecotone with spruce, while *Pinus mugo* is absent (see Sect. 7.5.3).

**Alpine belt** is not a continuous vegetation belt in the Czech mountains, although alpine vegetation types occur on the highest summits, especially in the Krkonoše Mts. They include open lichen-rich grasslands of *Festuca supina* or heathlands of *Calluna vulgaris*, as well as boulder fields with predominantly cryptogamic vegetation.

The numerical classification of the Czech landscapes based on vegetation (habitat) types recorded during the national project of habitat mapping (Divíšek et al. 2014) generally reflected the vegetation belts described here, except the last three belts which occur in small areas and were merged due to coarse spatial resolution of the analysis (Fig. 7.2). However, this analysis also emphasized the difference in the upper-colline and submontane landscapes between the Bohemian Massif and the Carpathian flysch zone.

## 7.4 Landscapes with a High Diversity of Vegetation Types

About 60% of the area of the Czech Republic is flat or gently undulating landscape at altitudes of 200–600 m, covered with a mosaic of forests, arable land, meadows and pastures. The mountain areas in the Bohemian Massif characteristically have rounded ridges and extensive plateaus, while the mountain areas on the flysch Carpathians are characterized by larger differences in altitudes, yet these flysch landscapes are in general much more gentle than the rugged landscapes in the Alps and Inner Carpathians. The gentle topography of Czech landscapes results in a relatively low diversity of vegetation types (low beta diversity). However, there are some restricted areas of high topographic heterogeneity, geological diversity or occurrence of bedrocks with sharply different effects on vegetation than the bedrocks that predominate in the wider landscape (Kučera 2005; Ložek 2011; Pánek and Hradecký 2016). In particular, they include:

**Deep river valleys of the Bohemian Massif** are incised in a gently undulating landscape formed of metamorphic and igneous rocks, mainly gneiss or granite. They are up to 200 m deep, with a v-shaped cross-section and numerous deeply entrenched meanders. They have a narrow, discontinuous floodplain, which only occurs on the inner banks of meanders or along some straight sections, being replaced by steep slopes adjacent directly to the river above the outer banks of meanders. Vegetation in these valleys is diverse and relatively well preserved in a natural or semi-natural state as this rugged terrain is unsuitable for agriculture and plantation forestry. Best examples of the natural/semi-natural vegetation are found in the Vltava valley in southern and central Bohemia (Jeník and Slavíková 1964; Zelený 2008) and adjacent lower sections of the Otava and Lužnice valleys, Berounka valley in western and central Bohemia (Sofron 1967; Kolbek et al. 1997, 1999, 2001, 2003), Sázava valley in central Bohemia, and valleys in south-western Moravia, especially of the Jihlava and Dyje rivers (Chytrý and Vicherek 1995, 1996, 2003; Tichý 1997; Chytrý and Grulich 2015; see Fig. 1.6 in Chap. 1, this book). These valleys are generally warmer than the surrounding landscape, but there are relatively cool sites on their north-facing lower slopes (Chytrý and Tichý 1998). Distribution of forest vegetation types in the valleys is strongly determined by topography, namely slope, aspect and position on the upper or lower slopes or at the bottom (Jeník and Slavíková 1964; Chytrý and Vicherek 1996; Tichý 1999; Zelený and Chytrý 2007). Heterogeneous mosaics of vegetation types occur especially along meandering sections of rivers (Fig. 7.3). South-facing upper slopes support thermophilous oak forests, whereas the opposite north-facing slopes are covered by beech or acidophilous oak forests. Lower parts of the valleys are covered by ravine forests on steeper slopes or oak-hornbeam forests on moderate slopes. Higher river terraces on the floors of the valleys are covered with oak-hornbeam forests, while there are riverine alder forests on the lower terraces. Open patches of grassland and scrub vegetation occur in some places at the upper edges of valley slopes, especially those with a southern aspect (Kučera and Mannová 1998). The diversity of flora and vegetation is locally enriched in places where there are outcrops of base-rich rocks

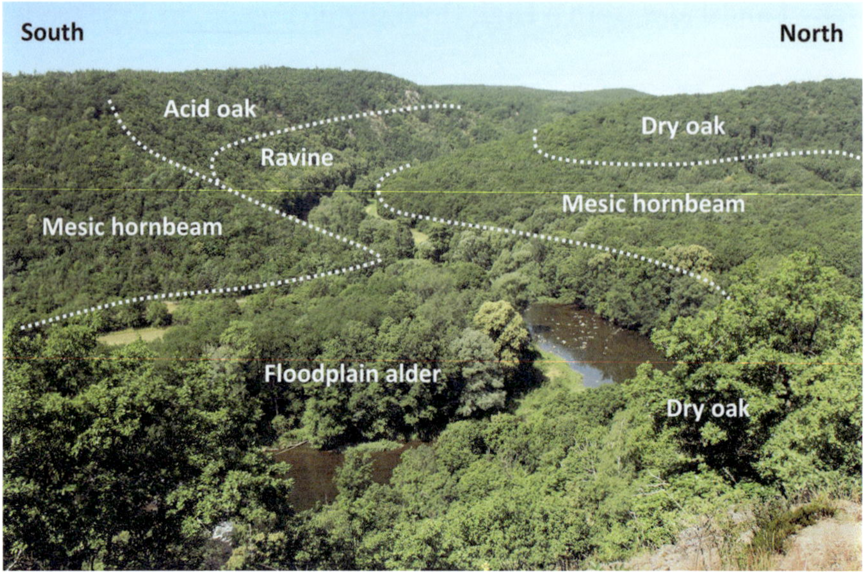


**Fig. 7.3** A typical distribution pattern of different vegetation types in a deep river valley of the Bohemian Massif (Dyje valley west of Šobes near Znojmo, south-western Moravia). Upper south-facing slopes are covered by dry thermophilous oak forests with *Quercus petraea* (*Quercion petraeae*), while the opposite upper north-facing slopes are covered by acidophilous oak forests, also dominated by *Q. petraea* (*Quercion roboris*). Steep lower slopes above the outer banks of meanders support ravine forests (*Tilio platyphylli-Acerion*), while gentler slopes are covered by mesic oak-hornbeam forests with *Q. petraea* and *Carpinus betulus* (*Carpinion betuli*). The discontinuous floodplain on the inner banks of meanders is either covered by natural floodplain forests with *Alnus glutinosa* (*Alnion incanae*) or by secondary vegetation of mesic meadows (*Arrhenatherion elatioris*). Credits for all photos in this chapter: M. Chytrý

such as metamorphic limestone, serpentinite, amphibolite or spilite. In many places river terraces have been deforested and converted into mesic meadows.

**Karst areas** occur locally in the colline and upper-colline belts in the Czech Republic. They are formed mainly of limestone of Silurian to Jurassic age. The largest areas with well-developed karst features include the Bohemian Karst south-west of Prague in central Bohemia, the Moravian Karst north of Brno in central Moravia and the Pavlov Hills in southern Moravia. These three areas differ in terms of precipitation, which is highest in the Moravian Karst and lowest in the Pavlov Hills (see Fig. 1.11 in Chap. 1, this book). Vegetation reflects the differences in humidity. The Moravian Karst is dominated by beech forests and contains small areas of oak-hornbeam forests, ravine forests and patches of thermophilous oak forests associated with dry grasslands on some south-facing upper slopes (Šmarda 1967). It is a plateau dissected by deep karst valleys that, because of topographic shading, contain patches of submontane-montane vegetation (Tichý 2015). In contrast, the Pavlov Hills is an isolated limestone ridge in the middle of a dry and warm forest-steppe area with steppes, some of them probably natural, thermophilous oak forests, oak-hornbeam forests and ravine forests; however, beech forests are absent (Horák 1969; Unar 2004; Danihelka et al. 2015a). The Bohemian Karst is intermediate between these two extremes, with a mosaic of different vegetation types ranging from steppes to beech forests (Skalický and Jeník 1974). The typical vegetation in all karst areas is *Sesleria caerulea* grassland with thermophilous or steppe species, occurring mainly on north-facing cliffs and limestone outcrops (Zlatník 1928).

**Sandstone pseudokarst** occurs at several sites at the margins of the area of Cretaceous sediments in northern, central and eastern Bohemia (see Fig. 1.4 in Chap. 1, this book). Most notable examples are in (1) Elbe Sandstones (Labské pískovce), including the Bohemian Switzerland (České Švýcarsko) National Park, in northern Bohemia, (2) Hradčany Cliffs in northern Bohemia (Sýkora 1970; Sádlo et al. 2011), (3) Kokořín area in central Bohemia (Kučera and Špryňar 1996; Sádlo 1996), (4) Bohemian Paradise (Český ráj) area in central and eastern Bohemia (Slavík 1977), (5) Broumov area including the Adršpach-Teplice Rocks in eastern Bohemia (Sýkora and Hadač 1984) and (6) Toulovcovy Maštale site in eastern Bohemia (Neuhäusl and Neuhäuslová-Novotná 1972). The altitude of these landscapes ranges from about 140 m (Elbe Sandstones) to about 600 m (Broumov area). Cretaceous sandstones are predominantly siliceous, acidic and nutrient-poor; only in a few places do they contain spatially restricted layers enriched with calcium carbonate, especially in the Hradčany Cliffs (Sádlo et al. 2011). Sandstone pseudokarst landscapes in their mature developmental phase, in the Czech Republic called 'rock cities', are made up of ~100 m deep, narrow canyons and dry gorges, which dissect sandstone plateaus, and isolated sandstone towers, protected from weathering by surface hardening caused by silica impregnation (Cílek and Kopecký 1998; Härtel et al. 2007). These landscapes contain a broad variety of vegetation types occurring on different landforms, although most of them are rather poor in species of seed plants, a feature generally typical of acidic bedrock. In contrast, flora of bryophytes and ferns is relatively rich (Härtel et al. 2007). Sandstone plateaus at lower altitudes may be locally covered with loess (e.g. in the Kokořín area) and such places support oak-hornbeam or thermophilous oak forests. However, the predominant vegetation in sandstone pseudokarst landscapes are acidophilous oak, beech or pine forests (alliances *Quercion roboris*, *Luzulo-Fagion sylvaticae* or *Dicrano-Pinion sylvestris*). Pine forests on rock tops, upper slopes and sandstone plateaus (Fig. 7.4) have been maintained by recurrent natural or human-induced fires for millennia (Novák et al. 2012, 2015; Adámek et al. 2015). In contrast, gorges are dominated by more mesic and fire-sensitive vegetation, especially acidophilous beech forests with spruce. Growing locally at the bottom of the gorges are alder galleries along the creeks and, in shaded deep gorges, wet spruce forests with montane plant species or small mires (Kuneš and Jankovská 2000), especially in the Broumov area. Patches of naturally open vegetation are found on the upper edges of sandstone cliffs and on steep rock faces. On shaded bottoms of deep gorges there are accumulations of sand, litter and moss polsters falling from the weathering cliffs, which provide substrate for fine-scale succession of bryophyte, dwarf-shrub and herbaceous vegetation (Zittová-Kurková 1984; Herben 1992; Sádlo 1996; Gutzerová and Herben 2001).

**Fig. 7.4** A sandstone pseudokarst area near Hrubá Skála in the Bohemian Paradise, eastern Bohemia, with rock tops and sandstone plateaus covered by *Pinus sylvestris* forests that are maintained by recurrent fires. The gorges between rocks have more mesic conditions, which support spruce and other mesophilous species

**Solitary volcanic hills** occur scattered across northern Bohemia, being most common in the České středohoří Mts. They are formed of either base-rich basalt or base-poor phonolite (or related trachytic rocks), both originated through Tertiary volcanic activity. Cretaceous sandstones or marlstones are usually found at their bases. Relative elevation of these hills above the surrounding landscape is often more than 250 m and altitudes of the summits are mostly between 400 and 700 m, although some are higher. For example, Milešovka, the highest mountain of the České středohoří Mts, reaches 837 m. Vegetation on volcanic hills depends on slope, aspect, position on lower or upper slope, bedrock (basalt versus phonolite/trachyte) and local occurrence of cliffs or boulder screes. Upper parts of the south-facing slopes of many hills are naturally treeless, especially in the precipitation-poor forest-steppe landscape in the south-western part of the České středohoří Mts, where south-facing slopes are covered by short-grass steppe (*Festucion valesiacae*) and north-facing slopes by woodland (Fig. 7.5; Slavíková et al. 1983). In areas with higher precipitation, located more to the east, especially in the highest part of the České středohoří Mts (Mt. Milešovka) and east of the Labe River, the natural vegetation of the volcanic hills is mainly beech, oak-hornbeam or ravine forest (*Fagion sylvaticae*, *Carpinion betuli* and *Tilio platyphylli-Acerion*), while treeless areas are small, usually restricted to south-facing cliffs near the hill tops (Kolbek 1978) with boulder screes below them (Cílek 1998). There are several light-demanding species with isolated occurrences in these treeless areas (Kolbek and Petříček 1972, 1979;

**Fig. 7.5** A landscape with solitary volcanic hills in the south-western part of the České středohoří Mts, northern Bohemia. *Stipa* steppe below the ridge of Raná hill contains an isolated occurrence of the continental steppe grass *Helictotrichon desertorum*. The conic hill Oblík in the background displays a pattern of the exposure-related forest-steppe, with woodland and shrubland on its north-facing slope (*left*) and steppe on the south-facing slope (*right*)

Petříček and Sýkora 1973; Sýkora 1979), which suggests a continuous existence of patches of open vegetation throughout the Holocene.

**Glacial cirques** are found at altitudes of 1000–1500 m in the highest mountain ranges in the Czech Republic: the Šumava and Sudetes (Krkonoše, Králický Sněžník and Hrubý Jeseník). They occur on massifs of siliceous rocks (phyllite, mica schist, gneiss and granite), on north-eastern to south-eastern slopes of the highest ridges or plateaus. Snow accumulated in these leeward positions in the colder periods of the Pleistocene, which resulted in the formation of small glaciers. Cirques in the Šumava Mts occur below the timberline and are covered mainly by spruce and beech forest and small patches of treeless vegetation, especially stands of the tall fern *Athyrium distentifolium* (Sofron and Štěpán 1971). At the bottoms of these cirques there are mesotrophic lakes, two of them with occurrence of quillworts (*Isoëtes echinospora* and *I. lacustris*; Husák et al. 2000). In contrast, cirques in the Sudetes occur from the montane belt with beech forests near their bottoms to the areas above the timberline. They are treeless in the central part due to periodical disturbances by avalanches (Fig. 7.6). The cirques are topographically heterogeneous with slopes of different aspect and inclination, cliffs, boulder screes and bottom areas with accumulation of fine soil. Soft-water springs are scattered on the slopes and there are small oucrops of base-rich rocks such as porphyry or metamorphic limestone in some of them. Due to their leeward position cirques are protected from strong wind

**Fig. 7.6** Upper part of the Velká Kotelní jáma, one of the glacial cirques in the Krkonoše Mts, with avalanche tracks covered by tall-forb vegetation, tall-fern stands of *Athyrium distentifolium* on boulder screes and *Pinus mugo* stands covering the areas adjacent to avalanche tracks

and have a warmer mesoclimate than the adjacent mountain slopes and summits (Jeník 1961). In winter there is deep snow cover, which protects the plants from frost; however, there are few areas where the snow covers the ground until summer (Jeník 1958; Hejcman et al. 2006a). Snow-bed vegetation is nearly absent except for some fragmentary stands with *Gnaphalium supinum* in open *Nardus stricta* grassland. Vegetation in the cirques in the Sudetes, especially in the Krkonoše and Hrubý Jeseník Mts, is very diverse (Jeník 1961; Jeník et al. 1980). There are various types of tall-herb subalpine vegetation dominated by forbs (alliance *Adenostylion alliariae*), grasses (*Calamagrostion villosae* and *Calamagrostion arundinaceae*) and ferns (*Dryopterido filicis-maris-Athyrion distentifolii*), *Vaccinium myrtillus* heaths (*Genisto pilosae-Vaccinion*), deciduous scrub of *Betula pubescens* subsp. *carpatica* and *Salix silesiaca* adapted to avalanche disturbance (*Salicion silesiacae*), *Pinus mugo* krummholz (*Pinion mugo*), rock-outcrop grasslands (*Agrostion alpinae*), shallow mires (*Caricion canescenti-nigrae*) and spring vegetation (*Swertio perennis-Dichodontion palustris*). Spruce and beech forests (*Piceion abietis*, *Fagion sylvaticae* and *Luzulo-Fagion sylvaticae*) grow on milder slopes of the cirques and on the moraines at the bottom. In the cirques of the Sudetes, larger lakes currently occur only on the Polish side of the Krkonoše Mts (see Chap. 1, Sect. 1.4, this book). The Sudetes cirques harbour several thermophilous species typical of low altitudes and peculiar plant communities composed of co-occurring alpine, subalpine and lowland species. *Calamagrostis arundinacea* grasslands on the warm slopes of cirques, which are particularly rich in these species, contain higher numbers of species than any other vegetation type in the Czech mountains (see Sect. 7.6

below). Jeník (1959, 1961) suggested that propagules of thermophilous species have been transported to the cirques from the western foothills of the mountain ranges by topographically modified wind currents ('theory of anemo-orographic systems'; see also Jeník 1997). Isolated occurrences of several species of continental distribution and a few endemics suggest the relict origin of these grasslands (Roleček et al. 2015).

**Lowland riverine landscapes** occur especially along the middle Labe River and lower courses of its tributaries Vltava and Ohře (Neuhäuslová-Novotná 1965; Kovář 1981, 1983) and along the middle and lower Morava and lower courses of other rivers in southern Moravia (Dyje, Jihlava and Svratka; Vicherek 1962b; Vicherek et al. 2000; Danihelka et al. 2015b). Floodplains of these rivers have been strongly modified by floods and associated accumulation of loamy sediments. The incidence of floods increased after the deforestation of submontane and montane areas in the Middle Ages (Opravil 1983; Štěrba et al. 2008; Kadlec et al. 2009; Ložek 2011). Typically the rivers flood after snowmelt in March–April and occasionally after heavy rainfall in summer or other periods of the year. The incidence of floods has declined during the last decades as a result of regulating the rivers. Lowland riverine landscapes contain a rich mosaic of aquatic vegetation in oxbows and pools (*Lemnetea* and *Potametea*), reed and tall-sedge marshes (*Phragmito-Magno-Caricetea*), productive meadows (*Deschampsion cespitosae*) that are inundated in spring but may dry out in late summer when the water table is more than 1 m below ground level (Balátová-Tuláčková 1968), softwood floodplain forests with *Salix alba* or, very rarely, also with *Populus nigra* (*Salicion albae*), and hardwood floodplain forests with *Fraxinus excelsior*, *Quercus robur*, *Ulmus laevis* and *U. minor*, and in southern Moravia also with *Fraxinus angustifolia* (*Alnion incanae*). Some terraces of the Labe River and lower Morava River (near Hodonín and Bzenec) are covered by acidic siliceous sand originating partly from Tertiary fluvial sedimentation and partly from Pleistocene aeolian sedimentation. Nowadays these sandy landscapes are covered mainly by pine forests, many of which are recent plantations, but there are also small remnants of acidophilous or thermophilous oak forests (Šmarda 1961; Chytrý and Horák 1997). Deforested areas in these landscapes support sandy grasslands (*Corynephorion canescentis*, *Armerion elongatae* and *Festucion vaginatae*; Klika 1931; Šmarda 1961). Besides lowland riverine landscapes, well-preserved complexes of floodplain vegetation also occur at mid-altitudes. Fast-running streams are rare in the Czech Republic and most rivers at mid-altitudes either flow in deep v-shaped valleys or, locally, in broad floodplains with numerous oxbows and alluvial pools with accumulations of organic sediment. Best examples of mid-altitude broad floodplains occur along the upper Lužnice River in the Třeboň Basin (Prach et al. 1996) and upper Vltava River in the Šumava Mts (Sádlo and Bufková 2002; Bufková et al. 2005).

**Serpentinite areas** occur in small patches in various areas of the Bohemian Massif, namely in western and south-western Bohemia, the Bohemian-Moravian Highlands and northern Moravia (see Fig. 1.7 in Chap. 1, this book). Serpentinite is a metamorphic rock occurring in association with other types of metamorphic rocks, particularly granulite. It has no causal links with specific topographic features, however, it affects vegetation through its extraordinary chemistry, namely a high content of magnesium and occurrence of heavy metals such as nickel, chromium and cobalt, which are toxic to many plant species (Proctor and Woodell 1975). Serpentinite occurring in a flat or gently undulating landscape is often covered with deep soil which moderates the effect of bedrock on the vegetation. However, when crossed by stream valleys or occurring on the tops of hills, serpentinite bedrock is exposed to erosion and covered by thin soil. In such situations the vegetation is markedly different from that in the surroundings. In forests, *Fagus sylvatica* and *Carpinus betulus* are replaced by *Pinus sylvestris*, and locally also by *Quercus petraea*. Canopy of pine and oak forests is rather open and it is possible that various light-demanding species have survived in them since the Early Holocene. In addition, the plant populations in these forests are isolated, which may have resulted in speciation of neoendemics such as *Cerastium alsinifolium* and *Minuartia smejkalii* (see Chap. 3, Sect. 3.4, this book). Best examples of specific serpentinite vegetation are found in the Slavkovský les Mts in western Bohemia, near Křemže in southern Bohemia, near the Želivka water reservoir in central Bohemia, in the middle Jihlava valley near Mohelno in south-western Moravia and near Raškov in northern Moravia. The serpentinite area near Mohelno (Suza 1928; Chytrý and Vicherek 1996; Daněk 2015) is the warmest and driest of all the serpentinite areas in the Czech Republic. Here the north-facing slopes are covered by forests of *Pinus sylvestris* with *Sesleria caerulea* and other montane basiphilous species (*Erico carneae-Pinion*). The south-facing slopes would be probably covered by the thermophilous oak forests (*Quercion petraeae*) in the natural landscape, but their deforestation and livestock grazing has resulted in the development of *Festuca-Stipa* steppe (*Alysso-Festucion pallentis*, *Festucion valesiacae*). Other serpentinite areas are covered mainly by pine forests (*Dicrano-Pinion sylvestris* and at Želivka partly also *Erico carneae-Pinion*), although some have been deforested and transformed into low-productive pastures. The coolest and wettest of these areas are in the Slavkovský les Mts (Hejtmánek 1954; Jeník 1994), where *Picea abies* frequently occurs in the understorey of species-poor *Pinus sylvestris* forest. Rock outcrops in all serpentinite areas are characterized by the occurrence of small ferns that are serpentinite specialists: *Asplenium adulterinum*, *A. cuneifolium* (Tájek et al. 2011) and at Mohelno, also *Notholaena marantae* (*Asplenion cuneifolii*; Vicherek 1970).

## 7.5 Vegetation Types

The brief overview of Czech vegetation types presented here focuses on the hierarchical levels of phytosociological classes and alliances. It follows the classification system accepted in the monograph Vegetation of the Czech Republic (Chytrý 2007, 2009, 2011, 2013), which is summarized in Table 7.1. More information, including classification to the level of associations and details on ecology, history and dynamics of particular vegetation types can be found in this monograph, as well as in more specialized, local or older studies cited therein. This monograph is not cited in the subsequent text dealing with particular vegetation types.

**Table 7.1** Phytosociological classification of Czech vegetation at the level of classes and alliances according to the monograph Vegetation of the Czech Republic (Chytrý 2007, 2009, 2011, 2013). Orders were omitted in order to keep the classification hierarchy at the national scale simple. See the monograph Vegetation of the Czech Republic for information on associations. Different names or concepts used in EuroVegChecklist (EVC, Mucina et al. 2016) are given in brackets

| |
|---|
| **Forests** |
| **LA. *Alnetea glutinosae* Br.-Bl. et Tüxen ex Westhoff et al. 1946** (including *Franguletea* Doing ex Westhoff in Westhoff et Den Held 1969) – Alder and willow carrs |
| LAA. *Alnion glutinosae* Malcuit 1929 – Alder carrs |
| LAB. *Salicion cinereae* Müller et Görs ex Passarge 1961 – Willow carrs |
| **LB. *Carpino-Fagetea* Jakucs 1967** (including *Alno glutinosae-Populetea albae* Fukarek et Fabijanić 1968) – Mesic deciduous broad-leaved forests |
| LBA. *Alnion incanae* Pawłowski et al. 1928 – Ash-alder alluvial forests |
| LBB. *Carpinion betuli* Issler 1931 – Oak-hornbeam forests |
| LBC. *Fagion sylvaticae* Luquet 1926 – Eutrophic beech forests |
| LBD. *Sorbo torminalis-Fagion sylvaticae* Passarge et Hofmann 1968 (in EVC included in *Fagion sylvaticae*) – Calcicole beech forests |
| LBE. *Luzulo-Fagion sylvaticae* Lohmeyer et Tüxen in Tüxen 1954 – Acidophilous beech forests |
| LBF. *Tilio platyphylli-Acerion* Klika 1955 (including *Melico-Tilion platyphylli* Passarge et Hofmann 1968) – Ravine forests |
| **LC. *Quercetea pubescentis* Doing Kraft ex Scamoni et Passarge 1959** – Thermophilous oak forests |
| LCA. *Quercion pubescenti-petraeae* Br.-Bl. 1932 – Peri-alpidic basiphilous thermophilous oak forests |
| LCB. *Aceri tatarici-Quercion* Zólyomi 1957 – Subcontinental forest-steppe oak forests |
| LCC. *Quercion petraeae* Issler 1931 – Acidophilous thermophilous oak forests |
| **LD. *Quercetea robori-petraeae* Br.-Bl. et Tüxen ex Oberdorfer 1957** – Acidophilous oak forests |
| LDA. *Quercion roboris* Malcuit 1929 – Acidophilous oak forests |
| **LE. *Erico-Pinetea* Horvat 1959** – Basiphilous submontane pine forests |
| LEA. *Erico carneae-Pinion* Br.-Bl. in Br.-Bl. et al. 1939 – Basiphilous montane pine forests of Central and South-eastern Europe |
| **LF. *Vaccinio-Piceetea* Br.-Bl. in Br.-Bl. et al. 1939** (including *Pyrolo-Pinetea sylvestris* Korneck 1974) – Boreal coniferous forests |
| LFA. *Festuco-Pinion sylvestris* Passarge et Hofmann 1968 – Basiphilous continental pine forests |
| LFB. *Dicrano-Pinion sylvestris* (Libbert 1933) Matuszkiewicz 1962 – Acidophilous boreo-continental pine forests |
| LFC. *Piceion abietis* Pawłowski et al. 1928 (syn.: *Piceion excelsae* Pawłowski et al. 1928) – Central European acidophilous spruce forests |
| LFD. *Vaccinio uliginosi-Pinion sylvestris* Passarge et Hofmann 1968 – Bog woodlands |
| **Scrub** |
| **KA. *Salicetea purpureae* Moor 1958** – Riparian willow scrub and willow-poplar forests |
| KAA. *Salicion triandrae* Müller et Görs 1958 – Willow scrub of loamy and sandy river banks |
| KAB. *Salicion elaeagno-daphnoidis* (Moor 1958) Grass in Mucina et al. 1993 – Willow scrub on river gravel accumulations |
| KAC. *Salicion albae* de Soó 1951 – Willow-poplar forests of lowland rivers |

(continued)

**Table 7.1** (continued)

| |
|---|
| **KB. *Rhamno-Prunetea* Rivas Goday et Borja Carbonell ex Tüxen 1962** (syn.: *Crataego-Prunetea* Tüxen 1962 and *Robinietea* Jurko ex Hadač et Sofron 1980) – Mesic and xeric scrub and *Robinia* groves |
| KBA. *Prunion fruticosae* Tüxen 1952 – Low xeric scrub |
| KBB. *Berberidion vulgaris* Br.-Bl. ex Tüxen 1952 – Tall mesic and xeric scrub |
| KBC. *Sambuco-Salicion capreae* Tüxen et Neumann ex Oberdorfer 1957 – Mesic scrub in forest clearings, canopy openings and disturbed sites |
| KBD. *Aegopodio podagrariae-Sambucion nigrae* Chytrý 2013 – Nitrophilous scrub in ruderal habitats |
| KBE. *Chelidonio majoris-Robinion pseudoacaciae* Hadač et Sofron ex Vítková in Chytrý 2013 – Black locust groves with nitrophilous species |
| KBF. *Balloto nigrae-Robinion pseudoacaciae* Hadač et Sofron 1980 – Black locust groves on dry sandy soils |
| KBG. *Euphorbio cyparissiae-Robinion pseudoacaciae* Vítková in Kolbek et al. 2003 – Low black locust groves and scrub at dry and warm sites with shallow soil |
| **KC. *Roso pendulinae-Pinetea mugo* Theurillat in Theurillat et al. 1995** – Subalpine krummholz vegetation |
| KCA. *Pinion mugo* Pawłowski et al. 1928 – Subalpine dwarf pine scrub |
| **Alpine and subalpine vegetation** |
| **AA. *Loiseleurio-Vaccinietea* Eggler ex Schubert 1960** – Alpine heathlands |
| AAA. *Loiseleurio procumbentis-Vaccinion* Br.-Bl. in Br.-Bl. et Jenny 1926 – Arcto-alpine dwarf-shrub vegetation |
| **AB. *Juncetea trifidi* Hadač in Klika et Hadač 1944** – Alpine grasslands on base-poor soil |
| ABA. *Juncion trifidi* Krajina 1933 – Wind-swept alpine grasslands on base-poor soil |
| ABB. *Nardo strictae-Caricion bigelowii* Nordhagen 1943 (syn.: *Nardo-Caricion rigidae* Nordhagen 1943) – Closed alpine grasslands on base-poor soil |
| **AC. *Elyno-Seslerietea* Br.-Bl. 1948** (in EVC Czech vegetation assigned to this class was reclassified to *Carici rupestris-Kobresietea bellardii* Ohba 1974) – Alpine grasslands on base-rich soil |
| ACA. *Agrostion alpinae* Jeník et al. 1980 – Species-rich rock-outcrop grasslands in the Sudetes cirques |
| **AD. *Mulgedio-Aconitetea* Hadač et Klika in Klika et Hadač 1944** (including *Betulo carpaticae-Alnetea viridis* Rejmánek ex Bouf et al. in Bouf et al. 2014) – Subalpine tall-forb and deciduous-shrub vegetation |
| ADA. *Calamagrostion villosae* Pawłowski et al. 1928 – Subalpine tall grasslands |
| ADB. *Calamagrostion arundinaceae* (Luquet 1926) Jeník 1961 (syn.: *Calamagrostion arundinaceae* (Luquet 1926) Oberdorfer 1957) – Subalpine grasslands with *Calamagrostis arundinacea* |
| ADC. *Salicion silesiacae* Rejmánek et al. 1971 – Subalpine deciduous scrub and woodland |
| ADD. *Adenostylion alliariae* Br.-Bl. 1926 – Subalpine tall-forb vegetation |
| ADE. *Dryopterido filicis-maris-Athyrion distentifolii* (Holub ex Sýkora et Štursa 1973) Jeník et al. 1980 – Subalpine tall-fern vegetation |
| **Rock and scree vegetation** |
| **SA. *Asplenietea trichomanis* (Br.-Bl. in Meier et Br.-Bl. 1934) Oberdorfer 1977** (including *Polypodietea* Jurko et Peciar ex Boscaiu et al. in Ratiu et al. 1966) – Vegetation of rocks, walls and stable screes |

(continued)

**Table 7.1** (continued)

| |
|---|
| SAA. *Cystopteridion* Richard 1972 – Vegetation of calcareous rock outcrops and walls |
| SAB. *Asplenion cuneifolii* Br.-Bl. ex Eggler 1955 (syn.: *Asplenion serpentini* Br.-Bl. et Tüxen ex Eggler 1955) – Vegetation of serpentinite outcrops |
| SAC. *Asplenion septentrionalis* Gams ex Oberdorfer 1938 – Vegetation of siliceous rock outcrops and talus slopes |
| SAD. *Androsacion alpinae* Br.-Bl. in Br.-Bl. et Jenny 1926 – Vegetation of siliceous talus slopes in subalpine and alpine belts |
| **SB. *Cymbalario muralis-Parietarietea judaicae* Oberdorfer 1969** (syn.: *Cymbalario-Parietarietea diffusae* Oberdorfer 1969) – Nitrophilous vegetation of walls |
| SBA. *Cymbalario muralis-Asplenion* Segal 1969 – Wall vegetation with neophytes of Mediterranean origin |
| **SC. *Thlaspietea rotundifolii* Br.-Bl. 1948** – Vegetation of mobile screes |
| SCA. *Stipion calamagrostis* Br.-Bl. et al. 1952 (syn.: *Stipion calamagrostis* Jenny-Lips ex Br.-Bl. 1950) – Vegetation of calcareous screes |
| SCB. *Galeopsion* Oberdorfer 1957 – Vegetation of siliceous screes |
| **Aquatic vegetation** |
| **VA. *Lemnetea* de Bolós et Masclans 1955** – Vegetation of free floating aquatic plants |
| VAA. *Lemnion minoris* de Bolós et Masclans 1955 – Vegetation of lemnids and free-floating aquatic ferns and liverworts |
| VAB. *Utricularion vulgaris* Passarge 1964 – Vegetation of bladderworts in mesotrophic to eutrophic water bodies |
| VAC. *Hydrocharition morsus-ranae* (Passarge 1964) Westhoff et den Held 1969 (syn.: *Stratiotion* Den Hartog et Segal 1964) – Vegetation of large free-floating vascular plants |
| **VB. *Potametea* Klika in Klika and Novák 1941** (syn.: *Potamogetonetea* Klika in Klika et Novák 1941) – Vegetation of aquatic plants rooted in the bottom |
| VBA. *Nymphaeion albae* Oberdorfer 1957 – Vegetation of aquatic plants rooted in the bottom and with leaves floating on the water surface |
| VBB. *Potamion* Miljan 1933 (syn.: *Potamogetonion* Libbert 1931 and *Potamogetonion graminei* Westhoff et Den Held 1969) – Vegetation of aquatic plants rooted in the bottom |
| VBC. *Batrachion fluitantis* Neuhäusl 1959 – Vegetation of aquatic plants in streams |
| VBD. *Ranunculion aquatilis* Passarge 1964 (syn.: *Ranunculion aquatilis* Passarge ex Theurillat in Theurillat et al. 2015) – Vegetation of aquatic plants in shallow water bodies with fluctuating water table |
| **VC. *Charetea* Fukarek ex Krausch 1964** – Vegetation of stoneworts |
| VCA. *Nitellion flexilis* Krause 1969 – Vegetation of stoneworts in calcium-poor water |
| VCB. *Charion globularis* Krausch 1964 (syn.: *Charion intermediae* Sauer 1937, *Charion vulgaris* (W. Krause et Lang 1977) W. Krause 1981 and *Charion canescentis* Krausch 1964) – Vegetation of stoneworts in calcium-rich or brackish water |
| **VD. *Littorelletea uniflorae* Br.-Bl. et Tüxen ex Westhoff et al. 1946** – Vegetation of oligotrophic water bodies |
| VDA. *Littorellion uniflorae* Koch ex Tüxen 1937 (syn.: *Lobelion dortmannae* Vanden Berghen 1964) – Submerged vegetation of oligotrophic water bodies |
| VDB. *Eleocharition acicularis* Pietsch ex Dierßen 1975 (syn.: *Littorellion uniflorae* Koch ex Klika 1935) – Vegetation of amphibious plants in shallow, oligotrophic to mesotrophic water bodies |

(continued)

**Table 7.1** (continued)

| |
|---|
| VDC. *Sphagno-Utricularion* Müller et Görs 1960 – Vegetation of oligotrophic pools with bladderworts |
| **Wetland vegetation** |
| **MA. *Isoëto-Nano-Juncetea* Br.-Bl. et Tüxen ex Br.-Bl. et al. 1952** – Vegetation of annual wetland herbs |
| MAA. *Eleocharition ovatae* Philippi 1968 (syn.: *Eleocharition soloniensis* Philippi 1968) – Vegetation of short-growing annual herbs on exposed bottoms of fisponds |
| MAB. *Radiolion linoidis* Pietsch 1973 – Vegetation of short-growing annual herbs on wet sand |
| MAC. *Verbenion supinae* Slavnić 1951 – Vegetation of annual herbs on base-rich exposed bottoms in warm areas |
| **MB. *Bidentetea tripartitae* Tüxen et al. ex von Rochow 1951** – Vegetation of annual nitrophilous wetland herbs |
| MBA. *Bidention tripartitae* Nordhagen ex Klika et Hadač 1944 – Nitrophilous vegetation of exposed bottoms and wet ruderal habitats |
| MBB. *Chenopodion rubri* (Tüxen 1960) Hilbig et Jage 1972 – Nitrophilous vegetation with *Chenopodium* and *Atriplex* in wet habitats |
| **MC. *Phragmito-Magno-Caricetea* Klika in Klika et Novák 1941** – Marsh vegetation |
| MCA. *Phragmition australis* Koch 1926 (syn.: *Phragmition communis* Koch 1926) – Fresh-water reed vegetation |
| MCB. *Meliloto dentati-Bolboschoenion maritimi* Hroudová et al. 2009 – Continental brackish marsh vegetation |
| MCC. *Eleocharito palustris-Sagittarion sagittifoliae* Passarge 1964 – Vegetation of large wetland herbs in habitats with periodical changes of water level |
| MCD. *Phalaridion arundinaceae* Kopecký 1961 – Reed and tall-sedge vegetation on river banks |
| MCE. *Glycerio-Sparganion* Br.-Bl. et Sissingh in Boer 1942 – Medium-tall reed stands along brooks and on floating islands |
| MCF. *Carici-Rumicion hydrolapathi* Passarge 1964 – Vegetation of wetland herbs on organic muddy sediments |
| MCG. *Magno-Caricion elatae* Koch 1926 – Tall-sedge vegetation in littoral zones of oligotrophic and mesotrophic water bodies |
| MCH. *Magno-Caricion gracilis* Géhu 1961– Tall-sedge vegetation in littoral zones of eutrophic water bodies |
| **Spring and mire vegetation** |
| **RA. *Montio-Cardaminetea* Br.-Bl. et Tüxen ex Klika et Hadač 1944** – Vegetation of springs |
| RAA. *Caricion remotae* Kästner 1941 – Vegetation of non-calcareous forest springs |
| RAB. *Lycopodo europaei-Cratoneurion commutati* Hadač 1983 – Vegetation of calcareous forest springs with tufa formation |
| RAC. *Epilobio nutantis-Montion fontanae* Zechmeister in Zechmeister et Mucina 1994 – Vegetation of subatlantic and submontane springs in open habitats |
| RAD. *Swertio perennis-Dichodontion palustris* Hadač 1983 (syn.: *Swertio perennis-Anisothecion squarrosi* Hadač 1983) – Vegetation of non-calcareous alpine and subalpine springs |

(continued)

**Table 7.1** (continued)

| |
|---|
| **RB. *Scheuchzerio palustris-Caricetea nigrae* Tüxen 1937** (syn.: *Scheuchzerio palustris-Caricetea fuscae* Tüxen 1937) – Vegetation of fens, transitional mires and bog hollows |
| RBA. *Caricion davallianae* Klika 1934 – Calcareous fens |
| RBB. *Sphagno warnstorfii-Tomentypnion nitentis* Dahl 1956 – Fens with calcicolous species and calcitolerant peat mosses |
| RBC. *Caricion canescenti-nigrae* Nordhagen 1937 (syn.: *Caricion fuscae* Koch 1926) – Sligthly acidic fens |
| RBD. *Sphagno-Caricion canescentis* Passarge (1964) 1978 – Acidic fens (transitional mires) |
| RBE. *Sphagnion cuspidati* Krajina 1933 (syn.: *Scheuchzerion palustris* Nordhagen ex Tüxen 1937) – Vegetation of bog hollows |
| **RC. *Oxycocco-Sphagnetea* Br.-Bl. et Tüxen ex Westhoff et al. 1946** – Bog vegetation |
| RCA. *Sphagnion magellanici* Kästner et Flössner 1933 (syn.: *Sphagnion medii* Kästner et Flössner 1933) – Continental and subcontinental bogs |
| RCB. *Oxycocco palustris-Ericion tetralicis* Nordhagen ex Tüxen 1937 – Oceanic and suboceanic bogs |
| RCC. *Oxycocco microcarpi-Empetrion hermaphroditi* Nordhagen ex Du Rietz 1954 – Boreal bogs |
| **Grasslands and heathlands below the timberline** |
| **TA. *Crypsietea aculeatae* Vicherek** 1973 – Vegetation of annual graminoids in saline habitats |
| TAA. *Cypero-Spergularion salinae* Slavnić 1948 – Inland salt marshes with annual halophilous grasses |
| **TB. *Thero-Salicornietea strictae* Tüxen in Tüxen et Oberdorfer 1958** – Vegetation of annual succulent halophytes |
| TBA. *Salicornion prostratae* Géhu 1992 – Inland salt marshes with annual succulent halophytes (now this vegetation type is extinct in the Czech Republic) |
| **TC. *Festuco-Puccinellietea* Soó ex Vicherek** 1973 – Saline grasslands |
| TCA. *Puccinellion limosae* Soó 1933 – Intermittently dry saline grasslands |
| TCB. *Juncion gerardii* Wendelberger 1943 – Mesic and wet saline grasslands |
| **TD. *Molinio-Arrhenatheretea* Tüxen 1937** – Meadows and mesic pastures |
| TDA. *Arrhenatherion elatioris* Luquet 1926 – Lowland to submontane mesic meadows |
| TDB. *Polygono bistortae-Trisetion flavescentis* Br.-Bl. et Tüxen ex Marschall 1947 (syn.: *Phyteumato-Trisetion* Ellmauer et Mucina Mucina 1993 and *Triseto flavescentis-Polygonion bistortae* Br.-Bl. et Tüxen ex Marschall 1947) – Montane mesic meadows |
| TDC. *Cynosurion cristati* Tüxen 1947 – Mesic pastures and perennial grasslands of trampled habitats |
| TDD. *Molinion caeruleae* Koch 1926 – Intermittently wet, nutrient-poor meadows |
| TDE. *Deschampsion cespitosae* Horvatić 1930 – Lowland floodplain meadows |
| TDF. *Calthion palustris* Tüxen 1937 – Wet tall-herb meadows |
| **TE. *Calluno-Ulicetea* Br.-Bl. et Tüxen ex Klika et Hadač 1944** – *Nardus* grasslands and heathlands |
| TEA. *Nardion strictae* Br.-Bl. 1926 – Subalpine *Nardus* grasslands |
| TEB. *Nardo strictae-Agrostion tenuis* Sillinger 1933 – Montane *Nardus* grasslands with alpine species |

(continued)

**Table 7.1** (continued)

| |
|---|
| TEC. *Violion caninae* Schwickerath 1944 – Submontane and montane *Nardus* grasslands |
| TED. *Nardo strictae-Juncion squarrosi* (Oberdorfer 1957) Passarge 1964 – Wet *Nardus* grasslands |
| TEE. *Euphorbio cyparissiae-Callunion vulgaris* Schubert ex Passarge in Scamoni 1963 (syn.: *Euphorbio-Callunion* Schubert ex Passarge 1964) – Dry lowland and colline heathlands |
| TEF. *Genisto pilosae-Vaccinion* Br.-Bl. 1926 – Submontane to subalpine *Vaccinium* heathlands |
| **TF. *Koelerio-Corynephoretea* Klika in Klika et Novák 1941** (including *Sedo-Scleranthetea* Br.-Bl. 1955) – Pioneer vegetation of sandy and shallow soils |
| TFA. *Corynephorion canescentis* Klika 1931 – Open sandy grasslands |
| TFB. *Thero-Airion* Tüxen ex Oberdorfer 1957 – Vegetation of annual grasses on sandy soils |
| TFC. *Armerion elongatae* Passarge 1964 – Closed sandy grasslands |
| TFD. *Hyperico perforati-Scleranthion perennis* Moravec 1967 – Submontane acidophilous vegetation of shallow soils |
| TFE. *Arabidopsion thalianae* Passarge 1964 (syn.: *Sedo albi-Veronicion dillenii* Korneck 1974) – Acidophilous vegetation of vernal therophytes and succulents |
| TFF. *Alysso alyssoidis-Sedion* Oberdorfer et Müller in Müller 1961 – Basiphilous vegetation of vernal therophytes and succulents |
| **TG. *Festucetea vaginatae* Soó ex Vicherek 1972** – Sandy steppes |
| TGA. *Festucion vaginatae* de Soó 1929 – Pannonian sandy steppe grasslands |
| **TH. *Festuco-Brometea* Br.-Bl. et Tüxen ex Soó 1947** (including *Trifolio-Geranietea sanguinei* Müller 1962) – Dry grasslands |
| THA. *Alysso-Festucion pallentis* Moravec in Holub et al. 1967 – Hercynian rock-outcrop vegetation with *Festuca pallens* |
| THB. *Bromo pannonici-Festucion pallentis* Zólyomi 1966 (syn.: *Bromo pannonici-Festucion csikhegyensis* Zólyomi 1966 corr. Mucina in Di Pietro et al. 2015) – Pannonian vegetation of limestone outcrops |
| THC. *Diantho lumnitzeri-Seslerion* (Soó 1971) Chytrý et Mucina in Mucina et al. 1993 – *Sesleria caerulea* grasslands |
| THD. *Festucion valesiacae* Klika 1931 – Narrow-leaved dry grasslands and short-grass steppes |
| THE. *Cirsio-Brachypodion pinnati* Hadač et Klika ex Klika 1951 (syn.: *Cirsio-Brachypodion pinnati* Hadač et Klika in Klika et Hadač 1944) – Subcontinental broad-leaved semi-dry grasslands and tall-grass steppes |
| THF. *Bromion erecti* Koch 1926 – Suboceanic broad-leaved semi-dry grasslands |
| THG. *Koelerio-Phleion phleoidis* Korneck 1974 – Acidophilous dry grasslands |
| THH. *Geranion sanguinei* Tüxen in Müller 1962 – Dry herbaceous fringe vegetation |
| THI. *Trifolion medii* Müller 1962 – Mesic herbaceous fringe vegetation |
| **Ruderal and weed vegetation** |
| **XA. *Polygono arenastri-Poëtea annuae* Rivas-Martínez 1975 corr. Rivas-Martínez et al. 1991** (syn.: *Polygono-Poetea annuae* Rivas-Mart. 1975) – Vegetation of trampled habitats |
| XAA. *Coronopodo-Polygonion arenastri* Sissingh 1969 (syn.: *Polygono-Coronopodion* Sissingh 1969) – Annual vegetation of dry trampled habitats |
| XAB. *Saginion procumbentis* Tüxen et Ohba in Géhu et al. 1972 – Annual vegetation of mesic trampled habitats |

(continued)

**Table 7.1** (continued)

| |
|---|
| **XB. *Stellarietea mediae* Tüxen et al. ex von Rochow 1951** (syn.: *Papaveretea rhoeadis* S. Brullo et al. 2001, *Sisymbrietea* Gutte et Hilbig 1975 and *Digitario sanguinalis-Eragrostietea minoris* Mucina et al. in Mucina et al. 2016) – Annual vegetation of arable land and ruderal habitats |
| XBA. *Caucalidion* von Rochow 1951 (syn.: *Caucalidion lappulae* von Rochow 1951) – Thermophilous weed vegetation of cereal fields on base-rich soils |
| XBB. *Veronico-Euphorbion* Sissingh ex Passarge 1964 – Basiphilous weed vegetation in root-crop fields |
| XBC. *Scleranthion annui* (Kruseman et Vlieger 1939) Sissingh in Westhoff et al. 1946 – Weed vegetation of cereal fields on acidic soils |
| XBD. *Arnoseridion minimae* Malato-Beliz et al. 1960 (in EVC included in *Scleranthion annui*) – Weed vegetation of cereal fields on nutrient-poor acidic soils |
| XBE. *Oxalidion fontanae* Passarge 1978 (syn.: *Oxalidion europeae* Passarge 1978) – Weed vegetation of cereal and root-crop fields in cool areas |
| XBF. *Spergulo arvensis-Erodion cicutariae* J. Tüxen in Passarge 1964 – Weed vegetation of dry sandy soils |
| XBG. *Atriplicion* Passarge 1978 – Ruderal vegetation of tall annual herbs |
| XBH. *Sisymbrion officinalis* Tüxen et al. ex von Rochow 1951 – Ruderal vegetation of winter-annual grasses |
| XBI. *Malvion neglectae* (Gutte 1972) Hejný 1978 – Ruderal vegetation of prostrate annual herbs on nutrient-rich soils |
| XBJ. *Salsolion ruthenicae* Philippi 1971 (syn.: *Salsolion ruthenicae* Philippi ex Oberdorfer 1983) – Annual ruderal vegetation of disturbed gravelly and sandy soils |
| XBK. *Eragrostion cilianensi-minoris* Tüxen ex Oberdorfer 1954 – Late-summer thermophilous ruderal and weed vegetation of sandy soils |
| **XC. *Artemisietea vulgaris* Lohmeyer et al. ex von Rochow 1951** – Xerophilous ruderal vegetation with biennial and perennial species |
| XCA. *Onopordion acanthii* Br.-Bl. et al. 1936 – Thermophilous archaeophyte-rich ruderal vegetation with biennial and perennial herbs |
| XCB. *Dauco carotae-Melilotion* Görs ex Rostański et Gutte 1971 – Ruderal vegetation with biennial and perennial herbs on stony and gravelly soils |
| XCC. *Convolvulo arvensis-Elytrigion repentis* Görs 1966 (syn.: *Convolvulo arvensis-Agropyrion repentis* Görs 1967) – Ruderal vegetation with perennial herbs on dry or intermittently dry soils |
| XCD. *Artemisio-Kochion prostratae* Soó 1964 – Relict vegetation of the Pleistocene loess steppes |
| XCE. *Arction lappae* Tüxen 1937 – Nitrophilous ruderal vegetation with biennial and perennial species in man-made habitats |
| **XD. *Galio-Urticetea* Passarge ex Kopecký** 1969 (in EVC included in *Epilobietea angustifolii*) – Nitrophilous perennial vegetation of wet to mesic habitats |
| XDA. *Senecionion fluviatilis* Tüxen ex Moor 1958 – Nitrophilous herbaceous fringes of floodplain forests |
| XDB. *Petasition hybridi* Sillinger 1933 (syn.: *Petasition officinalis* Sillinger 1933) – Vegetation of montane and submontane floodplains with *Petasites* |
| XDC. *Impatienti noli-tangere-Stachyion sylvaticae* Görs ex Mucina in Mucina et al. 1993 – Nitrophilous vegetation of forest fringes, canopy openings and clearings with perennial herbs |

(continued)

**Table 7.1** (continued)

| |
|---|
| XDD. *Geo urbani-Alliarion petiolatae* Lohmeyer et Oberdorfer in Görs et Müller 1969 – Nitrophilous vegetation of disturbed forest fringes with annual and biennial herbs |
| XDE. *Aegopodion podagrariae* Tüxen 1967 – Nitrophilous ruderal vegetation with broad-leaved perennial herbs |
| XDF. *Rumicion alpini* Scharfetter 1938 – Montane nitrophilous vegetation of broad-leaved herbs |
| **XE. *Epilobietea angustifolii* Tüxen et Preising ex von Rochow 1951** – Herbaceous vegetation of forest clearings and disturbed habitats in forest environments |
| XEA. *Fragarion vescae* Tüxen ex von Rochow 1951 (syn.: including *Epilobion angustifolii* Oberdorfer 1957) – Herbaceous vegetation at sites of disturbed forest |

## *7.5.1 Forests*

### 7.5.1.1 Forest History

Hypothetical distribution of natural forest vegetation in the Czech Republic is presented in the national maps of reconstructed natural (Mikyška et al. 1968–1972) and potential natural (Neuhäuslová et al. 1997) vegetation (Fig. 7.7). Both concepts describe the vegetation that would occur under twentieth century climate assuming no human disturbances, but while the former disregards past human-induced changes in substrates (e.g. building of ponds and water reservoirs, chanelling of rivers and mining pits), the latter takes them into account. According to these maps, lowlands in Moravia and the northern half of Bohemia would be dominated by oak-hornbeam forests, driest areas of northern and central Bohemia and southern Moravia by a mosaic of oak-hornbeam and thermophilous oak forest, mid-altitudes of western and southern Bohemia by acidophilous oak (partly also pine or fir) forests, submontane and montane areas by beech forests, the highest mountain areas by spruce forests and subalpine and alpine vegetation, and floodplains by alluvial forests. However, there is much uncertainty about actual species composition and distribution of potential natural vegetation types (Abraham et al. 2016), especially in the lowlands and at mid-altitudes, where forests have long been intensively exploited by humans and their present composition reflects past human impact (Nožička 1957). Human impact is responsible for fundamental differences in the historical development of forests in the lowland and colline belts and those at higher altitudes.

**Forests in the low-altitude areas**, settled by prehistoric farmers since the early Neolithic, were used as either coppices or wooded pastures. Coppice is an intensive management system in which trees are cut on a short rotation and left to regrow new shoots vegetatively from stumps or roots, forming multi-stemmed trees (Fig. 7.8a). The main purpose of coppicing was firewood production, but in many coppices some trees, especially oaks, were left uncut over several coppice cycles to produce

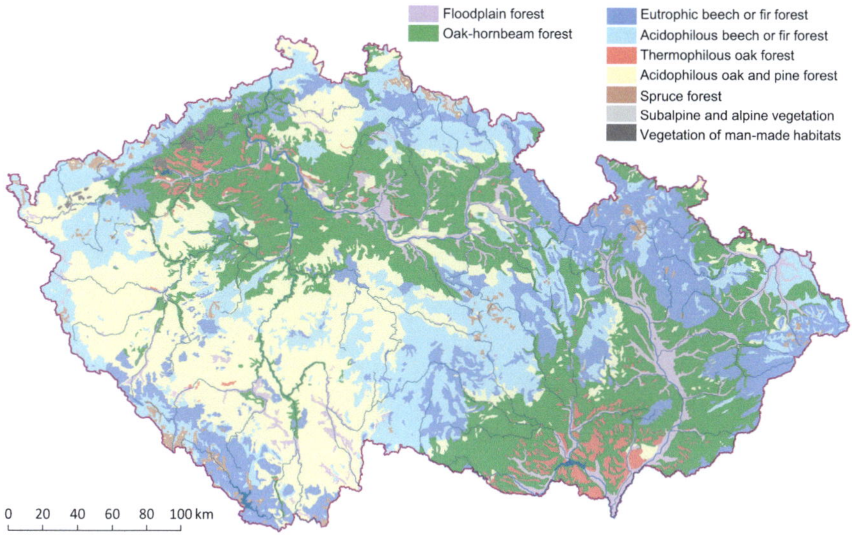


**Fig. 7.7** Potential natural vegetation according to Neuhäuslová et al. (1997), simplified

timber. These taller and older trees are called standards and the management system is called coppice-with-standards. In the Late Middle Ages coppices were widespread in areas below 500 m a.s.l. (Szabó et al. 2015) and rotation cycles shorter than 10 years were common (Szabó 2010). Coppicing favoured those trees that were good at regenerating vegetatively, especially *Carpinus betulus*. Oak recruitment was favoured by the opening up of the canopy that followed cutting (Altman et al. 2013).

Wooded pastures were mosaics of open woodland of single-stemmed trees and patches of grassland. In contrast to the single-purpose coppice management system, wooded pastures served multiple purposes, including domestic livestock grazing, fattening pigs on acorns and beech mast, hay making, harvesting leaf fodder from pollarded trees, collecting berries, mushrooms and oak galls, raking litter for animal bedding as well as a source of firewood and timber (Nožička 1957; Szabó and Hédl 2013). Wooded pastures were particularly suitable for light-demanding oak, providing suitable niches for its generative regeneration and supporting oak dominance in the canopy by maintaining an open woodland structure (Vera 2000). While coppices tended to be more common on nutrient-rich soils and steeper slopes, wooded pastures prevailed on less fertile soils and flat terrain, although this pattern was modified by socioeconomic factors (Szabó and Hédl 2013).

Both of these types of historical management were abandoned by the mid-twentieth century. For various socioeconomic reasons the rotation period of cop-

**Fig. 7.8** Legacies of historical forest management in lowland forests in southern Moravia: (**a**) Formerly coppiced lime (*Tilia platyphyllos*) on Děvín hill in the Pavlov Hills; (**b**) Old oaks (*Quercus robur*) in a hardwood floodplain forest on the Morava-Dyje floodplain near Lanžhot established in a former wooded pasture; when they die they are replaced by shade trees such as *Carpinus betulus*; (**c**) Trees established in wooded pastures are characterized by broadly stretching horizontal branches, as exhibited by the oaks (*Quercus cerris* and *Q. pubescens*) in Boří les Wood near Valtice

pices was being gradually extended, exceeding 30 years in the nineteenth and early twentieth century in Moravian coppices, and the ultimate reason for abandonment of coppicing was the replacement of firewood by fossil fuels (Machar 2009; Szabó 2010; Müllerová et al. 2014). After World War II foresters transformed coppices into high-forests by singling out the coppice stools. Wooded pastures gradually disappeared due to increasing state control of forest management after issuing forest regulation orders in the 1750s and 1760s. This led to gradual exclusion of domestic livestock from forests and separation of grassland pastures from wooded pastures (Nožička 1957), although this was a long-term process taking several decades, mainly motivated by the intensification of land management and increasing demand for hay. Diffuse boundaries between forests and grasslands changed to crisp boundaries as a result of introducing a so-called stable cadastre in the first half of the nineteenth century (Hejcman et al. 2013).

While coppices and wooded pastures provided a dynamic mosaic of different habitats with a high diversity of plants and invertebrates, their replacement by high-forests created a homogeneous landscape structure dominated by rather uniform areas of contiguous closed forest (Miklín and Čížek 2014). Abundance of nutrient demanding shade trees such as *Fraxinus excelsior*, *Tilia cordata* and *Acer campestre* gradually increased in many of these overgrown coppices, while oak tended to decline at some sites because of its inability to regenerate under the canopy of these trees (Fig. 7.8b, c; Hofmeister et al. 2004; Střeštík and Šamonil 2006; Janík et al. 2007; Müllerová et al. 2015). The resulting canopy closure led to the decline of many light-demanding and early successional species of plants and invertebrates, including endangered ones, which were typical of the former coppices and wooded pastures (Beneš et al. 2006; Hédl et al. 2010; Müllerová et al. 2015, Chudomelová et al. 2017). Also litter accumulation after cessation of litter raking may have contributed to depauperation of the herb layer (Vild et al. 2015).

Current plant species composition of the herb layer in abandoned coppices and wooded pastures comprises only a subset of species that occurred in original open woodlands, resulting in species-poor and spatially homogeneous communities (Kopecký et al. 2013). Some lowland woods are currently used as game preserves for wild ungulates, such as deer, mouflon or wild boar, which may partly counteract the succession towards closed-canopy forests after the cessation of coppicing or wood-pasture management. However, the main trend in the herb layer in such forests is the spread of ruderal, nutrient-demanding species (Chytrý and Danihelka 1993; Vild et al. 2017). Facing the current loss of biodiversity, nature conservationists and some foresters are increasingly interested in restoring coppices-with-standards and other historical forms of forest management in lowland nature reserves. An experimental restoration of coppice-with-standards in a lowland oak forest indicated that stand thinning can support a few oligotrophic light-demanding species, but at the same time it results in an increase in nutrient-demanding ruderal species, partly perhaps as a response to the accumulation of nitrogen resulting from several decades of abandonment and atmospheric deposition (Vild et al. 2013).

**Mid-altitude and montane forests** followed a different trajectory of historical development than the low-altitude forests. Negative archaeological evidence suggests that highlands and montane areas were scarcely populated before the medieval colonization, consequently their forests were probably in natural or climax state until the High and Late Middle Ages. During the colonization, which peaked in the thirteenth and fourteenth centuries, some of these forests, dominated by beech, spruce and fir, were clearcut to obtain agricultural land while the remaining stands

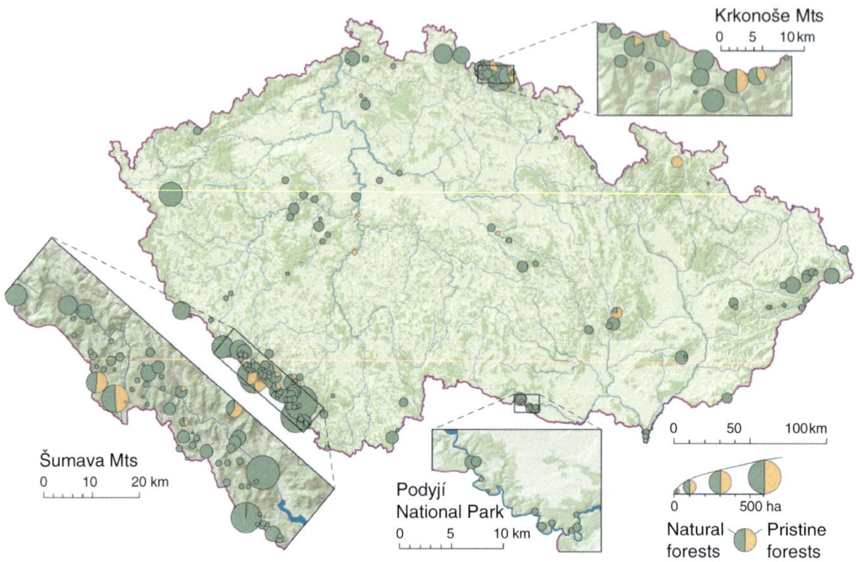


**Fig. 7.9** Pristine and natural old-growth forests larger than 10 ha (simplified from Winter et al. 2013). Circle sizes are proportional to the size of the forest stands

were used for selective timber logging, charcoal production from beech wood and livestock grazing (Nožička 1957). The spread of modern forestry practices in the first half of the nineteenth century resulted in clear-cutting and establishment of extensive even-aged plantations of the native coniferous trees *Picea abies* and *Pinus sylvestris*, and to some extent also of *Larix decidua*. Coniferous forests, most of them plantations, currently occupy the majority of forest land. Of the total area of present forests, 44.1% are occupied by spruce, 9.8% by pine and 3.3% by larch (Kučera et al. 2016).

The decline in the area of natural forests and their exploitation led some land owners to declare selected areas of mountain old-growth forests as nature reserves already in the nineteenth century. Two sites in the Novohradské hory Mts in southern Bohemia, Hojná Voda and Žofínský prales, protected by Count Buquoy since 1838, are among the oldest nature reserves in Europe. Currently most of the old-growth forests are protected in nature reserves or the strict-protection zones of national parks. An inventory of old-growth forests (http://bit.ly/2pjoMQs, 2012 data) yielded an estimate of 2467 ha of pristine (original) forest remaining in the Czech Republic, defined as forests with no logging or only selective logging that occurred more than 100 year ago, and no deadwood removal in the last 50 years. Natural forests (those with minor forestry interventions) were identified on a total area of 7525 ha and near-natural forests on an area of 19,575 ha. Most of the old-growth forests occur in the Šumava Mts and the Krkonoše Mts (Winter et al. 2013; Fig. 7.9). The central part of the Šumava Mts was settled rather late, in the mid-eighteenth century, probably because of the absence of alpine grassland that could

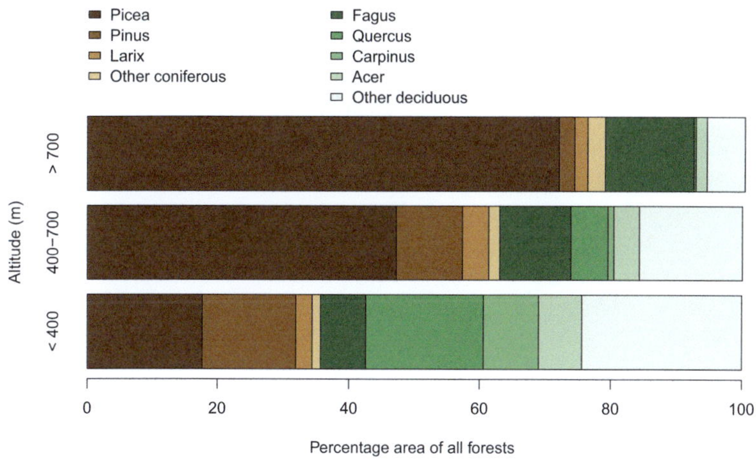


**Fig. 7.10** Percentages of different trees in the total area of current forests (including forest plantations) at three altitudinal ranges (data from the National Forest Inventory; Kučera et al. 2016)

be used as pastures. Extensive areas of spruce forest on the Šumava summits were subject to limited forestry interventions and their dynamics were dominated by large-scale natural disturbances (Čada et al. 2016). Currently the focus of nature conservation in mid- and high-altitude old-growth forests is on protecting the processes of natural ecosystem dynamics; this differs from the focus on protection of species diversity in low-altitude forests, which in most cases require conservation management that imitates historical forest management practices.

According to the National Forest Inventory made in the period of 2011–2015, the total area of forest in the Czech Republic is 2904 km$^2$, i.e. 36.8% of the area of the country. In a previous National Forest Inventory (2001–2004; ÚHÚL 2007) it was 2890 km$^2$. The increase is mainly due to natural successional processes and to a small extent due to human activities (Kučera et al. 2015). Of the total area of forest, 58.9% is composed of coniferous and 41.1% of broad-leaved deciduous trees (Kučera et al. 2016; Figs. 7.10 and 7.11).

### 7.5.1.2 Main Tree Species

Most Czech forests are dominated by a single species or codominated by two or three tree species (Svoboda 1953–1957; Neuhäuslová et al. 1998, 2001; Chytrý 2013). Superior competitors in zonal habitats include *Carpinus betulus* in the lowlands, *Fagus sylvatica* (and to some extent also *Abies alba* and *Picea abies*) at mid-altitude areas and in the mountains, and *Picea abies* at the highest altitudes. Other tree species are either confined to azonal (too wet or too dry) habitats or supported by forest management (Fig. 7.12).

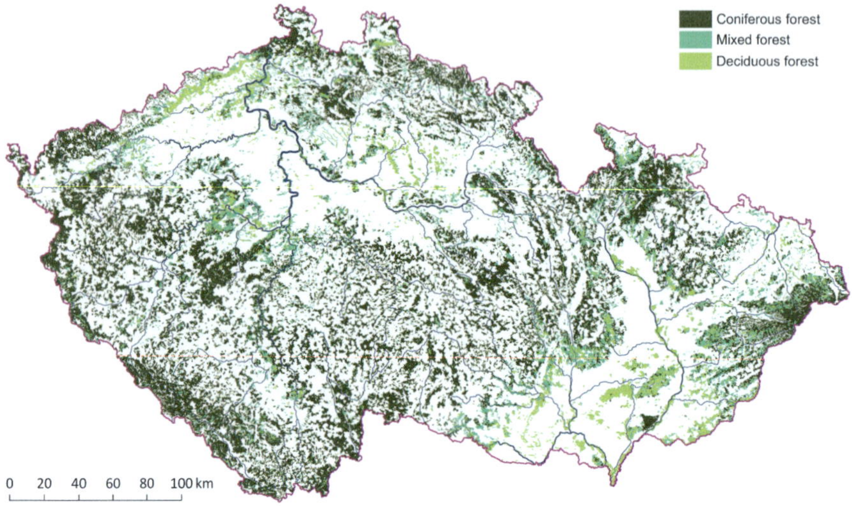


**Fig. 7.11** Forest cover in the Czech Republic (based on CORINE Land Cover 2000 data set; Bossard et al. 2000)

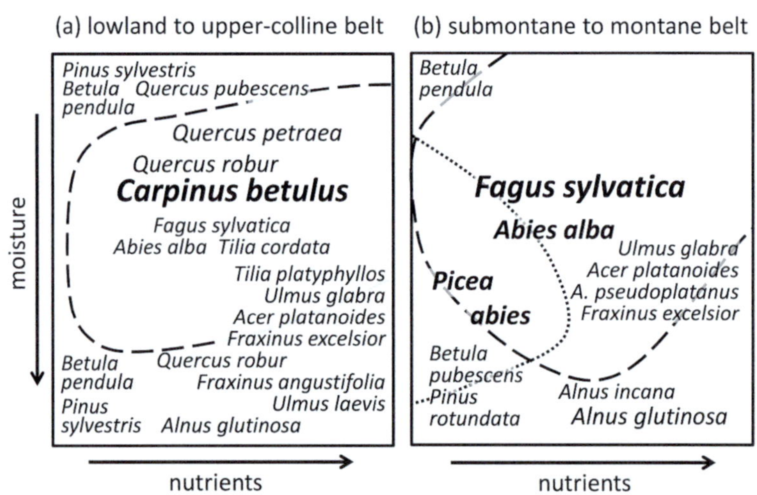


**Fig. 7.12** Schematic representation of the realized niches of the most common forest tree species in the Czech Republic, with respect to soil moisture and nutrient availability (Ellenberg-type ecograms): (**a**) lowland to upper-colline belt and (**b**) submontane to montane belt. Superior competitors are in larger *bold* letters and their realized niches are bounded by *broken* or *dotted lines*

**European beech** (*Fagus sylvatica*) is the most common deciduous tree in the country, occupying 10.3% of the total forested area (Kučera et al. 2016). It is a superior competitor on zonal, fertile soils because of its tall stature (it can be more than 40 m tall), dense canopy, life span of 200–300 years in undisturbed stands, shade tolerance when young (Svoboda 1953–1957) and intense root competition (Slavíková 1958). It does not tolerate flooded or water-logged sites, dry soils and areas where there is a risk of late frosts during the period of leaf flushing in spring (Ellenberg and Leuschner 2010). Nevertheless, it can grow on both acidic and base-rich soils. Beech forests or mixed forests of beech with spruce, fir, or both, are common in mid-altitude areas and the mountains but very rare in dry lowlands. However, beech forests can occasionally occur at altitudes of about 200 m. Rarity of beech in the lowlands may be partly due to historical management, e.g. its decline in coppiced forests in which hornbeam and other deciduous trees were favoured. Interestingly, former beech coppices are rare in the Czech Republic, although being rather common in other countries of Central Europe.

**European hornbeam** (*Carpinus betulus*) is widespread from the lowlands to the upper-colline belt, especially in the northern half of Bohemia and in Moravia and Silesia. Because of its shorter life span and shorter stature, it is a weaker competitor than beech. Nevertheless, it also forms a dense canopy like beech, which negatively affects recruitment of the light-demanding trees and makes it a superior competitor in areas where beech suffers from seasonal droughts or is injured by late frosts. It is most common on mesic nutrient-rich soils, but also grows on acidic or intermittently wet soils. Hornbeam was locally present in the lowlands of the eastern Czech Republic already in the Middle Holocene (Novák et al. 2017), however, it was the last species to expand its range in the current Czech territory during the Holocene. It spread in the Czech lowlands only in the Late Holocene and it was probably aided by forest management such as coppicing, because this tree easily resprouts from stumps. It migrated from the north-east (Pokorný 2002) and it is still absent from some parts of south-western Bohemia in spite of the presence of suitable habitats in this area.

**Oaks** (*Quercus* spp.) are long-living but light-demanding deciduous trees unable to reproduce generatively under the canopy of other broad-leaved trees. In undisturbed, non-managed forests, oaks would probably be confined to south-facing slopes with shallow soil, which are too dry for other broad-leaved trees. In spite of this, oaks are currently quite common in Czech forests, covering 7.9% of the forested land (Kučera et al. 2016), which is for most part a heritage of historical forest management. Oaks were preserved as standard trees in coppiced woodlands, planted for acorns or regenerated spotaneously from seed in wooded pastures. In today's forest reserves on mesic soils that are left to spontaneous succession, dying old individuals of oak are usually replaced by other species, e.g., hornbeam on mesic soils in the lowlands, beech at mid-altitudes and ash on floodplains but also on mesic soils (Hofmeister et al. 2004). *Quercus petraea* is drought-resistant and often occurs on slopes with shallow soil, both on acidic and base-rich bedrocks. *Quercus robur* occurs not only in floodplain forests along lowland rivers, but also on dry slopes, especially in south-western Bohemia. *Quercus pubescens* is a thermophilous species of sub-Mediterranean distribution, and in the Czech Republic is rather rare, confined to dry, base-rich soils in the driest areas in northern and central Bohemia and southern Moravia. *Quercus cerris* is a South-eastern European species reaching its distribution limit in southern Moravia. It is unclear whether it is a native species of the Czech flora, because it was also planted due to its rich and regular production of acorns. Other species of oak previously reported in the Czech Republic have an unclear taxonomic status (*Q. dalechampii*, *Q. polycarpa* and *Q. virgiliana*) or are most probably not native (*Q. frainetto*; Novák and Roleček 2013). The most frequently planted alien species of oak is the North American *Quercus rubra*, which is also spontaneously spreading (see Chap. 8, Sect. 8.10, this book).

**Noble hardwoods** (*Acer platanoides*, *A. pseudoplatanus*, *Fraxinus excelsior*, *Tilia cordata*, *T. platyphyllos* and *Ulmus glabra*) are confined to nutrient-rich soils that are both well supplied with water and well aerated. Leaf litter of these trees is rich in nutrients and decomposes rather quickly (1–2 years), therefore the base status and nutrient availability in the topsoil is rather high. Canopy shading by these trees is less intense than that by beech, hornbeam or fir, but stronger than by oak, and shade tolerance of their seedlings and saplings is positively correlated with the shading capacity of adult trees (Ellenberg and Leuschner 2010). Therefore these trees were superior competitors in the Middle Holocene, when they dominated landscapes at low and mid-altitudes, but were largely outcompeted from zonal habitats by beech, fir and hornbeam in the Late Holocene. Currently noble hardwoods are confined to ravines and lower slopes with scree accumulation, where they grow quickly, favoured by abundant moisture and nutrients, as opposed to beech and hornbeam, which fail to develop monodominant stands in such habitats. Within the group of noble hardwoods, *Tilia cordata* and *T. platyphyllos* are more thermophilous, having their optimum at lower altitudes, while *Ulmus glabra* and especially *Acer pseudoplatanus* are more tolerant of cold conditions and commonly occur in montane ravine forests. *Acer pseudoplatanus* locally grows in mixed stands with beech on precipitation-rich slopes in the mountains. *Fraxinus excelsior* is common not only in ravines and on talus slopes but also in floodplain forests along the middle and lower courses of rivers.

**Alders** (*Alnus* spp.) are typical trees of wet habitats. *Alnus glutinosa* is the most common tree in riverine forests at mid-altitudes and also dominates carrs in waterlogged depressions. In contrast, *Alnus incana* only occurs naturally on coarse well-aerated fluvial sediments (stones, gravel and sand) along mountain streams with fast-running water.

**Birches** (*Betula* spp.) are competitively weak, light-demanding but stress-tolerant trees. *Betula pendula* is common as an early successional species at deforested sites, including abandoned agricultural land, but it also grows in nutrient-poor, dry or wet habitats that are too stressful for other broad-leaved species. *Betula pubescens* subsp. *pubescens* is typical of nutrient-poor, wet or water-logged sites, usually marginal parts of mires or drained mires. *Betula pubescens* subsp. *carpatica* occurs in the mountains, being typical especially of the subalpine belt in the Sudetes, where it forms woodlands and scrub on and next to avalanche tracks in cirques.

**Poplars** (*Populus* spp.) are represented by *Populus tremula*, a stress-tolerant pioneer species ecologically similar to *Betula pendula*, but with extensive clonal growth, and two species that occur in lowland floodplain forests, *Populus nigra* and *P. alba*. The former is now very rare and the latter is a Southern European species considered as native only in southern Moravia, although it occurs also in Bohemia.

Several species or hybrids of alien poplars are planted: in particular, there are plantations of *Populus ×canadensis* in the potential habitats of floodplain hardwood forests and it is spontaneously spreading, especially in disturbed habitats (see Chap. 8, Sect. 8.10, this book).

**Willows** (*Salix* spp.) are represented mainly by shrubby species. Species with predominant arboreal growth include *Salix alba*, typical of lowland softwood floodplain forests, *Salix euxina*, which is common in riverine woodland and scrub at mid-altitudes, and *Salix caprea*, a species of early successional stages in disturbed forests or abandoned grasslands, especially on slightly wet mineral soils.

**Black locust** (*Robinia pseudoacacia*), a tree of North American origin, is the most common alien woody species in the Czech Republic, dominating in 0.6% of the total forested land (ÚHÚL 2007). It was extensively planted in the nineteenth century in warm areas as a source of honey, hard, resistant wood and for reducing soil erosion (Nožička 1957; Vítková et al. 2017; see also Chap. 8, Sect. 8.10, this book). Its ability to rapidly spread vegetatively by means of root suckers from planted stands to adjacent grasslands makes it a significant threat for biodiversity conservation.

Other broad-leaved deciduous tree species in the Czech Republic include *Fraxinus angustifolia*, a dominant tree in floodplain hardwood forests in southern Moravia, *Ulmus laevis*, a common species in floodplain hardwood forests, *Acer campestre* and *Ulmus minor*, subcanopy trees in lowland forests, *Sorbus* species (*S. aucuparia*, *S. aria* agg. and *S. torminalis* with related hybridogenous and apomictic taxa, and *S. domestica*), which are admixed in different types of forest, usually as subcanopy trees, and fruit trees or their native wild ancestors (especially *Malus domestica*, *M. sylvestris*, *Prunus avium*, *Pyrus communis* and *P. pyraster*), which are found as solitary individuals in lowland forests that were formerly managed as wooded pastures or coppices with standards.

**Norway spruce** (*Picea abies*) is tolerant of cool and long winters, but it requires moist or mesic conditions and occurs naturally either in areas of higher precipitation or at sites with high topographic moisture. It is adapted to acidic, nutrient-poor soils such as podzol or ranker on boulder screes, but it also grows in wet depressions and on organic soils of mires. Due to its shallow root system spruce is vulnerable to uprooting by wind. It dominates forests in the upper-montane belt and regularly occurs in mixed stands with beech and fir in the montane and submontane belt. In a warmer climate at low altitudes it is competitively inferior to broad-leaved deciduous trees, therefore its natural occurrence is restricted to wet organic soils, depressions accumulating cold air or shaded bottoms of deep valleys; in such habitats spruce occurs even at altitudes below 200 m (Mráz 1959). Spruce has been favoured since the Middle Ages by the selective logging of beech for charcoal production in montane and submontane forests. Archive sources indicate that it was rather common from the upper-colline belt to higher altitudes in the sixteenth to eighteenth centuries (Mráz 1959; Nožička 1972; Szabó et al. 2017). Pollen-based quantitative landscape reconstruction models even suggest that spruce may have been one of the most common trees at mid-altitudes of the Bohemian Massif continuously for approximately eight millenia (Abraham et al. 2016; Szabó et al. 2017). Since the early nineteenth century spruce has been extensively planted for timber, also at low altitudes, which has ultimately made it by far the most common tree species in the Czech landscape. Planting obscured its natural distribution pattern, which is a matter of contentious debate. Large areas of spruce forests died in the 1980s, especially in the Krušné and Jizerské hory Mts, and to a notable extent also in the Krkonoše and Moravskoslezské Beskydy Mts, as a result of atmospheric emissions of sulphur dioxide (Kubíková 1991). Large areas deforested due to air pollution were replanted with pollution-resistant North American *Picea pungens*, especially in the Krušné hory Mts, where it was planted on more than 8800 ha. However, since 2009 *P. pungens* is declining due to a massive outbreak of bud blight caused by the fungus *Gemmamyces piceae* (Černý et al. 2016) and is being replaced by the native *P. abies*. The native spruce is damaged only to a small extent by this pathogen and is no longer affected by air pollution, which was strongly reduced in the 1990s. Periodical decline of spruce stands at a more local scale is caused by outbreaks of bark beetle (*Ips typographus*), especially in plantations and forests weakened by air pollution or damaged by wind storms, but also in natural spruce forests in the mountains (Čada et al. 2016). Spruce tends to respond to the recent climate warming by shifting its optimum to higher altitudes, increasing growth rates at the altitudes near the timberline since the 1980s, but at the same time it is sensitive to more frequent drought events at mid-altitudes (Ponocná et al. 2016). This trend clearly calls for changes in forests management strategies.

**Silver fir** (*Abies alba*) grows in similar habitats as beech, especially on deep and not too dry soils; these two species often form mixed stands. Unlike beech, however, fir also occurs on heavy and wet mineral soils such as planosols or stagno-gleyic luvisols (Mráz 1959), but not on organic soils or dry soils in sandstone pseudokarst areas. Like beech, fir is sensitive to late spring frosts (Ellenberg and Leuschner 2010). Fir canopy is dense and there is limited light available to the understorey, however, fir saplings tolerate deep shade. Small fir individuals with thin stem can persist for dozens of years in the understorey and resume rapid growth when a canopy gap appears above them. This enables fir to compete successfully with fast-growing beech seedlings. Fir was very common in Czech forests between the Bronze Age and Middle Ages. Its expansion in forests at the expense of beech was probably aided by forest management, namely livestock grazing and litter raking (Málek 1983; Vrška et al. 2009). While deer frequently browse fir saplings, domestic livestock prefer saplings of broad-leaved trees. Raking removes the thick layer of slowly decomposing beech litter, which fir seedlings have difficulty penetrating. However, as a result of the forest regulations introduced in the mid-eighteen century, livestock was gradually removed from forest and its niche was filled by deer. Deer densities in the nineteenth to twentieth centuries considerably exceeded those typical of the Middle Ages, because large predators (wolf, bear and lynx) had been eradicated. Incidence of litter raking also steadily decreased. These changes have led to a decline in the abundance of fir since the nineteenth century, which may have been accelerated by air pollution in the twentieth century (Málek 1983; Vrška et al. 2009; Šebková et al. 2011; Janík et al. 2014).

**Scots pine** (*Pinus sylvestris*) has a very broad ecological niche, similar to that of birch. It occurs on soils that range from very wet to very dry, from acidic to basic, and from nutrient-poor to nutrient-rich. However, it is unable to regenerate when shaded by other trees. Therefore it is one of the weakest competitors among the Czech native trees. It is a pioneer species in disturbed habitats or on abandoned agricultural land, but it can also occur as a dominant tree species in successionally stable vegetation on forested peatlands, rock outcrops or serpentinite soils. In some lowland areas pine forests may have existed for millenia, probably supported or maintained by wild fires (Novák et al. 2012, 2015; Adámek et al. 2015). In some of its habitats, especially in sandstone pseudokarst areas such as the Elbe Sandstones, *P. sylvestris* has been recently outcompeted by *P. strobus*, an invasive species of North American origin (Hadincová et al. 1997; Pyšek et al. 2012a).

**Bog pine** (*Pinus uncinata* subsp. *uliginosa*) is endemic to Central Europe. It occurs on bogs in the upper-colline to montane belt especially in western and southern Bohemia, but it is also found in the Bohemian-Moravian Highlands and Hrubý Jeseník Mts (Businský 2009).

**European larch** (*Larix decidua*) is often used in forest plantations. It was widespread in the current Czech territory in the last glacial (Willis and van Andel 2004; Jankovská and Pokorný 2008) but it disappeared from most of its full-glacial range during the Holocene. A native population of larch survived throughout the Holocene in the Hrubý Jeseník Mts and its foothills, which is documented by finds of larch pollen grains in peat cores from this area in different periods of the Holocene (Dudová et al. 2012, 2014) and by written reports of the local use of larch wood in the sixteenth century (Nožička 1962). However, larch does not form any distinct type of natural forest in this area.

**Europan yew** (*Taxus baccata*) is a small tree occurring locally in the subcanopy of deciduous forests on steep slopes, especially in deep valleys.

#### 7.5.1.3 Main Types of Forest Vegetation

**Alder carrs** (class *Alnetea glutinosae*, alliance *Alnion glutinosae*; Fig. 7.13a) are forested wetlands dominated by *Alnus glutinosa* with a species-poor herb layer with tall sedges, especially *Carex acutiformis*, *C. elongata* and *C. riparia*. They occur on organic or gleyic soils in water-logged depressions, often near fishponds and in terrestrialized oxbows, from the lowlands to submontane areas (Douda 2008). Many current stands have developed from wet meadows and other types of abandoned wetland (Douda et al. 2009). Jeník (1980) proposed a model of natural autogenic cyclic succession of alder carrs with alternating forest and treeless stages. The pattern found in fossil pollen data from eastern Bohemia was interpreted as alternations of these two stages (Pokorný et al. 2000), however, it may have other causes than natural dynamics, e.g. human activities or changes in hydrological regime caused by external factors (Douda et al. 2009).

**Riparian willow-poplar forests** (class *Salicetea purpureae*, alliance *Salicion albae*) occur mainly along lowland rivers, where they are dominated by *Salix alba* or, rarely, by *Populus nigra*, and in southern Moravia *P. alba* is also present. Some stands can further be found along streams at mid-altitudes, where the dominant tree species is usually *Salix euxina* (Neuhäuslová 1987). These forests occur on fluvisols or gleyic soils on lower river terraces, which are flooded annually (Mezera 1956–1958). Because the rivers are now regulated and the ground-water table is lower, these softwood floodplain forests are changing into more mesic forest types such as hardwood floodplain forests (Vrška et al. 2006).

**Riparian alder forests** (class *Carpino-Fagetea*, alliance *Alnion incanae*) occur along medium-sized streams, brooks and around springs from the lowlands to the montane belt. Forests dominated by *Alnus incana* (Fig. 7.13b) form narrow galleries along fast-running mountain brooks with fluctuating water discharge. More extensive stands of *A. incana* occur on the montane floodplain of the upper Vltava River in the Šumava Mts (Sádlo and Bufková 2002). Forests of *Alnus glutinosa*, in places with an admixture of *Fraxinus excelsior* or other trees, occur on fluvisols or gleyic soils along streams from the mountains to the lowlands (Fig. 7.13c). Their herb layer is usually richer than that of upland forests in the same areas; it contains a mixture of species of alluvial habitats and mesophilous species that occur in adjacent forests (Douda 2008).

**Hardwood floodplain forests** (Fig. 7.13d–f) replace riparian alder forests on the floodplains of large lowland rivers. They are dominated by *Fraxinus excelsior*, *Quercus robur*, *Ulmus laevis*, *U. minor* and in southern Moravia also by the Southern European *Fraxinus angustifolia* (Mezera 1956–1958). As the species composition of the herb layer is similar to that of riparian alder forests (Douda 2008; Douda et al. 2016), both of these forest types are assigned to the same alliance (*Alnion incanae*). Charcoal analyses from numerous archaeological sites situated along Moravian lowland rivers indicate continuous dominance of oak, accompanied by ash and elm, in floodplain forests at least since the early Neolithic (Novák et al. 2017). The medieval deforestation of upland areas led to increased sedimentation in the floodplains (Kadlec et al. 2009), which may have caused nutrient enrichment and changes in the understorey of the floodplain forests, but their tree-layer composition probably did not change much. Although oaks were the most common trees in these forests for millenia, the ageing oak layer has been declining during recent decades. This process is accompanied by almost complete lack of oak regeneration and the space left by the oaks is being filled by other trees (Vrška et al. 2006). These dynamics suggest that historically the high abundance of oaks may have been caused by the past use of floodplains as wooded pastures and the current old oaks are a legacy from the times before the abandonment of this management. Elms declined dramatically after the epidemic of Dutch elm disease in the 1960s–1970s, but their populations are currently recovering (Dvořák et al. 2007). In the past the Czech hardwood floodplain forests were flooded almost every year in March–April, but the regulating of rivers, especially in the second half of the twentieth century, reduced the incidence of flooding and the hardwood floodplain forests have become more mesic. This has resulted in the spread of mesophilous tree species that are untypical of regularly flooded stands, especially *Acer campestre*, *Carpinus betulus* and *Tilia cordata.* In old-growth floodplain forests in southern Moravia, the first two of these species cur-

**Fig. 7.13** Examples of wet deciduous forest types in the Czech Republic: (**a**) *Alnus glutinosa* carr with *Carex elongata* near Polanka nad Odrou, north-eastern Moravia (*Carici elongatae-Alnetum glutinosae*, *Alnion glutinosae*); (**b**) montane riverine forest with *Alnus incana* in the Vydra valley, Šumava Mts (*Alnetum incanae*, *Alnion incanae*); (**c**) floodplain forest of middle stream courses with *Alnus glutinosa*, Tichá Orlice valley near Sudislav, eastern Bohemia (*Stellario nemorum-Alnetum glutinosae*, *Alnion glutinosae*); (**d**) hardwood floodplain forest of lower stream courses with *Fraxinus angustifolia* on the Dyje floodplain near Nové Mlýny, southern Moravia (*Fraxino pannonicae-Ulmetum*, *Alnion incanae*); (**e**, **f**) spring aspect of a hardwood floodplain forest of lower stream courses with the geophytes *Corydalis cava* (*left*) and *Galanthus nivalis* (*right*) (*Ficario vernae-Ulmetum campestris*, *Alnion incanae*)

rently have higher recruitment rates than the dominant *Fraxinus angustifolia*, which is more light-demanding (Janik et al. 2011, 2016a). These trends suggest that in the absence of regular flooding, hardwood floodplain forests are likely to develop towards broad-leaved forests dominated by mesophilous tree species.

**Oak-hornbeam forests** (class *Carpino-Fagetea*, alliance *Carpinion betuli*; Figs 7.14a and 7.15a) are widespread from the lowlands to the upper-colline belt. Dominant trees usually include *Carpinus betulus* and/or *Quercus petraea*, in some areas also *Q. robur*, and frequently there is an admixture of *Tilia cordata*. These forests have developed as a result of intensive management, especially by coppicing, since the Iron Age. Although they are no longer managed in this way, the tree species composition still reflects this influence. For example, the current distribution of oak-hornbeam forests in Moravia is roughly coincident with the extent of coppice management in the Late Middle Ages (Szabó et al. 2015). In the absence of management, oak would be probably outcompeted by hornbeam or noble hardwoods, and at higher altitudes, hornbeam by beech. During its postglacial migration, hornbeam did not reach some areas of southern and western Bohemia, where habitats suitable for oak-hornbeam forests are currently dominated by *Quercus robur* (Moravec 1964). Herb-layer species composition of oak-hornbeam forests is slightly different in the Bohemian Massif, Pannonian part of southern Moravia and Carpathian part of Moravia due to different phytogeographical influences (Neuhäuslová-Novotná 1964; Knollová and Chytrý 2004).

**Beech and mixed conifer-beech forests** (Figs 7.14b and 7.15b–c) are the predominant types of natural forest vegetation in the upper-colline to montane belts. At most sites there are currently mixed spruce-beech forests or monodominant stands of beech, but in the past mixed stands with fir or even pure fir forests were common at mid-altitudes (Rybníček and Rybníčková 1978; Kozáková et al. 2011). Nowadays fir-dominated forests are rare, occurring mainly in south-western Bohemia and northern and north-eastern Moravia (Boublík 2010). In many areas mixed conifer-beech and pure beech forests were replaced by spruce plantations. However, historical and palaeoecological data suggest that the proportion of spruce in these forests was high even before the establishment of spruce plantations in the early nineteenth century (Abraham et al. 2016; Szabó et al. 2017). The dynamics of old-growth beech or mixed beech forests have been studied in several nature reserves, e.g. Boubín, Milešice and Stožec in the Šumava Mts, Žofín and Hojná voda in the Novohradské hory Mts, Polom and Žákova hora in the Bohemian-Moravian Highlands and Razula, Salajka and Mionší in the Moravskoslezské Beskydy Mts (Průša 1985; Vrška et al. 2002, 2012; Janík et al. 2014; Král et al. 2014). The results of these studies indicate that these forests partly follow the classical model of fine-scale cyclic changes in the climax forest driven by canopy gap dynamics, with stages of growth, optimum (steady-state) and breakdown ('small developmental cycle'; Leibundgut 1993; Korpeľ 1995; Král et al. 2014). Old-growth forest at higher altitudes tend to have larger patches of the breakdown stage as a result of stronger disturbances by wind and other agents (Král et al. 2014). However, some of the changes observed recently in old-growth forests do not seem to be cyclical. Instead, they indicate long-term trends in forest development in response to cessa-

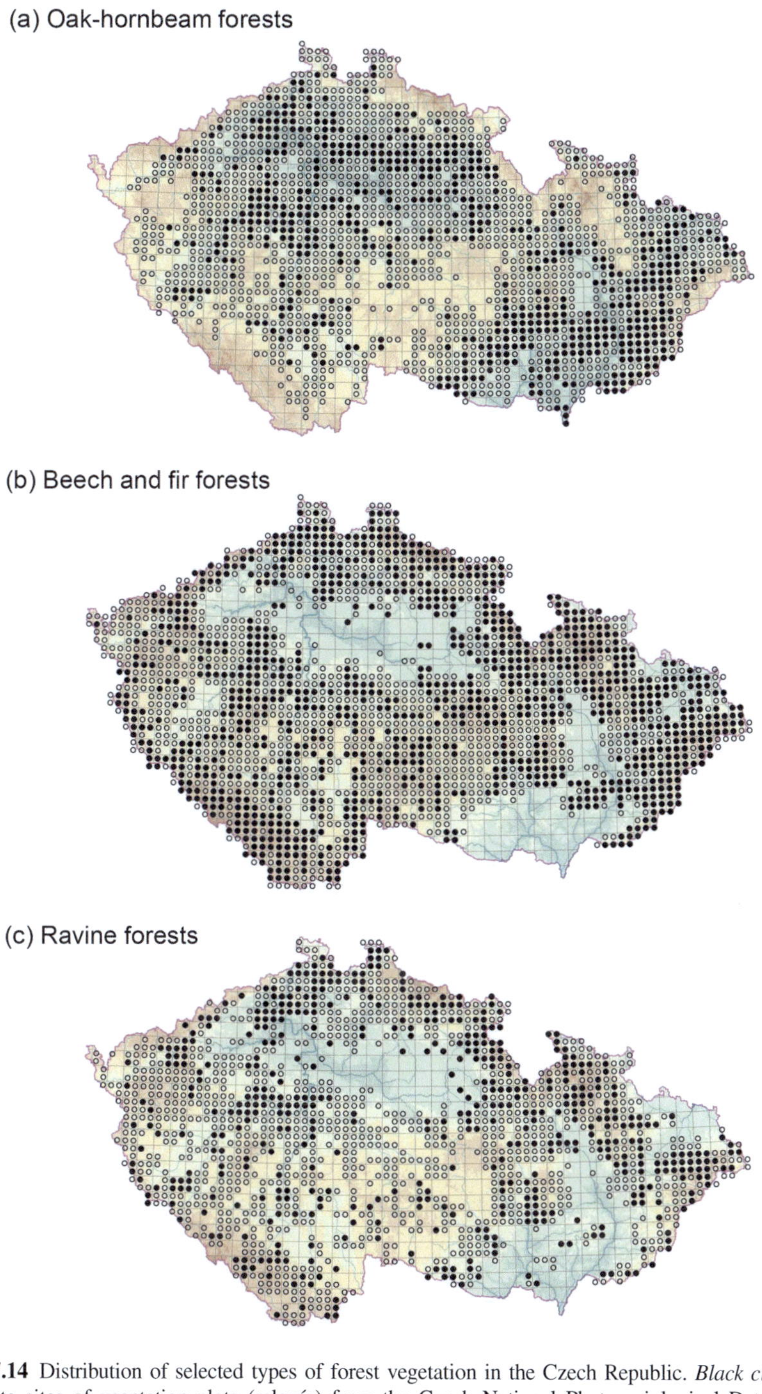


**Fig. 7.14** Distribution of selected types of forest vegetation in the Czech Republic. *Black circles* indicate sites of vegetation plots (relevés) from the Czech National Phytosociological Database (Chytrý and Rafajová 2003) and Database of Forest Typology of the Forest Management Institute (ÚHÚL) that correspond to the formal definitions of particular vegetation types proposed by Chytrý (2013). *Open circles* indicate records based on habitat mapping from 2001–2008 (Härtel et al. 2009), revised by Chytrý et al. (2010), which include fragmentary or transitional occurrences of individual vegetation types. Phytosociological alliances: (**a**) *Carpinion betuli*; (**b**) *Fagion sylvaticae*, *Sorbo torminalis-Fagion sylvaticae* and *Luzulo-Fagion sylvaticae*; (**c**) *Tilio platyphylli-Acerion*

**Fig. 7.15** Examples of mesic and dry deciduous forest types in the Czech Republic: (**a**) Oak-hornbeam forest (*Quercus petraea* and *Carpinus betulus*) near Čížov in south-western Moravia (*Galio sylvatici-Carpinetum betuli*, *Carpinion betuli*); (**b**) Old-growth beech forest in the Žákova hora National Nature Reserve near Cikháj, Bohemian-Moravian Highlands (*Galio odorati-Fagetum sylvaticae*, *Fagion sylvaticae*); (**c**) *Galium odoratum* and *Mercurialis perennis*, two typical species in the herb layer in a beech forest at Žákova hora; (**d**) Ravine forest on a slope with an accumulation of rock debris in the Dyje valley near Vranov nad Dyjí, south-western Moravia (*Aceri-Tilietum*, *Tilio platyphylli-Acerion*); (**e**) Thermophilous oak forest with *Quercus pubescens* on a calcareous flysch sandstone slope near Němčičky, southern Moravia (*Lithospermo purpurocaerulei-Quercetum pubescentis*, *Quercion pubescenti-petraeae*); (**f**) Acidophilous oak forest with *Quercus petraea* in the Ohře valley near Hory, western Bohemia (*Luzulo luzuloidis-Quercetum petraeae*, *Quercion roboris*)

tion of livestock grazing, increases in the population density of game since the eighteenth century or air pollution. For example, a steady increase in beech at the expense of fir and spruce has occurred at most of the sites studied (Vrška et al. 2012; Janík et al. 2014, 2016b).

Herb layers in Czech beech forests are remarkably poorer in species than those in beech forests in the Alps and central part of the Western Carpathians, which is partly due to predominance of poor acidic bedrocks in the Bohemian Massif (Ujházyová et al. 2016), partly due to historical legacies such as distance to the glacial refugia of temperate deciduous forests (Willner et al. 2009). The species composition of the herb layer in beech forests varies mainly in response to soil nutrient and base status (Willner et al. 2017a). These forests, belonging to the class *Carpino-Fagetea*, are accordingly assigned to the alliances *Fagion sylvaticae* (on eutrophic soils), *Sorbo torminalis-Fagion sylvaticae* (on calcareous soils; Boublík et al. 2007) and *Luzulo-Fagion sylvaticae* (on acidic soils). There is some evidence that atmospheric pollution in the second half of the twentieth century caused soil acidification and the decline in nutrient-demanding species in the herb layer in mountain beech forests, which resulted in the *Fagion sylvaticae* forests changing into *Luzulo-Fagion sylvaticae* forests (Hédl 2004; Hédl et al. 2011).

**Ravine forests** (class *Carpino-Fagetea*, alliance *Tilio platyphylli-Acerion*; Figs 7.14c and 7.15d) occur in ravines and on steep slopes, most commonly in deep river valleys. *Tilia cordata* and *T. platyphyllos*, with an admixture of *Carpinus betulus*, are more common in drier and warmer habitats at lower altitudes or on upper slopes, whereas *Acer pseudoplatanus*, *Fraxinus excelsior* and *Ulmus glabra* with an admixture of *Abies alba* and *Fagus sylvatica* are typical of wetter and cooler habitats at higher altitudes, on lower slopes or mountain tops that receive high precipitation. Dominant tree species of ravine forests have a good capacity of vegetative regeneration after disturbances caused by soil erosion or rock fall on steep slopes.

**Thermophilous oak forests** (Figs 7.15e and 7.16a) are dominated by *Quercus pubescens* in the driest and warmest habitats with base-rich soils, or *Q. petraea* in cooler habitats or on more acidic bedrock, or *Q. robur* on more mesic soils (Chytrý 1997; Roleček 2007). Natural stands are typically confined to the upper parts of south-facing slopes. They have an open canopy, a well-developed shrub layer and species-rich herb layer. However, there are thermophilous oak forests that developed as a result of historical management such as forest grazing (Mráz 1958). Abandonment of historical management by the mid-twentieth century caused the spread of trees with a denser canopy than oaks (e.g. *Fraxinus excelsior*; Hofmeister et al. 2004), decline of light-demanding, thermophilous species in the herb layer and the spread of species typical of mesic forests (Hédl et al. 2010; Kopecký et al. 2013). Thermophilous oak forests are assigned to the class *Quercetea pubescentis* and alliances *Quercion pubescenti-petraeae* (on slopes with shallow, base-rich soils, especially those on limestone, with some sub-Mediterranean species and frequent occurrence of *Quercus pubescens*), *Aceri tatarici-Quercion* (on loess and sandy soils in southern Moravian forest-steppe area, often on flat land, with continental floristic influence) and *Quercion petraeae* (on slopes of acidic rocks or on luvisols with a decalcified upper soil horizon, with dominance of *Quercus petraea* or *Q. robur*).

(a) Thermophilous oak forests

(b) Acidophilous oak forests

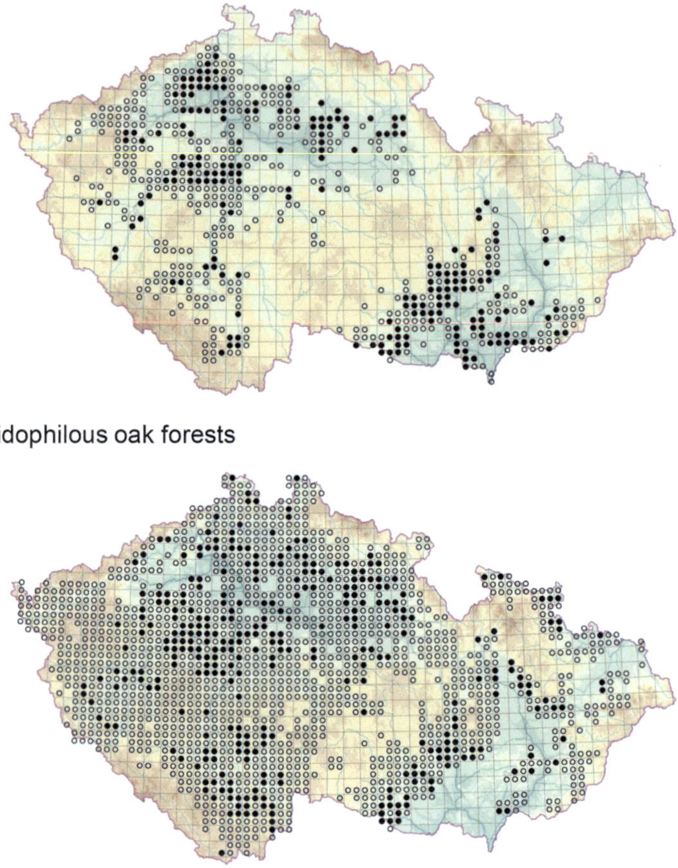

**Fig. 7.16** Distribution of oak forest vegetation: (**a**) *Quercetea pubescentis*; (**b**) *Quercetea robori-petraeae*. Symbols as in Fig. 7.14

**Acidophilous oak forests** (class *Quercetea robori-petraeae*, alliance *Quercion roboris*; Figs 7.16b and 7.15f) are dominated by *Quercus petraea* or *Q. robur*. In places *Betula pendula* or *Pinus sylvestris* occur in mixed stands with oaks. Their herb layer is species-poor, consisting of a few acidophilous species. These forests are confined to low-pH soils, which are usually mesic but can also be slightly dry or slightly wet (Neuhäusl and Neuhäuslová-Novotná 1967). In the Map of Potential Natural Vegetation of the Czech Republic (Neuhäuslová et al. 1997), acidophilous oak forests, together with mixed acidophilous forests with pine, fir and birch, are mapped over large areas at mid-altitudes of western and southern Bohemia, although few such forests currently occur in that area, which is largely covered by a mosaic of arable land and conifer plantations. Pollen data from the pre-medieval period indicate that these areas were mainly covered by forests of spruce, fir and beech with a relatively small amount of oak and birch (Pokorný 2002; Abraham et al. 2016).

**Pine forests** (Fig. 7.17a) are common at low and middle altitudes in the Czech Republic, but most of them are plantations of native *Pinus sylvestris*. Pine forests were common in the Late Glacial. The area under pine was greatly reduced by the Middle Holocene due to the spread of deciduous trees, but pine became more abundant again with the increase in human activity in the Late Holocene (Pokorný 2002, 2005, 2011; Kuneš et al. 2008, see also Fig. 6.4 in Chap. 6, this book). In some areas pine forests may have been maintained for millenia by periodical wild fires, which occur especially on hill tops and sunny slopes in sandstone pseudokarst areas (Novák et al. 2012, 2015; Adámek et al. 2015). Because of the weak competitive ability of *Pinus sylvestris*, pine forest can persist in undisturbed conditions only in stressful azonal habitats that are unsuitable for broad-leaved trees, especially on rock outcrops, serpentinite soils and bogs. There are several types of *Pinus sylvestris* forests in the Czech Republic with the diversity of the herb and moss layers depending on soil chemistry and moisture. Basiphilous pine forests of the class *Vaccinio-Piceetea*, alliance *Festuco-Pinion sylvestris*, containing several continental species typical of forest-steppe in their herb layer, occur on calcareous sandstone and marlstone, especially in northern Bohemia. Acidophilous pine forests (class *Vaccinio-Piceetea*, alliance *Dicrano-Pinion sylvestris*; Fig. 7.18a) occur in sandstone pseudokarst areas, on outcrops of igneous or metamorphic rocks in river valleys of the Bohemian Massif, and on sand dunes. However, in many cases they may have developed as a result of historical management. They are characterized by frequent occurrence of dwarf shrubs (*Vaccinium myrtillus* and *V. vitis-idaea*) and bryophytes of boreo-continental distribution. Pine forests on serpentinite at higher altitudes are similar to this boreo-continental type of pine forest, therefore they are also assigned to the alliance *Dicrano-Pinion sylvestris*. However, pine forests on serpentinite slopes at lower altitudes (near Želivka water reservoir in central Bohemia and in the Jihlava River valley near Mohelno in south-western Moravia) contain the basiphilous grass *Sesleria caerulea* and some species typical of basiphilous montane pine forests in the Alps and Carpathians; therefore they are assigned to the class *Erico-Pinetea* and alliance *Erico carneae-Pinion*.

**Spruce forests** (class *Vaccinio-Piceetea*, alliance *Piceion abietis*; Fig. 7.17b–c) are very common throughout the country, but most of them are plantations of native *Picea abies*. Natural pure spruce forests occur mainly in the upper-montane belt of high mountain ranges (Fig. 7.18b and 7.19) where they are maintained by large-scale disturbances (Box 7.1). They are also found at lower altitudes at the margins of bogs or in broad valleys where cold air accumulates and soil paludification occurs. They are characterized by occurrence of dwarf shrubs and bryophytes of boreo-continental distribution, but Central European mountain species, such as *Calamagrostis villosa*, are also common (Sofron 1981; Jirásek 1996a).

**Bog woodland** (class *Vaccinio-Piceetea*, alliance *Vaccinio uliginosi-Pinion sylvestris*; Figs 7.17c and 7.18c, d) occurs on the relatively drier parts of bogs and at their margins. Their dominant trees are *Betula pubescens* subsp. *pubescens*, *Picea abies*, *Pinus sylvestris* and *P. uncinata* subsp. *uliginosa*, and their herb and moss layers consist of a mixture of dwarf shrubs, herbs and bryophytes typical of either organic or mineral soils. *Betula pubescens* subsp. *pubescens* occurs especially in

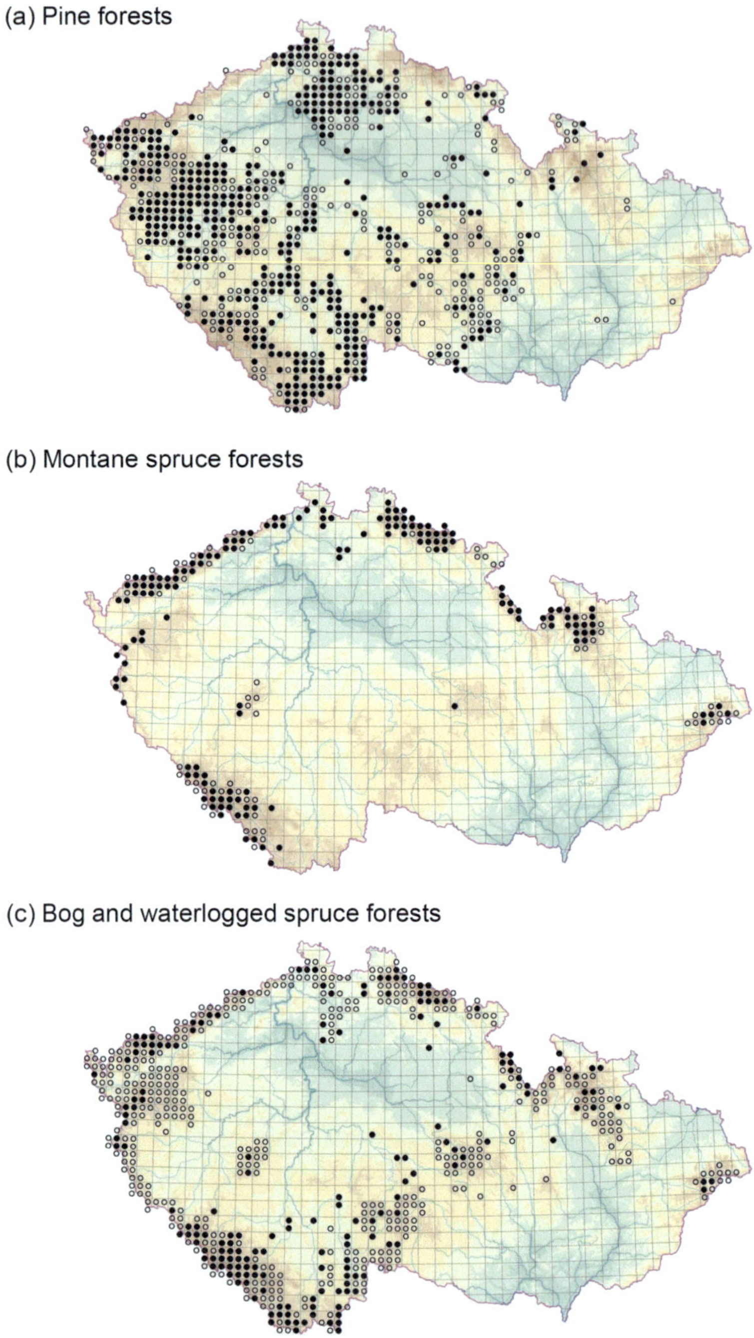


**Fig. 7.17** Distribution of selected types of coniferous forest vegetation: (**a**) *Erico carneae-Pinion*, *Festuco-Pinion sylvestris* and (mostly) *Dicrano-Pinion sylvestris*; (**b**) *Piceion abietis*, associations *Calamagrostio villosae-Piceetum abietis* and *Athyrio distentifolii-Piceetum abietis*; (**c**) *Piceion abietis*, associations *Equiseto sylvatici-Piceetum abietis* and *Soldanello montanae-Piceetum abietis*, and *Vaccinio uliginosi-Pinion sylvestris*, association *Vaccinio uliginosi-Piceetum abietis*. Some sites included in the maps may be plantations that resemble natural vegetation. Symbols as in Fig. 7.14

**Fig. 7.18** Examples of coniferous forest types in the Czech Republic: (**a**) *Pinus sylvestris* forest with *Vaccinium myrtillus* on quarzite outcrops on Goethe's Rock near Hazlov, western Bohemia (*Vaccinio myrtilli-Pinetum sylvestris*, *Dicrano-Pinion sylvestris*); (**b**) Upper-montane spruce forest near Boží Dar in the Krušné hory Mts, western Bohemia (*Calamagrostio villosae-Piceetum abietis*, *Piceion abietis*); (**c**) Bog woodland with *Pinus sylvestris* and the Central European endemic *Pinus uncinata* subsp. *uliginosa* on the Velké Dářko peat bog near Radostín in the Bohemian-Moravian Highlands (*Vaccinio-Pinetum montanae*, *Vaccinio uliginosi-Pinion sylvestris*); (**d**) Bog woodland with *Picea abies* on the Taiga peat bog near Kladská, Slavkovský les Mts, western Bohemia (*Vaccinio uliginosi-Piceetum abietis*, *Vaccinio uliginosi-Pinion sylvestris*)

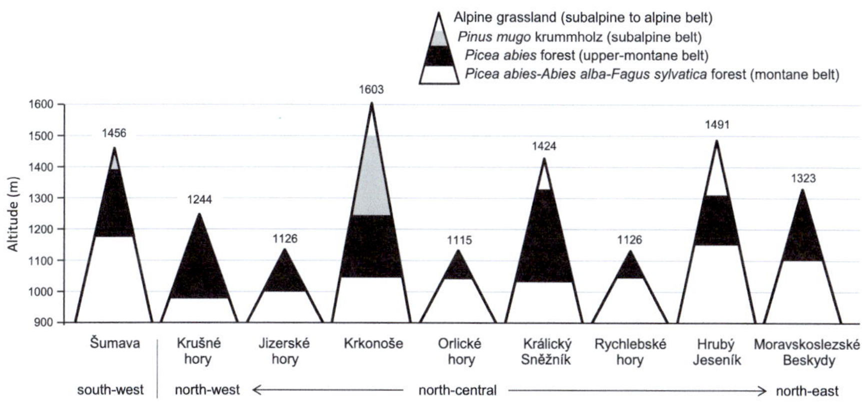


**Fig. 7.19** Altitudinal belts and timberline in the summit areas of the highest Czech mountain ranges. Altitudes of the highest peak in each range are given. The scheme represents entire mountain ranges including the parts ouside the borders of the Czech Republic (based on data from Jirásek 1996a; Neuhäuslová et al. 1997; Neuhäuslová 2001; Treml and Migoń 2015)

**Box 7.1 Natural Disturbances in Mountain Spruce Forests**

Several strong windstorms that occurred since the 1980s, particularly the violent Kyrill storm in 2007, caused windthrows affecting thousands of hectares in mountain spruce forests in the Šumava Mts and the adjacent Bayerischer Wald in Germany. Large-scale canopy dieback of old spruce stands was occurring since the mid-1990s due to outbreaks of bark beetle (*Ips typographus*), partly supported by storms and partly by a series of hot summers (Müller et al. 2008). As a large portion of the disturbed area was in the strict-protection zones of the Šumava and Bayerischer Wald National Parks, both Czech and German nature conservation authorities applied no-intervention management. Serious concerns about the appropriateness of this strategy were raised, because stand-replacing disturbances of such a large extent were felt to be catastrophic deviations from the natural forest dynamics by some managers and the public. Indeed traditional Central European forestry schools believed that fine-scale gap dynamics were the only type of natural disturbance in non-managed forests of temperate Europe, whereas they did not consider severe stand-replacing disturbances over large areas to be a part of the natural dynamics in these forests (e.g. Leibundgut 1993; Korpel' 1995). Some foresters therefore recommended salvage logging and forest restoration by planting.

Research motivated by these events challenged the traditional understanding of temperate forest dynamics. There are detailed historical records of stand-replacing forest disturbance that occured in the Šumava Mts in 1868–1870, also caused by violent windstorms followed by a bark beetle outbreak. Only about a third of forests older than 120 years survived that disturbance event (Brůna et al. 2013). Dendrochronological research in natural spruce forests in the summit areas above 1150 m a.s.l. in the Šumava Mts revealed that infrequent and severe disturbances of large extent occurred there also in the more distant past (Svoboda et al. 2012; Čada et al. 2013, 2016). Tree-ring analyses of old trees focused on patterns indicative of past stand disturbances,

(continued)

**Box 7.1** (continued)
namely abrupt and sustained growth increase, suggesting release of subcanopy trees from suppression by canopy trees, and rapid early growth, indicating tree origin in canopy gaps. Timing of these events was remarkably synchronized across sites and coincident with historical records of windstorms and bark-beetle outbreaks, suggesting that infrequent stand-replacing disturbances caused by these agents are an integral part of natural dynamics of spruce forests in the Šumava Mts. Similar evidence was simultaneously found for various forests in the Carpathians and other areas of temperate Europe (see Čada et al. 2016 for references). In the Šumava Mts severe large-scale disturbances occurring with mean recurrence intervals of less than 200 years prevent successional development towards mature spruce forests. Dendrochronologically recorded disturbances peaked in the 1820s, but became rare after 1880. In the absence of large disturbances during the twentieth century, spruce forests matured to the ages that are sensitive to windthrows and bark beetle attacks, contributing to the severity and large extent of the disturbances in the 1990s and 2000s (Brůna et al. 2013; Čada et al. 2016).

Forest regeneration after stand-replacing disturbance in non-intervention stands in the Šumava National Park is fast, with rapid regrowth of spruce, which germinates mainly on decaying logs, and *Sorbus aucuparia*, a typical subcanopy tree of mountain spruce forests (Jonášová and Prach 2004). Secondary succession in these forests includes little variation in plant species composition in spite of considerable change in stand structure. Dead wood provides habitats for saproxylic organisms and open areas support many species of invertebrates that are rare in closed forest (Müller et al. 2008). In contrast, salvage logging supports the spread of pioneer deciduous trees, thus changing the course of natural succession (Jonášová and Prach 2004). This experience provides strong support for no-intervention management after widthrows or bark-beetle attacks in protected areas.

(continued)

**Box 7.1** (continued)

Natural disturbances in the upper-montane natural spruce forests in the Šumava Mts: (**a**) The summit of Šumava on the Czech-Austrian border near Trojmezí (Dreieckmark, ~1350 m a.s.l.) in 2011: after the bark-beetle outbreak and large-scale windthrows caused by the Kyrill storm in 2007, the spruce forest on the Austrian side was logged (left), whereas non-intervention management was applied in the national park on the Czech side (right); (**b**) Windthrows in areas damaged by Kyrill; (**c**) Fast natural regeneration of spruce occurred below the standing trees killed by bark beetle in the non-managed forest moist habitats at lower altitudes, *Pinus sylvestris* in habitats in which the water table is regularly deeper than 30 cm below the soil surface and *Picea abies* in wet habitats in the mountains. *Pinus uncinata* subsp. *uliginosa* is typical of wet sites in the central parts of some bogs in the mountains and basins, especially in western and southern Bohemia (Neuhäusl 1972; Bastl et al. 2008), but it can also dominate bog woodlands in places where the water table is occasionally deeper than 30 cm. In areas affected by peat extraction and a subsequent drop in the ground-water table, *Pinus uncinata* subsp. *uliginosa* can be gradually replaced by spruce and birch (Unar et al. 2012).

**Black locust groves** are common in warm areas. For the most part, they are established plantations, but *Robinia pseudoacacia* has also spread spontaneously in abandoned grasslands and open oak or pine forests (Vítková et al. 2017). Being a legume, *Robinia* increases soil nitrogen availability through symbiotic fixation, which results in a dramatic change in the herb layer after *Robinia* is planted or invades. Therefore, *Robinia* is a significant threat especially to abandoned species-rich dry grasslands and open forests. In some schemes of vegetation classification, *Robinia* groves are included in the separate class *Robinietea pseudoacaciae* (Hadač and Sofron 1980; Vítková and Kolbek 2010; Mucina et al. 2016), but they also have features in common with the class of mesic to xeric scrub, *Rhamno-Prunetea*. In spite of large changes in the herb layer that occur after invasion by *Robinia*, the groves are distinctly differentiated according to habitat. Vítková and Kolbek (2010) distinguish three alliances: *Chelidonio majoris-Robinion pseudoacaciae* includes mesic stands in potential habitats of hornbeam or ravine forests, *Balloto nigrae-Robinion pseudoacaciae* includes high *Robinia* groves in dry lowland sandy habitats, and *Euphorbio cyparissiae-Robinion pseudoacaciae* is an alliance comprising *Robinia* woodlands or scrub on south-facing slopes with shallow soil in the potential habitats of thermophilous oak forests.

## 7.5.2 Scrub

Scrub vegetation is widespread and common in the intensively managed landscape of the Czech Republic, especially in successional habitats on abandoned grasslands and ex-arable land (Osbornová et al. 1990; Jírová et al. 2012). However, there is also scrub vegetation that is dependent on natural disturbance, such as riparian willow stands.

**Willow carrs** (class *Alnetea glutinosae*, alliance *Salicion cinereae*; Fig. 7.20a), in most cases dominated by *Salix cinerea* or *S. aurita*, occur in similar habitats as alder carrs, i.e. in water-logged depressions, often on organic soils. They are also spreading in abandoned wet meadows. At many sites, alder and willow carrs are linked both spatially and in terms of succession.

**Riparian willow scrub** commonly occurs along rivers from the upper to lower courses. This vegetation is maintained by periodic disturbances due to changes in the flow of the river, including damage by floating ice. The sandy or loamy sedi-

**Fig. 7.20** Examples of scrub vegetation in the Czech Republic: (**a**) Wet scrub of *Salix cinerea* near Prameny in the Slavkovský les Mts, western Bohemia (*Salicetum pentandro-auritae*, *Salicion cinereae*); (**b**) Willow scrub with *Salix euxina*, *S. purpurea*, *S. triandra* and *S. viminalis* along the Berounka River near Nižbor, central Bohemia (*Salicetum triandrae*, *Salicion triandrae*); (**c**) Low steppic scrub with *Prunus tenella* near Dolní Dunajovice, southern Moravia (*Prunetum tenellae*, *Prunion fruticosae*); (**d**) *Prunus spinosa* scrub in an agricultural landscape on flysch bedrock in the Carpathian foothills near Popice in southern Moravia (*Pruno spinosae-Ligustretum vulgaris*, *Berberidion vulgaris*)

ments that accumulate on the banks of most Czech rivers create a suitable habitat for *Salix triandra* and *S. viminalis* (Fig. 7.20b), and shrubs or trees of *S. euxina*. In gravelly places *S. purpurea* is more abundant. There are very few braided streams with extensive gravel beds in the Czech Republic, the most notable exceptions being the Ostravice and Morávka rivers in the northern part of the Moravskoslezské Beskydy Mts and their foothills in north-eastern Moravia (Eremiášová and Skokanová 2014). These streams, and some others in the same area, support scrub of *Salix elaeagnos* and *S. purpurea*, and rarely also *S. daphnoides* and *Myricaria germanica*. Riparian willow scrub is assigned to the same class as the riparian willow-poplar forests, *Salicetea purpureae*. Within this class, vegetation on sandy and loamy sediments is assigned to the alliance *Salicion triandrae* (Neuhäuslová 1985) and that on gravelly sediments to the alliance *Salicion elaeagno-daphnoidis*.

**Mesic and xeric scrub** is represented by several vegetation types in the Czech Republic. Low-growing scrub of the *Prunion fruticosae* alliance, dominated by the continental species *Prunus fruticosa* and *P. tenella* (Fig. 7.20c), is usually associated with dry grasslands in forest-steppe landscapes; the latter species only occurs at three sites in southern Moravia which represent the western limit of its distribution range. Tall dry to mesic scrub of the *Berberidion vulgaris* alliance occurs commonly on abandoned dry or mesic grasslands and at forest fringes (Fig. 7.20d). Dominant species include *Cornus mas*, *C. sanguinea*, *Corylus avellana*, *Cotoneaster integerrimus*, *Crataegus laevigata*, *C. monogyna*, *Euonymus europaeus*, *E. verrucosus*, *Ligustrum vulgare*, *Prunus mahaleb*, *P. spinosa*, *Rhamnus cathartica* and some species of *Rubus fruticosus* agg. A specific type of scrub, assigned to the alliance *Sambuco-Salicion capreae*, occurs in forest clearings, windthrow areas and other habitats of disturbed woodland. Typical dominant species of this type of scrub include *Corylus avellana*, *Populus tremula*, *Rubus idaeus* and some other *Rubus* species, *Salix caprea*, *Sambucus racemosa* and *Sorbus aucuparia*. Another type of tall scrub, assigned to the alliance *Aegopodio podagrariae-Sambucion nigrae*, occurs on nutrient-rich sites in warm areas, often in human settlements or agricultural landscapes. It is frequently dominated by *Sambucus nigra*, but alien shrub species are also common, including *Acer negundo*, *Lycium barbarum* and *Syringa vulgaris*.

### 7.5.3 Subalpine and Alpine Vegetation

Alpine timberline is reached only in three of the highest mountain groups in the Sudetes Mts in the north of the Czech Republic: Krkonoše (highest peak 1603 m a.s.l.), Králický Sněžník (1424 m) and Hrubý Jeseník (1491 m). There are two large treeless areas above the timberline in the Krkonoše Mts, one at the headwaters of the Labe in the western Krkonoše (~23 $km^2$) and the other in the headwaters of the Úpa in the eastern Krkonoše (~32 $km^2$). In addition, there are five smaller treeless areas. In the Králický Sněžník there is one small area above the timberline (~0.7 $km^2$) and in the Hrubý Jeseník there is one larger and four small such areas with a total area of ~10.5 $km^2$ (Treml and Migoń 2015).

The altitude of the treeline in Central European mountains north of the Alps increases from west to east by approximately 94 m per 100 km (Fig. 7.19), which reflects the rise in the temperature during the growing season with increasing continentality (Kašpar and Treml 2016). Treml and Migoń (2015) summarized the altitudes of the treeline ecotone based on standardized interpretation of aerial images from 2003 to 2005 (Fig. 7.21). Based on these data, the mean altitude of the timberline (i.e. the upper boundary of closed forest) is 1240 m in the Krkonoše and the mean altitude of treeline (defined as a line connecting the uppermost timberline positions on adjacent sections of the slope that differ in aspect) is 1278 m. The maximum altitude of the timberline is 1462 on the south-west facing slope of Mt. Sněžka and trees taller than 5 m have recently been found even at altitudes above 1500 m. The treeline ecotones in the Králický Sněžník and Hrubý Jeseník Mts are narrower than in the Krkonoše. The average altitudes of the timberlines in these eastern mountain groups in the Sudetes are 1320 m and 1302 m, respectively. The

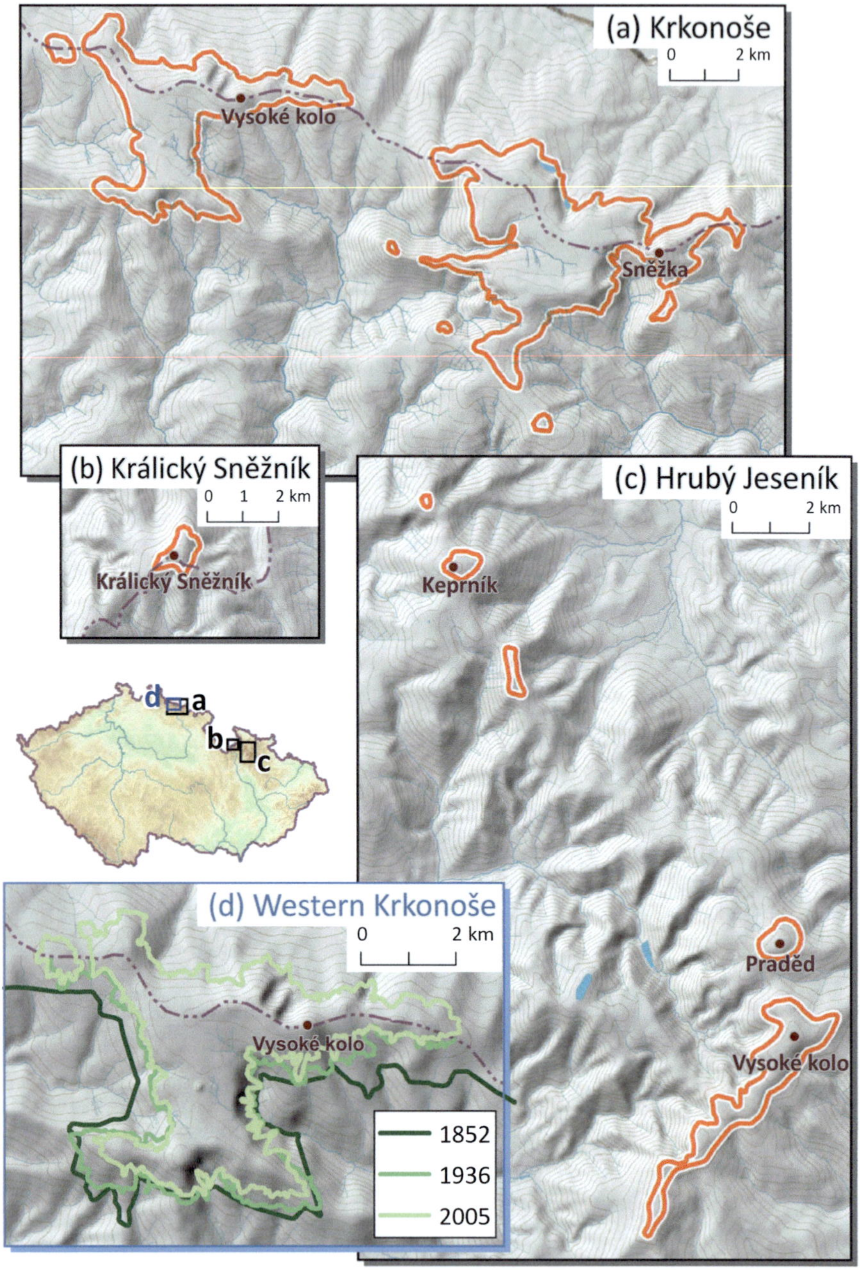


**Fig. 7.21** Treelines in the highest mountain groups in the Sudetes: (**a**) Krkonoše; (**b**) Králický Sněžník; (**c**) Hrubý Jeseník Mts; (**d**) altitudinal advance of the treeline on the Czech side of the western Krkonoše Mts between 1852 and 2005 (modified from Treml and Migoń 2015; note that timberline is shown instead of treeline on Mt. Liščí hora in the Krkonoše Mts)

treeline in the Hrubý Jeseník is on average at 1320 m and the highest timberline position at 1420 m on the west-facing slope of Mt. Praděd. In the glacial cirques timberline position is lowered by avalanches, in places going down to 900 m in the Krkonoše (Obří důl valley), 1220 m in the Králický Sněžník (Kotlina Moravy cirque) and 1070 m in the Hrubý Jeseník (Malá kotlina cirque). In addition to the Sudetes, timberline is also reached by Mt. Großer Arber on the Bavarian side of the Bohemian Forest (Bayerischer Wald), which has a small treeless area above ~1400 m, however, there is no area above the timberline in the Šumava Mts, which is the Czech side of the Bohemian Forest (Neuhäuslová 2001).

Following the terminology of Harsch and Bader (2011), the treeline ecotone in the Sudetes is of diffuse type with individual scattered trees or of island type with patches of clonally growing spruce (Vacek et al. 2012; Fig. 7.22). An abrupt timberline is caused by past human activities or, in glacial cirques, by avalanches. *Picea abies* is the only tree species in the treeline ecotone, with the exception of some individuals of *Acer pseudoplatanus*, *Sorbus aucuparia* and very rare occurrence of *Fagus sylvatica* (Jeník and Lokvenc 1962). The latter is present in the treeline ecotone on the south-west-facing slope of the Zlaté návrší ridge in the western Krkonoše, at altitudes of 1260–1370 m, which is the northernmost treeline ecotone formed by beech in Europe (Vacek and Hejcman 2012).

The hypothesis of continuous existence of treeless summit areas in the Sudetes throughout the Holocene is supported especially by the occurrence of well-developed patterned ground and other periglacial landforms that could not develop or remain preserved under forest cover (Sekyra et al. 2002; Treml et al. 2010), by the isolated localities of several light-demanding species of arctic or alpine grasslands and occurrence of several neoendemic species of *Hieracium* that could not survive on forested summits (Soukupová et al. 1995; Kaplan et al. 2016). Human activities extended the naturally treeless summit areas. Chalets were built at high altitudes in the Krkonoše from the fifteenth century onwards and the demand for firewood and pastures caused deforestation. However, Jeník and Lokvenc (1962) estimated that the Krkonoše timberline was lowered on average only by 20–30 m as a result of human activities. In the Hrubý Jeseník, the pollen and charcoal data and written sources indicate significant human activity on the summits already in the High Middle Ages and local effects even earlier (Jeník and Hampel 1992; Rybníček and Rybníčková 2004; Novák et al. 2010). Before human intervention the timberline in this mountain group was probably at 1400–1450 m a.s.l. and only restricted areas on the highest summits and exposed edges of summit plateaus were treeless (Rybníček and Rybníčková 2004; Treml et al. 2008; Novák et al. 2010). The artificial lowering of the timberline in the Krkonoše and Hrubý Jeseník Mts was partly reversed by afforestation with *Picea abies* in the nineteenth and the first half of the twentieth century.

A remarkable upward shift of the timberline has been observed in the Sudetes since the mid-twentieth century, characterized by upward migration of trees in the upper parts of the treeline ecotone and development of a closed-canopy spruce forest at the lower part of the ecotone. The annual rate of this upward advance has been 43 cm in the Krkonoše and 30 cm in the Hrubý Jeseník Mts (Treml and Chuman 2015; Fig. 7.21d). Detailed studies on climate change and land-use abandonment a

b

**Fig. 7.22** Timberlines in the Sudetes: (**a**) *Pinus mugo* krummholz belt between the upper-montane spruce forest and alpine grasslands in the Krkonoše Mts; (**b**) Diffuse treeline ecotone on Mt. Králický Sněžník, with groups of spruce and grassland patches; krummholz is absent here

revealed that the trigger for these changes was agricultural abandonment of mountain summits between the late nineteenth and mid-twentieth century rather than an increase in temperature (Treml et al. 2016).

**Krummholz** with *Pinus mugo* on mineral soils (class *Roso pendulinae-Pinetea mugo*, alliance *Pinion mugo*; Figs 7.22a and 7.23a) occurs only in the Krkonoše and Šumava Mts. It is absent from the Králický Sněžník and Hrubý Jeseník Mts (Fig. 7.22b), except for recent plantations, and is not recorded there even in the soil charcoal from the last 2000 years (Novák et al. 2010) or in written historical sources (Jeník and Hampel 1992). On the Czech side of the Šumava *Pinus mugo* occurs in

**Fig. 7.23** Examples of alpine and subalpine vegetation types in the Krkonoše Mts: (**a**) *Pinus mugo* patches and *Nardus stricta* grassland on a plateau above the timberline in the western Krkonoše (*Dryopterido dilatatae-Pinetum mugo*, *Pinion mugo*, and *Carici bigelowii-Nardetum strictae*, *Nardo strictae-Caricion bigelowii*); (**b**) Alpine heathland with *Calluna vulgaris* on Mt. Sněžka (*Avenello flexuosae-Callunetum vulgaris*, *Loiseleurio procumbentis-Vaccinion*); (**c**) Avalanche track in the Velká Kotelní jáma cirque with scrub of *Prunus padus* subsp. *borealis* (*Pado borealis-Sorbetum aucupariae*, *Salicion silesiacae*) and stands of the tall fern *Athyrium distentifolium* (*Adenostylo alliariae-Athyrietum distentifolii*, *Dryopterido filicis-maris-Athyrion distentifolii*)

small patches on boulder screes and rock outcrops below the timberline at altitudes of 1080–1360 m, but on the Bavarian side of this mountain range (Bayerischer Wald) it is also found in the treeline ecotone on Mt. Großer Arber at ~1400 m. In the Krkonoše krummholz is widespread in the treeline ecotone and above it, being most common at altitudes between 1230 and 1500 m. On boulder screes and avalanche tracks it descends as low as 1050 m. It declined slightly in abundance in the past as it was cut for firewood and removed in order to enlarge the grassland areas for grazing and haymaking, but was planted in some areas in the late nineteenth and twentieth century. Krummholz vegetation occurs on acidic soils in the Czech mountains and accordingly it is rather poor in species, with frequent occurrence of *Avenella flexuosa*, *Vaccinium myrtillus* and *V. vitis-idaea* (Jirásek 1996b; Šibík et al. 2010).

**Subalpine deciduous scrub or woodland** (class *Mulgedio-Aconitetea*, alliance *Salicion silesiacae*) occurs in cirques of the Krkonoše and Hrubý Jeseník Mts at the edges of avalanche tracks and on slopes where there are thick snowpacks. It is dominated by *Betula pubescens* subsp. *carpatica*, *Salix silesiaca* and *Sorbus aucuparia* and locally also by *Prunus padus* subsp. *borealis* and the Krkonoše endemic *Sorbus sudetica*. Subalpine tall forbs and tall ferns are common in the herb layer of this scrub or woodland.

**Alpine and subalpine grasslands and heathlands** above the timberline in the Sudetes are dominated by oligotrophic grasses adapted to acidic soils (Jeník 1961; Krahulec 1990a; Soukupová et al. 1995). Exposed wind-swept summit areas are covered by open grasslands of *Festuca supina* or heathlands of *Calluna vulgaris* (Fig. 7.23b), both with the boreo-alpine species *Carex bigelowii* and *Oreojuncus trifidus*, endemic *Hieracium* species and arcto-alpine lichens. These grasslands belong to the class *Juncetea trifidi*, alliance *Juncion trifidi*, and heathlands to the class *Loiseleurio-Vaccinietea*, alliance *Loiseleurio procumbentis-Vaccinion*. At less exposed sites protected by deep snow cover there are species-poor closed grasslands dominated by *Nardus stricta* (more common in the Krkonoše, Fig. 7.23a) or *Avenella flexuosa* (more common in the Hrubý Jeseník) and with occurrence of *Carex bigelowii* (class *Juncetea trifidi*, alliance *Nardo strictae-Caricion bigelowii*). These *Nardus* grasslands are the most common type of grassland on the summits of the Sudetes (Krahulec et al. 1997). The widespread distribution of *Nardus stricta* on the Krkonoše summits may have been a consequence of long-term cutting, grazing and the resulting nutrient depletion, which is suggested by the recent spread of tall grasses, such as *Calamagrostis villosa* and *Molinia caerulea*, in the former *Nardus* grasslands after cessation of management (Hejcman et al. 2006b, 2007, 2009). A species-rich type of *Nardus stricta* grassland (class *Calluno-Ulicetea*, alliance *Nardion strictae*) occurs locally at the timberline and on the slopes of cirques. *Vaccinium myrtillus* heathlands extend over large areas at sites with deep snowpacks on the leeward slopes, around the timberline and in open spaces among *Pinus mugo* bushes (class *Calluno-Ulicetea*, alliance *Genisto pilosae-Vaccinion*). On restricted outcrops of base-rich rocks (e.g. marble or porphyry) in the cirques of the Krkonoše and Hrubý Jeseník Mts, patches of species-rich basiphilous grassland with *Agrostis alpina* and *Festuca versicolor* occur (class *Elyno-Seslerietea*, alliance *Agrostion alpinae*).

**Subalpine tall-forb vegetation** (class *Mulgedio-Aconitetea*) occurs on treeless slopes and at the bottoms of the cirques, especially on avalanche tracks, along mountain streams and at wet and nutrient-rich sites in canopy openings around the timberline. Tall grasslands with *Calamagrostis villosa*, *Deschampsia cespitosa* and *Molinia caerulea* (alliance *Calamagrostion villosae*) occur on mesic soils on plateaus or gentle slopes. Grasslands of *Calamagrostis arundinacea* (alliance *Calamagrostion arundinaceae*) are typical of leeward steep slopes on avalanche tracks in the Krkonoše and Hrubý Jeseník Mts, which are protected by thick snow cover in winter but exposed to the sun and relatively warm in summer. These moderate mesoclimatic conditions are suitable for many herbaceous plant species, including thermophilous species typical of low-altitude grasslands and deciduous forests. This makes these *Calamagrostis arundinacea* grasslands the most species-rich vegetation type in the montane to alpine belt in the Czech Republic (Jeník 1961; Kočí 2001; see Sect. 7.6). Vegetation of tall broad-leaved dicot plants, such as *Adenostyles alliariae*, *Cicerbita alpina* and *Veratrum album* subsp. *lobelianum*, occurs in moist and nutrient-rich habitats in the subalpine belt, e.g. at seepage sites, along creeks and on bottoms of cirques (alliance *Adenostylion alliariae*). Stands of tall ferns, mainly *Athyrium distentifolium* and locally also *Dryopteris filix-mas*, are common on boulder screes in cirques (alliance *Dryopterido filicis-maris-Athyrion distentifolii*; Fig. 7.23c).

### 7.5.4 Rock and Scree Vegetation

Treeless rock outcrops, cliffs and screes occur mainly in river valleys of the Bohemian Massif, in karst and sandstone pseudokarst areas, on volcanic hills and in mountain cirques. As they are generally rare, small-sized and isolated in a forested landscape (Kubešová and Chytrý 2005), there are very few specialized or endemic species confined to these habitats in the Czech Republic, which sharply contrasts with a high number of habitat specialists and endemics on rock outcrops and screes in the Alps, Carpathians or mountain ranges of Southern Europe (Valachovič et al. 1997). Specialist species of rock outcrops and stable boulder screes in this country include small ferns such as *Asplenium* spp., *Cryptogramma crispa*, *Cystopteris fragilis*, *Gymnocarpium robertianum*, *Polypodium interjectum*, *P. vulgare*, *Trichomanes speciosum* (only gametophytes; Vogel et al. 1993) and *Woodsia ilvensis*, and some dicots such as *Saxifraga rosacea*. Some of these species also occur in man-made habitats such as stone quarries or walls; however, wall vegetation is poorer in species and contains a larger proportion of alien, annual, nutrient-demanding and ant-dispersed species than vegetation on natural rocks (Láníková and Lososová 2009; Lososová and Láníková 2010). Habitat specialists of mobile screes include *Epilobium dodonaei*, *Galeopsis angustifolia*, *G. ladanum*, *Gymnocarpium robertianum* and *Teucrium botrys* (Sádlo and Kolbek 1994).

Mobile screes composed of fine rock debris are mainly found in stone quarries in the Czech Republic and are rapidly stabilized by succession of perennial vegetation after the rock material has ceased to accumulate. The most common type of natural screes are stabilized talus slopes consisting of large boulders, which occur in river valleys of the Bohemian Massif and on volcanic hills in northern Bohemia. Some basalt talus slopes in the České středohoří Mts are characterized by an internal air circulation system (ventaroles) with winter exhalations of warm air at their upper edges and summer outflows of cold and humid air at their foot, at some sites even with summer ice holes (Kubát 1999). Consequently, the upper parts of these talus slopes locally support frost-sensitive species such as the Mediterranean or subtropical-suboceanic liverworts *Riccia ciliifera* and *Targionia hypophylla* on Boreč hill, while lower parts are often characterized by natural occurrence of spruce at low altitudes and a high diversity of bryophytes, many of them typical of boreal forests (Pilous 1959).

Vegetation on limestone and other calcareous rock outcrops is characterized by the ferns *Asplenium ruta-muraria*, *A. trichomanes* subsp. *quadrivalens*, *Cystopteris fragilis*, *Gymnocarpium robertianum* and *Polypodium interjectum* (class *Asplenietea trichomanis*, alliance *Cystopteridion*). Very similar vegetation also occurs on walls in which mortar was used as a binding material (Kolbek 1997; Duchoslav 2002). Serpentinite rocks host specialist ferns *Asplenium adulterinum* and *A. cuneifolium*. On serpentinite oucrops in the Jihlava valley near Mohelno (south-western Moravia), the Southern European-Western Asian fern *Notholaena marantae* also occurs at an isolated site (Vicherek 1970; class *Asplenietea trichomanis*, alliance *Asplenion cuneifolii*). The vegetation on siliceous rocks and stabilized talus slopes includes *Asplenium septentrionale*, *A. trichomanes* subsp. *trichomanes*, *Polypodium vulgare*, and at some sites also *Saxifraga rosacea* subsp. *sponhemica*, *S. r.* subsp. *steinmannii* and *Woodsia ilvensis* (class *Asplenietea trichomanis*, alliance *Asplenion septentrionalis*). Stands of the boreo-alpine fern *Cryptogramma crispa* (class *Asplenietea trichomanis*, alliance *Androsacion alpinae*) occur in a few places on siliceous screes and outcrops in cirques and on summit areas of high mountains. Some walls, especially at low altitudes, are covered by stands of *Cymbalaria muralis* and *Pseudofumaria lutea*, both neophytes of Mediterranean origin (class *Cymbalario muralis-Parietarietea judaicae*, alliance *Cymbalario muralis-Asplenion*). Unstable limestone screes occurring in karst areas, often in quarries, are characterized by *Galeopsis angustifolia*, *Gymnocarpium robertianum*, *Melica ciliata* and *Teucrium botrys* (class *Thlaspietea rotundifolii*, alliance *Stipion calamagrostis*). Species-poor vegetation with *Galeopsis ladanum* (class *Thlaspietea rotundifolii*, alliance *Galeopsion*) occurs locally on unstable screes of siliceous rocks.

### 7.5.5 Aquatic Vegetation

Various types of aquatic and wetland vegetation occur in fishponds, which are the most common type of still water bodies in the Czech Republic (Dykyjová and Květ 1978; see Fig. 1.13 in Chap. 1, this book). Although fishponds are artificial lakes, most of them have existed in the landscape for several centuries and many of them are of high conservation importance (Chytil et al. 1999). Other more common types of wetland habitats include oxbow lakes on floodplains and water bodies in abandoned quarries or loam, sand or gravel pits.

**Aquatic vegetation of eutrophic water** is best developed in small fishponds which are not used for intensive fish farming, alluvial pools or channels. The vegetation of free-floating aquatic plants (class *Lemnetea*) is composed of lemnids, such as *Lemna gibba*, *L. minor*, *L. trisulca* and *Spirodela polyrhiza*, or the aquatic liverworts *Riccia fluitans*, *R. rhenana* and *Ricciocarpos natans*. The rare species *Salvinia natans* and *Wolffia arrhiza* occur in this vegetation in the lowlands of north-eastern and southern Moravia, respectively (alliance *Lemnion minoris*). Bladderwort vegetation of eutrophic to mesotrophic water bodies is mostly composed of *Utricularia australis*, while the rare species *U. vulgaris* occurs only in the middle Labe floodplain (alliance *Utricularion vulgaris*). Larger free-floating aquatic plants include *Ceratophyllum demersum*, *C. submersum*, *Hydrocharis morsus-ranae* and *Stratiotes aloides* (alliance *Hydrocharition morsus-ranae*). Aquatic plants rooted in the bottom with leaves floating on the water surface include *Nuphar lutea*, *N. pumila*, *Nymphaea alba*, *N. candida*, *Nymphoides peltata*, *Persicaria amphibia* and *Trapa natans* (class *Potametea*, alliance *Nymphaeion albae*). Submerged aquatic plants occurring in still or slowly moving water are represented by various species of *Potamogeton*, *Myriophyllum spicatum*, *M. verticillatum*, *Najas marina*, *N. minor*, *Zannichellia palustris*, the neophyte *Elodea canadensis* (class *Potametea*, alliance *Potamion*) and stoneworts (*Chara* spp., *Nitella* spp. and *Tolypella* spp.; class *Charetea*, alliances *Charion globularis* and *Nitellion flexilis*). A few aquatic species may occur in monodominant stands in streams with a current of high or medium velocity: most commonly it is *Ranunculus fluitans*, *R. peltatus*, *Callitriche hamulata* or the moss *Fontinalis antipyretica* and less frequently *Ranunculus penicillatus* or *Myriophyllum alterniflorum* (class *Potametea*, alliance *Batrachion fluitantis*). *Ranunculus aquatilis*, *R. circinatus* and other species of this genus and in some places also *Callitriche hermaphroditica* and *Hottonia palustris* (class *Potametea*, alliance *Ranunculion aquatilis*) occur in still water bodies where the water level fluctuates and can occasionally be below the level of the bottom.

**Vegetation of oligotrophic and mesotrophic water** (class *Littorelletea uniflorae*) is rare in the Czech Republic, partly due to anthropogenic eutrophication of water bodies and partly because of the marginal position of this country with respect to the oceanic distribution of this vegetation. Two natural lakes in the Šumava Mts (Plešné, 1087 m a.s.l., and Černé, 1008 m a.s.l.) contain sparse low-productive monodominant stands of *Isoëtes echinospora* and *I. lacustris*, respectively (alliance *Littorellion uniflorae*; Husák et al. 2000, Čtvrtlíková et al. 2009). Some fishponds on acidic bedrock in submontane areas are habitats of a rare type of amphibious vegetation consisting of the perennial plants *Juncus bulbosus*, *Littorella uniflora* and *Pilularia globulifera*, or of more common vegetation with *Eleocharis acicularis* (alliance *Eleocharition acicularis*). Mire pools locally contain vegetation consisting of *Sparganium natans*, *Utricularia intermedia*, *U. minor* and *U. ochroleuca* (alliance *Sphagno-Utricularion*; Dítě et al. 2006).

### 7.5.6 Wetland vegetation

Like aquatic vegetation, the wetlands are mainly associated with fishponds and lowland river floodplains in the Czech Republic. There are two basic types of wetland: those with annual wetland herbs and marshes with tall perennials.

**Vegetation of annual wetland herbs** develops mainly on the bottoms of temporarily drained fishponds (Fig. 7.24a, b), although it is also found on fluvial muddy deposits along lowland rivers and in other habitats, including man-made ones. Traditional fishpond management included periodical summer draining at intervals of a few years, aimed at increasing pond productivity, suppressing fish parasites and reducing productive pond vegetation. Specialized annual wetland species germinate in the mud at the bottoms of drained fishponds, which remain saturated with water or very shallowly flooded for a period of several days or a few weeks after draining. In addition to the type of substrate on the pond bottom and climate, the species composition of this vegetation depends on timing and duration of pond draining (Šumberová et al. 2005). The intensification of fishpond management since the nineteenth century included increasing application of fertilizers, liming, fish feeding and reducing the periods when the ponds were drained to maximize fish production. From the mid-twentieth century, such intensive management was applied in most fishponds (Čítek et al. 1998; Šumberová 2003). Fishponds are rarely drained nowadays, but an alternative suitable habitat for annual wetland herbs is in fish-storage ponds (Šumberová et al. 2006). These are small ponds, often bounded by concrete walls, which are used for short-term storage of marketable fish and then drained (Fig. 7.24c). There are usually more plant species per unit area on the bottoms of fish-storage ponds than on the bottoms of ordinary fishponds, but because of a greater human impact, the former also contain more alien species (Šumberová et al. 2006). Vegetation of fishponds in submontane areas and basins with acidic bedrocks, but with nutrient-rich mud, includes *Carex bohemica*, *Coleanthus subtilis*, *Elatine triandra* and *Eleocharis ovata*, and that of fishponds in warmer areas or on more calcareous substrates additionally includes *Cyperus fuscus* (class *Isoëto-Nano-Juncetea*, alliance *Eleocharition ovatae*). On the exposed nutrient-poor sandy bottoms of fishpond margins or fish-storage ponds, or on wet sandy arable land in the southern Bohemian basins, a few last remnants can be found of vegetation with the rare species *Centunculus minimus*, *Hypericum humifusum*, *Illecebrum verticillatum*, *Juncus capitatus*, *J. tenageia*, *Pseudognaphalium luteoalbum*, *Radiola linoides* and *Tillaea aquatica* (class *Isoëto-Nano-Juncetea*, alliance *Radiolion linoidis*). Disturbed wet soils in warm lowlands, especially in southern Moravia, which are more base-rich than those at higher altitudes, are characterized by another group of rare species, including *Centaurium pulchellum*, *Cerastium dubium*, *Juncus ranarius*, *Lythrum hyssopifolia*, *Pulegium vulgare*, *Pulicaria vulgaris*, *Ranunculus sardous*, *Veronica anagalloides* and *V. catenata* (class *Isoëto-Nano-Juncetea*, alliance *Verbenion supinae*). In addition to this low-growing vegetation, stands of tall annual herbs (e.g. *Bidens cernuus*, *B. radiatus*, *B. tripartitus*, *Persicaria hydropiper*, *P. lapathifolia*, *Ranunculus sceleratus* and *Rumex maritimus*) also grow on the ini-

**Fig. 7.24** Examples of wetland vegetation in the Czech Republic: (**a**) Drained fishpond in which annual wetland vegetation is developing near Frymburk, south-western Bohemia; (**b**) Annual vegetation growing on the bottom of a drained fishpond with *Alisma plantago-aquatica*, *Persicaria lapathifolia*, *Ranunculus sceleratus*, *Rorippa palustris* and *Rumex maritimus* near Polanka nad Odrou, northern Moravia; (**c**) Storage fishponds near Tchořovice, south-western Bohemia; (**d**) Riverine reed vegetation with *Phalaris arundinacea* along the Berounka River near Račice, central Bohemia; (**e**) Tall-sedge vegetation dominated by *Carex rostrata* growing in mesotrophic water in Karhov fishpond near Studená, Bohemian-Moravian Highlands; (**f**) Tall-sedge vegetation dominated by *Carex riparia* in eutrophic water in an oxbow lake in the Dyje River floodplain near Lednice, southern Moravia

tially bare or frequently disturbed wet soils, which are rich in nutrients. If the bottom of the pond is exposed for a few months, these stands of tall-herbs constitute the next successional stage, which displaces the short-growing annual vegetation (class *Bidentetea tripartitae*; alliance *Bidention tripartitae*). On ammonium-rich and slightly saline wet soils, both in natural and man-made habitats, the vegetation includes nutrient-demanding *Chenopodiaceae* such as *Atriplex prostrata* subsp. *latifolia*, *Chenopodium ficifolium*, *C. glaucum* and *C. rubrum* (class *Bidentetea tripartitae*; alliance *Chenopodion rubri*).

**Marsh vegetation** (class *Phragmito-Magno-Caricetea*) occurs mainly in fishpond littoral zones (Dykyjová and Květ 1978) and river floodplains. Stands of *Phragmites australis* are the most common type of marsh vegetation. *Phragmites* is the strongest competitor among wetland herbs and has a broad ecological range, growing both in oligotrophic and eutrophic water bodies, and in water as deep as 2 m and at sites that are drained for a significant part of the growing season. Other wetland herbs dominate the vegetation only in habitats that are, for different reasons, unsuitable for *Phragmites* (Ellenberg and Leuschner 2010). For example, *Schoenoplectus lacustris* dominates in deeper water, *Typha angustifolia* and *T. latifolia* in nutrient-rich water bodies with frequently exposed bottom on which they can quickly regenerate from seed, *Glyceria maxima* in hypertrophic wetlands, *Acorus calamus* (alien to the Czech Republic) and *Sparganium erectum* in frequently disturbed littoral habitats, while shorter stands of *Equisetum fluviatile* occur at sites with a deep layer of organic sediment on the bottom. Dominance stands of the above-mentioned species are included in the alliance *Phragmition australis*. In the littoral zone of brackish water bodies or around mineral springs in lowland areas, marshes are dominated by halophilous species of *Bolboschoenus* (*B. maritimus* and *B. planiculmis*; Hroudová et al. 2009) or *Schoenoplectus tabernaemontani* (alliance *Meliloto dentati-Bolboschoenion maritimi*), but recently this vegetation has become rare in the Czech Republic. In wetlands with a fluctuating water level such as shallow littoral zones of fishponds, or fishponds in the years following draining, the vegetation consists of perennial or biennial herbs, e.g. *Alisma plantago-aquatica*, *Bolboschoenus yagara*, *Butomus umbellatus*, *Eleocharis palustris*, *Hippuris vulgaris*, *Oenanthe aquatica*, *Rorippa amphibia*, *Sagittaria sagittifolia*, *Scirpus radicans* and *Sparganium emersum* (alliance *Eleocharito palustris-Sagittarion sagittifoliae*). Riverine reeds are mostly dominated by *Phalaris arundinacea* (Fig. 7.24d), but in places that are less frequently flooded tall-sedge stands of *Carex buekii* also occur. The tall grass *Calamagrostis pseudophragmites* is typical of fluvial gravel beds, mainly in the Moravskoslezské Beskydy Mts in north-eastern Moravia and Silesia (Kopecký 1969; Kalníková and Eremiášová 2013; alliance *Phalaridion arundinaceae*). Short-growing small-sized marshes of *Berula erecta*, *Glyceria fluitans*, *G. notata*, *Leersia oryzoides* and *Nasturtium officinale* occur in small streams or ditches and on their banks, in fish-storage ponds and in fishpond littoral zones (alliance *Glycerio-Sparganion*). Another type of wetland vegetation develops in mesotrophic to dystrophic water bodies and wetlands in an advanced successional stage of terrestrialization, where organic sediment mixed with mud has accumulated. These stands may be dominated by *Calla palustris*, *Carex pseudocy-*

*perus* and *Cicuta virosa* (alliance *Carici-Rumicion hydrolapathi*). Tall-sedge stands generally develop in the later stages of wetland succession, in places where the accumulation of undecomposed litter of reed plants has raised the ground level and, as a consequence, flooding is shallower or shorter. There are two major types of tall-sedge marshes, mesotrophic (alliance *Magno-Caricion elatae*) and eutrophic (*Magno-Caricion gracilis*). The former includes especially stands dominated by *Carex rostrata* (Fig. 7.24e) and less commonly *C. appropinquata*, *C. diandra*, *C. elata* and *C. lasiocarpa*. Marshes with *Cladium mariscus*, which occur in a few calcium-rich fens in the lowland along the middle Labe River, are also mesotrophic. The latter, eutrophic type, includes especially the common stands with *Carex acuta* in fishpond littoral zones and on floodplains of large rivers, and also stands of *Carex acutiformis*, *C. disticha*, *C. paniculata*, *C. riparia* (Fig. 7.24f), *C. vesicaria* and *C. vulpina*.

### *7.5.7 Spring and Mire Vegetation*

Spring and mire vegetation occurs mainly in mid- to high-altitude areas in the Czech Republic, with the exception of calcareous fens, which are also found in the lowlands, especially in areas of central and eastern Bohemia with Cretaceous marlstone. Another exception is the basins around Doksy, Cheb and Třeboň (see the coniferous forest biome description above), which contain different types of acidic fens and bogs.

**Spring vegetation** (class *Montio-Cardaminetea*) with specialized short-growing vascular plants and abundant bryophytes occurs in small patches at seepage sites. Main factors determining its species composition are altitude, site insolation and calcium carbonate precipitation resulting in tufa formation. The most common type of spring vegetation occurs at soft-water seepage sites without tufa formation that are shaded by trees. Typical species of these forest springs include *Cardamine amara* subsp. *amara*, *C. a.* subsp. *austriaca*, *Carex remota*, *Chrysosplenium alternifolium* and *C. oppositifolium* (alliance *Caricion remotae*; Fig. 7.25a). Forest springs with tufa formation are mainly confined to the Bohemian Karst (Rivola 1982) and the Carpathian flysch zone in eastern Moravia. Their vegetation has a sparse herb layer, but the moss layer is well-developed and includes specialized bryophytes such as *Eucladium verticillatum*, *Palustriella commutata* and *Pellia endiviifolia* (alliance *Lycopodo europaei-Cratoneurion commutati*). Soft-water springs at open sites locally support specialized light-demanding vegetation with *Montia fontana* subsp. *amporitana* or *M. f.* subsp. *fontana*, occurring especially in the montane belt of higher mountain ranges (alliance *Epilobio nutantis-Montion fontanae*). Springs with tufa formation at open sites are found especially in the Carpathian flysch zone. They are characterized by the accumulation of organic sediment and their vegetation consequently corresponds to calcareous fens rather than to specialized spring vegetation. Remarkable types of spring vegetation occur around seepages in the subalpine belt of the Krkonoše and Hrubý Jeseník Mts;

**Fig. 7.25** Examples of spring and mire vegetation types in the Czech Republic: (**a**) Forest spring with *Carex remota* in the Žofínský prales old-growth forest, Novohradské hory Mts, southern Bohemia (*Caricetum remotae*, *Caricion remotae*); (**b**) Calcareous fen with *Schoenus ferrugineus* in the Hrabanovská černava mire near Lysá nad Labem, central Bohemia (*Junco subnodulosi-Schoenetum nigricantis*, *Caricion davallianae*); (**c**) Acidic fen with *Eriophorum angustifolium* in the Soos National Nature Reserve near Františkovy Lázně, western Bohemia (*Caricetum nigrae*, *Caricion canescenti-nigrae*); (**d**) *Andromeda polifolia* on a *Sphagnum magellanicum* hummock on the Jelení lázeň bog, Orlické hory Mts, eastern Bohemia (*Andromedo polifoliae-Sphagnetum magellanici*, *Sphagnion magellanici*); (**e**) A mosaic of open bog and forested bog with *Pinus uncinata* subsp. *uliginosa* and *Picea abies* on Taiga bog in the Slavkovský les Mts, western Bohemia (*Eriophoro vaginati-Sphagnetum recurvi* and *Ledo palustris-Pinetum uncinatae*, *Sphagnion magellanici*); (**f**) Mountain bog with *Pinus mugo* in the Rolavská vrchoviště National Nature Reserve, Krušné hory Mts, western Bohemia (*Vaccinio uliginosi-Pinetum mugo*, *Sphagnion magellanici*)

dominant species include *Allium schoenoprasum*, *Cardamine amara* subsp. *opicii*, *Swertia perennis* and mosses *Dichodontium palustre*, *Philonotis seriata* and *Pohlia wahlenbergii* (alliance *Swertio perennis-Dichodontion palustris*; Hadač 1983).

**Vegetation of fens and transitional mires** (class *Scheuchzerio palustris-Caricetea nigrae*) occurs in habitats that are permanently saturated with water and poor in nitrogen and phosphorus, in which undecomposed litter of sedges, other herbs and mosses accumulates to form a peat layer. Although some fens and transitional mires existed in the territory of the present Czech Republic already in the Late Glacial and remained in the landscape throughout the Holocene (Sádlo 2000), most of them developed at sites of former forested seepages or carrs after deforestation, which took place especially during the high-medieval colonization of mid-altitude areas (Rybníček and Rybníčková 1974; Rybníčková 1974; Rybníčková et al. 2005; Hájková et al. 2012). As a result, they contain fewer relict species than old fens that have existed continuously since the Late Glacial or Early Holocene, for example those in the central part of the Western Carpathians in Slovakia (Hájek et al. 2011). This vegetation varies along a 'poor-rich gradient', i.e. from acidic, base-poor and species-poor types to species rich calcareous fens (Hájek et al. 2002, 2006; Peterka et al. 2014, 2017). Calcareous fens with short sedges (*Carex davalliana*, *C. flacca*, *C. flava*, *C. nigra* and *C. panicea*) and several species of specialist herbs and mosses such as *Bryum pseudotriquetrum* and *Campylium stellatum* (alliance *Caricion davallianae*) occur mainly in the Carpathian flysch zone and in the lowland areas with Cretaceous marlstone in eastern Bohemia (Figs. 7.25b and 7.26a), although in the latter area many fens have been drained. Other fen types occur mostly at higher altitudes (Fig. 7.26b). At some mid-altitude sites, especially in the Bohemian-Moravian Highlands, fen vegetation contains a mixture of calcicolous and acidophilous species, including calcium-tolerant peat mosses (e.g. *Sphagnum contortum*, *S. teres* and *S. warnstorfii*; alliance *Sphagno warnstorfii-Tomentypnion nitentis*). The most common type of Czech fen vegetation is moderately rich in bases, with a substrate of pH 5–6, and is usually rather rich in species. It consists of short sedges, grasses, dicots and a rich moss layer with *Aulacomnium palustre*, *Sarmentypnum exannulatum*, *Straminergon stramineum* and peat mosses, e.g. *Sphagnum palustre* and *S. subsecundum* (alliance *Caricion canescenti-nigrae*; Fig. 7.25c). Acidic fens (transitional mires) at mid-altitude sites, which are very poor in calcium despite being fed by ground water rather than rain water, support species-poor vegetation of sedges (*Carex canescens*, *C. echinata*, *C. lasiocarpa*, *C. nigra* and *C. rostrata*) and mosses (*Polytrichum commune*, *Sphagnum fallax*, *S. flexuosum* and *S. palustre*; alliance *Sphagno-Caricion canescentis*). In the course of mire succession they develop from the above-mentioned fen types. Bog hollows with continuous or nearly continuous moss layer (mainly *Sphagnum cuspidatum* and *Warnstorfia fluitans*) and sparse herb layer with *Carex limosa*, *C. rostrata*, *Eriophorum angustifolium* and *Scheuchzeria palustris* are extremely poor in nutrients and bases, nevertheless, their structure is similar to fens (alliance *Sphagnion cuspidati*). Many fens have been destroyed by draining in the past. At present, their remnants are mostly protected in nature reserves, but their plant species composition tends to change in response to nutrient enrichment and water level changes.

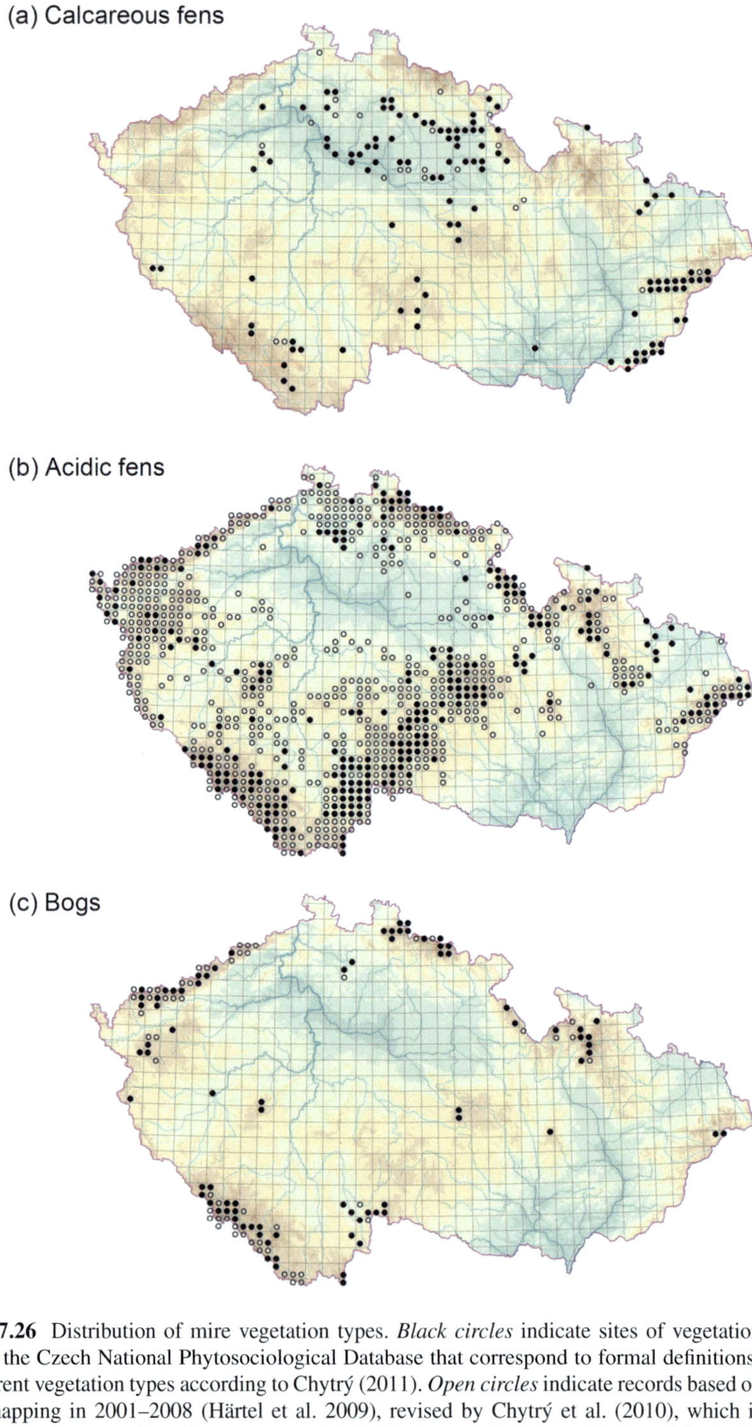


**Fig. 7.26** Distribution of mire vegetation types. *Black circles* indicate sites of vegetation plots from the Czech National Phytosociological Database that correspond to formal definitions of the different vegetation types according to Chytrý (2011). *Open circles* indicate records based on habitat mapping in 2001–2008 (Härtel et al. 2009), revised by Chytrý et al. (2010), which include fragmentary or transitional occurrences of individual vegetation types. Phytosociological classes and alliances: (**a**) *Caricion davallianae*; (**b**) *Sphagno warnstorfii-Tomentypnion nitentis*, *Caricion canescenti-nigrae* and *Sphagno-Caricion canescentis*; (**c**) *Oxycocco-Sphagnetea*

Comparisons of historical and recent vegetation-plot records from the Třeboň Basin and Bohemian-Moravian Highlands indicate a decline in specialized species, especially those of rich fens, and conversely, spread of competitive, nutrient demanding species of both vascular plants and bryophytes (Hájek et al. 2015; Navrátilová et al. 2017).

**Bogs** (class *Oxycocco-Sphagnetea*; Figs 7.25d–f and 7.26c) are the most advanced mire type developed through the process of autogenic succession. In the Czech Republic, they are represented by raised bogs, which are elevated above the adjacent terrain. Therefore they are entirely fed by rainwater (ombrotrophic mires) and extremely poor in nutrients and bases. They are dominated by specialist peat mosses (e.g. *Sphagnum fallax*, *S. flexuosum*, *S. magellanicum*, *S. rubellum* and *S. russowii*) and dwarf shrubs (*Calluna vulgaris*, *Empetrum hermaphroditum*, *E. nigrum*, *Rhododendron tomentosum*, *Vaccinium myrtillus*, *V. uliginosum* and *V. vitis-idaea*). Some bogs are characterized by a distinct microtopography with dry hummocks and wet hollows. Continental and subcontinental bogs (alliance *Sphagnion magellanici*) can be open, especially in precipitation-rich areas, but bogs at the sites where the water level is occasionally lower during dry summer are covered by a sparse tree layer with pine: *Pinus sylvestris* dominates bog woodlands at low altitudes, namely in the Doksy area and Třeboň Basin, whereas the Central European endemic *P. uncinata* subsp. *uliginosa* occurs in mountainous areas (but also in the Třeboň Basin), frequently in a mixture with *P. sylvestris*. At high altitudes in some Czech mountain ranges (Novohradské hory, Šumava, Krušné hory, Jizerské hory and Krkonoše), shrubby *Pinus mugo* becomes dominant (Fig. 7.25f). A rare type of bog vegetation includes suboceanic mires with *Trichophorum cespitosum* and *Sphagnum papillosum* (alliance *Oxycocco palustris-Ericion tetralicis*); they occur in the Jizerské hory Mts, the most precipitation-rich area in the Czech Republic. Unlike most other bogs in the Czech Republic, they do not have a hummock-and-hollow microtopography. A boreal type of mires with arctic-boreal species such as *Betula nana*, *Rubus chamaemorus* and *Vaccinium microcarpum* occurs in the upper-montane to subalpine belts in the Krkonoše, Šumava and Krušné hory Mts. Peat mosses typical of this vegetation include *Sphagnum compactum* and *S. fuscum* (alliance *Oxycocco microcarpi-Empetrion hermaphroditi*). In the past large areas of bogs were destroyed by peat extraction, but there has been a tendency for peatland to recover by spontaneous succession at some of these sites (Konvalinková and Prach 2010). Even now undisturbed bogs are endangered by the lowering of the ground-water table and atmospheric nutrient deposition (Hájková et al. 2011a). Summit bogs in the Sudetes were also polluted by aerial liming applied in the early 1990s, which may have promoted decomposition and associated release of phosphorus (Poulíčková et al. 2013). Yet, it seems that these factors had little effect on the species composition in Sudetes summit bogs, even though the local levels of atmospheric nitrogen deposition are among the highest in Europe (Hájková et al. 2011a).

## *7.5.8 Grasslands and Heathlands Below the Timberline*

### 7.5.8.1 Grassland History and Management

Grasslands were widespread in the current Czech territory in the Pleistocene, when the dry and cold climate limited the development of forest. Pleistocene landscape was rather open and covered by a mosaic of grassland, tundra and woodland (Kuneš et al. 2008; see Chap. 6, this book). Pleistocene cold steppe, which has been reconstructed for the full and late glacial periods in lowland areas, based on both fossil pollen (e.g. Rybníčková and Rybníček 1972, 2014; Kuneš et al. 2008) and mollusc records (Ložek 2001, 2011), was a precursor of current steppe grasslands. Putative relicts of the steppe flora during the coldest periods include *Agropyron pectinatum*, *Bassia prostrata* and *Taraxacum serotinum* (Horsák et al. 2015), currently occurring rarely on eroded loess slopes with discontinuous vegetation cover in southern Moravia (alliance *Artemisio-Kochion prostratae*). However, it is possible that many, if not most, species of Czech dry steppes are Pleistocene relicts (see Kajtoch et al. 2016). Pleistocene stream valleys probably also contained various species that are currently typical of wet meadows.

With climatic amelioration in the Early Holocene, the area of grassland was reduced due to the spread of forest. The key question is whether grassland survived the period of a few centuries between the beginning of the humid period in the Middle Holocene and onset of Neolithic farming ~7600 years before present, which could have reversed the spread of forest. Continuous persistence of specialist steppe snails in the fossil record, preservation of chernozem soils and current occurrence of many plant species of continental steppe (some of them with disjunct ranges) in these areas suggest the continuous persistence of steppe vegetation in forest-steppe areas of northern Bohemia and southern Moravia throughout the Holocene, not only on rock outcrops, but also on deeper soils (Sádlo et al. 2005; Ložek 2011; Pokorný et al. 2015). A landscape reconstruction model based on pollen data for the dry area in southern Moravia indicates that the proportion of open land was continuously about 20% even in the wettest period of the Middle Holocene (Kuneš et al. 2015; Abraham et al. 2016). The driest types of steppe grassland may have existed at some sites throughout the Holocene independently of human influence, especially on south-facing slopes in the forest-steppe areas in northern and central Bohemia and southern Moravia, and on rock outcrops (Ložek 2011). It is possible that grassland existed continuously also at the peripheries of these dry regions, supported by human management, whereas precipitation-rich areas at higher altitudes became nearly continuously forested (Hájková et al. 2011b; Hájek et al. 2016). In addition to human activities, grasslands may have been partly maintained also by large wild herbivores, which lived in Central Europe until the Modern Period (auroch, wild horse and European bison) and beaver (Vera 2000).

In spite of the likely historical continuity of grassland, most areas and vegetation types of grasslands and heathlands below the alpine timberline in the Czech Republic are secondary and developed in artificially deforested areas under long-term management. This is indicated by shrub and tree encroachment after their abandonment. According to their management, grasslands are divided into meadows and pastures. In Central Europe, the term 'meadow' is traditionally used for grasslands mown for hay; once abandoned, in most cases they are overgrown by shrubs and trees and ultimately replaced by forest. Meadows are confined to mesic and wet habitats that are naturally nutrient-rich (such as in floodplains) or artifically fertilized. In contrast, all types of soils from wet to dry and from oligotrophic to eutrophic have traditionally been used as pastures. Some grasslands are cut in spring and then grazed in summer and autumn (Hejcman et al. 2013).

The productivity of most grasslands in the prehistoric landscape was probably low except for those in floodplains. The oldest archaeological finds of scythes are from the fifth century BC (Beranová and Kubačák 2010) and it is therefore assumed that from the Neolithic to the early Iron Age secondary grasslands were maintained almost exclusively by grazing. The main source of winter fodder for livestock was not hay from grassland but leaf fodder from pollarded trees. Hay making introduced in the Iron Age was much more effective than leaf fodder collecting in terms of the amount of dry matter collected per unit time. This was necessary for the maintenance of expanding areas of arable land, which were managed using cattle as draught animals, and cattle require good winter feeding (Hejcman et al. 2013). It is likely that already in the Iron Age some differentiation was made between pastures, usually on less productive soils, and mown grasslands on more productive soils, but the alternation of grazing and mowing at single sites was probably common. The extent of secondary grasslands and heathlands significantly increased with medieval colonization and deforestation of mid-altitude areas, nevertheless the extensive management continued into the Modern Period. The existence of hay meadows in the Middle Ages is documented in written sources and paintings, from archaeological finds of long scythes and indirectly by several finds in archaeological contexts of *Arrhenatherum elatius*, a tall grass typical of productive hay meadows (Hejcman et al. 2013).

Long-term grazing and hay-making led to continuous nutrient depletion of grassland ecosystems. Grasslands were not fertilized except for areas close to some farms with stalling for cattle (Krahulec et al. 1997; Semelová et al. 2008; Hejcman et al. 2013). Therefore, they were generally not very productive, enabling only one hay cut annually, often in late spring or summer (Jongepierová 2008). Productive grasslands suitable for two or even three hay cuts annually were mainly in floodplains, where nutrients were supplied naturally by floods.

Hay meadows on mesic soils, although being a very typical component of the Central European landscape and composed mainly of native species, is a relatively modern vegetation type. Increasing population in the eighteenth century required increased crop production, which started to be achieved by application of farmyard manure to arable fields. The demand for farmyard manure caused a shift in animal husbandry from free grazing to livestock enclosure in paddocks and barns. This in turn increased the demand for hay, which resulted in separation of land used for hay making and grazing, and subsequent differentiation of meadow and pasture vegetation types. Sowing of legumes was introduced to increase grassland fertility through symbiotic nitrogen fixation, resulting in higher yields of herbage and more farmyard manure, which could be occasionally also used to fertilize meadows. More extensive application of fertilizers to meadows was associated with the introduction of mineral fertilizers since the second half of the nineteenth century (Poschlod et al. 2009; Hejcman et al. 2013; Poschlod 2015). The nutrient status and species composition of fertilized meadows changed from the originally oligotrophic, nutrient depleted grasslands with short grasses such as *Agrostis capillaris* and *Festuca rubra* and continental species belonging to the relict Early Holocene species pool (Blažková 1979; Roleček et al. 2015) to mesotrophic or eutrophic types. This change in nutrient status was associated with the spread of tall, nutrient demanding grasses and dicot herbs, most notably *Arrhenatherum elatius*, which is now a common dominant species in eutrophic meadows, but is considered to be alien (archaeophyte) to Central Europe (Poschlod et al. 2009; Pyšek et al. 2012a). Other species forming the current species pool in these eutrophic meadows probably came mainly from various floodplain habitats, open woodlands and tall-grass steppes.

Merging small private farms into large cooperative and state-run farms enforced by the communist government in the 1950s and increased use of mineral fertilizers (Bičík and Jančák 2009) led to intensive management of meadows in productive and easily accessible areas, whereas grasslands in marginal areas and less productive habitats, where hay-making or grazing became unprofitable, were abandoned (see Fig. 1.14b in Chap. 1, this book). Large areas of formerly managed grasslands were also abandoned in areas originally with German-speaking inhabitants who were forced to leave the country after World War II (Kopecký and Vojta 2009; Vojta and Drhovská 2012). At the same time, large areas of wet meadows were drained in an attempt to obtain more productive agricultural land. All of these activities led to a dramatic decline in grassland species diversity. Intensive grassland management with plowing and seeding with forage grasses, fertilizer input and several hay cuts annually was introduced across large areas. This led to the development of species-poor plant communities with a few species capable of fast regrowth after cutting, e.g. *Alopecurus pratensis*, *Dactylis glomerata*, *Festuca arundinacea*, *F. pratensis* and *Phleum pratense*, which are currently the most common grassland type in the Czech Republic (Hejcman et al. 2012).

Profound changes in grassland management occurred after the political changes in 1989 and subsequent break-up of the subsidized cooperative and state-run farms. The populations of domestic livestock were greatly reduced. For example, the number of cattle decreased by ~60% between 1990 and 2005 and stabilized only later on (Czech Statistical Office; www.czso.cz). Consequently, there was a reduced demand for forage and less economic interest in grassland management. Large areas of grassland were abandoned, which resulted in the spread of tall competitive plant species, such as *Alopecurus pratensis*, *Elymus repens*, *Phalaris arundinacea* and *Urtica dioica* in wet grasslands (Prach 2008), *Arrhenatherum elatius* in dry grasslands (Dostálek and Frantík 2012; Holub et al. 2012), *Bistorta officinalis* in mountain grasslands (Pecháčková and Krahulec 1995), *Calamagrostis epigejos* in various grasslands from wet to dry (Sedláková and Fiala 2001; Stránská 2004; Holub et al. 2012; Klimeš et al. 2013), *Calamagrostis villosa* and *Molinia caerulea* in subalpine grasslands (Hejcman et al. 2006b, 2009), *Molinia arundinacea* in semi-dry grasslands (Klimeš et al. 2013) and shrub and tree encroachment in most grassland types (e.g. Osbornová et al. 1990; Stránská 2004; Dostálek and Frantík 2012; Jírová et al. 2012). Increasing dominance of these competitive species usually causes a decrease in species richness.

While many grasslands were abandoned, extensive areas of arable land were converted to grassland in the 1990s as a result of socioeconomic changes (Bičík and Krupková 2009; see Fig. 1.14b in Chap. 1, this book). In some cases, especially in warm and dry lowland and colline areas, spontaneous grassland succession on ex-arable land resulted in dry or semi-dry grasslands of considerable conservation importance as they became habitats of endangered species, especially if there were propagule sources in remnant ancient dry grasslands in the neigbourhood (Osbornová et al. 1990; Jongepierová et al. 2004; Jírová et al. 2012; Knappová et al. 2012; Sojneková and Chytrý 2015). Dry grasslands of conservation importance are spontaneously developing also in some abandoned stone quarries (Novák and Konvička 2006). Some species-rich grasslands were restored using commercial and regional seed mixtures. Very successful restoration projects were conducted especially in the White Carpathians, where ~500 ha of species-rich semi-dry grassland were restored using a regional seed mixture (Michley et al. 2012; Prach et al. 2013a). Non-sown target species from the adjacent ancient grasslands also gradually became established in the restored grassland, in addition to sown species (Prach et al. 2015). Experimental transplantation of small blocks of meadow turf was also used in the White Carpathians as an additional means of enhancing the restoration of species-rich grassland (Mudrák et al. 2017).

Some areas of grassland are currently under subsidized extensive management. An increasingly applied low-cost management method is mulching, i.e. cutting the sward but not removing the herbage. Clippings are cut into smaller pieces and left on the ground to decompose. Mulched grasslands tend to lose some nutrients because of easier decomposition of mulched biomass and subsequent nutrient leaching, but they lose less nutrients than grasslands cut for hay in which the biomass is regularly removed (Pavlů et al. 2016). In low-productive submontane and montane mesic grasslands dominated by *Festuca rubra* mulching does not seem to lead to a loss of plant species richness or significant change in species composition compared to traditional cutting of hay (Mašková et al. 2009; Gaisler et al. 2013), therefore it is also used for grassland maintenance in some protected areas. In contrast, in more productive grasslands, such as submontane mesic grasslands with *Dactylis glomerata* and *Festuca pratensis*, mulching can lead to a decrease in plant species richness relative to hay cutting, although mulched grasslands still support higher numbers of species per unit area than abandoned grasslands (Pavlů et al. 2016).

Preferred option in protected areas is the reintroduction of traditional site-specific management, either hay making (e.g. Jongepierová 2008; Pavlů et al. 2011; Klimeš et al. 2013) or grazing (Krahulec et al. 2001; Pavlů et al. 2007; Hejcman et al. 2008; Dostálek and Frantík 2012), depending on the area and vegetation type. In most cases cutting is done by machinery and large areas of grasslands are cut homogeneously within a short period of time. This is different from the traditional cutting by scythes, which was done over a longer period and created structurally more diverse mosaic of cut and uncut grasslands. To allow for seed ripening and to protect invertebrates, strips or patches of uncut grassland are sometimes left in the cut grassland areas; such mixed management seems to result in a higher diversity of both plants and invertebrates (Bonari et al. 2017). Grazing in protected areas is usually done by cattle, sheep or goats. A promising option for larger protected areas is the rewilding using de-domesticated horses, auroch-like cattle and European bison, i.e. large herbivores that were dwelling in the Central European landscape from the Early Holocene up to the Modern Period, but were extirpated by humans. As these animals are well adapted to year-round-grazing with minimum input from managers, they are able to keep the landscape open and sustain grassland at a very low cost (Dostál et al. 2012, 2014). Introduction of these herbivores into abandoned military areas in the Labe lowland near Milovice and Benátky nad Jizerou in central Bohemia started in 2015.

#### 7.5.8.2 Main Types of Grassland Vegetation

**Meadows and mesic pastures** (class *Molinio-Arrhenatheretea*) are the most typical grassland type in the Czech Republic. The most common meadow type in the Czech Republic is mesic meadows dominated by the tall grass *Arrhenatherum elatius* (alliance *Arrhenatherion elatioris*; Fig. 7.27a), which occur from lowland to montane belt, although at higher altitudes the abundance of *Arrhenatherum* decreases while shorter grasses become dominant, in particular *Agrostis capillaris*, *Festuca rubra* and *Trisetum flavescens*. These meadows are usually cut twice a year with occasional aftermath grazing. At high altitudes, especially in the Krušné hory, Jizerské hory, Krkonoše and Orlické hory Mts, mesic meadows are dominated by the above mentioned short grasses and contain several species that are absent in low-altitude meadows, e.g. *Arabidopsis halleri*, *Cirsium heterophyllum*, *Geranium sylvaticum*, *Meum athamanticum*, *Poa chaixii* and *Silene dioica* (alliance *Polygono bistortae-Trisetion flavescentis*). Because of the shorter growing season and lower productivity in the mountains, these meadows are usually cut only once a year. Mesic pastures with *Lolium perenne*, *Plantago lanceolata*, *P. major* subsp. *major*, *Trifolium pratense* and *T. repens* (alliance *Cynosurion cristati*) are also common in this country, especially at mid-altitudes. Many of them are poor in species and their structure and species composition is similar to lawns in city parks, which are mulched several times a year, or to trampled grasslands in human settlements. More species-rich mesic pastures are found especially in the Carpathian flysch zone (Rozbrojová et al. 2010), where they are related to less intensive management. Meadows on low-productive soils, which are wet in spring but dry in summer, are often dominated by *Molinia arundinacea* or *M. caerulea* (alliance *Molinion caeruleae*; Havlová 2006; Fig. 7.27b). Due to their low productivity and the late phenology of the dominant species, they used to be cut once a year, in summer. They occur mainly in the highlands in the central part of the Bohemian Massif (Brdy Mts and Bohemian-Moravian Highlands) and basins of southern Bohemia, but they can also be found on raised areas on the floodplains of large lowland rivers. However, in the last few decades the area of these meadows has been strongly reduced due to eutrophication. Wet meadows with natural fertilizer input through spring flooding are

**Fig. 7.27** Examples of mesic and wet grassland types in the Czech Republic: (**a**) Eutrophic mesic grassland with *Leucanthemum vulgare* agg. and *Trifolium pratense*, Brno-Řečkovice, southern Moravia (*Pastinaco sativae-Arrhenatheretum elatioris*, *Arrhenatherion elatioris*); (**b**) Mesotrophic intermittently wet grassland with *Betonica officinalis* near Březka, eastern Bohemia (*Molinietum caeruleae*, *Molinion caeruleae*); (**c**) Sligtly acidophilous eutrophic wet meadow with *Angelica sylvestris* and *Cirsium palustre* (*Angelico sylvestris-Cirsietum palustris*, *Calthion palustris*) near Kvilda, Šumava Mts, south-eastern Bohemia; (**d**) Nutrient-poor grassland around granite outcrops in an agricultural landscape near Kadov, southern Bohemia (*Festuco capillatae-Nardetum strictae*, *Violion caninae*); (**e**) Dry heathland with *Calluna vulgaris* on Kraví hora hill near Znojmo, southern Moravia (*Euphorbio cyparissiae-Callunetum vulgaris*, *Euphorbio cyparissiae-Callunion vulgaris*); (**f**) Mesic heathland with the dwarf shrubs *Calluna vulgaris*, *Erica carnea*, *Vaccinium myrtillus* and *V. vitis-idaea* on a serpentinite outcrop at Křížky site in the Slavkovský les Mts, western Bohemia (*Vaccinio-Callunetum vulgaris*, *Genisto pilosae-Vaccinion*)

divided into the alliances *Deschampsion cespitosae* and *Calthion palustris*. The former alliance includes meadows in broad floodplains, which are flooded in spring but dry out in summer, especially along lowland rivers such as the Labe, lower Dyje and lower Morava (Balátová-Tuláčková 1968). These meadows are dominated by grasses, in particular *Alopecurus pratensis*, and used to be cut three times a year (Honsová et al. 2007). The latter alliance comprises meadows along small streams or at seepage sites (Fig. 7.27c). They are usually slightly flooded in spring, the water table remains high in summer and the soil is wet all year round. These meadows are dominated by tall dicot herbs, especially different species of *Cirsium*, depending on soil chemistry (Hájek and Hájková 2004). They used to be cut two times annually. Abandoned meadows of *Calthion palustris* are usually dominated by the tall forb *Filipendula ulmaria*.

***Nardus stricta* grasslands** (Fig. 7.28a) occur on nutrient-poor acidic soils over granite, gneiss, schist and similar hard and nutrient-poor bedrocks. Due to their low productivity, they were traditionally used for livestock grazing, although occasionally they were mown once a year or every second year in summer. As a result of abandonment since the mid-twentieth century, coupled with nutrient deposition, many *Nardus* grasslands were replaced by mesic grasslands with a higher biomass production (Krahulec et al. 1997). *Nardus stricta* grasslands occur from the submontane belt to areas above the timberline, and their species composition varies with altitude. The most widespread type with some thermophilous species (class *Calluno-Ulicetea*, alliance *Violion caninae*; Fig. 7.27d) is found in the submontane to montane belt. In the montane belt of the Krkonoše Mts, *Nardus* grasslands contain a mixture of submontane and subalpine species (class *Calluno-Ulicetea*, alliance *Nardo strictae-Agrostion tenuis*; Krahulec 1990b; Krahulec et al. 1997). *Nardus* grasslands occurring around the timberline in the Krkonoše, Králický Sněžník and Hrubý Jeseník Mts contain subalpine species while submontane species are absent (class *Calluno-Ulicetea*, alliance *Nardion strictae*). *Nardus* grasslands above the timberline in these mountain ranges are species-poor and contain arcto-alpine species such as *Carex bigelowii* (class *Juncetea trifidi*, alliance *Nardo-Caricion bigelowii*; see above). Species-poor *Nardus* grasslands with *Juncus squarrosus* develop on wet acidic soils in the montane and submontane belts (class *Calluno-Ulicetea*, alliance *Nardo strictae-Juncion squarrosi*).

**Heathlands** are dominated by *Calluna vulgaris*, *Vaccinium myrtillus* and *V. vitis-idaea*, and locally also codominated by *Arctostaphylos uva-ursi* (northern, central and southern Bohemia), *Erica carnea* (western and southern Bohemia) and *Genista pilosa* (south-western Moravia). Other dwarf shrubs that are typical of Atlantic heathlands of Western Europe are absent in the Czech Republic because of its more continental climate with winter frost. Czech heathlands occur on nutrient-poor bedrocks such as granite, gneiss, schist, sandstone or sand, usually on shallow soil. For the most part, they are secondary vegetation that developed in deforested areas that were disturbed and depleted of nutrients by grazing, cutting or burning, and locally perhaps also by sod-cutting (Sedláková and Chytrý 1999). Small patches of primary heathlands are restricted to rock outcrops and more extensive primary heathlands occur above the timberline. Dry lowland heathlands that occur espe-

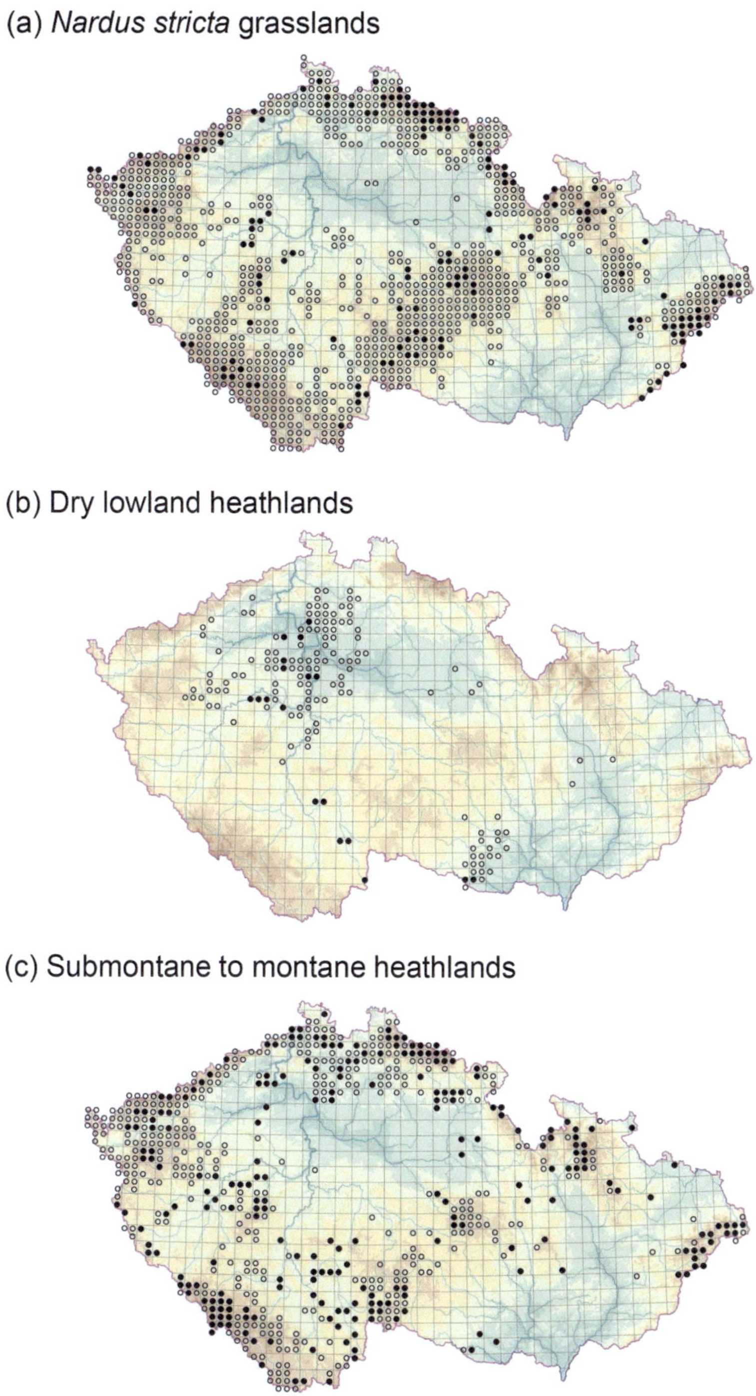


**Fig. 7.28** Distribution of selected types of grassland and heathland vegetation. *Black circles* indicate sites of vegetation plots from the Czech National Phytosociological Database that correspond to the formal definitions of different vegetation types according to Chytrý (2007). *Open circles* indicate records based on habitat mapping in 2001–2008 (Härtel et al. 2009), revised by Chytrý et al. (2010), which include fragmentary or transitional occurrences of individual vegetation types. Phytosociological alliances and classes: (**a**) *Nardion strictae*, *Nardo strictae-Agrostion tenuis* and *Violion caninae*; (**b**) *Euphorbio cyparissiae-Callunion vulgaris*; (**c**) *Genisto pilosae-Vaccinion*

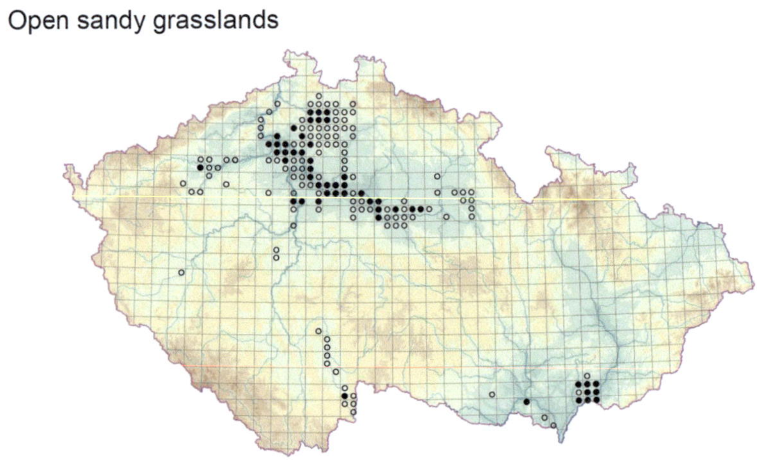


**Fig. 7.29** Distribution of sandy grasslands (*Corynephorion canescentis*) (see Fig. 7.28 for details)

cially in central Bohemia and south-western Moravia are dominated by *Calluna vulgaris*, while *Vaccinium* species are absent; they are rich in thermophilous continental species typical of dry grasslands (class *Calluno-Ulicetea*, alliance *Euphorbio cyparissiae-Callunion vulgaris*; Figs 7.27e and 7.28b). Submontane to montane heathlands are usually codominated by *Calluna vulgaris*, *Vaccinium myrtillus* and *V. vitis-idaea*, while thermophilous continental species are absent (Geringhoff and Daniëls 1998). Extensive stands dominated by *Vaccinium myrtillus* occur above the timberline in the Krkonoše, Králický Sněžník and Hrubý Jeseník Mts, in places where thick snow cover provides winter protection to dwarf shrubs (class *Calluno-Ulicetea*; alliance *Genisto pilosae-Vaccinion*; Figs 7.27f and 7.28c).

**Pioneer vegetation of sandy and shallow soils** occurs in lowland sandy areas (Fig. 1.4 in Chap. 1, this book) and on outcrops of poorly weathered rocks, especially granite, gneiss and schist of the Bohemian Massif, Cretaceous sandstone or Palaeozoic to Jurassic limestone. Most types of this vegetation occur in habitats of potential forests and are maintained by disturbances. Sand accumulations of both fluvial and aeolian origin in the Czech Republic are siliceous, with a low pH. Species-poor sandy grasslands dominated by the subatlantic bunch grass *Corynephorus canescens*, with abundant mosses and lichens, occur mainly in the Doksy area of northern Bohemia, along the Labe River and in the sandy area near the town of Hodonín in south-eastern Moravia (class *Koelerio-Corynephoretea*, alliance *Corynephorion canescentis*; Figs 7.29 and 7.30a). The latter area also hosts species-rich continental (Pannonian) sandy-steppe vegetation dominated by *Festuca psammophila* subsp. *dominii* and *Stipa borysthenica* and containing several species with a continental distribution (class *Festucetea vaginatae*, alliance *Festucion vaginatae*). In some areas, especially in the lowland and colline belt, small patches of vegetation with the short annual grasses *Aira praecox* (Černý et al. 2007) and *Vulpia*

**Fig. 7.30** Examples of dry grassland types in the Czech Republic: (**a**) *Corynephorus canescens* sandy grassland near Bzenec, southern Moravia (*Corniculario aculeatae-Corynephoretum canescentis*, *Corynephorion canescentis*); (**b**) Acidophilous rock-outcrop vegetation with *Gagea bohemica* and ephemeral vernal annuals on Bouchal hill near Ivančice, southern Moravia (*Festuco-Veronicetum dillenii*, *Arabidopsion thalianae*); (**c**) A limestone outcrop with *Aurinia saxatilis* and *Festuca pallens* on Růžový vrch hill in the Pavlov Hills, southern Moravia (*Festuco pallentis-Aurinietum saxatilis*, *Alysso-Festucion pallentis*); (**d**) Dry steppic grassland with *Stipa capillata* near Hnojnice in the České středohoří Mts, northern Bohemia (*Festuco valesiacae-Stipetum capillatae*, *Festucion valesiacae*); (**e**) Spring view of a semi-dry grassland with *Adonis vernalis* near Kurdějov, southern Moravia (*Polygalo majoris-Brachypodietum pinnati*, *Cirsio-Brachypodion pinnati*); (**f**) Saline grassland with *Tripolium pannonicum* subsp. *pannonicum* and *Inula britannica* near Sedlec, southern Moravia (*Scorzonero parviflorae-Juncetum gerardii*, *Juncion gerardii*)

*myuros* occur in disturbed sandy or gravelly habitats (class *Koelerio-Corynephoretea*, alliance *Thero-Airion*). At less frequently disturbed sites, lowland sandy grasslands become more closed and narrow-leaved fescues (*Festuca brevipila*, *F. ovina* and *F. rubra*) dominate; sand specialists (e.g. *Armeria elongata* subsp. *elongata*) are also common in these grasslands (class *Koelerio-Corynephoretea*, alliance *Armerion elongatae*). In the colline to submontane belt of the Bohemian Massif, an analogue of the *Festuca*-dominated lowland sandy grasslands is the vegetation with *Festuca ovina* and short herbs adapted to very shallow nutrient-poor soils with a low pH, e.g. *Jasione montana*, *Pilosella officinarum*, *Rumex acetosella* and *Scleranthus perennis*. These short swards with abundant mosses and lichens (*Ceratodon purpureus*, *Polytrichum piliferum* and *Cladonia* spp.) occur mainly in the southern half of Bohemia and in south-western Moravia (class *Koelerio-Corynephoretea*, alliance *Hyperico perforati-Scleranthion perennis*; Moravec 1967). Vegetation with short-lived vernal annuals and succulents (especially *Sedum acre*, *S. album* and *S. sexangulare*) occur in dry and warm areas, partly on rock outcrops, partly in disturbed patches within dry grasslands where cover of perennial herbs is reduced. The species composition of this vegetation differs between acidic bedrock and limestone, the former being characterized by *Arabidopsis thaliana*, *Gagea bohemica*, *Myosotis stricta*, *Veronica dillenii* and *V. verna* (class *Koelerio-Corynephoretea*, alliance *Arabidopsion thalianae*; Fig. 7.30b) and the latter by *Acinos arvensis*, *Alyssum alyssoides*, *Arabidopsis thaliana*, *Cerastium pumilum*, *Erophila spathulata*, *Medicago minima*, *Poa bulbosa*, *Saxifraga tridactylites*, *Thlaspi perfoliatum* and *Veronica praecox* (class *Koelerio-Corynephoretea*, alliance *Alysso alyssoidis-Sedion*).

**Dry grasslands** (class *Festuco-Brometea*) vary in species composition and productivity depending on the availability of water, which in turn depends on soil depth and precipitation. At a higher hierarchical level, they can be divided into rocky grasslands, dry grasslands and semi-dry grasslands, corresponding to the orders *Stipo-Festucetalia*, *Festucetalia valesiacae* and *Brometalia erecti* (Willner et al. 2017b). Dry grasslands on outcrops of various types of poorly weathered rocks occur mainly in northern and central Bohemia and southern Moravia (Fig. 7.31a). Most of these grasslands are dominated by the narrow-leaved tussocky grass *Festuca pallens*, which occurs both on siliceous rocks and limestone (alliance *Alysso-Festucion pallentis*; Fig. 7.30c). On the limestone outcrops in the Pavlov Hills of southern Moravia, *Festuca pallens* is accompanied by sub-Mediterranean species such as *Fumana procumbens*, *Melica ciliata*, *Poa badensis* and *Teucrium montanum* (alliance *Bromo pannonici-Festucion pallentis*). *Festuca pallens* grasslands are confined to south-facing slopes, whereas north-facing slopes on limestone outcrops are covered by *Sesleria caerulea* grasslands (alliance *Diantho lumnitzeri-Seslerion*). *Sesleria caerulea* and some other species growing in the same grasslands (e.g. *Biscutella laevigata* and *Saxifraga paniculata*) are typical of the montane to subalpine belt of the limestone Alps and Carpathians. Their occurrence in the colline belt in northern and central Bohemia and southern Moravia is probably a relict of a broader distribution of these species at lower altitudes in the Pleistocene. However, at low altitudes these species are mixed with species of continental steppes and sub-Mediterranean rock-outcrop grasslands. Short-grass continental steppe (the

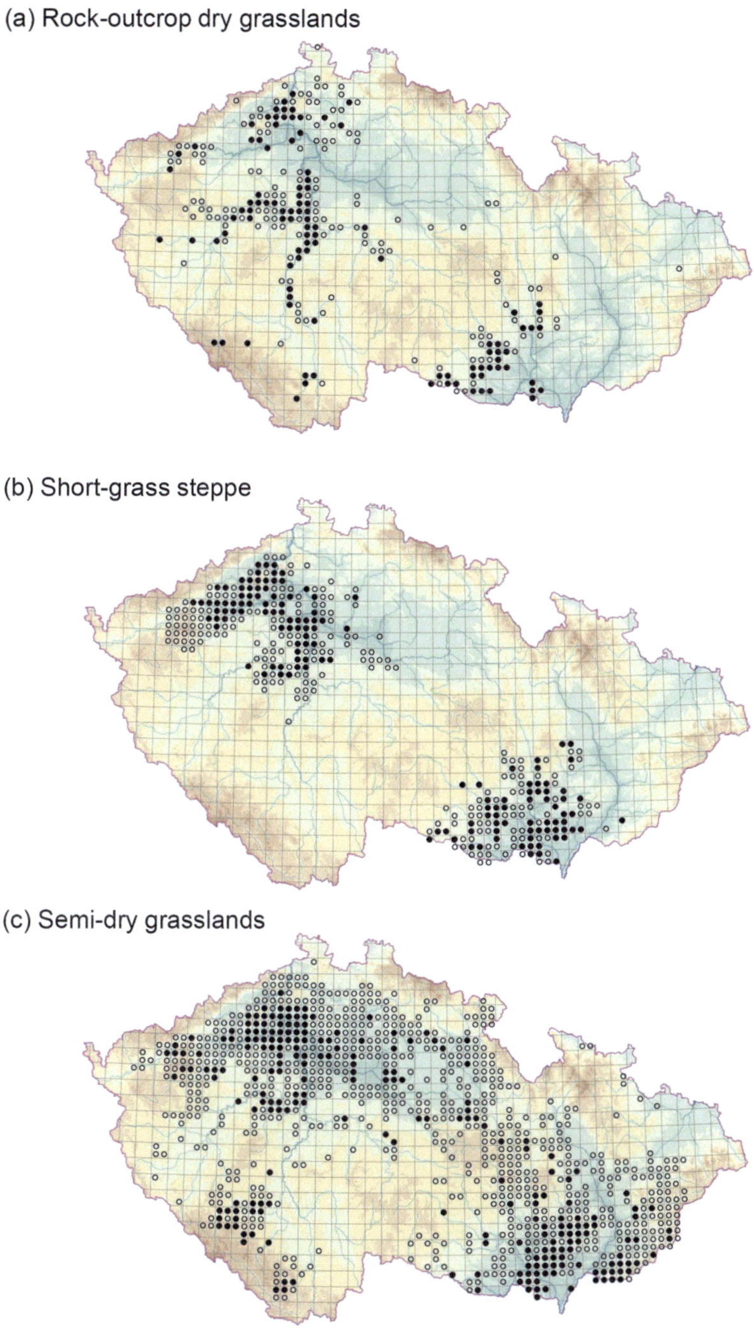


**Fig. 7.31** Distribution of dry grassland types: (**a**) *Alysso-Festucion pallentis*, *Bromo pannonici-Festucion pallentis* and *Diantho lumnitzeri-Seslerion*; (**b**) *Festucion valesiacae*; (**c**) *Cirsio-Brachypodion pinnati* and *Bromion erecti* (see Fig. 7.28 for details)

'true steppe' in Russian literature) occurs on both shallow (rendzina or ranker) or deep (chernozem) soils in the driest areas of northern and central Bohemia and southern Moravia. Dominant species include *Carex humilis*, *Festuca valesiaca*, *Stipa capillata*, *S. pennata*, *S. pulcherrima* and other feather grasses and at slightly more mesic sites also *Festuca rupicola* (alliance *Festucion valesiacae*; Kolbek 1975, 1978; Dúbravková et al. 2010; Figs 7.30d and 7.31b). More mesic soils on flat land or gentle slopes in the driest area in northern Bohemia and southern Moravia support patches of *Stipa tirsa* steppe. This species is the most moisture-demanding of the Czech feather grasses (Rychnovská and Úlehlová 1975). It may have dominated extensive steppes on chernozem in the dry lowlands before these were converted to arable land. Nowadays only small fragmentary stands of *S. tirsa* steppe are preserved, although extensive stands still existed in the foothills of the White Carpathians in south-eastern Moravia in the first half of the twentieth century (Sillinger 1929; Podpěra 1930). Disturbed steppe sites on deep soil (e.g. on erosion-prone steep slopes, around rabbit colonies or on ex-arable land) at some sites in southern Moravia harbour competitively poor continental steppe species such as *Astragalus exscapus*, *Crambe tataria*, *Iris pumila* and *Taraxacum serotinum*. Short-grass steppes have traditionally been used as pastures for livestock, usually mixed herds of different kinds of animals grazing on common land. However, some of these steppes can be natural, existing independenly of human management. Semi-dry grasslands (Figs 7.30e and 7.31c) are usually dominated by the broad-leaved rhizomatous grass *Brachypodium pinnatum*, but in some areas other species can also be important, e.g. *Bromus erectus*, *Koeleria pyramidata* (at higher altitudes), *Dorycnium pentaphyllum* agg. and *Inula ensifolia* (both in southern Moravia) and *Sesleria caerulea* (marlstone slopes in northern Bohemia). Based on their species composition, these grasslands are divided into suboceanic and continental types, correspoding respectively to the alliances *Bromion erecti* and *Cirsio-Brachypodion pinnati* (Illyés et al. 2007). In the Czech Republic the former occurs mainly in the upper-colline belt, i.e. areas with a wetter, more oceanic climate, whereas the latter is found in dry lowland and colline areas in the northern half of Bohemia and in southern Moravia. *Cirsio-Brachypodion pinnati* grasslands correspond to tall-grass steppe or 'meadow steppe' of Russian literature. Semi-dry grasslands are characterized by a higher biomass production than other types of dry grasslands, therefore many of them used to be mown for hay once a year, usually in summer (Jongepierová 2008), but some of them were also grazed by livestock. These grasslands are among the most species-rich vegetation types in the Czech Republic, both in terms of their regional species pool (Sádlo et al. 2007) and local species richness (Chytrý et al. 2015; see Sect. 7.6). Dry grasslands on acidic soils (alliance *Koelerio-Phleion phleoidis*) are poorer in species than semi-dry grasslands. They occur in dry and warm areas of the Bohemian Massif with poorly weathered rocks, especially in central Bohemia and south-western Moravia. Dominant species are mainly graminoids such as *Carex humilis*, *Festuca ovina*, *F. rupicola* and *Helictotrichon pratense* (Chytrý et al. 1997). The herbaceous vegetation of forest fringes is a specific type of dry grassland occurring in ecotonal habitats at the interface of forest and grassland. These communities are composed of both grassland species and species of forest

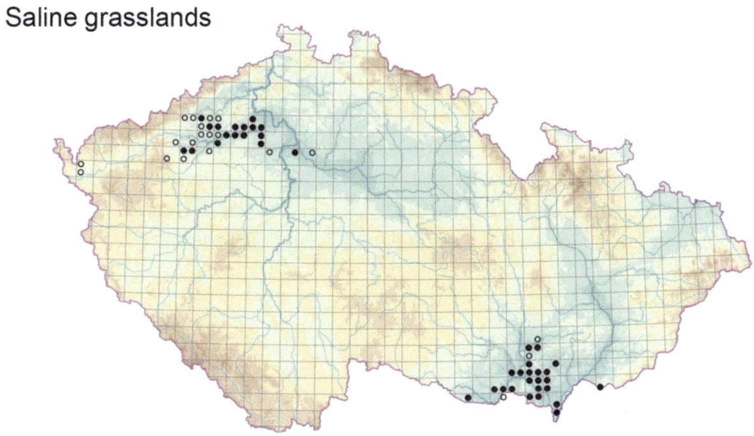


**Fig. 7.32** Distribution of saline grasslands (*Festuco-Puccinellietea*) (see Fig. 7.28 for details)

herb layer, and dominated by broad-leaved dicots. Fringe vegetation dominated by *Dictamnus albus*, *Geranium sanguineum* or *Peucedanum cervaria* develops in dry and warm areas, often in an ecotone between dry grassland and thermophilous oak forest or xeric scrub (alliance *Geranion sanguinei*). In contrast, fringe vegetation with *Melampyrum nemorosum* or *Trifolium medium* develops in slightly cooler and wetter areas, typically at the transition between semi-dry grassland or mesic meadow and oak-hornbeam forest (alliance *Trifolion medii*).

**Inland saline vegetation** (Figs. 7.30f and 7.32) is found locally in dry lowlands in both areas of the forest-steppe biome within the Czech Republic. It has stronger biogeographical links with the inland saline vegetation in South-eastern and Eastern Europe than with coastal salt marshes (Eliáš et al. 2013). It used to be most diverse and rich in obligate halophytes in southern Moravia (Šmarda 1953; Vicherek 1962a, 1973), where solonchak (but not solonetz) soils existed in areas with salt-rich Miocene sediments, around mineral springs rich in calcium sulphate and on the shores of natural shallow saline lakes, which existed in the lowlands south-east of Brno but were drained in the first half of the nineteenth century. Most historical occurrences of saline vegetation have disappeared because of the lowering of ground-water table after draining and the regulation of rivers, or conversion to arable land or productive grassland (Grulich 1987). Some types of saline vegetation have disappeared entirely from the country, especially those occurring on soils with a very high salt concentration or in seasonally flooded habitats. In particular, species-poor vegetation with the succulent annual halophytes *Salicornia perennans* and *Suaeda prostrata* (class *Thero-Salicornietea strictae*, alliance *Salicornion prostratae*) occurred at some sites in southern Moravia but has not been observed since the 1980s (Vicherek 1973; Grulich 1987). Also vegetation of annual halophytic grasses *Crypsis aculeata* and *C. schoenoides* (class *Crypsietea aculeatae*, alliance *Cypero-Spergularion salinae*), formerly recorded at several sites in southern Moravia (Vicherek 1973), has declined considerably, although it still persists at a few sites. Last remnants of saline grasslands with *Carex distans*, *C. otrubae*, *C. secalina*, *Juncus gerardii*, *Lotus tenuis*, *Melilotus dentatus*, *Plantago maritima*, *Puccinellia distans*, *Pulegium vulgare*, *Scorzonera parviflora*, *Spergularia media*, *Taraxacum bessarabicum*, *Trifolium fragiferum* and *Tripolium pannonicum* subsp. *pannonicum* (class *Festuco-Puccinellietea*, alliances *Puccinellion limosae* at drier sites and *Juncion gerardii* at wetter sites) occur at a few sites in southern Moravia, and impoverished stands that lack several species of specialized halophytes of continental distribution also occur in northern and central Bohemia (Toman 1988).

## 7.5.9 Ruderal and Weed Vegetation

The structure and species composition of the ruderal and weed vegetation in the Czech Republic mainly depends on the frequency and severity of disturbance. In frequently disturbed habitats in human settlements, along roads and on arable land, this vegetation is more open and composed mainly of annual plants. Where disturbance is less frequent or less severe, ruderal vegetation becomes denser and biennial and perennial plants prevail.

In the pre-Neolithic landscape, ruderal vegetation was probably spatially restricted, confined to disturbed nutrient-rich sites such as sediment accumulations on floodplains, places where animals gathered and camps of Palaeolithic or Mesolithic hunter-gatherers. In the present landscape of the Czech Republic the native flora still makes up a considerable proportion of the vegetation in man-made habitats, accounting in most cases for ~35–40% of the regional species pool and 40–70% of species in individual stands of annual vegetation, and ~ 60% of the regional species pool and 55–80% of the species in individual stands of perennial ruderal vegetation (Chytrý et al. 2005; Sádlo et al. 2007; Pyšek et al. 2012b). Human-mediated spread of alien species, initially mainly from Southern Europe and the Middle East, started in the Neolithic. These pre-Modern Period invaders (archaeophytes, defined as alien species that were introduced before 1500 AD; Pyšek et al. 2012a) are best represented in weed communities on arable land, where they represent on average 35–55% of plant species at individual sites, and in annual ruderal vegetation, where their mean proportion is 25–50%. However, in other types of ruderal vegetation they usually make up less than 35%. Aliens that arrived in the Modern Period, many of them from the Americas, eastern Asia and other continents (neophytes), make up 15–25% of the regional species pool of ruderal and weed vegetation. However, as most of them are rare, their percentage representation in individual vegetation stands is usually less than 10% of the total number of species (Chytrý et al. 2005; Sádlo et al. 2007; Simonová and Lososová 2008; Pyšek et al. 2012b).

**Weed vegetation of arable land** includes a high proportion of species from Southern Europe and the Middle East, which were introduced into Central Europe since approximately 7500 years before present, together with the Middle Eastern crops such as wheat, barley, pea, poppy and flax. Primitive agrotechniques used in the early Neolithic probably permitted many perennial species, mostly belonging to native flora, to occur as weeds on agricultural land, but with the improvements in ploughing technology from the late Neolithic onwards annual species became more prominent (Holzner and Immonen 1982; Kühn 1994). In the Middle Ages three-field crop rotation was introduced with spring cereals (*Triticum*), winter cereals (*Avena*, *Hordeum* and *Secale*) and fallow in the three sequential seasons. This system was gradually abandoned at the end of eighteenth century, when it was replaced by crop rotation with legumes (*Medicago sativa* and *Trifolium pratense*), cereals and root crops (namely potato and since the nineteenth century also beet). Maize planting was introduced in warm lowland areas in the twentieth century. New agro-techniques needed for planting these new crops favoured new species of weeds. Neophytic weeds, mainly of American origin, spread especially in cultures of root crops and maize but have remained less significant in cereal cultures; nevertheless, archaeophytic weeds have always been more important than neophytic weeds in all kinds of crops (Pyšek et al. 2005a; Lososová and Grulich 2009).

In the 1950s small arable fields in former Czechoslovakia were merged into extensive fields managed by large cooperative or state-owned farms, which introduced intensive agrotechniques relying on new crop cultivars, pesticides and mineral fertilizers (Bičík and Jančák 2009). Improved techniques of seed separation led to a decline in species adapted to seeding with crop, e.g. *Agrostemma githago*, *Bromus secalinus*, *Bupleurum rotundifolium* and *Ranunculus arvensis* (Kornaś 1988). Local species richness of weed vegetation also decreased during the second half of the twentieth century (Pyšek et al. 2005b) and its composition changed. Archaeophytes, especially rare ones, tended to decrease in abundance, while neophytes increased (Pyšek et al. 2005a; Lososová and Simonová 2008). In some weed species (e.g. *Amaranthus retroflexus*, *Chenopodium album* and *Echinochloa crus-galli*) herbicide-resistant populations evolved and spread over large areas of arable land (Chodová et al. 2004). No-till farming with reduced soil disturbance, used on some farms since the 1990s, supported the spread of perennial weeds, most notably *Artemisia vulgaris*, *Cirsium arvense* and *Elymus repens* (Mikulka and Kneifelová 2005). Generalist species adapted to modern agrotechniques are predominant in current weed vegetation, with *Chenopodium album* agg., *Cirsium arvense*, *Elymus repens*, *Fallopia convolvulus* and *Viola arvensis* being the most common (Lososová et al. 2008). Nowadays species-rich weed communities or those containing specialist weed species are found mainly at the field margins. However, recent introduction of organic farming in some places has had a positive effect on survival of some rare species of weeds (Kolářová et al. 2013).

Composition of Czech weed vegetation depends mainly on altitude (which includes the joint effects of lower temperature, higher precipitation and more acidic soils at higher altitudes), crop type (cereals vs. root crops), time of year and type of management, i.e. conventional or organic farming (Lososová et al. 2004; Kolářová et al. 2014). The percentage of annuals and aliens among weed species generally decreases with altitude. In early spring, when the cover of crops is still sparse, there are many small short-lived vernal annual weeds, such as *Arabidopsis thaliana*, *Erophila verna*, *Veronica hederifolia*, *V. sublobata*, *V. triloba* and *V. triphyllos*. As the crop cover increases and becomes denser, these short species disappear and the weed community then consists mainly of taller weeds that germinated in the previous autumn or early in the spring (e.g. *Capsella bursa-pastoris*, *Caucalis platycarpos*, *Centaurea cyanus*, *Consolida regalis*, *Fallopia convolvulus* and *Sinapis arvensis*). These cool-season weeds make up a large proportion of the weeds in cereal fields. In contrast, hoeing or selective tillage in root crops provides space for late-germinating thermophilous weeds, such as *Chenopodium polyspermum*, *Echinochloa crus-galli*, *Mercurialis annua* and *Setaria pumila*. Populations of these weeds develop in stubble fields after the cereal harvest (Lososová et al. 2006).

Classification of the weed vegetation in the Czech Republic reflects the variation in altitude, crop type and seasonal differences between spring and summer (Kropáč 2006; Lososová et al. 2006). All types of weed vegetation are assigned to the class *Stellarietea mediae*. The weed communities in cereal crops differ between warmer and drier lowland areas, where the soils tend to be base-rich (alliance *Caucalidion*), and cooler and wetter mid-altitude areas, where the soils are usually acidic (alliance *Scleranthion annui*). Suboceanic weed communities in cereal fields on very poor sandy soils or wet podzols occurred rarely in the western part of Bohemia and their last locality is preserved in the Třeboň Basin of southern Bohemia (alliance *Arnoseridion minimae*). Weed communities in root crops growing in warm, dry and base-rich habitats (alliance *Veronico-Euphorbion* on loamy or clayey soils and alliance *Spergulo arvensis-Erodion cicutariae* on sandy soils) also differ from those in cool, wet and base-poor habitats (alliance *Oxalidion fontanae*).

**Ruderal vegetation** is very variable depending on the frequency and intensity of disturbance, soil nutrients, soil moisture and climate. However, like in weed vegetation, the diversity of ruderal vegetation is also declining in terms of number of vegetation types, species richness and number of archaeophytes (Pyšek et al. 2004). Most frequently disturbed sites with exposed bare soil are occupied by ruderal communities of annual species. These communities are dominated by either summer annuals of the genera *Atriplex*, *Chenopodium* or *Sisymbrium* (class *Stellarietea mediae*, alliance *Atriplicion*) or winter-annual grasses such as *Bromus sterilis*, *B. tectorum* or *Hordeum murinum* (class *Stellarietea mediae*, alliance *Sisymbrion officinalis*). Vegetation with low-growing archaeophytic annuals (e.g. *Anthemis cotula*, *Malva neglecta*, *M. pusilla*, *Mercurialis annua* and *Urtica urens;* class *Stellarietea mediae*, alliance *Sisymbrion officinalis*) occurs especially at trampled sites in village yards and fowl runs, but it has been declining due to village urbanization since the second half of the twentieth century. Annual ruderal vegetation occurring on nutrient-poor sandy or gravelly soils is characterized by *Dysphania botrys*, *Plantago arenaria* and *Salsola tragus* subsp. *tragus* (class *Stellarietea mediae*, alliance *Salsolion ruthenicae*). On dry and nutrient-rich sandy or gravelly soils, both in ruderal (often trampled) habitats and on arable land, this vegetation is dominated by species with $C_4$ carbon fixation pathway such as *Cynodon dactylon*, *Digitaria sanguinalis*, *Eragrostis minor*, *Panicum capillare* and *Portulaca oleracea* (class *Stellarietea mediae*, alliance *Eragrostion cilianensi-minoris*). Annual or short-lived species (e.g. *Lepidium ruderale*, *Matricaria discoidea*, *Poa annua* and *Polygonum arenastrum*) are also dominant in trampled habitats (class *Polygono arenastri-*

*Poëtea annuae*; Simonová 2008). At lower altitudes and drier sites, *Polygonum arenastrum* prevails and in places is accompanied by thermophilous species, such as *Eragrostis minor* or *Sclerochloa dura* (alliance *Coronopodo-Polygonion arenastri*), while at high altitudes or wetter sites, more perennial species (e.g. *Lolium perenne*, *Plantago major* subsp. *major* and *Trifolium repens*) and bryophytes occur in trampled habitats along with annual vascular plants.

In ruderal habitats with less frequent or weaker disturbance, biennial and perennial species are more common and eventually become dominant, although annual species are also well represented. On dry base-rich soils, e.g. on loess, limestone, and also on building rubble or in waste places, archaeophytes prevail, including *Artemisia absinthium*, *Carduus acanthoides*, *Onopordum acanthium*, *Reseda lutea* and *R. luteola* (class *Artemisietea vulgaris*, alliance *Onopordion acanthii*). Currently this vegetation type is declining due to village urbanization. On dry and nutrient-poor anthropogenic substrates, often with a high content of gravel or dross, widespread biennial and perennial ruderal species are common, including *Artemisia vulgaris*, *Cirsium arvense*, *Daucus carota*, *Elymus repens*, *Melilotus albus*, *M. officinalis*, *Solidago canadensis*, *Tanacetum vulgare* and *Tussilago farfara* (class *Artemisietea vulgaris*, alliance *Dauco carotae-Melilotion*). On slopes with loamy soils subject to occasional landslides or solifluction, especially in dry and warm areas, semi-natural to ruderal herbaceous vegetation occurs that is dominated by species with extensive root or rhizome systems, e.g. *Bromus inermis*, *Convolvulus arvensis* and *Elymus repens* (class *Artemisietea vulgaris*, alliance *Convolvulo arvensis-Elytrigion repentis*). On mesic soils, infrequently disturbed ruderal vegetation is dominated by medium-tall to tall perennial dicots such as *Arctium lappa*, *A. tomentosum*, *Ballota nigra*, *Chenopodium bonus-henricus*, *Conium maculatum* or *Sambucus ebulus* (class *Artemisietea vulgaris*, alliance *Arction lappae*). In mesic to wet, infrequently disturbed habitats, large native umbellifers such as *Aegopodium podagraria*, *Anthriscus nitidus*, *A. sylvestris*, *Chaerophyllum aromaticum*, *C. aureum* and *C. bulbosum* dominate (class *Galio-Urticetea*, alliance *Aegopodion podagrariae*). Such habitats are frequently invaded by large perennial broad-leaved neophytes, including *Helianthus tuberosus*, *Heracleum mantegazzianum*, *Reynoutria* ×*bohemica*, *R. japonica*, *Solidago canadensis*, *S. gigantea* and *Symphyotrichum lanceolatum* (see Chap. 8, Sect. 8.10, this book). Similar habitats in the Jizerské hory Mts, Krkonoše and Orlické hory Mts locally support monodominant stands of the neophyte *Rumex alpinus* (class *Galio-Urticetea*, alliance *Rumicion alpini*).

In natural or semi-natural ecotonal or naturally disturbed habitats on nutrient-rich or wet soils, there is productive and dense vegetation dominated by dicot herbs, with a floristic composition similar to ruderal vegetation in man-made habitats. Along lowland and mid-altitudinal streams and channels, this vegetation is characterized by a frequent occurrence of *Epilobium hirsutum*, *Rubus caesius* and herbaceous lianas such as *Calystegia sepium*, *Cuscuta europaea*, *Echinocystis lobata*, *Fallopia dumetorum*, *Humulus lupulus* and *Silene baccifera* (class *Galio-Urticetea*, alliance *Senecionion fluviatilis*). The invasive alien *Impatiens glandulifera* is common in this vegetation. Herbaceous vegetation dominated by *Petasites hybridus* occurs along mountain streams; on some gravel accumulations in the

Moravskoslezské Beskydy Mts, this species is replaced by *P. kablikanus* (class *Galio-Urticetea*, alliance *Petasition hybridi*). Mesic, nutrient-rich habitats at the edges of forest and in natural canopy openings (e.g. on screes, landslides and at windthrow sites) are dominated by tall dicot herbs such as *Aruncus dioicus*, *Eupatorium cannabinum*, *Impatiens noli-tangere*, *Lunaria rediviva*, *Parietaria officinalis*, *Stachys sylvatica* and *Urtica dioica* (class *Galio-Urticetea*, alliance *Impatienti noli-tangere-Stachyion sylvaticae*). In warmer areas, similar habitats are dominated by annual or short-lived species, e.g. *Alliaria petiolata*, *Anthriscus cerefolium*, *Chaerophyllum temulum*, *Chelidonium majus*, *Galium aparine* and *Torilis japonica* (class *Galio-Urticetea*, alliance *Geo urbani-Alliarion petiolatae*). In forest clearings, areas deforested by wildfires, insect outbreaks, wind storms or air pollution, in canopy gaps and forest fringes, soils are usually poor in bases but with temporarily increased availability of nitrogen. Herbaceous vegetation in such habitats usually contains a mixture of acidophilous and nitrophilous species, but also species of the former forest herb layer (class *Epilobietea angustifolii*, alliance *Fragarion vescae*). Dominant species include tall grasses (e.g. *Calamagrostis arundinacea*, *C. epigejos* and, in the mountains, *C. villosa*), tall dicot herbs (e.g. *Epilobium angustifolium* and *Senecio ovatus*), tall ferns (e.g. *Athyrium filix-femina* and *Pteridium aquilinum*) and medium-tall to dwarf shrubs (e.g. *Rubus idaeus* and *Vaccinium myrtillus*).

## 7.6 Species Richness in Different Vegetation Types

There are non-random differences in the mean number of plant species per unit area (local species richness, alpha diversity) among vegetation types. These differences can be partly due to the environmental conditions in habitats of particular vegetation types, especially stress and disturbance, partly due to interspecific interactions, especially competition, and partly due to species-pool effects (Grime 1979; Zobel 1997). Regional species pools are subsets of the species in regional floras that can occur in specific habitats. The sizes of the regional species pools, which depend on evolutionary and biogeographical history, usually influence the local species richness in different vegetation types. Sádlo et al. (2007) assigned 2347 taxa of Czech vascular flora to regional species pools of 88 habitat types, which correspond to the main vegetation types in this country. This compilation indicates that the largest species pools, containing from 500 to nearly 800 species, exist for mesic to xeric scrub, dry herbaceous fringe vegetation, mesic and dry grasslands, some types of ruderal vegetation, some types of deciduous forests and pine or larch plantations.

The existence of a national database of vegetation plots and national vegetation classification makes it possible to compare local species richness of individual vegetation types in the Czech Republic. Vegetation databases contain sets of plot records that were sampled non-systematically, which can affect the estimates of mean species richness (e.g. Chytrý 2001). However, possible biases tend to be much smaller than the real differences between vegetation types if datasets of many plots from large areas and many vegetation types are analyzed. Therefore, if the species counts recorded in plots of different sizes are recalculated to a single size using the species–area relationship (e.g. Rosenzweig 1995), the patterns in species richness across vegetation types are realistic.

Mean numbers of species of vascular plants occurring in phytosociological alliances of Czech vegetation, calculated from a geographically stratified set of 27,852 vegetation plots, are shown in Fig. 7.33. Because forests (and to some extent also scrub) are sampled using larger plots than those used for sampling open vegetation, species numbers were recalculated to areas of 100 $m^2$ for forests and scrub and 10 $m^2$ for the other vegetation types. Among forests and scrub, the thermophilous oak forests are the most species-rich, whereas coniferous forests on acidic soils and subalpine *Pinus mugo* scrub are the poorest vegetation types. Among open vegetation types, the highest species richness is recorded in grasslands, especially in the semi-dry grasslands of the *Bromion erecti* and *Cirsio-Brachypodion pinnati* alliances, intermittently wet, nutrient-poor meadows of the *Molinion caeruleae* alliance, and subalpine grasslands of the *Calamagrostion arundinaceae* alliance. Some types of arable weed vegetation and calcareous fens also contain high numbers of species in small plots. In contrast, most types of aquatic vegetation and bogs are very poor in species (Table 7.2).

Recent comparative studies showed that semi-dry grasslands in the White Carpathians of south-eastern Moravia hold several world records in fine-scale species richness of vascular plants (Wilson et al. 2012; Chytrý et al. 2015). Most of these grasslands belong to the association *Brachypodio pinnati-Molinietum arundinaceae*, which has a transitional species composition between the alliances *Bromion erecti* and *Cirsio-Brachypodion pinnati*; Fig. 7.34). Typically there are about 50 species of vascular plants in 1 $m^2$ and more than 100 species in 100 $m^2$ in these grasslands (Merunková et al. 2012), but local maxima are much higher (Table 7.3). This extraordinary species richness probably results from a combination of several factors conducive to high species richness, which co-occur in this area:

- Geographical position on the transition between the Pannonian lowlands and the Carpathians means that species of both lowland steppic and mid-altitude mesic flora are likely to co-occur there.
- Continuous existence of these grasslands for millenia, possibly throughout the Holocene, has resulted in the accumulation of many species at individual sites (Hájková et al. 2011b; Hájek et al. 2016).
- Large current area of these grasslands reduces the chance of random species extinctions caused by the habitat island effect (Michalcová et al. 2014). Both the hypothesis of continuous existence of this habitat and the hypothesis of reduced species extinction are supported by isolated occurrences of several relict species in these grasslands (Hájková et al. 2011b; Roleček et al. 2015).
- Heterogeneous land cover (a mosaic of grasslands, forests, scrub patches, solitary trees and small wetlands around springs) exists on relatively homogeneous flysch bedrock, which gives rise to deep, mesic to sligtly dry soils. As a result, most species from the local species pool can grow at almost any site in the

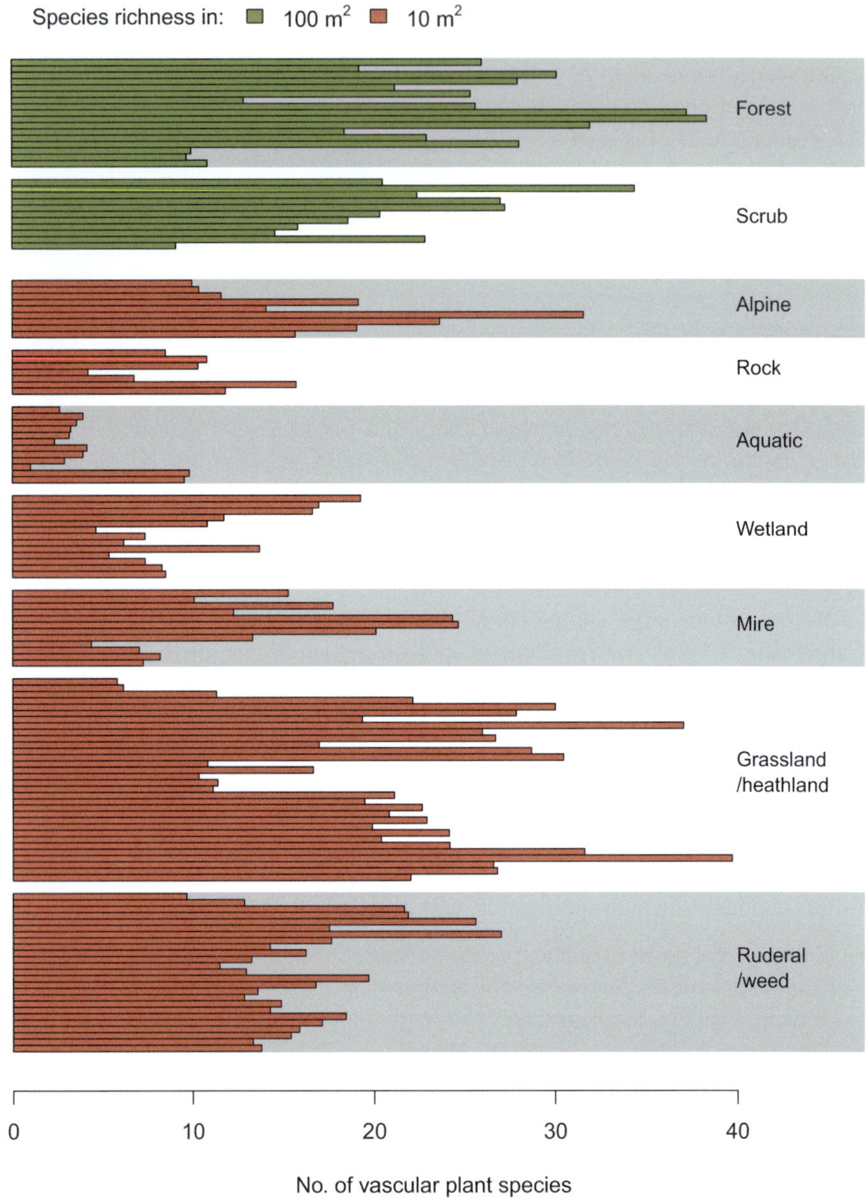


**Fig. 7.33** Mean number of vascular plant species recorded in small areas in different vegetation types of the Czech Republic. Vegetation types are defined as phytosociological alliances and shown in the same order as in Table 7.1. Species numbers were counted in a geographically stratified data set of 27,852 vegetation plots recorded in the Czech Republic that were assigned to associations of the national vegetation classification by the expert system described by Chytrý (2013). Species numbers recorded in plots of various sizes were recalculated for a size of 100 m² (for forests and scrub) and 10 m² (for other formations) using the species–area relationship calculated separately for these two groups of formations using the equation $S = c \cdot A^z$, where $S$ is the number of species, $A$ is the area, and $c$ and $z$ are parameters. See Table 7.2 for information on the most species-rich and most species-poor alliances within individual formations

**Table 7.2** Examples of the phytosociological alliances that are richest or poorest in the number of vascular plant species in individual formations. The mean numbers of vascular plant species in 100 $m^2$ (for forests and scrub) and 10 $m^2$ (for other formations) are given in brackets. See Table 7.1 for more information on the alliances and Fig. 7.33 for explanation of the data analysis

| |
|---|
| **Forests** |
| Richest: LCB *Aceri tatarici-Quercion* (38.3 species/100 $m^2$), LCA *Quercion pubescenti-petraeae* (37.2), LCC *Quercion petraeae* (31.9), LBA *Alnion incanae* (30.1), LBB *Carpinion betuli* (28.0) |
| Poorest: LFC *Piceion abietis* (9.7), LFB *Dicrano-Pinion sylvestris* (10.0), LFD *Vaccinio uliginosi-Pinion sylvestris* (10.9) |
| **Scrub** |
| Richest: KAB *Salicion elaeagno-daphnoidis* (34.4 species/100 $m^2$), KBB *Berberidion vulgaris* (27.3), KBA *Prunion fruticosae* (27.0) |
| Poorest: KCA *Pinion mugo* (9.1) |
| **Alpine and subalpine vegetation** |
| Richest: ADB *Calamagrostion arundinaceae* (31.6 species/10 $m^2$), ADC *Salicion silesiacae* (23.6) |
| Poorest: AAA *Loiseleurio procumbentis-Vaccinion* (10.0), ABA *Juncion trifidi* (10.4) |
| **Rock and scree vegetation** |
| Richest: SCA *Stipion calamagrostis* (15.7 species/10 $m^2$) |
| Poorest: SAD *Androsacion alpinae* (4.2) |
| **Aquatic vegetation** |
| Richest: VDB *Eleocharition acicularis* (9.8 species/10 $m^2$), VDC *Sphagno-Utricularion* (9.6) |
| Poorest: VDA *Littorellion uniflorae* (1) |
| **Wetland vegetation** |
| Richest: MAA *Eleocharition ovatae* (19.3 species/10 $m^2$), MAB *Radiolion linoidis* (17.0), MAC *Verbenion supinae* (16.6) |
| Poorest: MCA *Phragmition australis* (4.6), MCE *Glycerio-Sparganion* (5.4) |
| **Spring and mire vegetation** |
| Richest: RBB *Sphagno warnstorfii-Tomentypnion nitentis* (24.7 species/10 $m^2$), RBA *Caricion davallianae* (24.3) |
| Poorest: RBE *Sphagnion cuspidati* (4.4) |
| **Grasslands and heathlands below the timberline** |
| Richest: THF *Bromion erecti* (39.7 species/10 $m^2$), TDD *Molinion caeruleae* (37.0), THE *Cirsio-Brachypodion pinnati* (31.6), TEC *Violion caninae* (30.5), TDA *Arrhenatherion elatioris* (30.0) |
| Poorest: TAA *Cypero-Spergularion salinae* (5.8), TBA *Salicornion prostratae* (6.2), TEF *Genisto pilosae-Vaccinion* (10.3) |
| **Ruderal and weed vegetation** |
| Richest: XBE *Oxalidion fontanae* (27.0 species/10 $m^2$), XBC *Scleranthion annui* (25.6) |
| Poorest: XAA *Coronopodo-Polygonion arenastri* (9.7) |

**Fig. 7.34** Large areas of mesic to semi-dry meadows with scattered trees and shrubs in the White Carpathian Mts in south-eastern Moravia hold several world records in the number of vascular plant species observed in small plots. For example, 82 species in 1 $m^2$ and 109 species in 16 $m^2$ were recorded in the Porážky meadow shown in this picture

landscape, and forest species and species of spring habitats spread into grasslands. This generates low beta diversity (different sites have similar species composition) but high alpha diversity (many species occur at each site; Michalcová et al. 2014).

- Although soils are homogeneous across the landscape, they are probably ecologically heterogeneous locally, with spatiotemporal heterogeneity in the availability of water.
- Most environmental factors associated with these grasslands (e.g. productivity, moisture, nutrient availability and soil pH) have intermediate values, with no extremes, which is suitable for a large proportion of species of the regional flora (Merunková et al. 2012).
- Grasslands are regularly mown once a year and not fertilized, which creates an intermediate disturbance regime that prevents potentially dominant species from developing a large biomass and outcompeting weaker species (Klimeš et al. 2000, 2013).
- Another factor that may limit potentially dominant species is the occasional summer droughts in areas with deep soils over flysch. They can be severe enough to reduce biomass of potential dominant species but insufficient to exclude competitively weak species (Klimeš 2008).

**Table 7.3** Examples of the maximum numbers of species of vascular plants recorded in plots of small size at different grassland sites in the White Carpathians, at altitudes between 380 and 580 m (data from Chytrý et al. 2015, updated)

| Plot size | No. of species | Site | Remark |
|---|---|---|---|
| 0.1 $m^2$ | 43 | Čertoryje National Nature Reserve | World record |
| 0.25 $m^2$ | 45 | Hutě Nature Reserve | Czech record |
| 1 $m^2$ | 82 | Porážky National Nature Reserve | Czech record |
| 4 $m^2$ | 88 | Čertoryje National Nature Reserve | Czech record |
| 16 $m^2$ | 109 | Porážky National Nature Reserve | Czech record |
| 25 $m^2$ | 116 | Čertoryje National Nature Reserve | World record |
| 49 $m^2$ | 131 | Čertoryje National Nature Reserve | World record |
| 100 $m^2$ | 133 | Čertoryje National Nature Reserve | Czech record |

## 7.7 Current Vegetation Changes

The vegetation classification presented above and summarized in detail by Chytrý (2007, 2009, 2011, 2013) is mainly based on data collected and concepts developed in the twentieth century. However, during the last few decades the vegetation in the Czech Republic was profoundly affected by socioeconomic and environmental changes. A national assessment of the Natura 2000 habitats identified that the most important impacts and threats include eutrophication, biological invasions and natural succession, followed by inappropriate management of forests, overgrazing by game, draining of wet habitats, cessation of grazing and mowing, afforestation of open habitats and drying out (Chobot 2016).

Sádlo and Pokorný (2003) suggested that the recent changes in the landscape and vegetation are of comparable magnitude as the major changes that occurred after the introduction of Neolithic agriculture and after colonization of the mid-altitude areas in the Middle Ages. They estimated that the greatest changes in Czech vegetation, marking the transition from the pre-1950 state to the current state, occurred in the 1970s–1980s, but a comprehensive quantitative analysis of these changes is missing. The recent change is associated with a dramatic decline in small-scale disturbances caused by agricultural and forestry management, which used to occur in numerous rural subdivisions, mainly due to the activity of small land owners and village populations. Such landscape management, typical of the period between 1850 and 1950, changed dramatically after agricultural collectivization in the 1950s. Many areas of grassland that were formerly mown or grazed were abandoned and former coppices and wooded pastures were converted to high-forest or replaced by forest plantations. Current landscapes are characterized by infrequent but strong disturbances of a large spatial extent, which favour fast-spreading ruderal (including alien) species. Extensive disturbed areas are often left to spontaneous succession (Novák and Konvička 2006; Konvalinková and Prach 2010; Trnková et al. 2010; Prach et al. 2013b, 2014b, 2016). Abandoned arable land and grasslands continuously develop into shrubland and woodland (Kopecký and Vojta 2009; Jírová et al. 2012; Vojta and Drhovská 2012; Prach et al. 2014a, b, 2016).

Nowadays there is a much smaller export of nutrients from grasslands and forests than a century ago. Nutrient accumulation in ecosystems is further enhanced by atmospheric nutrient deposition and widespread use of agricultural fertilizers. Exceedance of critical loads for eutrophication due to nitrogen deposition in the Czech Republic is close to the highest in Europe (European Environment Agency 2010). Although the air pollution in terms of both nitrogen oxides and ammonia has been decreasing slightly since the early 1990s (European Environment Agency 2015), open forests and oligotrophic grasslands typical of the first half of the twentieth century, with light-demanding and stress-tolerant species, are changing to species-poor vegetation dominated by nutrient-demanding, highly competitive species (e.g. Prach 2008; Hédl et al. 2010; Dostálek and Frantík 2012; Holub et al. 2012; Verheyen et al. 2012; Kopecký et al. 2013). In addition to increasing the nitrogen content of soil, atmospheric deposition also results in soil acidification (Hédl et al. 2011).

The pattern of plant migrations has also changed with an increasing importance of long-distance dispersal, supported by international trade and extensive transport of various commodities. On average, four new alien plant species arrive in the Czech Republic every year (Pyšek et al. 2012a), mostly of American or Asian origin, and many of them invade natural vegetation outside man-made habitats (Chytrý et al. 2005; see also Chap. 8, this book). The effect of global warming on vegetation and species migrations is currently much less pronounced than that of changes in land-use, even in the alpine treeline ecotone (Treml et al. 2016), but its importance is likely to increase in the future.

Nature conservation is trying to reverse or at least halt these changes by applying measures such as subsidized cutting and grazing, especially in protected areas, which include four national parks, 26 protected landscape areas and 2580 small-scale protected areas (national nature reserves, nature reserves, national nature monuments and nature monuments; http://drusop.nature.cz, April 2016 data; see Chap. 9, Sect. 9.5, this book). However, although some nature conservation measures have been successful, it may be very hard or impossible to halt some of the current changes in vegetation, especially outside the protected areas and in protected areas of small size.

**Acknowledgements** I thank Jiří Danihelka, Petr Pyšek and Marinus Werger for critical reading and comments on a previous version of this chapter, Jan Divíšek, Václav Treml and Tomáš Vrška for the original GIS data from their research, Ondřej Hájek for preparing the maps, Jan Divíšek for an R script for recalculating species richness based on the species–area relationship, Kryštof Chytrý for preparing the photographic plates, Pavel Dřevojan and Zuzana Sixtová for editing the list of references, Tony Dixon for proof-reading the English text and Irena Axmanová for her help with proof correcting. This work was supported by the Czech Science Foundation (project no. 14-36079G, Centre of Excellence Pladias).

## References

Abraham V, Kuneš P, Petr L, Svitavská Svobodová H, Kozáková R, Jamrichová E, Švarcová MG, Pokorný P (2016) A pollen-based quantitative reconstruction of the Holocene vegetation updates a perspective on the natural vegetation in the Czech Republic and Slovakia. Preslia 88:409–434

Adámek M, Bobek P, Hadincová V, Wild J, Kopecký M (2015) Forest fires within a temperate landscape: a decadal and millennial perspective from a sandstone region in Central Europe. Forest Ecology and Management 336:81–90

Altman J, Hédl R, Szabó P, Mazůrek P, Riedl V, Müllerová J, Kopecký M, Doležal J (2013) Tree-rings mirror management legacy: dramatic response of standard oaks to past coppicing in Central Europe. PLoS One 8:e55770

Balátová-Tuláčková E (1968) Grundwasserganglinien und Wiesengesellschaften (Vergleichende Studie der Wiesen aus Südmähren und der Südwestslowakei). Přírodovědné práce ústavů Československé akademie věd v Brně 2(2):1–37

Bastl M, Burian M, Kučera J, Prach K, Rektoris L, Štech M (2008) Central European pine bogs change along an altitudinal gradient. Preslia 80:349–363

Beneš J, Čížek O, Dovala J, Konvička M (2006) Intensive game keeping, coppicing and butterflies: the story of Milovicky Wood, Czech Republic. Forest Ecology and Management 237:353–365

Beranová M, Kubačák A (2010) Dějiny zemědělství v Čechách a na Moravě [The history of agriculture in Bohemia and Moravia]. Libri, Praha

Bičík I, Jančák V (2009) Vývoj spotřeby průmyslových hnojiv [Development of the consumption of industrial fertilizers]. In: Hrnčiarová T, Mackovčin P, Zvara I (eds) Atlas krajiny České republiky [Landscape atlas of the Czech Republic]. Ministerstvo životního prostředí České republiky, Praha and Výzkumný ústav Silva Taroucy pro krajinu a okrasné zahradnictví, Průhonice, p 172

Bičík I, Krupková L (2009) Změna výměry trvalých travních porostů (1845–2000) [Change of permanent grassland area (1845–2000)]. In: Hrnčiarová T, Mackovčin P, Zvara I (eds) Atlas krajiny České republiky [Landscape atlas of the Czech Republic]. Ministerstvo životního prostředí České republiky, Praha and Výzkumný ústav Silva Taroucy pro krajinu a okrasné zahradnictví, Průhonice, p 88

Blažková D (1979) Das *Potentillo albae-Festucetum rubrae* – eine Reliktgesellschaft der vorintensiven Landwirtschaft. Preslia 51:47–69

Bohn U, Neuhäusl R, Gollub G, Hettwer C, Neuhäuslová Z, Raus T, Schlüter H, Weber H (eds) (2000–2003) Karte der natürlichen Vegetation Europas. Maßstab 1: 2 500 000. Teil 1–3/Map of the natural vegetation of Europe. Scale 1: 2 500 000. Part 1–3. Landwirtschaftsverlag, Münster

Bonari G, Fajmon K, Malenovský I, Zelený D, Holuša J, Jongepierová I, Kočárek P, Konvička O, Uřičář J, Chytrý M (2017) Management of semi-natural grasslands benefiting both plant and insect diversity: the importance of heterogeneity and tradition. Agriculture, Ecosystems and Environment 246:243–252

Bossard M, Feranec J, Otahel J (2000) CORINE land cover technical guide – addendum 2000. European Environment Agency, Copenhagen

Boublík K (2010) Formalized classification of the vegetation of *Abies alba*-dominated forests in the Czech Republic. Biologia 65:822–831

Boublík K, Petřík P, Sádlo J, Hédl R, Willner W, Černý T, Kolbek J (2007) Calcicolous beech forests and related vegetation in the Czech Republic: a comparison of formalized classifications. Preslia 79:141–161

Breckle SW (2002) Walter's vegetation of the earth. The ecological systems of the geo-biosphere. Springer, Berlin

Brůna J, Wild J, Svoboda M, Heurich M, Müllerová J (2013) Impacts and underlying factors of landscape-scale, historical disturbance of mountain forest identified using archival documents. Forest Ecology and Management 305:294–306

Bufková I, Prach K, Bastl M (2005) Relationships between vegetation and environment within the montane floodplain of the upper Vltava River (Šumava National Park, Czech Republic). Silva Gabreta 11(Suppl 2):1–78

Businský R (2009) Borovice blatka v novém pojetí [A new concept in bog pine]. Zprávy České botanické společnosti 44:35–43

Čada V, Svoboda M, Janda P (2013) Dendrochronological reconstruction of the disturbance history and past development of the mountain Norway spruce in the Bohemian Forest, Central Europe. Forest Ecology and Management 295:59–68

Čada V, Morrissey RC, Michalová Z, Bače R, Janda P, Svoboda M (2016) Frequent severe natural disturbances and non-equilibrium landscape dynamics shaped the mountain spruce forest in central Europe. Forest Ecology and Management 363:169–178

Černý T, Petřík P, Boublík K, Kolbek J (2007) Vegetation with *Aira praecox* in the Czech Republic compared to its variability in Western Europe. Phytocoenologia 37:115–134

Černý K, Pešková V, Soukup F, Havrdová L, Strnadová V, Zahradník D, Hrabětová M (2016) *Gemmamyces* bud blight of *Picea pungens*: a sudden disease outbreak in Central Europe. Plant Pathology 65:1267–1278

Chobot K (ed) (2016) Druhy a přírodní stanoviště. Hodnotící zprávy o stavu v České republice 2013 [Species and natural habitats. Evaluation reports about the status in the Czech Republic 2013]. Agentura ochrany přírody a krajiny ČR, Praha

Chodová D, Mikulka J, Kočová M, Salava J (2004) Origin, mechanism and molecular basis of weed resistance to herbicides. Plant Protection Science 40:151–168

Chudomelová M, Hédl R, Zouhar V, Szabó P (2017) Open oakwoods facing modern threats: will they survive the next fifty years? Biological Conservation 210:163–173

Chytil J, Hakrová P, Hudec K, Husák Š, Jandová J, Pellantová J (eds) (1999) Mokřady České republiky – přehled vodních a mokřadních lokalit ČR [Wetlands of the Czech Republic – overview of aquatic and wetland localities of the CR]. Český ramsarský výbor, Mikulov

Chytrý M (1997) Thermophilous oak forests in the Czech Republic: syntaxonomical revision of the *Quercetalia pubescenti-petraeae*. Folia Geobotanica et Phytotaxonomica 32:221–258

Chytrý M (2001) Phytosociological data give biased estimates of species richness. Journal of Vegetation Science 12:439–444

Chytrý M (ed) (2007) Vegetace České republiky 1. Travinná a keříčková vegetace [Vegetation of the Czech Republic 1. Grassland and heathland vegetation]. Academia, Praha

Chytrý M (2009) Vegetace České republiky 2. Ruderální, plevelová, skalní a suťová vegetace [Vegetation of the Czech Republic 2. Ruderal, weed, rock and scree vegetation]. Academia, Praha

Chytrý M (ed) (2011) Vegetace České republiky 3. Vodní a mokřadní vegetace [Vegetation of the Czech Republic 3. Aquatic and wetland vegetation]. Academia, Praha

Chytrý M (ed) (2013) Vegetace České republiky 4. Lesní a křovinná vegetace [Vegetation of the Czech Republic 4. Forest and scrub vegetation]. Academia, Praha

Chytrý M, Danihelka J (1993) Long-term changes in the field layer of oak and oak-hornbeam forests under the impact of deer and mouflon. Folia Geobotanica et Phytotaxonomica 28:225–245

Chytrý M, Grulich V (2015) Podyjí National Park. In: Chytrý M, Danihelka J, Michalcová D (eds) Botanical excursions in Moravia, Field guide for the 58th IAVS Symposium. Masaryk University, Brno, pp 37–64

Chytrý M, Horák J (1997) Plant communities of the thermophilous oak forests in Moravia. Preslia 68(1996):193–240

Chytrý M, Rafajová M (2003) Czech National Phytosociological Database: basic statistics of the available vegetation-plot data. Preslia 75:1–15

Chytrý M, Tichý L (1998) Phenological mapping in a topographically complex landscape by combining field survey with an irradiation model. Applied Vegetation Science 1:225–232

Chytrý M, Vicherek J (1995) Lesní vegetace Národního parku Podyjí/Thayatal. Die Waldvegetation des Nationalparks Podyjí/Thayatal. Academia, Praha

Chytrý M, Vicherek J (1996) Přirozená a polopřirozená vegetace údolí řek Oslavy, Jihlavy a Rokytné [Natural and semi-natural vegetation of the Oslava, Jihlava and Rokytná river valleys]. Přírodovědný sborník Západomoravského muzea v Třebíči 22:1–125

Chytrý M, Vicherek J (2003) Travinná, keříčková a křovinná vegetace Národního parku Podyjí/Thayatal [Grassland, heathland, and scrub vegetation of the Podyjí/Thayatal National Park]. Thayensia 5:11–84

Chytrý M, Mucina L, Vicherek J, Pokorny-Strudl M, Strudl M, Koó AJ, Maglocký Š (1997) Die Pflanzengesellschaften der westpannonischen wergstrauchheiden und azidophilen Trockenrasen. Dissertationes Botanicae 277:1–108

Chytrý M, Pyšek P, Tichý L, Knollová I, Danihelka J (2005) Invasions by alien plants in the Czech Republic: a quantitative assessment across habitats. Preslia 77:339–354

Chytrý M, Kučera T, Kočí M, Grulich V, Lustyk P (eds) (2010) Katalog biotopů České republiky [Habitat catalogue of the Czech Republic], Ed, vol 2. Agentura ochrany přírody a krajiny ČR, Praha

Chytrý M, Dražil T, Hájek M, Kalníková V, Preislerová Z, Šibík J, Ujházy K, Axmanová I, Bernátová D, ... Vymazalová M (2015) The most species-rich plant communities in the Czech Republic and Slovakia (with new world records). Preslia 87:217–278

Cílek V (1998) Ekologie sutí Českého středohoří. Eduard Brabec. Kritická recenze a přehled výsledků [The ecology of taluses of the České středohoří Mts. Eduard Brabec. Critical review and the most important results]. In: Cílek V, Kopecký J (eds) Pískovcový fenomén: klima, život a reliéf [Sandstone phenomenon: climate, relief and life]. Zlatý kůň, Praha, pp 49–59

Cílek V, Kopecký J (1998) Pískovcový fenomén: klima, reliéf a život [Sandstone phenomenon: climate, life and relief]. Zlatý kůň, Praha

Čítek J, Krupauer V, Kubů F (1998) Rybnikářství [Fishpond management], 2nd edn. Informatorium, Praha

Čtvrtlíková M, Vrba J, Znachor P, Hekera P (2009) The effects of aluminium toxicity and low pH on the early development of *Isoëtes echinospora*. Preslia 81:135–149

Daněk P (2015) Mohelno Serpentinite Steppe. In: Chytrý M, Danihelka J, Michalcová D (eds) Botanical excursions in Moravia, Field guide for the 58th IAVS Symposium. Masaryk University, Brno, pp 65–74

Danihelka J, Grulich V, Chytrý M (2015a) Pavlov Hills. In: Chytrý M, Danihelka J, Michalcová D (eds) Botanical excursions in Moravia, Field guide for the 58th IAVS Symposium. Masaryk University, Brno, pp 117–138

Danihelka J, Grulich V, Šumberová K (2015b) Dyje-Morava floodplain near Lanžhot. In: Chytrý M, Danihelka J, Michalcová D (eds) Botanical excursions in Moravia, Field guide for the 58th IAVS Symposium. Masaryk University, Brno, pp 163–174

Dítě D, Navrátilová J, Hájek M, Valachovič M, Pukajová D (2006) Habitat variability and classification of *Utricularia* communities: comparison of peat depressions in Slovakia and the Třeboň basin. Preslia 78:331–343

Divíšek J, Chytrý M, Grulich V, Poláková L (2014) Landscape classification of the Czech Republic based on the distribution of natural habitats. Preslia 86:209–231

Dostál D, Jirků M, Konvička M, Čížek L, Šálek M (2012) Návrat zubra evropského (*Bison bonasus*) do České republiky: Potenciální přínosy a perspektivní lokality. Česká krajina, Kutná Hora

Dostál D, Konvička M, Čížek L, Šálek M, Robovský J, Horčičková E, Jirků M (2014) Divoký kůň (*Equus ferus*) a pratur (*Bos primigenius*): klíčové druhy pro formování české krajiny. Česká krajina, Kutná Hora

Dostálek J, Frantík T (2012) The impact of different grazing periods in dry grasslands on the expansive grass *Arrhenatherum elatius* L. and on woody species. Environmental Management 49:855–861

Douda J (2008) Formalized vegetation classification of alder carr and floodplain forests in the Czech Republic. Preslia 80:199–224

Douda J, Čejková A, Douda K, Kochánková J (2009) Development of alder carr after abandonment of wetland grasslands during the last 70 years. Annals of Forest Science 66(712):1–13

Douda J, Boublík K, Slezák M, Biurrun I, Nociar J, Havrdová A, Doudová J, Aćić S, Brisse H, ... Zimmermann NE (2016) Vegetation classification and biogeography of European floodplain forests and alder carrs. Applied Vegetation Science 19:147–163

Dúbravková D, Chytrý M, Willner W, Illyés E, Janišová M, Kállayné Szerényi J (2010) Dry grasslands in the Western Carpathians and the northern Pannonian Basin: a numerical classification. Preslia 82:165–221

Duchoslav M (2002) Flora and vegetation of stony walls in east Bohemia (Czech Republic). Preslia 74:1–25

Dudová L, Hájková P, Buchtová H, Opravilová V (2012) Formation, succession and landscape history of Central-European summit raised bogs: A multiproxy study from the Hrubý Jeseník Mts. The Holocene 23:230–242

Dudová L, Hájková P, Opravilová V, Hájek M (2014) Holocene history and environmental reconstruction of a Hercynian mire and surrounding mountain landscape based on multiple proxies. Quaternary Research 82:107–120

Dvořák M, Tomšovský M, Jankovský L, Novotný D (2007) Contribution to identify the causal agents of Dutch elm disease in the Czech Republic. Plant Protection Science 43:142–145

Dykyjová D, Květ J (eds) (1978) Pond littoral ecosystems. Springer Verlag, Berlin

Eliáš P Jr, Sopotlieva D, Dítě D, Hájková P, Apostolova I, Senko D, Melečková Z, Hájek M (2013) Vegetation diversity of salt-rich grasslands in Southeast Europe. Applied Vegetation Science 16:521–537

Ellenberg H, Leuschner C (2010) Vegetation Mitteleuropas mit den Alpen in ökologischer, dynamischer und historischer Sicht, 6th edn. Verlag Eugen Ulmer, Stuttgart

Eremiášová R, Skokanová H (2014) Response of vegetation on gravel bars to management measures and floods: case study from the Czech Republic. Ekológia 33:274–285

European Environment Agency (2010) The European environment. State and outlook 2010. Air pollution. European Environment Agency, Copenhagen

European Environment Agency (2015) Air pollution fact sheet 2014. Czech Republic. European Environment Agency, Copenhagen

Gaisler J, Pavlů V, Pavlů L, Hejcman M (2013) Long-term effects of different mulching and cutting regimes on plant species composition of *Festuca rubra* grassland. Agriculture, Ecosystems and Environment 178:10–17

Geringhoff H, Daniëls FJA (1998) Vegetationskundliche Untersuchungen zu Zwergstrauch-Heiden in Gebirgslagen der Tschechischen Republik. Tuexenia 18:103–117

Grime JP (1979) Plant strategies and vegetation processes. Wiley, Chichester

Grulich V (1987) Slanomilné rostliny na jižní Moravě [Halophilous plants in southern Moravia]. Český svaz ochránců přírody, Břeclav

Gutzerová N, Herben T (2001) Patch dynamics and local succession in a sandstone area with frequent disturbance. Journal of Vegetation Science 12:533–544

Hadač E (1983) A survey of plant communities of springs and mountain brooks in Czechoslovakia. Folia Geobotanica et Phytotaxonomica 13:126–163

Hadač E, Sofron J (1980) Notes on syntaxonomy of cultural forest communities. Folia Geobotanica et Phytotaxonomica 15:245–258

Hadincová V, Dobrý J, Hanzélyová D, Härtel H, Herben T, Krahulec F, Kyncl J, Moravcová L, Šmilauer P, Šmilauerová M (1997) Invazní druh *Pinus strobus* v Labských pískovcích [Invasion of *Pinus strobus* in the Labské pískovce sandstone area]. Zprávy České botanické společnosti, Materiály 14:63–79

Hais M, Chytrý M, Horsák M (2016) Exposure-related forest-steppe: A diverse landscape type determined by topography and climate. Journal of Arid Environments 135:75–84

Hájek M, Hájková P (2004) Environmental determinants of variation in Czech *Calthion* wet meadows: a synthesis of phytosociological data. Phytocoenologia 34:33–54

Hájek M, Hekera P, Hájková P (2002) Spring fen vegetation and water chemistry in the Western Carpathian flysch zone. Folia Geobotanica 37:205–224

Hájek M, Horsák M, Hájková P, Dítě D (2006) Habitat diversity of central European fens in relation to environmental gradients and an effort to standardise fen terminology in ecological studies. Perspectives in Plant Ecology, Evolution and Systematics 8:97–114

Hájek M, Horsák M, Tichý L, Hájková P, Dítě D, Jamrichová E (2011) Testing a relict distributional pattern of fen plant and terrestrial snail species at the Holocene scale: a null model approach. Journal of Biogeography 38:742–755

Hájek M, Jiroušek M, Navrátilová J, Horodyská E, Peterka T, Plesková Z, Navrátil J, Hájková P, Hájek T (2015) Changes in the moss layer of Czech fens indicate early succession triggered by nutrient enrichment. Preslia 87:279–301

Hájek M, Dudová L, Hájková P, Roleček J, Moutelíková J, Jamrichová J, Horsák M (2016) Contrasting Holocene environmental histories may explain patterns of species richness and rarity in a Central European landscape. Quaternary Science Reviews 133:48–61

Hájková P, Hájek M, Rybníček K, Jiroušek M, Tichý L, Králová Š, Mikulášková E (2011a) Long-term vegetation changes in bogs exposed to high atmospheric deposition, aerial liming and climate fluctuation. Journal of Vegetation Science 22:891–904

Hájková P, Roleček J, Hájek M, Horsák M, Fajmon K, Polák M, Jamrichová E (2011b) Prehistoric origin of the extremely species-rich semi-dry grasslands in the Bílé Karpaty Mts (Czech Republic and Slovakia). Preslia 83:185–204

Hájková P, Horsák M, Hájek M, Lacina A, Buchtová H, Pelánková B (2012) Origin and contrasting succession pathways of the Western Carpathian calcareous fens revealed by plant and mollusc macrofossils. Boreas 41:690–706

Harsch MA, Bader MY (2011) Tree form – a potential key to understanding treeline dynamics. Global Ecology and Biogeography 20:582–596

Härtel H, Cílek V, Herben T, Jackson A, Williams R (eds) (2007) Sandstone landscapes. Academia, Praha

Härtel H, Lončáková J, Hošek M (eds) (2009) Mapování biotopů v České republice. Východiska, výsledky, perspektivy [Habitat mapping in the Czech Republic. Background, results, perspectives]. Agentura ochrany přírody a krajiny ČR, Praha

Havlová M (2006) Syntaxonomical revision of the *Molinion* meadows in the Czech Republic. Preslia 78:87–102

Hédl R (2004) Vegetation of beech forests in the Rychlebské Mountains, Czech Republic, re-inspected after 60 years with assessment of environmental changes. Plant Ecology 170:243–265

Hédl R, Kopecký M, Komárek J (2010) Half a century of succession in a temperate oakwood: from species-rich community to mesic forest. Diversity and Distributions 16:267–276

Hédl R, Petřík P, Boublík K (2011) Long-term patterns in soil acidification due to pollution in forests of the Eastern Sudetes Mountains. Environmental Pollution 159:2586–2593

Hejcman M, Dvořák IJ, Kociánová M, Pavlů V, Nežerková P, Vítek O, Rauch O, Jeník J (2006a) Snow depth and vegetation pattern in a late-melting snowbed analyzed by GPS and GIS in the Giant Mountains, Czech Republic. Arctic, Antarctic and Alpine Research 38:90–98

Hejcman M, Pavlů V, Hejcmanová P, Gaisler J, Hakl J, Rauch O (2006b) Farmer decision making and its effect on subalpine grassland succession in the Giant Mts., Czech Republic. Acta Societatis Botanicorum Poloniae 75:165–174

Hejcman M, Klaudisová M, Štursa J, Pavlů V, Schellberg J, Hejcmanová P, Hakl J, Rauch O, Vacek S (2007) Revisiting a 37 years abandoned fertilizer experiment on *Nardus* grassland in the Czech Republic. Agriculture, Ecosystems and Environment 118:231–236

Hejcman M, Žáková I, Bílek M, Bendová P, Hejcmanová P, Pavlů V, Stránská M (2008) Sward structure and diet selection after sheep introduction on an abandoned grassland in the Giant Mts., Czech Republic. Biologia 63:506–514

Hejcman M, Klaudisová M, Hejcmanová P, Pavlů V, Jones M (2009) Expansion of *Calamagrostis villosa* in sub-alpine *Nardus stricta* grassland: cessation of cutting management or high nitrogen deposition? Agriculture, Ecosystems and Environment 129:91–96

Hejcman M, Strnad L, Hejcmanová P, Pavlů V (2012) Response of plant species composition, biomass production and biomass chemical properties to high N, P and K application rates in *Dactylis glomerata*- and *Festuca arundinacea*-dominated grassland. Grass and Forage Science 67:488–506

Hejcman M, Hejcmanová P, Pavlů V, Beneš J (2013) Origin and history of grasslands in Central Europe – a review. Grass and Forage Science 68:345–363

Hejtmánek J (1954) Vřesovcové bory v Císařském lese [*Erica* pine forests in the Císařský les Mts]. Ochrana přírody 9:70–76

Herben T (1992) Ecotone – basic building unit of sandstone ecosystems of northern Bohemia. Ekológia 11:287–291

Hofmeister J, Mihaljevič M, Hošek J (2004) The spread of ash (*Fraxinus excelsior*) in some European oak forests: an effect of nitrogen deposition or successional change? Forest Ecology and Management 203:35–47

Holub J, Jirásek V (1967) Zur Vereinheitlichung der Terminologie in der Phytogeographie. Folia Geobotanica et Phytotaxonomica 2:91–113

Holub P, Tůma I, Záhora J, Fiala K (2012) Different nutrient use strategies of expansive grasses *Calamagrostis epigejos* and *Arrhenatherum elatius*. Biologia 67:673–680

Holzner W, Immonen R (1982) Europe: an overview. In: Holzner W, Numata M (eds) Biology and ecology of weeds, Geobotany 2. W. Junk, The Hague, pp 203–226

Honsová D, Hejcman M, Klaudisová M, Pavlů V, Kocourková D, Hakl J (2007) Species composition of an alluvial meadow after 40 years of applying nitrogen, phosphorus and potassium fertilizer. Preslia 79:245–258

Horák J (1969) Waldtypengruppen der Pavlovské kopce (Pollauer Berge). Přírodovědné práce ústavů Československé akademie věd v Brně 3:1–40

Horsák M, Chytrý M, Hájková P, Hájek M, Danihelka J, Horsáková V, Ermakov N, German DA, Kočí M, ... Valachovič M (2015) European glacial relict snails and plants: environmental context of their modern refugial occurrence in southern Siberia. Boreas 44:638–657

Hroudová Z, Hrivnák R, Chytrý M (2009) Classification of inland *Bolboschoenus*-dominated vegetation in Central Europe. Phytocoenologia 39:205–215

Husák Š, Vöge M, Weilner C (2000) *Isoëtes echinospora* and *I. lacustris* in the Bohemian Forest lakes in comparsion with other European sites. Silva Gabreta 4:245–252

Illyés E, Chytrý M, Botta-Dukát Z, Jandt U, Škodová I, Janišová M, Willner W, Hájek O (2007) Semi-dry grasslands along a climatic gradient across Central Europe: vegetation classification with validation. Journal of Vegetation Science 18:835–846

Janík D, Vrška T, Šamonil P, Unar P, Adam D, Hort L, Král K (2007) Struktura a ekologie doubrav Národního parku Podyjí na příkladu lokality Lipina [Structure and ecology of oak woods in the Podyjí National Park as exemplified by the Lipina locality]. Thayensia 7:175–206

Janik D, Adam D, Vrška T, Hort L, Unar P, Král K, Šamonil P, Horal D (2011) Field maple and hornbeam populations along a 4-m elevation gradient in an alluvial forest. European Journal of Forest Research 130:197–208

Janík D, Adam D, Hort L, Král K, Šamonil P, Unar P, Vrška T (2014) Tree spatial patterns of *Abies alba* and *Fagus sylvatica* in the Western Carpathians over 30 years. European Journal of Forest Research 133:1015–1028

Janík D, Adam D, Hort L, Král K, Šamonil P, Unar P, Vrška T (2016a) Patterns of *Fraxinus angustifolia* in an alluvial old-growth forest after declines in flooding events. European Journal of Forest Research 135:215–228

Janík D, Král K, Adam D, Hort L, Samonil P, Unar P, Vrska T, McMahon S (2016b) Tree spatial patterns of *Fagus sylvatica* expansion over 37 years. Forest Ecology and Management 375:134–145

Jankovská V (1992) Vegetationsverhältnisse und Naturumwelt des Beckens Jestřebská kotlina am Ende des Spätglazials und im Holozän (Doksy-Gebiet). Folia Geobotanica et Phytotaxonomica 27:137–148

Jankovská V, Pokorný P (2008) Forest vegetation of the last full-glacial period in the Western Carpathians (Slovakia and Czech Republic). Preslia 80:307–324

Jeník J (1958) Geobotanická studie lavinového pole v Modrém dole v Krkonoších [Geobotanical study of an avalanche field in Modrý důl valley in the Krkonoše Mts]. Acta Universitatis Carolinae, Biologica 5(1):49–95

Jeník J (1959) Kurzgefasste Übersicht der Theorie der anemo-orographischen Systeme. Preslia 31:337–357

Jeník J (1961) Alpinská vegetace Krkonoš, Králického Sněžníku a Hrubého Jeseníku [Alpine vegetation of the Krkonoše, Králický Sněžník and Hrubý Jeseník]. Nakladatelství Československé akademie věd, Praha

Jeník J (1980) Struktura slatinné olšiny (*Carici elongatae-Alnetum*) v regresivní fázi [Structure of an alder carr (*Carici elongatae-Alnetum*) in a regression phase]. In: Zborník referátov z 3. zjazdu Slovenskej botanickej spoločnosti při SAV. Slovenská botanická spoločnosť, Zvolen, pp 53–57

Jeník J (1994) Serpentine vegetation in Slavkov Forest, Western Bohemia. Novitatis Botanicae Universitatis Carolinae 8(1993–1994):51–62

Jeník J (1997) Anemo-orographic systems in the Hercynian Mts. and their effects on biodiversity. Acta Universitatis Wratislaviensis 1950, Ser C 4:9–21

Jeník J, Hampel R (1992) Die waldfreien Kammlagen des Altvatergebirges. Geschichte und Ökologie. Mährisch-Schlesischer Sudetengebirgsverein, Kirchheim

Jeník J, Lokvenc T (1962) Die alpine Waldgrenze im Krkonoše Gebirge. Rozpravy Československé akademie věd, Řada matematických a přírodních věd 72(1):1–65

Jeník J, Slavíková J (1964) Střední Vltava a její přehrady z hlediska geobotanického [Middle Vltava River and its dams from the geobotanical viewpoint]. In: Jeník J (ed) Vegetační problémy při budování vodních děl [Vegetation issues related to water reservoir construction]. ČSAV, Praha, pp 67–100

Jeník J, Soukupová L (1992) Microtopography of subalpine mires in the Krkonoše Mountains, the Sudetes. Preslia 64:313–326

Jeník J, Bureš L, Burešová Z (1980) Syntaxonomic study of vegetation in Velká kotlina cirque, the Sudeten mountains. Folia Geobotanica et Phytotaxonomica 15:1–28

Jiménez-Alfaro B, Chytrý M, Rejmánek M, Mucina L (2014) The number of vegetation types in European countries: major determinants and extrapolation to other regions. Journal of Vegetation Science 25:863–872

Jirásek J (1996a) Společenstva přirozených smrčin České republiky [Natural spruce forest communities in the Czech Republic]. Preslia 67(1995):225–259

Jirásek J (1996b) Společenstva kosodřeviny (*Pinus mugo*) v České republice [Communities of dwarf pine (*Pinus mugo*) in the Czech Republic]. Preslia 68:1–12

Jírová A, Klaudisová A, Prach K (2012) Spontaneous restoration of target vegetation in old-fields in a central European landscape: a repeated analysis after three decades. Applied Vegetation Science 15:245–252

Jonášová M, Prach K (2004) Central-European mountain spruce (*Picea abies* (L.) Karst.) forests: regeneration of tree species after a bark beetle outbreak. Ecological Engineering 23:15–27

Jongepierová I (ed) (2008) Louky Bílých Karpat [Grasslands of the White Carpathian Mountains]. ZO ČSOP Bílé Karpaty, Veselí nad Moravou

Jongepierová I, Jongepier JW, Klimeš L (2004) Restoring grassland on arable land: an example of a fast spontaneous succession without weed-dominated stages. Preslia 76:361–369

Kadlec J, Grygar T, Světlík I, Ettler V, Mihaljevič M, Diehl JF, Beske-Diehl S, Svitavská-Svobodová H (2009) Morava River floodplain development during the last millennium, Strážnické Pomoraví, Czech Republic. The Holocene 19:499–509

Kajtoch Ł, Cieślak E, Varga Z, Paul W, Mazur MA, Sramkó G, Kubisz D (2016) Phylogeographic patterns of steppe species in Eastern Central Europe: a review and the implications for conservation. Biodiversity and Conservation 25:2309–2339

Kalníková V, Eremiášová R (2013) Rozšíření třtiny pobřežní (*Calamagrostis pseudophragmites* (Haller f.) Koeler) na řece Ostravici [Distribution of *Calamagrostis pseudophragmites* (Haller f.) Koeler on the Ostravice River]. Acta Carpathica Occidentalis 4:3–14

Kaplan Z, Danihelka J, Lepší M, Lepší P, Ekrt L, Chrtek J Jr, Kocián J, Prančl J, Kobrlová L, Hroneš M, Šulc V (2016) Distributions of vascular plants in the Czech Republic. Part 3. Preslia 88:459–544

Kašpar J, Treml V (2016) Thermal characteristics of alpine treelines in Central Europe north of the Alps. Climate Research 68:1–12

Klika J (1931) O rostlinných společenstvech a jejich sukcesi na obnažených písečných půdách lesních ve středním Polabí [On plant communities and their succession on bare sandy forest soils in the Middle Labe area]. Sborník Československé akademie zemědělské 6:277–302

Klimeš L (2008) Druhové bohatství luk [Species diversity of grasslands]. In: Jongepierová I (ed) Louky Bílých Karpat [Grasslands of the White Carpathian Mountains]. ZO ČSOP Bílé Karpaty, Veselí nad Moravou, pp 89–94

Klimeš L, Jongepierová I, Jongepier JW (2000) The effect of mowing on a previously abandoned meadow: a ten-year experiment. Příroda 17:7–24

Klimeš L, Hájek M, Mudrák O, Dančák M, Preislerová Z, Hájková P, Jongepierová I, Klimešová J (2013) Effects of changes in management on resistance and resilience in three grassland communities. Applied Vegetation Science 16:640–649
Knappová J, Hemrová L, Münzbergová Z (2012) Colonization of central European abandoned fields by dry grassland species depends on the species richness of the source habitats: a new approach for measuring habitat isolation. Landscape Ecology 27:97–108
Knollová I, Chytrý M (2004) Oak-hornbeam forests of the Czech Republic: geographical and ecological approaches to vegetation classification. Preslia 76:291–311
Kočí M (2001) Subalpine tall-forb vegetation (*Mulgedio-Aconitetea*) in the Czech Republic: syntaxonomical revision. Preslia 73:289–331
Kolářová M, Tyšer L, Soukup J (2013) Impact of site conditions and farming practices on the occurrence of rare and endangered weeds on arable land in the Czech Republic. Weed Research 53:489–498
Kolářová M, Tyšer L, Soukup J (2014) Weed vegetation of arable land in the Czech Republic: Environmental a management factors determining weed species composition. Biologia 69:443–448
Kolbek J (1975) Die *Festucetalia valesiacae*-Gesellschaften im Ostteil des Gebirges České středohoří (Böhmisches Mittelgebirge). 1. Die Pflanzengesellschaften. Folia Geobotanica et Phytotaxonomica 10:1–57
Kolbek J (1978) Die *Festucetalia valesiacae*-Gesellschaften im Ostteil des Gebirges České středohoří (Böhmisches Mittelgebirge). 2. Synökologie, Sukzession und syntaxonomische Ergänzungen. Folia Geobotanica et Phytotaxonomica 13:235–303
Kolbek J (1997) Plant communities on walls in the Czech Republic – preliminary notes. Zprávy České botanické společnosti, Materiály 15:61–67
Kolbek J, Petříček V (1972) Vegetační poměry státní přírodní rezervace Sedlo [Vegetation of the State Nature Reserve Sedlo]. Československá ochrana prírody 13:125–166
Kolbek J, Petříček V (1979) Vegetace Malého a Velkého Bezdězu a její vztah k Českému Středohoří [Vegetation of Malý and Velký Bezděz hills and its relationships to the České Středohoří Mts]. Sborník Severočeského muzea, Přírodní vědy 11:5–95
Kolbek J, Blažková D, Husová M, Moravec J, Neuhäuslová Z, Sádlo J (1997) Potential natural vegetation of the Biosphere reserve Křivoklátsko. Academia, Praha
Kolbek J, Blažková D, Břízová E, Kučera T, Ložek V, Rybníček K, Rybníčková E, Rydlo J (1999) Vegetace Chráněné krajinné oblasti a Biosférické rezervace Křivoklátsko 1. Vývoj krajiny a vegetace, vodní, pobřežní a luční společenstva [Vegetation of the Protected Landscape Area and Biosphere Reserve Křivoklátsko 1. Development of the landscape and vegetation, aquatic, river bank and meadow communities]. Agentura ochrany přírody a krajiny ČR and Botanický ústav AV ČR, Praha
Kolbek J, Neuhäuslová Z, Sádlo J, Dostálek J, Havlíček P, Husáková J, Kučera T, Kropáč Z, Lecjaksová S (2001) Vegetace Chráněné krajinné oblasti a Biosférické rezervace Křivoklátsko 2. Společenstva skal, strání, sutí, primitivních půd, vřesovišť, termofilních lemů a synantropní vegetace [Vegetation of the Protected Landscape Area and Biosphere Reserve Křivoklátsko 2. Plant communities of the rocks, screes, primitive soils, heathlands, thermophilous margins and ruderal sites]. Academia, Praha
Kolbek J, Bílek O, Boublík K, Brabec J, Černý T, Härtel H, Husová M, Jelínek J, Kučera T, ... Vítková M (2003) Vegetace Chráněné krajinné oblasti a Biosférické rezervace Křivoklátsko 3. Společenstva lesů, křovin, pramenišť, balvanišť a acidofilních lemů [Vegetation of the Protected Landscape Area and Biosphere Reserve Křivoklátsko 3. Plant communities of the forests, shrubs, forest springs, boulder screes and acidophilous fringes]. Academia, Praha
Konvalinková P, Prach K (2010) Spontaneous succession of vegetation in mined peatlands: a multi-site study. Preslia 82:423–435
Kopecký K (1969) Klassifikationsvorschlag der Vegetationsstandorte an den Ufern der tschechoslowakischen Wasserläufe unter hydrologischen Gesichtspunkten. Archiv für Hydrobiologie 66:326–347
Kopecký M, Vojta J (2009) Land use legacies in post-agricultural forests in the Doupovské Mountains, Czech Republic. Applied Vegetation Science 12:251–260

Kopecký M, Hédl R, Szabó P (2013) Non-random extinctions dominate plant community changes in abandoned coppices. Journal of Applied Ecology 50:79–87
Kornaś J (1988) Speirochore Ackerwildkräuter: von ökologischer Spezialisierung zum Ausssterben. Flora 180:83–91
Korpeľ Š (1995) Die Urwälder der Westkarpaten. Gustav Fischer Verlag, Stuttgart
Kovář P (1981) The grassland communities of the Southeastern Basin of the Labe River 1. Syntaxonomy. Folia Geobotanica et Phytotaxonomica 16:1–43
Kovář P (1983) The grassland communities of the Southeastern Basin of the Labe River 2. Synecology Folia Geobotanica et Phytotaxonomica 18:161–187
Kozáková R, Šamonil P, Kuneš P, Novák J, Kočár P, Kočárová R (2011) Contrasting local and regional Holocene histories of *Abies alba* in the Czech Republic in relation to human impact: evidence from forestry, pollen and anthracological data. The Holocene 21:431–444
Krahulec F (1990a) Alpine vegetation of the Králický Sněžník Mts. (The Sudeten Mts.). Preslia 62:307–322
Krahulec F (1990b) *Nardo-Agrostion* communities in the Krkonoše and West Carpathians Mts. Folia Geobotanica et Phytotaxonomica 25:337–347
Krahulec F, Blažková D, Balátová-Tuláčková E, Štursa J, Pecháčková S, Fabšičová M (1997) Louky Krkonoš: rostlinná společenstva a jejich dynamika [Grasslands of the Krkonoše Mountains: plant communities and their dynamics]. Opera Corcontica 33(1996):1–250
Krahulec F, Skálová H, Herben T, Hadincová V, Wildová R, Pecháčková S (2001) Vegetation changes following sheep grazing in abandoned mountain meadows. Applied Vegetation Science 4:97–102
Král K, McMahon SM, Janík D, Adam D, Vrška T (2014) Patch mosaic of developmental stages in central European natural forests along vegetation gradient. Forest Ecology and Management 330:17–28
Kropáč Z (2006) Segetal vegetation in the Czech Republic: synthesis and syntaxonomical revision. Preslia 78:123–209
Kubát K (1999) Luftströmung in den Blockhalden des Böhmischen Mittelgebirges als ein mikroklimatischer Faktor. Decheniana, Beihefte 37:81–84
Kubešová S, Chytrý M (2005) Diversity of bryophytes on treeless cliffs and talus slopes in a forested central European landscape. Journal of Bryology 27:35–46
Kubíková J (1991) Forest dieback in Czechoslovakia. Vegetatio 93:101–108
Kučera T (2000) Výškové vegetační stupně používané v české botanické literatuře [Altitudinal belts in the Czech botanical literature]. Zprávy České botanické společnosti 35:109–112
Kučera T (2005) Koncept ekologických fenoménů v interpretaci středoevropské vegetace [Ecological phenomena concept: the interpretation of Central European vegetation]. Malacologica Bohemoslovaca 3:47–77
Kučera T, Mannová V (1998) Srovnávací studie křivoklátských pleší [Comparative study of bald habitats in the Křivoklátsko Biosphere Reserve]. Sborník Západočeského muzea v Plzni, Příroda 97:1–48
Kučera T, Špryňar P (1996) Flóra a vegetace Kokořínského dolu [Flora and vegetation of Kokořínský důl valley]. Příroda 7:181–235
Kučera M, Adolt R, Kratěna L, Kohn I, Fejfar J, Závodský J, Piškytlová K, Čech Z (2015) Výstupy Národní inventarizace lesů uskutečněné v letech 2011–2015. 1. Plocha lesa [Outputs of the National Forest Inventory realized in 2011–2015. 1. Forest area]. Lesnická práce 94(12): Appendix
Kučera M, Adolt R, Kohn I, Piškytlová K, Kratěna L, Fejfar J, Závodský J, Čech Z (2016) Výstupy Národní inventarizace lesů uskutečněné v letech 2011–2015. 7. Zastoupení dřevin [Outputs of the National Forest Inventory realized in 2011–2015. 7. Representation of woody plants]. Lesnická práce 95(6): Appendix
Kühn F (1994) Veränderungen der Unkrautflora von Mähren während der Entwicklung der Landwirtschaft. Naturschutz und Landschaftspflege in Brandenburg, Sonderheft 1994(1):8–13
Kuneš P, Jankovská V (2000) Outline of Late Glacial and Holocene vegetation in a landscape with strong geomorphological gradients. Geolines 11:112–114

Kuneš P, Pelánková B, Chytrý M, Jankovská V, Pokorný P, Petr L (2008) Interpretation of the last-glacial vegetation of eastern-central Europe using modern analogues from southern Siberia. Journal of Biogeography 35:2223–2236
Kuneš P, Svobodová-Svitavská H, Kolář J, Hajnalová M, Abraham V, Macek M, Tkáč P, Szabó P (2015) The origin of grasslands in the temperate forest zone of east-central Europe: long-term legacy of climate and human impact. Quaternary Science Reviews 116:15–27
Láníková D, Lososová Z (2009) Rocks and walls: natural *versus* secondary habitats. Folia Geobotanica 44:263–280
Leibundgut H (1993) Europäische Urwälder. Paul Haupt, Bern
Lososová Z, Grulich V (2009) Chorological spectra of arable weed vegetation types in the Czech Republic. Phytocoenologia 39:235–252
Lososová Z, Láníková D (2010) Differences in trait compositions between rocky natural and artificial habitats. Journal of Vegetation Science 21:520–530
Lososová Z, Simonová D (2008) Changes during the 20th century in species composition of synanthropic vegetation in Moravia (Czech Republic). Preslia 80:291–305
Lososová Z, Chytrý M, Cimalová Š, Kropáč Z, Otýpková Z, Pyšek P, Tichý L (2004) Weed vegetation of arable land in Central Europe: gradients of diversity and species composition. Journal of Vegetation Science 15:415–422
Lososová Z, Chytrý M, Cimalová Š, Otýpková Z, Pyšek P, Tichý L (2006) Classification of weed vegetation of arable land in the Czech Republic and Slovakia. Folia Geobotanica 41:259–273
Lososová Z, Chytrý M, Kühn I (2008) Plant attributes determining the regional abundance of weeds on central European arable land. Journal of Biogeography 35:177–187
Ložek V (2001) Molluscan fauna from the loess series of Bohemia and Moravia. Quaternary International 76–77:141–156
Ložek V (2011) Po stopách pravěkých dějů. O silách, které vytvářely naši krajinu [Tracing the prehistoric events. On the forces that formed our landscape]. Dokořán, Praha
Machar I (2009) Coppice-with-standards in floodplain forests – a new subject for nature protection. Journal of Forest Science 55:306–311
Málek J (1983) Problematika ekologie jedle bělokoré a jejího odumírání [Ecology of silver fir and its dieback]. Studie ČSAV 1983(11):1–112
Mašková Z, Doležal J, Květ J, Zemek F (2009) Long-term functioning of a species-rich mountain meadow under different management regimes. Agriculture, Ecosystems and Environment 132:192–202
Merunková K, Preislerová Z, Chytrý M (2012) White Carpathian grasslands: can local ecological factors explain their extraordinary species richness? Preslia 84:311–325
Mezera A (1956–1958) Středoevropské nížinné luhy I–II [Central European lowland floodplain forests I–II]. Státní zemědělské nakladatelství, Praha
Michalcová D, Chytrý M, Pechanec V, Hájek O, Jongepier JW, Danihelka J, Grulich V, Šumberová K, Preislerová Z, ... Zelený D (2014) High plant diversity of grasslands in a landscape context: a comparison of contrasting regions in central Europe. Folia Geobotanica 49:117–135
Miklín J, Čížek L (2014) Erasing a European biodiversity hot-spot: open woodlands, veteran trees and mature forests succumb to forestry intensification, succession, and logging in a UNESCO Biosphere Reserve. Journal of Nature Conservation 22:35–41
Mikulka J, Kneifelová M (eds) (2005) Plevelné rostliny [Weed plants]. Profi Press, Praha
Mikyška R, Deyl M, Holub J, Husová M, Moravec J, Neuhäusl R, Neuhäuslová-Novotná Z (1968–1972) Geobotanická mapa ČSSR 1. České země [Geobotanical map of the CSSR 1. Bohemian lands]. Vegetace ČSSR, Ser A, 2:1–204. Academia, Praha [Explanatory text 1968, map sheets 1968–1972]
Mitchley J, Jongepierová I, Fajmon K (2012) Regional seed mixtures for the re-creation of species-rich meadows in the White Carpathian Mountains: results of a 10-yr experiment. Applied Vegetation Science 15:253–263
Moravec J (1964) Differenzierung der Pflanzengesellschaften des *Carpinion* Issler 1931 emend. Oberd. 1953 durch Migration in Südwestböhmen. Preslia 36:165–177

Moravec J (1967) Zu den azidophilen Trockenrasengesellschaften Südwestböhmens und Bemerkungen zur Syntaxonomie der Klasse Sedo-Scleranthetea. Folia Geobotanica et Phytotaxonomica 2:137–178

Mráz K (1958) Beitrag zur Kenntnis der Stellung des *Potentillo-Quercetum*. Archiv für Forstwesen 7:703–728

Mráz K (1959) Příspěvek k poznání původnosti smrku a jedle ve vnitrozemí Čech [Contribution to knowledge of natural occurrence of spruce and fir in inland Bohemia]. Práce Výzkumných ústavů lesnických ČSR 17:135–180

Mucina L, Bültmann H, Dierßen K, Theurillat JP, Raus T, Čarni A, Šumberová K, Willner W, Dengler J, ... Tichý L (2016) Vegetation of Europe: Hierarchical floristic classification system of vascular plant, bryophyte, lichen, and algal communities. Applied Vegetation Science 19 (Suppl 1):3–264

Mudrák O, Fajmon K, Jongepierová I, Doležal J (2017) Restoring species-rich meadow by means of turf transplantation: long-term colonization of ex-arable land. Applied Vegetation Science 20:62–73

Müller J, Bußler H, Goßner M, Rettelbach T, Duelli P (2008) The European spruce bark beetle *Ips typographus* in a national park: from pest to keystone species. Biodiversity and Conservation 17:2979–3001

Müllerová J, Szabó P, Hédl R (2014) The rise and fall of traditional forest management in southern Moravia: a history of the past 700 years. Forest Ecology and Management 331:104–115

Müllerová J, Hédl R, Szabó P (2015) Coppice abandonment and its implications for species diversity in forest vegetation. Forest Ecology and Management 343:88–100

Navrátilová J, Hájek M, Navrátil J, Hájková P, Frazier RJ (2017) Convergence and impoverishment of fen communities in a eutrophicated agricultural landscape of the Czech Republic. Applied Vegetation Science 20:225–235

Neuhäusl R (1972) Subkontinental Hochmoore und ihre Vegetation. Studie ČSAV 1972(13):1–121

Neuhäusl R, Neuhäuslová-Novotná Z (1967) Syntaxonomische Revision der azidophilen Eichen- und Eichenmischwälder im westlichen Teile der Tschechoslowakei. Folia Geobotanica et Phytotaxonomica 2:1–41

Neuhäusl R, Neuhäuslová-Novotná Z (1972) Bory pískovcových Maštalí u Proseče a jejich kontaktní společenstva [Pine forests of sandstone Maštale near Proseč and their contact communities]. Preslia 44:254–269

Neuhäuslová Z (1985) *Salicetum triandro-viminalis* – společenstvo křovitých vrb na březích českých a moravských toků [*Salicetum triandro-viminalis* – a community of shrubby willows on the banks of Czech and Moravian streams]. Preslia 57:313–333

Neuhäuslová Z (1987) Společenstva vrby bílé a vrby křehké v České socialistické republice [Communities of *Salix alba* and *Salix fragilis* in the Czech Socialist Republic]. Preslia 59:25–50

Neuhäuslová Z (2001) Mapa potenciální přirozené vegetace Národního parku Šumava [The map of potential natural vegetation of the Šumava National Park]. Silva Gabreta Suppl 1: 1–189

Neuhäuslová Z, Moravec J, Chytrý M, Sádlo J, Rybníček K, Kolbek J, Jirásek J (1997) Mapa potenciální přirozené vegetace České republiky 1 : 500 000 [Map of potential natural vegetation of the Czech Republic 1: 500 000]. Botanický ústav AV ČR, Průhonice

Neuhäuslová Z, Blažková D, Grulich V, Husová M, Chytrý M, Jeník J, Jirásek J, Kolbek J, Kropáč Z, ... Sádlo J (1998) Mapa potenciální přirozené vegetace České republiky. Textová část [Map of potential natural vegetation of the Czech Republic. Explanatory text]. Academia, Praha

Neuhäuslová Z, Moravec J, Chytrý M, Ložek V, Rybníček K, Rybníčková E, Husová M, Grulich V, Jeník J, ... Wild J (2001) Potential natural vegetation of the Czech Republic. Braun-Blanquetia 30:1–80

Neuhäuslová-Novotná Z (1964) Zur Charakteristik der *Carpinion*-Gesellschaften in der Tschechoslowakei. Preslia 36:38–54

Neuhäuslová-Novotná Z (1965) Waldgesellschaften der Elbe- und Egerauen. Vegetace ČSSR, Ser A, 1:387–497, 509–517. Academia, Praha

Novák J, Konvička M (2006) Proximity of valuable habitats affects succession patterns in abandoned quarries. Ecological Engineering 26:113–122
Novák P, Roleček J (2013) Dub balkánský (*Quercus frainetto*) – nová lokalita na jihozápadní Moravě a zhodnocení výskytu v České republice [New record of *Quercus frainetto* in SW Moravia and notes on its occurrence in the Czech Republic]. Zprávy České botanické společnosti 48:1–7
Novák J, Petr L, Treml V (2010) Late-Holocene human-induced changes to the extent of alpine areas in the East Sudetes, Central Europe. The Holocene 20:895–905
Novák J, Sádlo J, Svobodová-Svitavská H (2012) Unusual vegetation stability in a lowland pine forest area (Doksy region, Czech Republic). The Holocene 22:947–955
Novák J, Svoboda J, Šída P, Prostředník J, Pokorný P (2015) A charcoal record of Holocene woodland succession from sandstone rock shelters of North Bohemia (Czech Republic). Quaternary International 366:25–36
Novák J, Abraham V, Kočár P, Petr L, Kočárová R, Nováková K, Houfková P, Jankovská V, Vaněček Z (2017) Middle- and upper-Holocene woodland history in central Moravia (Czech Republic) reveals biases of pollen and anthracological analysis. The Holocene 27:349–360
Nožička J (1957) Přehled vývoje našich lesů [Historical overview of our forests]. Státní zemědělské nakladatelství, Praha
Nožička J (1962) Jesenický modřín: původní jeho výskyt a zavádění modřínu v českých zemích [Larch of the Jeseník Mts: its native occurrence and introduction of larch in the Bohemian lands]. Krajské nakladatelství, Ostrava
Nožička J (1972) Původní výskyt smrku v českých zemích [Original occurrence of spruce in the Bohemian lands]. Státní zemědělské nakladatelství, Praha
Olson DM, Dinerstein E, Wikramanayake ED, Burgess ND, Powell GVN, Underwood EC, D'Amico JA, Itoua I, Strand HE, ... Kassem KR (2001) Terrestrial ecoregions of the world: a new map of life on Earth. Bioscience 51:933–938
Opravil E (1983) Údolní niva v době hradištní [River floodplain in the Early Middle Ages]. Studie Archeologického ústavu ČSAV v Brně 11(2):1–78
Osbornová J, Kovářová M, Lepš J, Prach K (eds) (1990) Succession in abandoned fields: studies in Central Bohemia, Czechoslovakia. Kluwer, Dordrecht
Pánek T, Hradecký J (eds) (2016) Landscapes and landforms of the Czech Republic. Springer, Cham
Pavlů L, Pavlů V, Gaisler J, Hejcman M, Mikulka J (2011) Effect of long-term cutting versus abandonment on the vegetation of a mountain hay meadow (*Polygono-Trisetion*) in Central Europe. Flora 206:1020–1029
Pavlů L, Gaisler J, Hejcman M, Pavlů V (2016) What is the effect of long-term mulching and traditional cutting regimes on soil and biomass chemical properties, species richness and herbage production in *Dactylis glomerata* grassland? Agriculture, Ecosystems and Environment 217:13–21
Pavlů V, Hejcman M, Pavlů L, Gaisler J (2007) Restoration of grazing management and its effect on vegetation in an upland grassland. Applied Vegetation Science 10:375–382
Pecháčková S, Krahulec F (1995) Efficient nitrogen economy – key to the success of *Polygonum bistorta* in an abandoned mountain meadow. Folia Geobotanica et Phytotaxonomica 30:211–222
Peterka T, Plesková Z, Jiroušek M, Hájek M (2014) Testing floristic and environmental differentiation of rich fens on the Bohemian Massif. Preslia 86:337–366
Peterka T, Hájek M, Jiroušek M, Jiménez-Alfaro B, Aunina L, Bergamini A, Dítě D, Felbaba-Klushyna L, Graf U, ... Chytrý M (2017) Formalized classification of European fen vegetation at the alliance level. Applied Vegetation Science 20:124–142
Petříček V, Sýkora T (1973) Státní přírodní rezervace Ralsko [State Nature Reserve Ralsko]. Ochrana přírody 28:152–155
Pilous Z (1959) Mechorosty Státní přírodní rezervace Borečský vrch v Českém Středohoří [Bryophytes of the State Nature Reserve Borečský Hill in the České Středohoří Mts]. Ochrana přírody 14:97–99

Podpěra J (1930) Vergleichende Studien über das *Stipetum stenophyllae*. Veröffentlichungen des Geobotanischen Institutes Rübel in Zürich 6:191–210
Pokorný P (2002) Palaeogeography of forest trees in the Czech Republic around 2000 BP: methodical approach and selected results. Preslia 74:235–246
Pokorný P (2005) Role of man in the development of Holocene vegetation in Central Bohemia. Preslia 77:113–128
Pokorný P (2011) Neklidné časy. Kapitoly ze společných dějin přírody a lidí [Unstable times. Chapters from the common history of nature and humans]. Dokořán, Praha
Pokorný P, Klimešová J, Klimeš L (2000) Late Holocene history and vegetation dynamics of a floodplain alder carr: a case study from eastern Bohemia, Czech Republic. Folia Geobotanica 35:43–58
Pokorný P, Chytrý M, Juřičková L, Sádlo J, Novák J, Ložek V (2015) Mid-Holocene bottleneck for central European dry grasslands: Did steppe survive the forest optimum in northern Bohemia, Czech Republic? The Holocene 25:716–726
Ponocná T, Spyt B, Kaczka R, Büntgen U, Treml V (2016) Growth trends and climate responses of Norway spruce along elevational gradients in East-Central Europe. Trees 30:1633–1646
Poschlod P (2015) Geschichte der Kulturlandschaft. Ulmer, Stuttgart
Poschlod P, Baumann A, Karlik P (2009) Origin and development of grasslands in Central Europe. In: Veen P, Jefferson R, de Smidt J, van der Straaten J (eds) Grasslands in Europe of high nature value. KNNV Publishing, Zeist, pp 15–25
Poulíčková A, Hájková P, Kintrová K, Baťková R, Czudková M, Hájek M (2013) Tracing decadal environmental change in ombrotrophic bogs using diatoms from herbarium collections and transfer functions. Environmental Pollution 179:201–209
Prach K (2008) Vegetation changes in a wet meadow complex during the past half-century. Folia Geobotanica 43:119–130
Prach K, Jeník J, Large ARG (eds) (1996) Floodplain ecology and management. The Lužnice River in the Třeboň Biosphere Reserve, Central Europe. SPB Academic Publishing, The Hague
Prach K, Jongepierová I, Řehounková K (2013a) Large-scale restoration of dry grasslands on ex-arable land using a regional seed mixture: establishment of target species. Restoration Ecology 21:33–39
Prach K, Lencová K, Řehounková K, Dvořáková H, Jírová A, Konvalinková P, Mudrák O, Novák J, Trnková R (2013b) Spontaneous vegetation succession at different central European mining sites: a comparison across seres. Environmental Science and Pollution Research 20:7680–7685
Prach K, Jírová A, Doležal J (2014a) Pattern of succession in old-field vegetation at a regional scale. Preslia 86:119–130
Prach K, Řehounková K, Lencová K, Jírová A, Konvalinková P, Mudrák O, Študent V, Vaněček Z, Tichý L, ... Pyšek P (2014b) Vegetation succession in restoration of disturbed sites in Central Europe: the direction of succession and species richness across 19 seres. Applied Vegetation Science 17:193–200
Prach K, Fajmon K, Jongepierová I, Řehounková K (2015) Landscape context in colonization of restored dry grasslands by target species. Applied Vegetation Science 18:181–189
Prach K, Tichý L, Lencová K, Adámek M, Koutecký T, Sádlo J, Bartošová A, Novák J, Kovář P, ... Řehounková K (2016) Does succession run towards potential natural vegetation? An analysis across seres. Journal of Vegetation Science 27:515–523
Proctor J, Woodell SRJ (1975) The ecology of serpentine soils. Advances in Ecological Research 9:255–366
Průša E (1985) Die böhmischen und mährischen Urwälder – ihre Struktur und Ökologie. Vegetace ČSSR, Ser A, 15:1–577. Academia, Praha
Pyšek P, Chocholoušková Z, Pyšek A, Jarošík V, Chytrý M, Tichý L (2004) Trends in species diversity and composition of urban vegetation over three decades. Journal of Vegetation Science 15:781–788
Pyšek P, Jarošík V, Chytrý M, Kropáč Z, Tichý L, Wild J (2005a) Alien plants in temperate weed communities: prehistoric and recent invaders occupy different habitats. Ecology 86:772–785

Pyšek P, Jarošík V, Kropáč Z, Chytrý M, Wild J, Tichý L (2005b) Effects of abiotic factors on species richness and cover in Central European weed communities. Agriculture, Ecosystems and Environment 109:1–8
Pyšek P, Chytrý M, Pergl J, Sádlo J, Wild J (2012a) Plant invasions in the Czech Republic: current state, introduction dynamics, invasive species and invaded habitats. Preslia 84:575–629
Pyšek P, Danihelka J, Sádlo J, Chrtek J Jr, Chytrý M, Jarošík V, Kaplan Z, Krahulec F, Moravcová L, ... Tichý L (2012b) Catalogue of alien plants of the Czech Republic (2nd edition): checklist update, taxonomic diversity and invasion patterns. Preslia 84:155–255
Rivola M (1982) Vegetace středočeských pěnovců [Vegetation of central Bohemian tufa formations]. Preslia 54:329–339
Roleček J (2007) Formalized classification of thermophilous oak forests in the Czech Republic: what brings the Cocktail method? Preslia 79:1–21
Roleček J, Hájek M, Karlík P, Novák J (2015) Reliktní vegetace na mezických stanovištích [Relict vegetation on mesic sites]. Zprávy České botanické společnosti 50:201–245
Rosenzweig ML (1995) Species diversity in space and time. Cambridge University Press, Cambridge
Rozbrojová Z, Hájek M, Hájek O (2010) Vegetation diversity of mesic meadows and pastures in the West Carpathians. Preslia 82:307–332
Rybníček K, Rybníčková E (1974) The origin and development of waterlogged meadows in central part of the Šumava foothills. Folia Geobotanica et Phytotaxonomica 9:45–70
Rybníček K, Rybníčková E (1978) Palynological and historical evidence of virgin coniferous forests at middle altitudes in Czechoslovakia. Vegetatio 36:95–103
Rybníček K, Rybníčková E (2004) Pollen analyses of sediments from the summit of the Praděd range in the Hrubý Jeseník Mts (Eastern Sudetes). Preslia 76:331–347
Rybníčková E (1974) Die Entwicklung der Vegetation und Flora im südlichen Teil der Böhmisch-Mährischen Höhe während des Spätglazials und Holozäns. Vegetace ČSSR, Ser A, 7:1–163. Academia, Praha
Rybníčková E, Rybníček K (1972) Erste Ergebnisse paläogeobotanischer Untersuchungen des Moores bei Vracov, Südmähren. Folia Geobotanica et Phytotaxonomica 7:285–308
Rybníčková E, Rybníček K (2014) Palaeovegetation in the Pavlovské vrchy hills region (South Moravia, Czech Republic) around 25 000 BP: the Bulhary core. Vegetation History and Archaeobotany 23:719–728
Rybníčková E, Hájková P, Rybníček K (2005) The origin and development of spring fen vegetation and ecosystems – palaeogeobotanical results. In: Poulíčková A, Hájek M, Rybníček K (eds) Ecology and palaeoecology of spring fens of the West Carpathians. Palacký University Press, Olomouc, pp 29–62
Rychnovská M, Úlehlová B (1975) Autökologische Studie der tschechoslowakischen *Stipa*-Arten. Vegetace ČSSR, Ser A, 8:1–171. Academia, Praha
Sádlo J (1996) Náčrt vegetace CHKO Kokořínsko [Outline of vegetation of Protected Landscape Area Kokořínsko]. Příroda 7:143–167
Sádlo J (2000) Původ travinné vegetace slatin v Čechách: sukcese kontra cenogeneze [The origin of grassland vegetation of fen peats in the Czech Republic: succession versus coenogenesis]. Preslia 72:495–506
Sádlo J, Bufková I (2002) Vegetace Vltavského luhu na Šumavě a problém reliktních praluk [Vegetation of the Vltava river alluvial plain in the Šumava Mts (Czech Republic) and the problem of relict primary meadows]. Preslia 74:67–83
Sádlo J, Kolbek J (1994) Náčrt nelesní vegetace sutí kolinního až montánního stupně České republiky [An outline of the non-forest vegetation of stony debris in colline to montane belts in the Czech Republic]. Preslia 66:217–236
Sádlo J, Pokorný P (2003) Rostlinné expanze a vývoj krajiny v holocenní perspektivě [Range expansions in plant and landscape development in a Holocene perspective]. Zprávy České botanické společnosti, Materiály 19:5–16

Sádlo J, Pokorný P, Hájek P, Dreslerová D, Cílek V (2005) Krajina a revoluce. Významné přelomy ve vývoji kulturní krajiny Českých zemí [Landscape and revolution. Milestones in the development of cultural landscape of the Bohemian lands]. Malá Skála, Praha

Sádlo J, Chytrý M, Pyšek P (2007) Regional species pools of vascular plants in habitats of the Czech Republic. Preslia 79:303–321

Sádlo J, Petřík P, Boublík K, Rychtařík P, Šímová I (2011) Diverzita rostlinstva Hradčanských stěn (Dokesko) a její příčiny [Habitats, vegetation and flora of the Hradčanské stěny rocks (Doksy region, northern Bohemia): causes of diversity]. Zprávy České botanické společnosti 46:17–38

Šebková B, Šamonil P, Janík D, Adam D, Král K, Vrška T, Hort L, Unar P (2011) Spatial and volume patterns of an unmanaged submontane mixed forest in Central Europe: 160 years of spontaneous dynamics. Forest Ecology and Management 262:873–885

Sedláková I, Chytrý M (1999) Regeneration patterns in a Central European dry heathland: effects of burning, sod-cutting and cutting. Plant Ecology 143:77–87

Sedláková I, Fiala K (2001) Ecological problems of degradation of alluvial meadows due to expanding *Calamagrostis epigejos*. Ekológia 20 (Suppl 3):226–233

Sekyra J, Kociánová M, Štursová H, Kalenská J, Dvořák I, Svoboda M (2002) Frost phenomena in relationship to mountain pine. Opera Corcontica 39:69–114

Semelová V, Hejcman M, Pavlů V, Vacek S, Podrázský V (2008) The Grass Garden in the Giant Mts (Czech Republic): residual effect of long-term fertilization after 62 years. Agriculture, Ecosystems and Environment 123:337–342

Šibík J, Šibíková I, Kliment J (2010) The subalpine *Pinus mugo*-communities of the Carpathians with a European perspective. Phytocoenologia 40:155–188

Sillinger P (1929) Bílé Karpaty. Nástin geobotanických poměrů se zvláštním zřetelem ke společenstvům rostlinným [Bílé Karpaty Mts. A geobotanical outline with special reference to plant communities]. Rozpravy Královské české společnosti nauk, Třída mathematicko-přírodovědecká, Nová řada, 8(3): 1–71

Simonová D (2008) Vegetation of trampled habitats in the Czech Republic: a formalized phytosociological classification. Phytocoenologia 38:177–191

Simonová D, Lososová Z (2008) Which factors determine plant invasions in man-made habitats in the Czech Republic? Perspectives in Plant Ecology, Evolution and Systematics 10:89–100

Skalický V (1988) Regionálně fytogeografické členění [Regional phytogeographical division]. In: Hejný S, Slavík B, Chrtek J, Tomšovic P, Kovanda M (eds) Květena České socialistické republiky 1 [Flora of the Czech Socialist Republic 1]. Academia, Praha, pp 103–121

Skalický V, Jeník J (1974) Květena a vegetační poměry Českého krasu z hlediska ochrany přírody [Flora and vegetation of the Bohemian Karst from the perspective of nature conservation]. Bohemia Centralis 3:101–140

Slavík B (1977) Floristicko-fytogeografická charakteristika Českého ráje z hlediska ochrany přírody [Floristic-phytogeographical characterization of Český ráj from the perspective of nature conservation]. Bohemia Centralis 5:43–123

Slavíková J (1958) Einfluss den Buche (*Fagus silvatica* L.) als Edifikator auf die Entwicklung der Krautschicht in Buchenphytozönosen. Preslia 30:19–42

Slavíková J, Molíková M, Rejmánek M, Rydlo J, Studnička M, Studničková I, Suchara I, Štolcová-Březinová J (1983) Ecological and vegetational differentiation of solitary conic hill. Vegetace ČSSR, Ser A, 13:1–221. Academia, Praha

Šmarda F (1961) Rostlinná společenstva území přesypových písků lesa Doubravy u Hodonína [Plant communities of the mobile sand area in Doubrava Wood near Hodonín]. Práce Brněnské základny Československé akademie věd 33(1):1–56

Šmarda J (1953) Halofytní květena jižní Moravy [Halophytic flora of southern Moravia]. Práce Moravskoslezské akademie věd přírodních 25(3):121–166

Šmarda J (1967) Vegetační poměry Moravského krasu. (Příspěvek k řešení bioindikace krasového reliéfu) [Vegetation of Moravian Karst. (A contribution to bioindication of karst terrain)]. Československá ochrana prírody 3: 139–168, 5: 139–164

Sofron J (1967) Lesní a křovinná společenstva údolí střední Berounky [Forest and scrub communities of the middle Berounka valley]. Sborník Západočeského muzea v Plzni, Příroda 1:20–37
Sofron J (1981) Přirozené smrčiny západních a jihozápadních Čech [Natural spruce forests of western and south-western Bohemia]. Studie ČSAV 1981(7):1–127
Sofron J, Štěpán J (1971) Vegetace šumavských karů [Vegetation of the Šumava cirques]. Rozpravy Československé akademie věd, Řada matematických a přírodních věd 81(1): 1–58
Sojneková M, Chytrý M (2015) From arable land to species-rich semi-natural grasslands: Succession in abandoned fields in a dry region of central Europe. Ecological Engineering 77:373–381
Soukupová L, Kociánová M, Jeník J, Sekyra J (eds) (1995) Arctic-alpine tundra in the Krkonoše, the Sudetes. Opera Corcontica 32:5–88
Štěrba O, Měkotová J, Bednář V, Šarapatka B, Rychnovská M, Kubíček F, Řehořek V (2008) Říční krajina a její ekosystémy [Riverine landscape and its ecosystems]. Univerzita Palackého, Olomouc
Stránská M (2004) Successional dynamics of *Cynosurus* pasture after abandonment in Podkrkonoší. Plant, Soil and Environment 50:364–370
Střeštík S, Šamonil P (2006) Ecological valence of expanding European ash (*Fraxinus excelsior* L.) in the Bohemian Karst (Czech Republic). Journal of Forest Science 52:293–305
Šumberová K (2003) Veränderungen in der Teichwirtschaft und ihr Einfluß auf die Vegetation in der Tschechischen Republik. Mit Beispielen von *Isoëto-Nanojuncetea-*, *Isoëto-Littorelletea-* und *Bidentetea*-Arten im Becken von Třeboň (Wittingauer Becken). Mitteilungen des Badischen Landesvereins für Naturkunde und Naturschutz, Neue Folge 18:7–24
Šumberová K, Horáková V, Lososová Z (2005) Vegetation dynamics on exposed pond bottoms in the Českobudějovická basin (Czech Republic). Phytocoenologia 35:421–448
Šumberová K, Lososová Z, Fabšičová M, Horáková V (2006) Variability of vegetation of exposed pond bottoms in relation to management and environmental factors. Preslia 78:235–252
Suza J (1928) Geobotanický průvodce serpentinovou oblastí u Mohelna na jihozápadní Moravě (ČSR) [Geobotanical guide to the serpentinite area near Mohelno in south-western Moravia (CSR)]. Rozpravy České akademie věd a umění, Třída II, Matematicko-přírodnická, 37(31): 54–57
Svoboda P (1953–1957) Lesní dřeviny a jejich porosty. Část I.–III. [Forest woody plants and their stands. Part I.–III.]. Státní zemědělské nakladatelství, Praha
Svoboda M, Janda P, Nagel TA, Fraver S, Rejzek J, Bače R (2012) Disturbance history of an old-growth sub-alpine *Picea abies* stand in the Bohemian Forest, Czech Republic. Journal of Vegetation Science 23:86–97
Sýkora T (1970) Lesní společenstva jihozápadní části Hradčanské plošiny [Forest communities of the south-western part of the Hradčanská Plateau]. Studie ČSAV 1970(7):9–43
Sýkora T (1979) Botanická inventarizace CHÚ a popis anemo-orografického systému Milešovky v Českém středohoří [Botanical survey of the protected area and description of the anemo-orographic system of Mt. Milešovka in České středohoří Mts]. Stipa 4:34–79
Sýkora T, Hadač E (1984) Příspěvek k fytogeografii Adršpašsko-Teplických skal [Contribution to the phytogeography of the Adršpach-Teplice rock complex]. Preslia 56:359–376
Szabó P (2010) Driving forces of stability and change in woodland structure: a case-study from the Czech lowlands. Forest Ecology and Management 259:650–656
Szabó P, Hédl R (2013) Socio-economic demands, ecological conditions and the power of tradition: past woodland management decisions in a Central Europen landscape. Landscape Research 38:243–261
Szabó P, Müllerová J, Suchánková S, Kotačka M (2015) Intensive woodland management in the Middle Ages: spatial modelling based on archival data. Journal of Historical Geography 48:1–10
Szabó P, Kuneš P, Svobodová-Svitavská H, Švarcová MG, Křížová L, Suchánková S, Müllerová J, Hédl R (2017) Using historical ecology to reassess the conservation status of coniferous forests in Central Europe. Conservation Biology 31:150–160
Tájek P, Bucharová A, Münzbergová Z (2011) Limitation of distribution of two rare ferns in fragmented landscape. Acta Oecologica 37:495–502

Tichý L (1997) Lesní vegetace údolí Dyje v okolí Vranovské přehrady a mapa potenciální přirozené vegetace [Forest vegetation of the Dyje valley in the area of Vranov Reservoir (SW Moravia) and the map of potential natural vegetation]. Zprávy České botanické společnosti, Materiály 15:109–130

Tichý L (1999) Predictive modeling of the potential natural vegetation pattern in the Podyjí National Park, Czech Republic. Folia Geobotanica 34:243–252

Tichý L (2015) Moravian Karst. In: Chytrý M, Danihelka J, Michalcová D (eds) Botanical excursions in Moravia, Field guide for the 58th IAVS Symposium. Masaryk University, Brno, pp 85–108

Tolasz R, Míková T, Valeriánová A, Voženílek V (eds) (2007) Atlas podnebí Česka [Climate atlas of Czechia]. Český hydrometeorologický ústav, Praha and Univerzita Palackého v Olomouci, Olomouc

Toman M (1988) Beiträge zum xerothermen Vegetationskomplex Böhmens. II. Die Salzflora Böhmens und ihre Stellung zur Xerothermvegetation. Feddes Repertorium 99:205–235

Treml V, Chuman (2015) Ecotonal dynamics of the altitudinal forest limit are affected by terrain and vegetation structure variables: an example from the Sudetes Mountains in Central Europe. Arctic, Antarctic, and Alpine Research 47:133–146

Treml V, Migoń P (2015) Controlling factors limiting timberline position and shifts in the Sudetes: a review. Geographia Polonica 88:55–70

Treml V, Jankovská V, Petr L (2008) Holocene dynamics of the alpine timberline in the High Sudetes. Biologia 63:73–80

Treml V, Krizek M, Engel Z (2010) Classification of patterned ground based on morphometry and site characteristics: a case study from the High Sudetes, Central Europe. Permafrost and Periglacial Processes 21:67–77

Treml V, Šenfeldr M, Chuman T, Ponocná T, Demková K (2016) Twentieth century treeline ecotone advance in the Sudetes Mountains (Central Europe) was induced by agricultural land abandonment rather than climate change. Journal of Vegetation Science 27:1209–1221

Trnková R, Řehounková K, Prach K (2010) Spontaneous succession of vegetation on acidic bedrock in quarries in the Czech Republic. Preslia 82:333–343

ÚHÚL (2007) Národní inventarizace lesů v České republice 2001–2004. Úvod, metody, výsledky [National forest inventory in the Czech Republic 2001–2004. Introduction, methods, results]. ÚHÚL, Brandýs nad Labem

Ujházyová M, Ujházy K, Chytrý M, Willner W, Čiliak M, Máliš F, Slezák M (2016) Diversity of beech forest vegetation in the Eastern Alps, Bohemian Massif and the Western Carpathians. Preslia 88:435–457

Unar J (2004) Xerotermní vegetace Pavlovských vrchů [Xerothermic vegetation of the Pavlov Hills]. Sborník Přírodovědeckého klubu v Uherském Hradišti, Suppl 11:1–140

Unar P, Janík D, Souček J, Vrška T, Adam D, Král K, Hort L (2012) The *Pinus rotundata* Link bog forests on mined peat bogs – is the conservation of undisturbed edge an effective tool for its protection? Polish Journal of Ecology 60:707–715

Vacek S, Hejcman M (2012) Natural layering, foliation, fertility and plant species composition of a *Fagus sylvatica* stand above the alpine timberline in the Giant (Krkonoše) Mts., Czech Republic. European Journal of Forest Research 131:799–810

Vacek S, Hejcmanová P, Hejcman M (2012) Vegetative reproduction of *Picea abies* by artificial layering at the ecotone of the alpine timberline in the Giant (Krkonoše) Mountains, Czech Republic. Forest Ecology and Management 263:199–207

Valachovič M, Dierssen K, Dimopoulos P, Hadač E, Loidi J, Mucina L, Rossi G, Valle Tendero F, Tomaselli M (1997) The vegetation on screes – a synopsis of higher syntaxa in Europe. Folia Geobotanica et Phytotaxonomica 32:173–192

Vera FWM (2000) Grazing ecology and forest history. CABI Publishing, Wallingford

Verheyen K, Baeten L, De Frenne P, Bernhardt-Römermann M, Brunet J, Cornelis J, Decocq G, Dierschke H, Eriksson O, ... Verstraeten G (2012) Driving factors behind the eutrophication signal in understorey plant communities of deciduous temperate forests. Journal of Ecology 100:352–365

Vicherek J (1962a) Rostlinná společenstva jihomoravské halofytní vegetace [Plant communities of southern Moravian halophytic vegetation]. Spisy přírodovědecké fakulty University J. E. Purkyně v Brně, Řada L17, 430: 65–96

Vicherek J (1962b) Typy fytocenos aluviální nivy dolního Podyjí se zvláštním zaměřením na společenstva luční [Phytocoenosis types in the lower Dyje floodplain with special reference to meadow communities]. Folia Facultatis Scientiarum Naturalium Universitatis Purkynianae Brunensis, Biologia 3(5):1–113

Vicherek J (1970) Ein Beitrag zur Syntaxonomie der Felsspalten- und Rissenpflanzengesellschaften auf Serpentin in Mitteleuropa. Folia Facultatis Scientiarum Naturalium Universitatis Purkynianae Brunensis, Biologia 11(3):83–89

Vicherek J (1973) Die Pflanzengesellschaften der Halophyten- und Subhalophytenvegetation der Tschechoslowakei. Vegetace ČSSR, Ser A, 5:1–200. Academia, Praha

Vicherek J, Antonín V, Danihelka J, Grulich V, Gruna B, Hradílek Z, Řehořek V, Šumberová K, Vampola P, Vágner A (2000) Flóra a vegetace na soutoku Moravy a Dyje [Flora and vegetation at the confluence of the Morava and Dyje rivers]. Masarykova univerzita, Brno

Vild O, Roleček J, Hédl R, Kopecký M, Utínek D (2013) Experimental restoration of coppice-with-standards: Response of understorey vegetation from the conservation perspective. Forest Ecology and Management 310:234–241

Vild O, Kalwij JM, Hédl R (2015) Effects of simulated historical tree litter raking on the understorey vegetation in a central European forest. Applied Vegetation Science 18:569–578

Vild O, Hédl R, Kopecký M, Szabó P, Suchánková S, Zouhar V (2017) The paradox of long-term ungulate impact: increase of plant species richness in a temperate forest. Applied Vegetation Science 20:282–292

Vítková M, Kolbek J (2010) Vegetation classification and synecology of Bohemian *Robinia pseudacacia* stands in a Central European context. Phytocoenologia 40:205–241

Vítková M, Müllerová J, Sádlo J, Pergl J, Pyšek P (2017) Black locust (*Robinia pseudoacacia*) beloved and despised: A story of an invasive tree in Central Europe. Forest Ecology and Management 384:287–302

Vogel JC, Jeßen S, Gibby M, Jermy AC, Ellis L (1993) Gametophytes of *Trichomanes speciosum* (*Hymenophyllaceae*: *Pteridophyta*) in Central Europe. Fern Gazette 14:227–232

Vojta J, Drhovská L (2012) Are abandoned wooded pastures suitable refugia for forest species? Journal of Vegetation Science 23:880–891

Vrška T, Hort L, Adam D, Odehnalová P, Horal D (2002) Dynamika vývoje pralesovitých rezervací v České republice. Svazek I. Českomoravská vrchovina – Polom, Žákova hora [Developmental dynamics of virgin forest reserves in the Czech Republic. Volume I. The Českomoravská vrchovina Upland – Polom, Žákova hora Mt.]. Academia, Praha

Vrška T, Adam D, Hort L, Odehnalová P, Horal D, Král K (2006) Dynamika vývoje pralesovitých rezervací v České republice. Svazek 2. Lužní lesy – Cahnov-Soutok, Ranšpurk, Jiřina [Developmental dynamics of virgin forest reserves in the Czech Republic. Volume II. Floodplain forests – Cahnov-Soutok, Ranšpurk, Jiřina]. Academia, Praha

Vrška T, Adam D, Hort L, Kolář T, Janík D (2009) European beech (*Fagus sylvatica* L.) and silver fir (*Abies alba* Mill.) rotation in the Carpathians – a developmental cycle or a linear trend induced by man? Forest Ecology and Management 258:347–356

Vrška T, Šamonil P, Unar P, Hort L, Adam D, Král K, Janík D (2012) Dynamika vývoje pralesovitých rezervací v České republice III. Šumava a Český les – Diana, Stožec, Boubínský prales, Milešický prales [Developmental dynamics of virgin forest reserves in the Czech Republic III. Šumava Mts and Český les Mts – Diana, Stožec, Boubín virgin forest, Milešice virgin forest]. Academia, Praha

Willis KJ, van Andel TH (2004) Trees or no trees? The environments of central and eastern Europe during the Last Glaciation. Quaternary Science Reviews 23:2369–2387

Willner W, Di Pietro R, Bergmeier E (2009) Phytogeographical evidence for post-glacial dispersal limitation of European beech forest species. Ecography 32:1011–1018
Willner W, Jiménez-Alfaro B, Agrillo E, Biurrun I, Campos JA, Čarni A, Casella L, Csiky J, Ćušterevska R, ... Chytrý M (2017a) Classification of European beech forests: a Gordian Knot? Applied Vegetation Science 20:494–512
Willner W, Kuzemko A, Dengler J, Chytrý M, Bauer N, Becker T, Biţă-Nicolae C, Botta-Dukát Z, Čarni A, ... Janišová M (2017b) A higher-level classification of the Pannonian and western Pontic steppe grasslands (Central and Eastern Europe). Applied Vegetation Science 20:143–158
Wilson JB, Peet RK, Dengler J, Pärtel M (2012) Plant species richness: the world records. Journal of Vegetation Science 23:796–802
Winter S, Vrška T, Begehold H (2013) Forest naturalness as a key to forest biodiversity preservation. In: Kraus D, Krumm F (eds) Integrative approaches as an opportunity for the conservation of forest biodiversity. European Forest Institute, pp 52–63
Zelený D (2008) Lesní vegetace v údolí Vltavy severně od Zlaté Koruny (okres Český Krumlov) [Forest vegetation in the Vltava river valley north of Zlatá Koruna (district of Český Krumlov)]. Zprávy České botanické společnosti 43:111–169
Zelený D, Chytrý M (2007) Environmental control of vegetation pattern in deep river valleys of the Bohemian Massif. Preslia 79:205–222
Zittová-Kurková J (1984) Bryophyte communities of sandstone rocks in Bohemia. Preslia 56:125–152
Zlatník A (1928) Études écologiques et sociologiques sur le *Sesleria coerulea* et le *Seslerion calcariae* en Tchécoslovaquie. Rozpravy Královské České společnosti nauk, Třída mathematicko-přírovědecká, Nová řada 8(1):1–115
Zobel M (1997) The relative role of species pools in determining plant species richness: an alternative explanation of species coexistence? Trends in Ecology and Evolution 12:266–269

# Chapter 8
# Plant Invasions in the Czech Republic

**Petr Pyšek, Milan Chytrý, Jan Pergl, Jiří Sádlo, and Jan Wild**

**Abstract** At present there are 1454 alien taxa (species, subspecies, varieties, hybrids and cultivars) of vascular plants recorded in the Czech Republic, among them 350 archaeophytes, introduced since the beginning of the Neolithic until the end of the Middle Ages, and 1104 neophytes, introduced in the Modern Period. Of the total number, 985 (67.7%) taxa are classified as casual, 408 (28.1%) as naturalized but non-invasive and 61 (4.2%) as invasive. Aliens make up 33.1% of the total plant diversity recorded in this country, or 14.4% of the permanently present flora. The highest levels of invasion of plant communities are recorded in cities and villages and their surroundings, floodplains of large rivers, disturbed landscapes in the north, and agricultural landscapes and forest plantations in the warm lowlands, especially in southern, central and eastern Bohemia. The habitats and vegetation types harbouring the highest percentages of alien species in the Czech Republic are generally those with a high level of disturbance or with a fluctuating input of nutrients.

## 8.1 Introduction

Thorough studies of plants in human-made habitats have been a traditional topic of Czech botany since the 1970s (e.g. Hejný et al. 1973, 1979; Jehlík and Hejný 1974; Jehlík 1998). The systematic research into the biogeography and ecology of alien

P. Pyšek (✉)
Institute of Botany, The Czech Academy of Sciences, 252 43 Průhonice, Czech Republic

Department of Ecology, Faculty of Science, Charles University,
Viničná 7, 128 44 Praha 2, Czech Republic
e-mail: petr.pysek@ibot.cas.cz

M. Chytrý
Department of Botany and Zoology, Masaryk University,
Kotlářská 2, 611 37 Brno, Czech Republic
e-mail: chytry@sci.muni.cz

J. Pergl • J. Sádlo • J. Wild
Institute of Botany, The Czech Academy of Sciences, 252 43 Průhonice, Czech Republic
e-mail: jan.pergl@ibot.cas.cz; saadlo@volny.cz; jan.wild@ibot.cas.cz



M. Chytrý et al. (eds.), *Flora and Vegetation of the Czech Republic*, Plant and Vegetation 14, DOI 10.1007/978-3-319-63181-3_8

plants in this country was triggered by the publication of the *Catalogue of alien plants of the Czech Republic* (Pyšek et al. 2002b, updated by Pyšek et al. 2012b). In the last decade, a wide array of issues were addressed at the regional scale, such as the role of species traits in determining species invasiveness (e.g. Pyšek et al. 2009, 2011a; Kubešová et al. 2010; Moravcová et al. 2010), patterns of habitat invasibility (Chytrý et al. 2005, 2008a, 2009b; Pyšek et al. 2005; Simonová and Lososová 2008; Lososová and Cimalová 2009; Lososová and Grulich 2009), including invasions of nature reserves (Pyšek et al. 2002b, 2003a), as well as topics related to pathways of introduction (Pergl et al. 2016c), impact and risk assessment (Křivánek and Pyšek 2006; Chytrý et al. 2009b; Hejda et al. 2009a; Pyšek et al. 2011b, 2012c; Pergl et al. 2016d) and case studies of individual invasive species (see Pyšek et al. 2012a for an overview, and Hejda 2013; Horáčková et al. 2014; Čuda et al. 2015, 2016; Reif et al. 2016; Skálová et al. 2017; Vítková et al. 2017 for examples of recent research).

Here we (i) review patterns in the diversity of the alien flora in the Czech Republic and the historical dynamics of introductions at time scales of centuries; (ii) summarize available information on the patterns of invasion across landscapes and habitats in this country; and (iii) provide fact sheets of invasive neophytes in the Czech Republic including information on their impact and distribution. A comprehensive analysis of the structure and composition of the Czech alien flora, including other characteristics not reported here, such as taxonomic patterns, life histories, incidence in plant communities, habitat niche and pathways of introduction, can be found in the two editions of the *Catalogue of alien plants of the Czech Republic* (Pyšek et al. 2002b, 2012b), as well as other summary papers addressing plant invasions in this country (Chytrý et al. 2005, 2009b; Pyšek et al. 2012a).

## 8.2 Patterns in the Diversity of the Alien Flora

Based on a recent update, the alien flora recorded in the Czech Republic consists of 1454 taxa (species, subspecies, varieties, hybrids and cultivars) of vascular plants (see Pyšek et al. 2012b; their Appendix 2 for the list of taxa). This figure is slightly higher than 1434 alien taxa reported by Danihelka (2013), referred to in other chapters of this book, because the latter source did not count varieties and cultivars. The list of taxa (including varieties and cultivars) published decade earlier (Pyšek et al. 2002b) reported 1378 taxa. This increase by 76 taxa was not only due to newly arrived taxa recorded during the last decade but also due to a thorough exploration and taxonomic re-evaluation of the literature, herbaria and other sources (Pyšek et al. 2012b). The 1454 recorded taxa consist of 350 archaeophytes (plants introduced since the beginning of the Neolithic until the end of the Middle Ages; see Holub and Jirásek 1967; Pyšek et al. 2002b, 2004b for definitions) and 1104 neophytes (plants introduced in the Modern Period). Danihelka (2013) reports 345 taxa of archaeophytes and 1089 of neophytes. These two groups differ markedly in the numbers of taxa in particular categories along the introduction–naturalization–invasion continuum (INIC; following the concept of Richardson et al. 2000; Blackburn

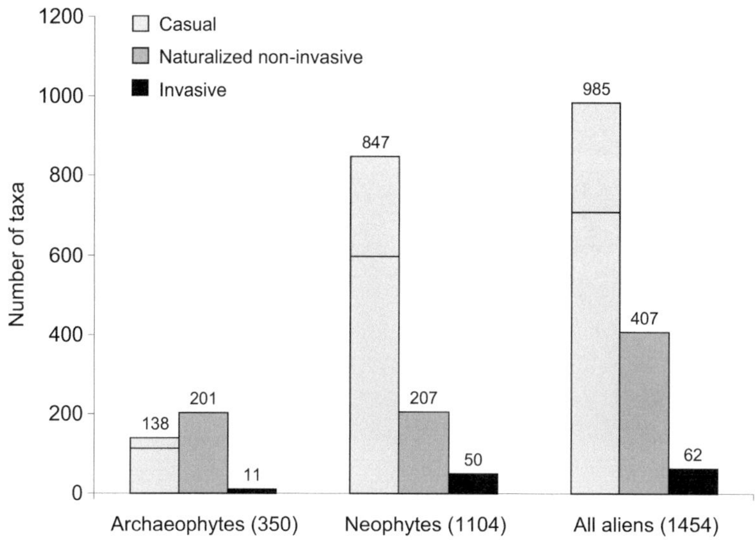


**Fig. 8.1** Numbers of alien taxa in the Czech flora, shown separately for all aliens and the two subgroups distinguished with respect to their residence time, archaeophytes and neophytes. Taxa are classified according to the stage they have reached along the introduction–naturalization–invasion continuum (INIC), which describes how species proceed in the invasion process in terms of overcoming geographical, environmental and biotic barriers (Richardson et al. 2000; Blackburn et al. 2011). Within each group, taxa are divided into casual, naturalized but non-invasive, and invasive; the *upper part* of the bar representing casuals indicates the taxa that have vanished (taken from Pyšek et al. 2012a)

et al. 2011; Fig. 8.1; Table 8.1). While the numbers of naturalized but non-invasive archaeophytes and neophytes are similar (201 vs 207), there are more invasive taxa among the neophytes (50 vs 11). The markedly higher total taxonomic diversity of neophytes is due to the much higher number of casual taxa (847 vs 138; Fig. 8.1). Consequently, expressed in relative terms, the ratio of naturalized and casual species is the opposite in both groups: the majority of archaeophytes, but only a minority of neophytes are naturalized (archaeophytes: 60.6% vs 39.4% of casuals; neophytes: 23.3% vs 76.7%; Table 8.1). The percentage of naturalized taxa among all aliens increases, particularly for neophytes, if only plants assumed to be currently present are taken into account. If taxa that are considered to have vanished (i.e. recorded only once or a few times in the past and not observed for the last 25 years; see Pyšek et al. 2012b) are excluded, the opposite pattern remains the same (Table 8.1).

This difference in species richness of the two historical groups of alien species results partly from the fact that the number of archaeophytes is by definition (introduced up to 1500 A.D.) no longer increasing (Pyšek and Jarošík 2005). Even more relevant for species richness is that the archaeophytes we observe today are winners in the invasion process that has lasted for millennia and we have no information on the frequency of failures in the past (Pyšek et al. 2012b). In fact, current invasions

**Table 8.1** Percentages of the different groups of alien taxa, based on their residence-time categories and invasion status, in the flora of the Czech Republic. Percentages of taxa in particular groups are shown for (a) the complete alien flora with all taxa ever recorded considered, and (b) taxa that may be assumed to occur at present (i.e. excluding the 277 taxa that have vanished from the complete flora). Note that 'invasive' is a subgroup of 'naturalized', therefore the total number of naturalized taxa is the sum of 'naturalized non-invasive', and 'invasive' (taken from Pyšek et al. 2012a)

| | (a) All taxa (%) | | | | (b) Currently present taxa (%) | | | |
|---|---|---|---|---|---|---|---|---|
| | Casual | Naturalized non-invasive | Invasive | Naturalized total | Casual | Naturalized non-invasive | Invasive | Naturalized total |
| Archaeophytes | 39.4 | 57.4 | 3.2 | 60.6 | 34.6 | 62.0 | 3.4 | 65.4 |
| Neophytes | 76.7 | 18.7 | 4.6 | 23.3 | 69.9 | 24.1 | 6.0 | 30.1 |
| All aliens | 67.7 | 28.0 | 4.3 | 32.3 | 60.1 | 34.6 | 5.3 | 39.9 |

by neophytes tell a story of the archaeophytes' past, at least to some extent (Pyšek et al. 2011a). There is no reason to believe that invasions by archaeophytes were in principle different from modern invasions of neophytes in terms of introductions-and-failures, and booms-and-busts (sensu Williamson 1996; Blackburn et al. 2011) or that the dynamics were not similar to that recorded for neophytes. However, the intensity and frequency of these phenomena were probably smaller than today due to different pathways, lower propagule pressure, a smaller extent of human-induced disturbance and absence of invaders from overseas. The assumption of similar dynamics in the past is reflected in some archaeophytes being labelled as post-invasive; this category included species that now have a stable, or even decreasing population and do not invade new, modern habitat types, but their population dynamics and type of occurrence suggests that they might have been dominant in the vegetation in the past (see Pyšek et al. 2002b; Medvecká et al. 2012).

If archaeophytes and neophytes are merged into a single group of aliens, 67.7% of the taxa in the total Czech alien flora (985) are classified as casual and 32.3% (469) as naturalized. Separating the latter group according to the more advanced stages of the INIC indicates that 408 taxa (28.1% of the total alien flora) are currently naturalized but non-invasive, and 61 taxa (4.2%) are invasive (Fig. 8.1, see overview at the end of this chapter).

These figures correspond reasonably well with results from other European countries that have compiled complete checklists of their alien plants (see Lambdon et al. 2008; van Kleunen et al. 2015 for overviews). However, including a complete record of all the casual taxa on national or regional lists is still an exception rather than a rule. The comparability is also limited by the variation in the approaches to national checklists that may differ mainly in how criteria for naturalized or invasive taxa are applied, and whether only neophytes or also archaeophytes are included. Considering only neophytes (archaeophytes cannot be assessed because of missing information on unsuccessful casuals), the naturalization rate of 23.3% (i.e. the percentage of taxa in the total pool of introduced aliens that became naturalized) for the

Czech flora corresponds well with that recorded in other European countries such as 22.7% for Belgium (Verloove 2006), 20.5% for Hungary (Balogh et al. 2004), 25.4% for Austria (Essl and Rabitsch 2002) and 26.8% for Slovakia (Medvecká et al. 2012).

A comprehensive account of the alien flora in Slovakia (Medvecká et al. 2012) provides an excellent opportunity to compare the patterns of plant invasions in two neighbouring countries with shared culture and history, which were part of the same state until 1992, yet differ mainly in topographic heterogeneity and land use and hence in opportunities for invasion. A brief comparison of the two most recent catalogues (Medvecká et al. 2012; Pyšek et al. 2012b) indicates that overall, disproportionally fewer taxa are considered naturalized in the Czech Republic (28.1% of all aliens) than in Slovakia (39.1%), the difference being mainly due to archaeophytes (19.5% vs 39.4%). The proportions of casual, naturalized and invasive neophytes in the Czech Republic and Slovakia are very similar (75.7% vs 73.2%, 18.7% vs 22.8% and 4.5% vs 4.0%, respectively), indicating that in both checklists the transitions along the INIC are reasonably well captured.

## 8.3 Regional Distributions of Alien and Invasive Plants

Maps based on the percentage of alien species in 35 habitat types in the Czech Republic (Fig. 8.2), derived from quantitative data for ~19,000 vegetation plots (Chytrý et al. 2009b), show that both archaeophytes and neophytes are most common in lowland agricultural regions and urban areas, but sparsely represented in mountainous areas. At middle altitudes, agricultural areas are more invaded than forested areas. Outside agricultural and urban areas, high levels of invasion are found especially in lowland sandy areas and river corridors. The regional levels of invasion by neophytes follow a similar pattern whether expressed for all terrestrial habitats (Fig. 8.2a) or only natural and semi-natural habitats (Fig. 8.2b), but are generally lower in the latter, regardless of whether the relative numbers or covers of neophytes in vegetation plots are used (Chytrý et al. 2009b).

In addition to these maps that reflect the distribution of all aliens at the plant community scale, Pyšek et al. (2012a) published a map based on 26,462 floristic records of species that are considered invasive, which shows that the most invaded are the surroundings of big cities, floodplains of large rivers, post-mining disturbed landscapes in the northern part of the country, and warm lowlands in east-central Bohemia and southern Moravia (Fig. 8.2c). The hotspots of the occurrence of all aliens and currently spreading invaders are closely correlated. This confirms that the pattern recorded at the regional scale in the Czech Republic is similar to that previously highlighted at the continental scale in Europe by Chytrý et al. (2012); these authors report a strong positive relationship between the mean level of invasion by all aliens recorded in individual 50 km mapping grid cells and the percentage of the total number of the worst invasive plant species occurring in Europe present in these cells.

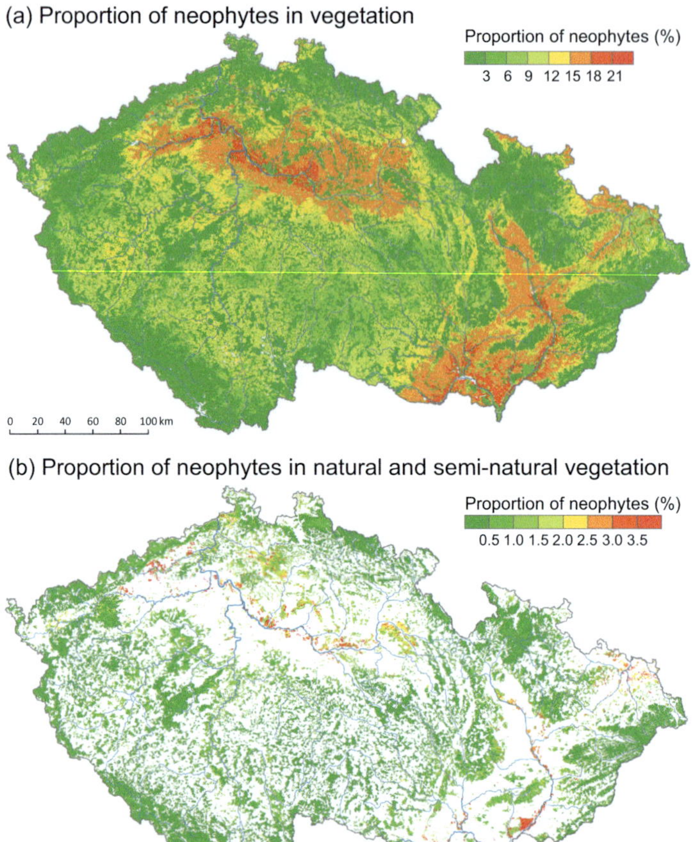


**Fig. 8.2** Regional levels of invasion by neophytes in the Czech Republic. (**a**, **b**) Level of invasion by all neophytes regardless their invasion status in all vegetation types and separately in natural and semi-natural vegetation only. The values are percentage numbers of alien taxa among all taxa interpolated by generalized linear models for particular combinations of land-cover types and altitudes. (**c**) Levels of invasion based on the occurrence of 40 invasive neophytes. Percentages of invasive neophyte taxa present in grid cells of 6 × 10 min were interpolated using the inverse distance weighted interpolator applied to 12 neighbouring points (power parameter = 2). The location of cities and towns is indicated, with symbol size corresponding to their size (data were taken from Chytrý et al. 2009b: maps a, b, and Pyšek et al. 2012a: map c, and redrawn by O. Hájek)

## 8.4 Contribution of Aliens to Plant Diversity in the Czech Republic

The data available for both native (Danihelka et al. 2012) and alien flora (Pyšek et al. 2012b) in the Czech Republic make it possible to precisely estimate the enrichment of the national flora by alien species (Pyšek et al. 2003b). Considering all the taxa ever recorded in the national flora, i.e. including extinct natives and vanished aliens, gives totals of 1454 alien and 2945 native taxa and indicates that the former make up 33.1% (Fig. 8.3). Of this value, 8.0% are attributed to archaeophytes and 25.1% to neophytes. The total percentage in the flora is the figure usually reported in plant invasion literature. For European countries for which complete lists of casual species are available, the percentage of all aliens among the total flora ranges widely from rather low values in southern countries, such as 13% in Italy (Celesti-Grapow et al. 2009) and 12% in Spain (Sanz-Elorza 2004; cited by Celesti-Grapow et al. 2009), to those comparable with the Czech Republic recorded in Central European countries. These include e.g. Austria (27.0%, Essl and Rabitsch 2002), Hungary (26.6%, Balogh et al. 2004; number of native taxa from Winter et al. 2009) and Poland (27.3%, Tokarska-Guzik 2005; number of native taxa from Winter et al. 2009). The highest percentages of alien taxa are reported from the more northerly and westerly located regions in Europe, e.g. 35.3% in Estonia (Ööpik et al. 2008), 41.0% in Belgium (Verloove 2006; number of native taxa from Winter et al. 2009) and 53.4% in the UK (Lambdon et al. 2008), where higher percentages of aliens are partly attributed to the relatively species-poor native floras.

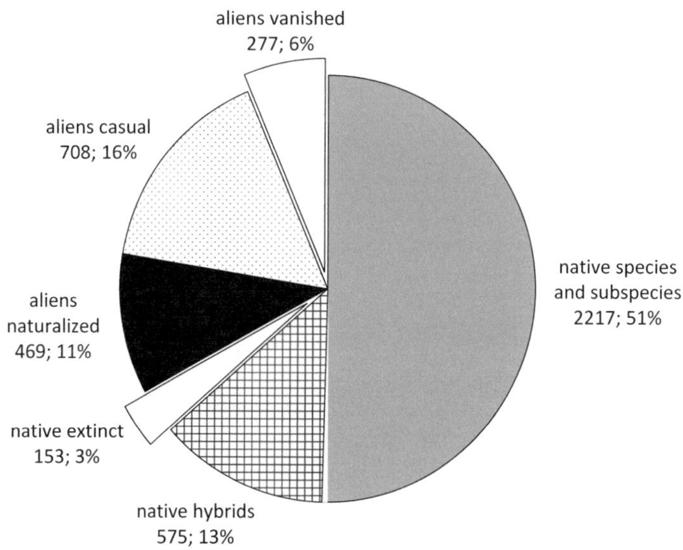


**Fig. 8.3** Composition of the flora of the Czech Republic in terms of its native and alien status, categories of aliens, and historical and current presence of taxa in this country (data on the numbers in the native flora are from Kaplan 2012 and Danihelka et al. 2012, those on the alien flora from Pyšek et al. 2012b)

Similar estimates in the literature, however, depend on which taxa are included in the assessment of total plant diversity (e.g. subspecies, hybrids), how thorough and long-term is the recording of casual aliens, or whether or not archaeophytes are included, or distinguished at all (see Pyšek et al. 2002b; Williamson 2002; Lambdon et al. 2008 for discussion). This can be demonstrated using the Czech flora; excluding hybrids or taxa no longer present (extinct natives and vanished aliens) shifts the percentage of aliens to 36.4% or 29.7%, respectively. This is because hybridization among native species is more frequent (19.5%) than hybridization involving alien species (6.6%), hence the percentage increases. The opposite is true for extinct and vanished taxa, as extinctions account for 5.2% of the total native flora, but vanished taxa for 19.1% of the alien flora.

As the variation in the reported figures results not only from the composition of floras but also from other factors that introduce biases, invasion histories and regions can only be rigorously compared keeping these potentially biasing factors in mind. Again, neighbouring Slovakia is a country for which such a comparison is possible based on a recent checklist of alien species compiled using similar criteria. In this country alien taxa make up 21.5% of the total number ever recorded in its territory, of which 6.6% are attributed to archaeophytes and 14.9% to neophytes (Medvecká et al. 2012). The overall contribution of alien taxa is therefore markedly lower than the above estimate for the Czech Republic. Another possible comparison is of the taxa that are permanently present in the two countries, i.e. excluding extinct natives and including only naturalized alien taxa (Fig. 8.3). This measure yields a figure of 14.4% alien contribution to the permanent plant diversity in the Czech Republic, 6.5% and 7.9% attributed to archaeophytes and neophytes, respectively. For Slovakia, 373 alien taxa permanently present (206 naturalized archaeophytes and 167 naturalized neophytes; Medvecká et al. 2012) make up 10.3% of the contribution of alien taxa to the total plant diversity, a figure again by ~50% smaller than in the Czech Republic (Pyšek et al. 2012b).

Since both countries have similar sizes, climates, demographies and current macroeconomic parameters, i.e. factors that are known to determine alien species richness at a large scale (e.g. Vilà and Pujadas 2001; Pyšek et al. 2010), and a comparable intensity of floristic research, the explanation of these differences needs to be sought mainly in topography, land cover and land-use history. First, large parts of Slovakia consist of mountainous areas of the Western Carpathians, which are less invaded because of the general tendency of mountainous areas to be less invaded than areas at lower altitudes (Becker et al. 2005; Alexander et al. 2011; McDougall et al. 2011). Second, the Bohemian lands were industrialized already in the second half of the nineteenth century, while Slovakia was industrialized only after World War II. This earlier start of industry in the area that is now the Czech Republic undoubtedly intensified transportation and introduction of alien species.

## 8.5 Dynamics of Invasions Over Time

Pyšek et al. (2012b) assigned the year of the first record for 771 neophytes (i.e. 70% of the total number recorded) and showed that there has been a steady trend of approximately four new alien arrivals per year since the beginning of the nineteenth century without any distinct acceleration or deceleration. If the dynamics based on taxa with known year of the first record are projected to the total neophyte flora, the total number of estimated neophytes would reach 1264 by 2050 should the current trend persist (see Pyšek et al. 2012b for details). These trends suggest that the number of alien species recorded in the Czech Republic will be increasing at a similar rate in the near future, which corresponds to the trend reported for Europe (Lambdon et al. 2008; Hulme et al. 2009).

Displaying the dynamics of arrivals of alien neophytes separately by invasion status reveals that the rate of increase in the cumulative number of naturalized taxa somewhat decelerated in the second half of the twentieth century, while casuals exhibit an opposite trend, with a steep increase since the 1950s (Fig. 8.4). This acceleration is due to landscapes becoming more suitable for invasions after the dramatic changes in land-use that occurred after World War II (Müllerová et al. 2005; Williamson et al. 2005) but also, to some extent, reflects increased interest in alien plant research since the 1970s (Pyšek et al. 2002b, 2011b). However, the rapidly increasing numbers of neophytes are a warning because naturalized or even invasive species recruit from casuals with a delay of decades, due to lag phenomena and invasion debt (Kowarik 1995; Aikio et al. 2010; Essl et al. 2011).

The dynamics of arrivals also convincingly demonstrate the role of residence time in invasions; this is evident if naturalized species are categorized into naturalized but not (yet) invasive and invasive. Not only is the current distribution of neophytes positively related to minimum residence time, with those present for a longer time being more widely distributed or more abundant (Pyšek and Jarošík 2005;

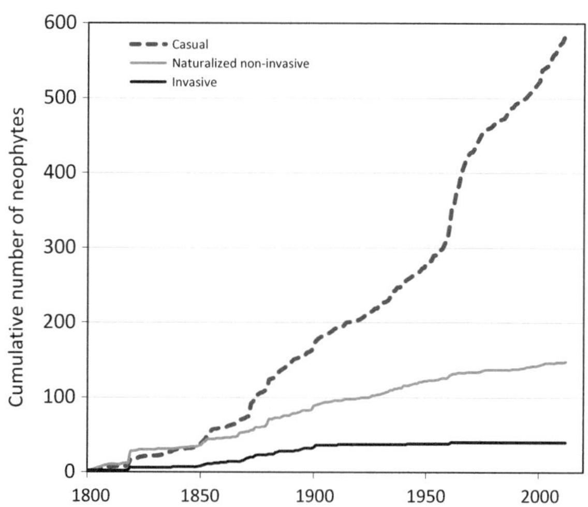


**Fig. 8.4** Cumulative numbers of neophytes reported up to the given year shown separately for casual, naturalized non-invasive, and invasive species. Based on taxa for which the year of the first record is available (n = 771) (data from Pyšek et al. 2012a)

Rejmánek et al. 2005), but also residence time affects the invasion status of alien species. Of the neophytes currently classified as invasive in the Czech Republic (Pyšek et al. 2012b), 50% were introduced prior to 1872, i.e. earlier than naturalized but non-invasive (1886) and much earlier than casual neophytes (1956), or 90% of invasive, naturalized but non-invasive, and casual taxa were introduced before 1901, 1968 and 2001, respectively (Pyšek et al. 2012a).

The rates of species introductions differ not only based on invasion status but also on origin. An analysis of the source of alien taxa in the Czech flora by Pyšek et al. (2012b) revealed that the Mediterranean area (covering parts of Southern Europe, Northern Africa and Western Asia from Turkey and Israel to Afghanistan) is the main source not only of archaeophytes, 52.7% of which arrived from there, but also of neophytes (28.7%; Fig. 8.5). For neophytes, the other most important sources are extra-Mediterranean parts of Europe, contributing 19.9% of taxa, North America (16.7%) and extra-Mediterranean Asia (14.2%; Fig. 8.5). The dynamics of introduction of neophytes from these main donor regions indicate that the arrivals from the Mediterranean area and extra-Mediterranean Europe proceeded at the same speed until approximately the 1870s, after which introductions from the Mediterranean region became more frequent. In general, species native to more distant areas in Asia and North America arrived later (Fig. 8.6). In relative terms, however, the introductions of the extra-Mediterranean European species were the fastest: 50% of all currently known taxa from this region were in this country by 1895, compared to 1926 for the Mediterranean species, 1935 for North American species and 1958 for Asian species (Pyšek et al. 2012a).

The estimates of the long-term introduction dynamics of archaeophytes reveal that 35.2% of presently known taxa were introduced in the Neolithic/Chalcolithic period (5500–2200 years BC) and more than half (52.7%) are thought to have been

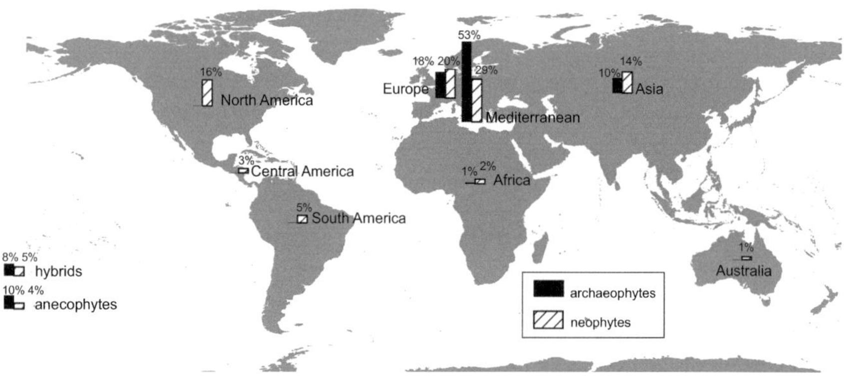


**Fig. 8.5** Regions from which alien plants in the Czech Republic originated. The height of the bars reflects the percentage of the total number of taxa that are alien in the Czech Republic and native to particular regions (shown on top of the bar). The Mediterranean region includes parts of Southern Europe, Northern Africa and Western Asia from Turkey and Israel to Afghanistan. Europe, Asia and Africa refer to their parts other than those in the Mediterranean region in this delimitation (from Pyšek et al. 2012a)

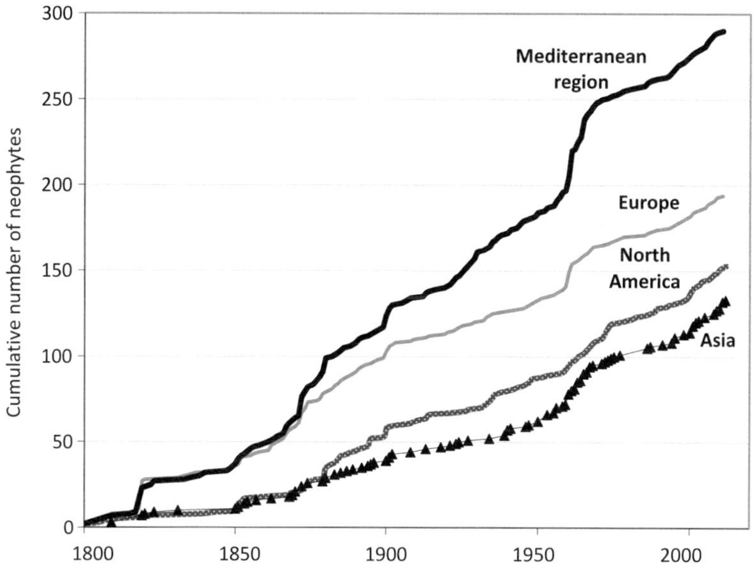


**Fig. 8.6** Arrivals of alien neophytes in the Czech Republic from four main donor regions, shown as increase in the cumulative number of taxa originating from a given region. If a taxon originated from more than one region, only a fraction was attributed to each, e.g. ½ for a taxon originating from two regions (taken from Pyšek et al. 2012a)

present by the end of the Bronze Age, ~750 years BC (Pyšek et al. 2003b). The temporal dynamics of archaeophytes in the Czech flora also reflects a clear effect of residence time, detectable after centuries to millenia since the start of their invasions. Taxa that are most widespread at present were introduced earlier than those currently less widely distributed (Pyšek and Jarošík 2005).

## 8.6 Invasive Taxa in the Czech Republic

In the checklist of alien flora in the Czech Republic 61 taxa are labelled as invasive (Pyšek et al. 2012b). This work followed the definition of an 'invasive species' as one that forms self-replacing populations over many life cycles, produces reproductive offspring, often in very large numbers at considerable distances from the parent population or site of introduction, and has the potential to spread over long distances (Richardson et al. 2000; Pyšek et al. 2004b; Blackburn et al. 2011). In addition to this definition, Pyšek et al. (2012b) introduced the metapopulation criterion to separate invasive species from naturalized, in order to account for the historical population dynamics of the respective taxa, and classified the invasion status based on the population history viewed from the current perspective, i.e. the state in which the populations of a given species exist at present. Therefore, some taxa previously considered invasive are now classified as naturalized, reflecting the 'boom-and-bust phenomenon' (sensu Williamson 1996; Strayer et al. 2017). Another principle adopted was that of the highest stage achieved at the population level, reflecting that individual populations of an alien species may occur in a region at different stages of the INIC (e.g. Essl et al. 2009; Richardson and Pyšek 2012). Therefore, if some populations of a species reach the invasion stage, the species is classified as invasive (see Pyšek et al. 2012b for details on the approach).

Among the taxa currently considered as invasive there are 11 archaeophytes (*Angelica archangelica* subsp. *archangelica*, *Arrhenatherum elatius*, *Atriplex sagittata*, *Cirsium arvense*, *Conium maculatum*, *Digitaria ischaemum*, *Echinochloa crus-galli*, *Eragrostis minor*, *Portulaca oleracea* subsp. *oleracea*, *Prunus cerasifera* and *Stellaria pallida*) and 50 neophytes. Invasive neophytes occur in a wide range of habitats (Figs. 8.7 and 8.8) and are addressed in detail in Sect. 8.6, where the information on their ecology, invasion history, current distribution, trends and impact is summarized.

## 8.7 Patterns of Plant Invasions Across Landscapes and Habitats in the Czech Republic

Important insights into the patterns of invasion of different habitats and plant communities are gained by combining the knowledge on alien flora with approaches of vegetation ecology (Pyšek and Chytrý 2014). Plant invasion patterns across landscapes, habitats or vegetation types can be quantified in terms of the level of invasion, measured as the number of alien species, or the percentage of aliens to all species, per unit area. The level of invasion is influenced by habitat invasibility (i.e. its vulnerability to invasion; Rejmánek 1989) and propagule pressure of alien species (Lonsdale 1999; Chytrý et al. 2008a). The latter can be understood either as the number of propagules arriving at a given location (propagule pressure sensu stricto) or the number of species (colonization pressure; Lockwood et al. 2009). Technically, habitat invasibility can be quantified by separating the effect of propagule pressure from the level of invasion measured at particular sites (Lonsdale 1999; Chytrý et al. 2008a). However, it is difficult to obtain an exact measurement of propagule pressure. Therefore studies of invasibility are limited to the use of proxy variables such as the type of land use in the surroundings of the sites sampled. In the Czech Republic such a study indicated a rather strong relationship between the level of invasion and habitat invasibility: generally, invasible habitats tend to be invaded to a high level and vice versa, although there are exceptions (Chytrý et al. 2008a). Unfortunately, for habitats restricted to areas with limited propagule pressure, it cannot be decided by observational studies whether they are invasible or resistant to invasions.

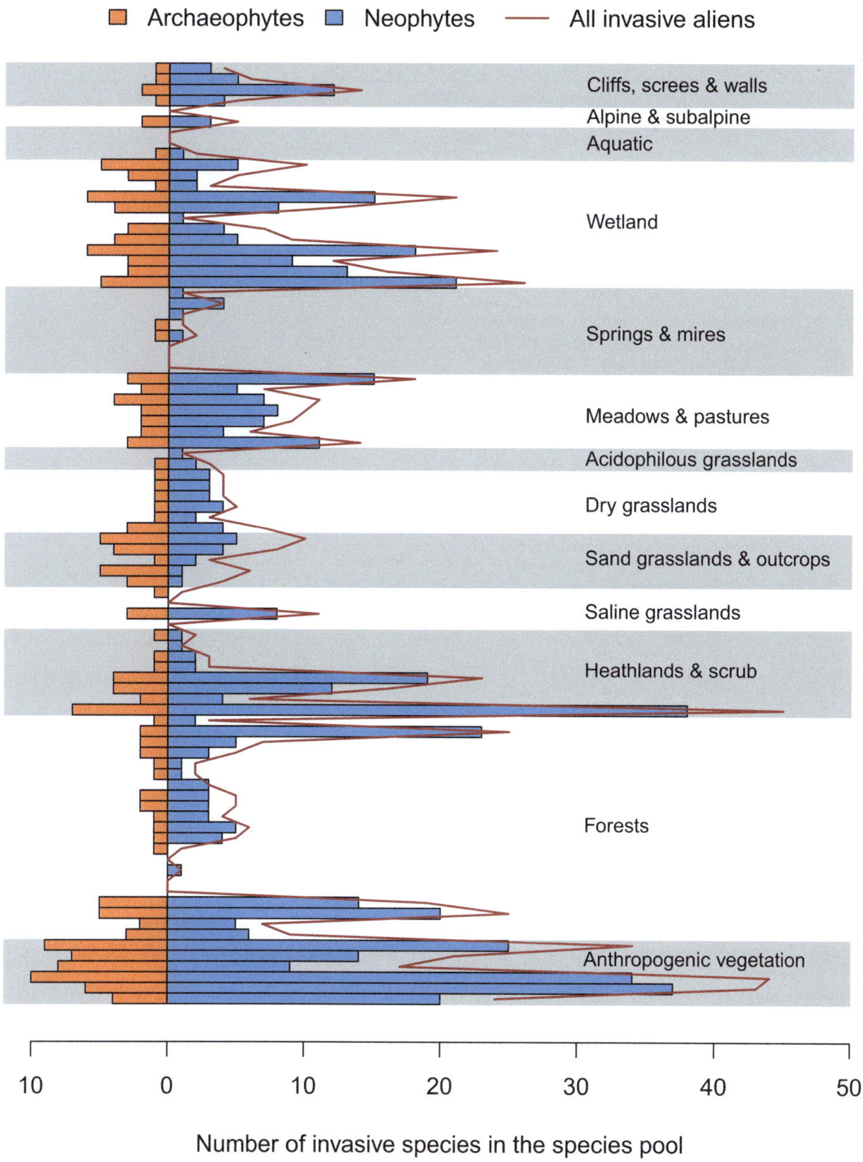


**Fig. 8.7** Numbers of invasive archaeophytes (n = 11) and invasive neophytes (n = 50) in different habitats (n = 88) in the Czech Republic, based on taxa listed in this chapter (data from Sádlo et al. 2007 and Pyšek et al. 2012a)

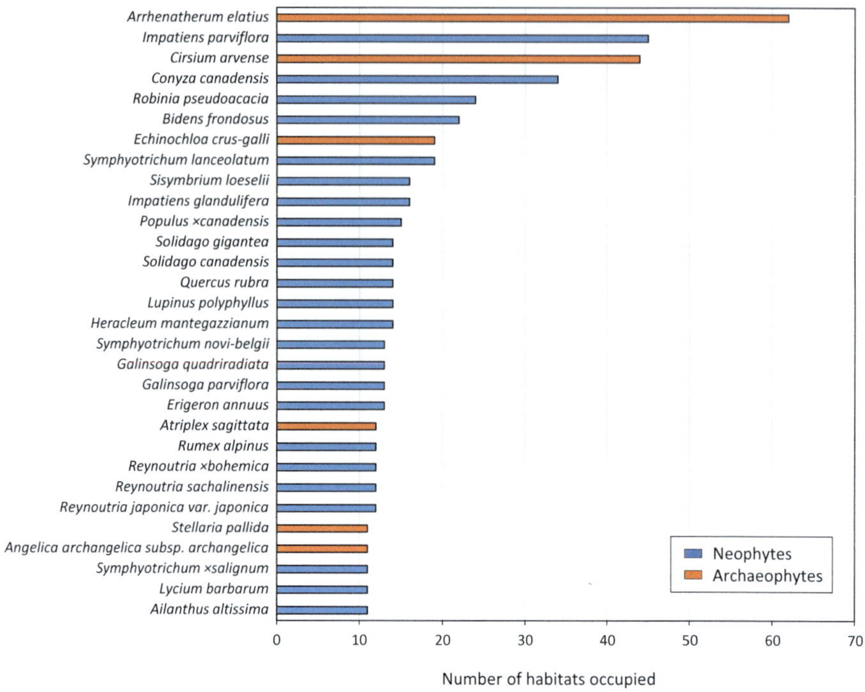


**Fig. 8.8** Numbers of habitats invaded by archaeophytes and neophytes that are considered invasive in the Czech Republic. The classification of habitats follows that used in Sádlo et al. (2007; n = 88). Only taxa occurring in more than 10 habitats are shown (based on data in Pyšek et al. 2012a)

## *8.7.1 Level of Invasion and Altitude*

A decrease in the level of plant invasions at high altitudes is a globally consistent pattern. Generally, alien floras in mountain areas are subsets of alien floras in the surrounding lowlands, the former consisting of species with the broadest climatic tolerance, which continuously spread from the foothill regions (Alexander et al. 2011; McDougall et al. 2011). For example, Pyšek et al. (2011c) demonstrated that neophytes in the Czech Republic were most often introduced into areas located at 250–400 m a.s.l., from where they subsequently spread to higher altitudes. This is supported by the fact that species with earlier first records (thus, presumably introduced earlier) tend to have a broader altitudinal range. The altitudinal pattern of invasion can be explained by the low-altitude filter effect, limited landscape connectivity in the mountains and gene flow from lowland populations to peripheral populations at high altitudes, which would hinder the development of adaptations to high-altitude environments (Becker et al. 2005). In addition, the available total land area decreases with altitude, which may further contribute to lower levels of invasion, e.g. by reducing the diversity and extent of suitable habitats and thereby reducing the frequency of invasion foci for further spread.

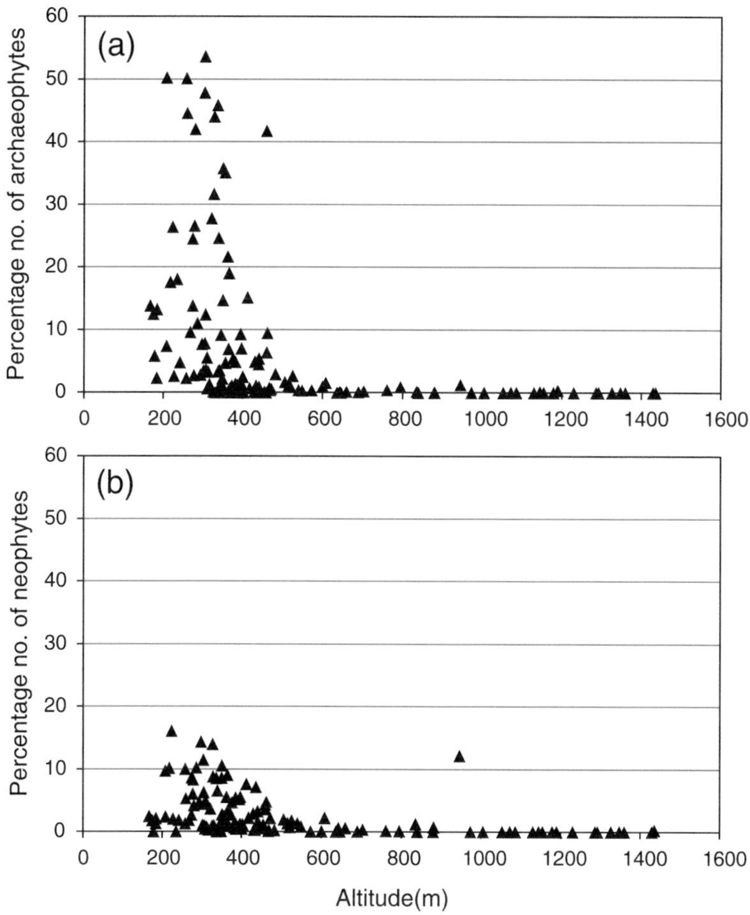


**Fig. 8.9** Percentage of (**a**) archaeophytes and (**b**) neophytes, relative to the number of all species recorded in vegetation plots, plotted against altitude. Values are means for phytosociological alliances recognized in the Czech Republic. Percentages of alien species are from Fig. 8.10 and altitudes were taken from the vegetation plots used as the data source for that figure. Note that the altitudinal range in the Czech Republic is 115–1603 m. The outlier on the right side of the lower graph is the alliance *Rumicion alpini* (taken from Pyšek et al. 2012a)

Although the altitudinal range in the Czech Republic (115–1603 m) is not very broad, the decrease in the level of invasion with altitude is distinct and well documented (Fig. 8.9). It has been demonstrated in comparative studies of landscape segments, e.g. using a transect through a landscape in southern Bohemia that also included an altitudinal gradient (Mihulka 1998), Czech nature reserves (Pyšek et al. 2002a) and grid cells of flora mapping in the White Carpathian Mts in south-eastern Moravia (Otýpková et al. 2011). A possible explanation of this pattern might be that higher altitudes generally include smaller areas of invasible habitats. However, analyses performed within individual habitats, using vegetation plots, also detect a decreasing trend in the level of invasion with altitude, e.g. for weed communities on arable land (Lososová et al. 2004; Pyšek et al. 2005), ruderal vegetation (Simonová and Lososová 2008) and most other habitat types in the Czech Republic except those with narrow altitudinal ranges (Chytrý et al. 2009b). In most contexts, neophytes respond to altitude more strongly than archaeophytes, the former being more distinctly concentrated in the lowlands.

### *8.7.2 Level of Invasion Across Habitats*

Previous research (Chytrý et al. 2005) identified arable land and anthropogenic ruderal vegetation as the habitat types harbouring the highest percentages of both archaeophytes and neophytes. Deciduous broad-leaved forest plantations (of e.g. *Populus ×canadensis*, *Quercus rubra* and *Robinia pseudoacacia*) are also frequently invaded, especially by neophytes. These habitats not only have the largest percentage of alien species in small plots but also the largest regional pools of alien species, i.e. species occurring in the Czech Republic and adapted to these habitats (Sádlo et al. 2007). Levels of invasion for more finely defined vegetation types, basically at the level of phytosociological alliances, are quantified for ruderal vegetation (Simonová and Lososová 2008), weed vegetation (Lososová and Grulich 2009) and forests (Chytrý et al. 2009b).

An analysis of the habitat species pools as defined in Sádlo et al. (2007) revealed that there is a strong phylogenetic signal in the invasion process for both the invaders and the invaded vegetation (Lososová et al. 2015). The numbers of alien species in particular habitat types in the Czech Republic are related both to the phylogenetic structure of plant communities occurring in these habitat types and the phylogenetic position of the invading alien species. Frequently disturbed herbaceous vegetation with strong phylogenetic clustering is more invaded than other vegetation types, possibly due to disturbance acting as an environmental filter. However, alien species not only invade more massively those community types that are phylogenetically clustered but also increase the post-invasion degree of phylogenetic clustering of those communities as they tend to be from the same lineages as native species. By this process, invaded community types become even more phylogenetically clustered. These findings support the hypothesis that the relatedness of invaders to native species promotes invasion because they share adaptations to the same environments. However, such trends are not detected for phylogenetically more diverse and less invaded communities, such as forests (Lososová et al. 2015).

Here we provide a new quantification of the level of invasion for plant communities in the Czech Republic at the level of phytosociological alliances, following the classification system accepted in the monograph Vegetation of the Czech Republic (Chytrý 2007–2013) and the most recent edition of the Catalogue of alien species of the Czech Republic (Pyšek et al. 2012b; Fig. 8.10). It is based on a stratified-random selection of 20,830 vegetation-plot records (relevés) from the Czech National Phytosociological Database (Chytrý and Rafajová 2003) and prepared using the

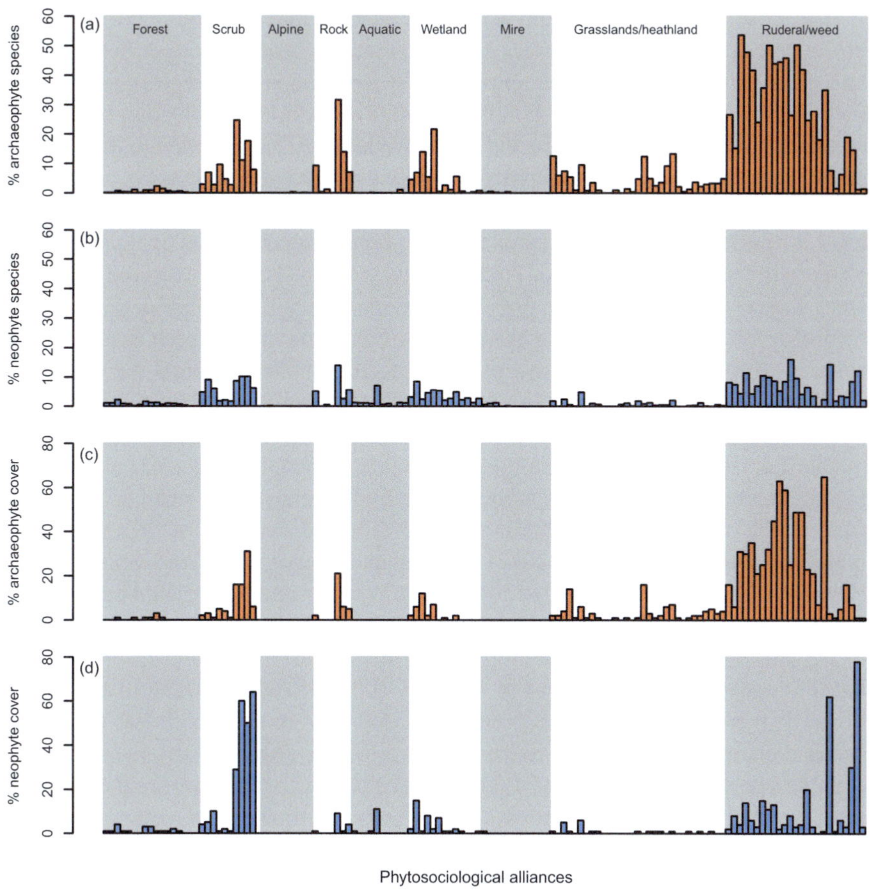


**Fig. 8.10** Mean percentages and mean summed covers of archaeophytes and neophytes, relative to all species recorded in vegetation plots, calculated for phytosociological alliances in the Czech Republic. The graphs are based on 20,830 plots of all vegetation types from the Czech National Phytosociological Database, excluding those made before 1980 and in plot sizes deviating from standards for a given vegetation type. The plots were classified to phytosociological associations using an expert system for automated classification developed as a part of the project Vegetation of the Czech Republic (see Chytrý 2007–2013 for details; based on data presented in Pyšek et al. 2012a: Table 3; see this paper for more details)

same methods as the previous quantification (Chytrý et al. 2005), but it uses more finely defined vegetation units and newer plots sampled since 1980. Compared to previous analyses (Chytrý et al. 2005, 2009b; Simonová and Lososová 2008; Lososová and Grulich 2009), the new results report slightly lower numbers of archaeophytes and higher numbers of native species in some cases, which is mainly due to reclassification of 41 species considered as archaeophytes by Pyšek et al. (2002b) to native in Pyšek et al. (2012b). However, the general trend remains the same as reported in previous studies.

The most invaded habitats and vegetation types of the Czech Republic, in terms of the percentages of alien species they harbour, are generally either disturbed or have fluctuating inputs of resources, especially nutrients and in some cases water or light. A comparison with corresponding habitats in other European countries shows that such habitats are also among the most invaded elsewhere, although the composition of alien floras is very different among European regions (Chytrý et al. 2008b). In contrast, habitats with limited fluctuation of resource availability such as dry, wet and saline grasslands, base-rich fens and broad-leaved deciduous woodlands appear to be rather resistant to invasion, although they can be invaded to some extent in areas subject to high propagule pressure (Chytrý et al. 2008a). Similar patterns are recorded in comparative studies of invasions of habitats in the Czech Republic and eastern North America (Kalusová et al. 2014, 2015). This supports the theory of fluctuating resource availability as an important driver of community invasibility (Davis et al. 2000).

Sites, areas or habitats with a high percentage of neophytes usually also have a high percentage of archaeophytes, as shown by the analyses of plant species lists from Czech nature reserves (Pyšek et al. 2002a), grid mapping of flora at a landscape scale (Otýpková et al. 2011) and vegetation plots from the whole Czech Republic (Chytrý et al. 2005). However, there is some variation around this general trend. Archaeophytes are often more numerous in treeless vegetation on dry to mesic soils, while neophytes are more common in disturbed woody vegetation, wetlands or aquatic habitats (Chytrý et al. 2005, 2008b). On arable land in the Czech Republic there are proportionally more archaeophytes in areas with less precipitation and on drier soils, such as chernozem or rendzina, although neophytes are also common there (Pyšek et al. 2005). This pattern obviously results from the fact that most archaeophytes originate from dry areas in the Mediterranean region, including the Middle East (Pyšek et al. 2012b), and are therefore preadapted to dry open habitats (di Castri 1989; Pyšek and Jarošík 2005).

The incidence of alien species (their number, percentage of all species, or cover) is generally higher in early successional stages and decreases with successional age (Rejmánek 1989). This pattern is recorded in different successional seres in the Czech Republic, for example in abandoned fields in dry areas of the Bohemian Karst (central Bohemia; Rejmánek 1989) or southern Moravia (Sojneková and Chytrý 2015), southern Bohemian bogs disturbed by peat extraction (Bastl et al. 1997), disused sand or gravel pits across the country (Bastl et al. 1997; Řehounková and Prach 2008) and some other early-successional habitats (Prach et al. 2008). However, the course of succession can be changed if strongly competitive alien species occur near the successional site. For example, Řehounková and Prach (2008) report that the occurrence of mature *Robinia pseudoacacia* trees within 100 m of a disused sand or gravel pit resulted in the formation of *Robinia* groves during spontaneous succession in the pit.

The most invaded areas in the Czech Republic are agricultural landscapes with predominantly arable fields, cities and villages, and forest plantations in the lowlands. River floodplains, especially in the lowlands, are also strongly invaded (Vymyslický 2001; Matějček 2008; Kalusová 2009; Kalníková 2012; Šenová and Matějček 2013). Mid- and high-altitude landscapes are less invaded, especially if they are forested (Chytrý et al. 2009b). In a European context, the Czech Republic belongs to a strongly invaded area of Western and Central Europe, which is characterized by on average higher levels of invasion than the boreal zone of Northern Europe or the Mediterranean and sub-Mediterranean zones in Southern Europe (Chytrý et al. 2009a).

Habitat type is the most important determinant of the level of plant invasion in the Czech Republic, followed by altitude (and associated effects of climate) and variation in propagule pressure (Chytrý et al. 2008a). However, the patterns in the invasion of habitats also depend on the original habitats in which the introduced species occur in their native ranges. Hejda et al. (2009b) show that most neophytes occurring in the Czech Republic originate from various types of ruderal vegetation, dry grasslands, broad-leaved deciduous woodlands, moist and wet grasslands including tall-herbaceous vegetation, cliffs and rock outcrops, arable land and mesic grasslands. Especially those neophytes that originate from riverine habitats, eroded slopes and avalanche tracks are the most likely to become invasive, once they are introduced into Central Europe (Hejda et al. 2009b).

The Czech flora was recently used for investigating the biogeographic aspects of habitat invasions, in particular the role of habitats as sources of species that are invasive on other continents. Dostál et al. (2013) demonstrated that the native species in the Czech flora that originate from more productive habitats or have a broader niche are more successful invaders elsewhere in the world. Outside their native ranges, Central European species invade habitats that are similar to their habitats in Central Europe (Kalusová et al. 2013, 2014). European habitats that are important donors of alien species for other continents experience the highest levels of invasion by alien species from other regions (reciprocal species-pool effect; Kalusová et al. 2014). Hejda et al. (2015), in their global analysis of habitat affinities of alien species in their native ranges, took into account the direction of invasion and showed that European grassland species are much more successful as world-wide invaders than are grassland species from other continents in invading Europe. Conversely, New World wetland species invading Europe are more successful than *vice versa*. These authors also showed that successful invaders are adapted to a broad spectrum of successional phases in their native range, ranging from grasslands to forests. A comparison of the level of invasion in corresponding habitats in the Czech Republic and eastern North America demonstrates that although the pattern in the level of invasion is very similar in both areas, North American habitats are more invaded at the habitat scale and to some extent also at the plot scale (Kalusová et al. 2015). Interestingly, the direction of invasion between continents also affects how strong the impact on native biota will be. Hejda et al. (2017) show that European invaders have more profound impacts in North America than North American invaders in Europe.

### 8.7.3 Changes in the Level of Invasion over Time

There are distinct temporal trends in the percentages of alien species in Czech vegetation, but these trends are opposite in archaeophytes compared to neophytes. Lososová and Simonová (2008) compared data on the ruderal and weed flora in Moravia between the early twentieth century and the turn of the twenty-first century, showing a decrease in archaeophytes and increase in neophytes. The same trends for these two groups of aliens were confirmed by analyses of data spanning a few decades in the second half of the twentieth century, e.g. for arable weed vegetation throughout the Czech Republic (Lososová et al. 2004; Pyšek et al. 2005), ruderal vegetation in the city of Plzeň (Pyšek et al. 2004a; here the increase in neophytes was non-significant) and road side vegetation in the Orlické hory Mts and their foothills (Dostálek et al. 2016). These trends will probably continue in the future, although the future spread of alien species will mainly depend on changing land use. European scenarios of future changes in the level of invasion project smaller increases in Central Europe than in North-western Europe (Chytrý et al. 2012). The mean level of invasion across landscapes may even decrease slightly in regions where large areas of arable land are abandoned, leading to a subsequent decrease in alien species during succession.

## 8.8 Impact, Management and Legislation

In recent years, an important new avenue of research into biological invasions has focused on describing, classifying and categorizing impacts of alien biota (see Vilà et al. 2010, 2011; Pyšek et al. 2012c for the most comprehensive global accounts on plants). A system for classifying alien species according to the magnitude of their environmental impacts, based on the mechanisms of impact, has been proposed (EICAT; Blackburn et al. 2014) and recently adopted as an official tool by IUCN, similar to their Red List scheme. A GISS (Generic Impact Scoring System) scheme based on semi-quantitative scores assigned to impact categories allows not only environmental but also socioeconomic impacts to be assessed (Kumschick et al. 2015; Nentwig et al. 2016). Using GISS, Rumlerová et al. (2016) assessed the impacts of 128 alien plant species in Europe, of which 86 are known to occur in the Czech Republic (Table 8.2).

Among the 20 taxa invasive in Europe with the highest scores of summary impact, there are 11 that are also invasive in the Czech Republic; these plants pose the greatest threat to the natural environment. The scores for environmental impacts are generally higher than those for socioeconomic impacts, reflecting that to some extent, rigorous data on the latter are still rather scarce. Table 8.2 shows several taxa with potentially great impacts, both environmental and socioeconomic, that are however not realized due to their casual occurrence in the Czech Republic, mostly due to the unsuitable climate.

The Czech Republic is still lacking a comprehensive assessment of the ecological and economic impacts of its alien plants. Studies rigorously testing ecological

**Table 8.2** Twenty alien species occurring in the Czech Republic that rank highest based on their potential impacts in Europe. Taken from Rumlerová et al. (2016) who assigned scores for all 86 species assessed for Europe and occurring in Czech Republic. Summary scores across six types of environmental and socioeconomic impacts are used as the measure, with each type scored 1–5 (see Nentwig et al. 2016 for details). Environmental impacts include (1) direct impact on plants; (2) impact on animals; (3) indirect impacts on other species; (4) transmission of diseases and parasites; (5) impact through hybridization; and (6) impact on ecosystems. Socioeconomic impacts concern (1) agricultural production; (2) animal production; (3) forestry; (4) human infrastructure; (5) human health; and (6) human social life. Species are ranked according to the sum of their impacts. Their status in the Czech Republic is given in brackets (based on Pyšek et al. 2012b); invasive species are in *bold*

| Species | Environmental | Socioeconomic | Sum |
|---|---|---|---|
| *Eichhornia crassipes* (cas) | 16 | 13 | 29 |
| *Elodea canadensis* (nat) | 15 | 8 | 23 |
| ***Reynoutria japonica* (inv)** | 12 | 9 | 21 |
| ***Heracleum mantegazzianum* (inv)** | 13 | 8 | 21 |
| ***Robinia pseudoacacia* (inv)** | 11 | 9 | 20 |
| ***Solidago canadensis* (inv)** | 14 | 4 | 18 |
| ***Ambrosia artemisiifolia* (inv)** | 9 | 8 | 17 |
| ***Prunus serotina* (inv)** | 12 | 5 | 17 |
| *Ambrosia trifida* (cas) | 10 | 6 | 16 |
| *Arctotheca calendula* (cas) | 7 | 9 | 16 |
| ***Conyza canadensis* (inv)** | 8 | 8 | 16 |
| *Elaeagnus angustifolia* (cas) | 11 | 5 | 16 |
| ***Ailanthus altissima* (inv)** | 11 | 4 | 15 |
| *Rosa rugosa* (cas) | 12 | 3 | 15 |
| *Elodea nuttallii* (cas) | 8 | 6 | 14 |
| ***Impatiens glandulifera* (inv)** | 10 | 4 | 14 |
| ***Lupinus polyphyllus* (inv)** | 11 | 3 | 14 |
| *Oxalis pes-caprae* (cas) | 9 | 5 | 14 |
| ***Bidens frondosus* (inv)** | 7 | 6 | 13 |
| *Datura stramonium* (nat) | 5 | 8 | 13 |

impacts under local conditions are few (Hejda et al. 2009a; Horáčková et al. 2014; Jandová et al. 2014) and similar to the GISS-based one presented above (Table 8.2), but are a good beginning. A thorough nationwide assessment of ecological and economic impacts of alien plants in the Czech Republic is one of the most important future tasks for both researchers and practitioners. A more systematic scoring of impacts is currently underway and urgently needed in order to provide state authorities and responsible management bodies with a knowledge base that can be used to effectively mitigate the problem of plant invasion at the national level. In terms of costs incurred by invasions of alien species, information is only available for small (mostly protected) areas and specific eradication measures and management efforts. These data relate to individual invasive species, such as *Pinus strobus* in the Bohemian Switzerland National Park, for which a management cost of almost 1.5 million EUR was estimated for the period of 2000–2016 (including costs associated with management of the locally non-native *Larix decidua*; see Chap. 9, Sect. 9.7.3, this book). Another example is the management of six major invasive taxa subjected to control in the Bohemian Paradise (Český ráj) Landscape Protected Area in 2003, where the cost reached 450,000 CZK (Křivánek 2006). A study of *Heracleum mantegazzianum* estimates that the total annual economic impact of this species in the Czech Republic is 2.5 million CZK (Linc 2012).

The adoption by the European Union of Regulation 1143/2014 on invasive alien species (IAS; European Union 2014) is currently the most important policy instrument related to biological invasions in Europe. The backbone of this regulation is a list of species alien to the EU and identified as invasive based on a detailed evidence-based risk assessment. The species listed are banned from being imported, traded, possessed, bred, transported, used and released into the environment (Genovesi et al. 2014). In February 2016, the European Commission adopted a first list of IAS of Union concern consisting of 37 species, selected from the 65 species for which fully compliant risk assessments were available (Roy et al. 2015). Among the 13 plant species on this list, only *Eichhornia crassipes* and *Heracleum persicum* are recorded from the Czech Republic, but both only as casuals. Although the selection of species for this list has been criticized as being influenced by political interests (Pergl et al. 2016b), this Regulation is an important milestone for an invasive species policy in Europe. In the Czech Republic, the key legislative instrument for biodiversity conservation is Act no. 114/1992, which restricts deliberate introductions of non-native species into the wild. Another important legislative tool is Act no. 326/2004 on phytosanitary care, which focuses on weeds and sets the obligation to minimize the impact of alien species on nature. There are also other acts related to particular sectors such as forestry (no. 289/1995), agriculture (no. 334/1992), water resources (no. 254/2001) or genetically modified organisms (no. 78/2004) that concern alien species and their introductions (Šíma 2008).

## 8.9 National Black List of Invasive Plants

Intensive research focused on the categorization of alien plant species in the Czech Republic based on their invasiveness and the recent beginning of systematic scoring of impacts mentioned above (Kumschick et al. 2015; Rumlerová et al. 2016), yielded robust knowledge that allowed the preparation of the first national Black, Grey and Watch Lists of invasive biota (Pergl et al. 2016d). The Black List classifies species to three categories defined on the basis of the magnitude of the impact and pathway of introduction that affect the opportunities for action; for each category management measures are recommended that can be used by land managers, policy makers and other stakeholders. Species with a low impact, but for which some level of management and regulation is desirable, are included on the Grey List. Potentially dangerous species occurring in European countries with similar climates, as well as those introduced in the past but with currently no known wild populations in the Czech Republic, are included on the Watch List. In total, there are 78 plant species on the Black List, 47 on the Grey List, and 25 on the Watch List (Table 8.3).

**Table 8.3** Categories of the Black List with indications of recommended management, handling restrictions, examples of species and criteria for classifying that are derived from environmental and socio economic impact, population status and distribution of the target species (based on Pergl et al. 2016b)

| | Grouping criteria | Population status and distribution | Recommended local management | Handling and release restrictions | No. | Species |
|---|---|---|---|---|---|---|
| BL1 | High environmental and socio economic impact | Abundant, occurring in a wide range of habitats throughout the country | Complete eradication, disposal of abandoned plantations | No release, application of trade regulations | 2 | *Ambrosia artemisiifolia*, *Heracleum mantegazzianum* |
| BL2 | Moderate to massive environmental impact; depending on human actions that promote species spread | Often found as remnants of planting in gardens and plantations, usually with a wide distribution | Stratified approach; promoting alternative native species; permit only in areas with low conservation value; disposal of abandoned plantations | No release, legislative regulations of trade, handling, and planting | 49 | *Azolla filiculoides*, *Acer negundo*, *Ailanthus altissima*, *Allium paradoxum*, *Amorpha fruticosa*, *Arrhenatherum elatius*, *Asclepias syriaca*, *Beta vulgaris Altissima* Group, *Buddleja davidii*, *Colutea arborescens*, *Cornus alba*, *C. sericea*, *Cytisus scoparius*, *Echinocystis lobata*, *Echinops exaltatus*, *E. sphaerocephalus*, *Fallopia aubertii*, *Fraxinus pennsylvanica*, *Galega officinalis*, *Galeobdolon argentatum*, *Helianthus ×laetiflorus*, *H. pauciflorus*, *H. tuberosus*, *Impatiens glandulifera*, *Laburnum anagyroides*, *Lupinus polyphyllus*, *Lycium barbarum*, *Parthenocissus inserta*, *P. quinquefolia*, *Physocarpus opulifolius*, *Phytolacca acinosa*, *Pinus nigra*, *P. strobus*, *Populus ×canadensis*, *P. balsamifera*, *Prunus cerasifera*, *P. serotina*, *Pyracantha coccinea*, *Quercus rubra*, *Reynoutria ×bohemica*, *R. japonica*, *R. sachalinensis*, *Rhus typhina*, *Robinia pseudoacacia*, *Rudbeckia laciniata*, *Solidago canadensis*, *S. gigantea*, *Symphoricarpos albus*, *Symphyotrichum novi-belgii*, *Telekia speciosa* |
| BL3 | Moderate to massive environmental impact; results of unintentional introductions | Usually with a wide distribution resulting from spontaneous spread | Stratified approach; due to spontaneous spread there is no need to tolerate in any area | No release | 27 | *Abutilon theophrasti*, *Alopecurus myosuroides*, *Amaranthus albus*, *A. powellii*, *A. retroflexus*, *Bunias orientalis*, *Cannabis sativa* var. *spontanea*, *Cirsium arvense*, *Conium maculatum*, *Consolida hispanica*, *Conyza canadensis*, *Cuscuta campestris*, *Digitaria ischaemum*, *Echinochloa crus-galli*, *Galinsoga parviflora*, *G. quadriradiata*, *Iva xanthiifolia*, *Orobanche minor*, *Oxalis corniculata* var. *corniculata*, *O. dillenii*, *Panicum miliaceum* subsp. *agricola*, *Portulaca oleracea*, *Rumex alpinus*, *R. longifolius* subsp. *sourekii*, *Senecio inaequidens*, *Setaria faberi*, *S. verticillata* |

An important feature of the Black List is that it adopts a stratified approach to management, which reflects the local and regional context of a particular invasion and allows the formulation of the optimal strategy for each species. An example is the management of *Robinia pseudoacacia* (black locust) in the Czech Republic, the planting of which can be allowed in areas where the stands are not an imminent threat to the landscape, but should be prohibited, and extant stands eradicated, from sites with a nature conservation value, such as dry grasslands (Vítková et al. 2017). The stratified approach thus discriminates where and when the management of alien species is needed and efficient, and where the eradication is neither effective nor necessary, e.g. in urban and suburban areas (Pergl et al. 2016d). The Black, Grey and Watch Lists have served as a basis for implementing the methodology of monitoring alien species by the Ministry of Environment of the Czech Republic (Pergl et al. 2016a).

## 8.10 Accounts of Invasive Taxa in the Czech Republic

In the following section, factsheets of invasive neophytes in the Czech Republic are presented, as classified in Pyšek et al. (2012b). The distribution maps use the grid cells of the Central European mapping system (KFME, Kartierung der Flora Mitteleuropas; Niklfeld 1998; Schönfelder 1999) and are based on data from the Pladias database (www.pladias.cz). Only occurrences outside cultivation are included in the maps. Maps are not shown for taxa for which insufficient information on distribution is available, which is mostly due to the low reliability of records for taxonomically difficult taxa that are easily mistaken in the field. Note that the distribution maps are generally somewhat underestimated since they are not the result of systematic mapping of invasive species, but for the taxa presented they reflect the patterns in their geographic distribution reasonably well. A complete list of the references from which the information for these accounts was extracted can be found in Pyšek et al. (2012a). For all species these accounts are also partly based on the Flora of the Czech Republic (Hejný et al. 1988, 1990, 1992; Slavík et al. 1995, 1997, 2000, 2004; Štěpánková et al. 2010).

***Acer negundo*** (*Sapindaceae*, Fig. 8.11a) is a dioecious tree reaching a height of up to 20 m; in the Czech Republic it is not as tall being only up to 10 m in height. It is native to eastern and central parts of North America. Its moderate resistance to flooding combined with tolerance to water deficits enable it to occur in a wide range of habitats ranging from wetlands and floodplain forests to relatively dry forests and grasslands. The introduced range covers Europe, parts of Asia and Australia. The species was imported into Europe as an ornamental tree in the seventeenth century, with its first record in the UK dating from 1688. It became a popular garden tree due to its fast growth in the first few years. In the second half of the nineteenth century it was planted in parks, along roads, later on in wind-breaks and shelter belts. The

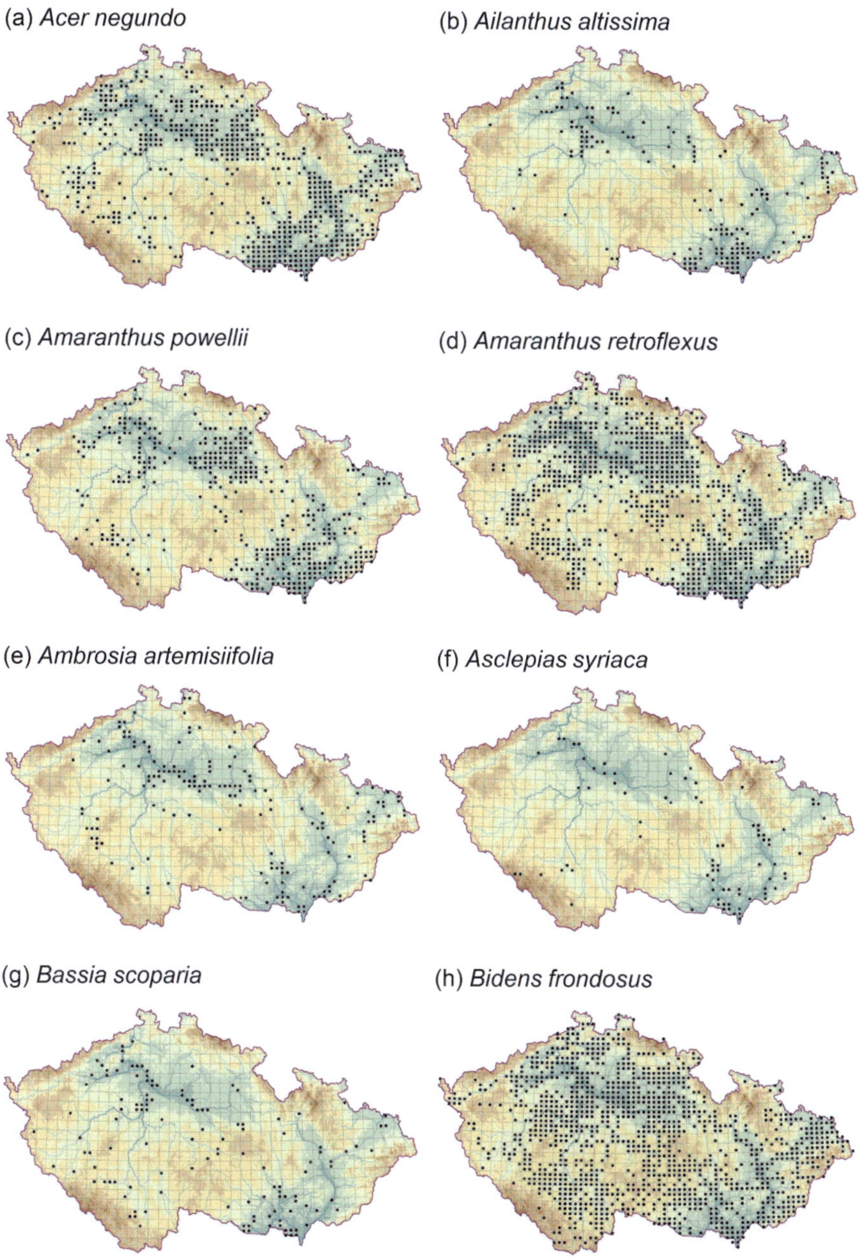


**Fig. 8.11** Distribution of selected invasive plant species in the Czech Republic. Species distribution maps in this chapter are based on data stored in the Pladias database (www.pladias.cz, accessed in January 2017), which is used with the permission of the original data providers (Institute of Botany CAS, Masaryk University, Brno, the Nature Conservation Agency of the Czech Republic and Czech Botanical Society), except for *Ambrosia artemisiifolia* which is based on the map published by Skálová et al. (2017). Maps of commonly planted woody species are not shown because their records in the database often relate to planted individuals. All the maps were prepared by O. Hájek

first record of spontaneous occurrence in the wild in the Czech Republic is probably from the bank of the Labe River near Neratovice in 1875. It is commonly grown, with a variety of cultivars on the market. The species is planted in urban areas, industrial zones and wind-breaks, where the planted trees give rise to new populations originating from self-seeding. It spreads by seed and easily resprouts from stumps. It prefers moist open habitats with a sufficient supply of organic nutrients on clayey or sandy soils. Above 350 m a.s.l. it only occasionally reproduces by seed. It also forms spontaneous invasive populations along large rivers, where it invades disturbed floodplain forests (especially along the Morava and lower Dyje rivers), and in areas reclaimed after coal mining. Except in these plantations, it rarely occurs as a strong dominant with a high cover, and often occurs together with other neophytes such as *Fraxinus pennsylvanica*, *Helianthus tuberosus*, *Robinia pseudoacacia*, *Solidago* spp. and *Symphyotrichum* spp. It is recorded from seven habitat types. The invasion is most pronounced in warm areas of the country, namely in southern and central Moravia, where it is common, develops monodominant stands and spreads rapidly. *Acer negundo* is ranked among the 10 most invasive woody species in the world. It has negative impacts due to its pollen causing allergenic reactions in humans and competition with other species at invaded sites.

***Ailanthus altissima*** (*Simaroubaceae*, Figs. 8.11b and 8.12a) is a fast-growing deciduous tree, ~20 m tall, reproducing both vegetatively by root suckers and by seed (up to 325,000 seeds per year are produced by a single tree). It is native to East Asia (China and Korea) and has been introduced worldwide, including Australia and the Pacific Islands. This species was introduced into Europe in 1784 or earlier (some sources suggest an introduction date of ca 1750 into the UK) as an ornamental tree. In the Czech Republic the first record of planting comes from 1813 and in the wild from a forest near Veltrusy, central Bohemia, in 1874. The species is highly tolerant of polluted air and poor soils, and is markedly resistant to varying temperatures, and different levels of humidity, light and moisture and is thus able to grow in stressful habitats. At present it is spreading in large cities and their suburbs and industrial zones where it forms large stands independent of original plantings. Beside urban habitats, it occurs in shrub and grassland vegetation, especially in warmer areas; it is recorded from 11 habitat types (Fig. 8.8). It is spread by wind, water and contaminated soil. As a rather weak competitor it is only able to establish in sparse vegetation, although after establishment it quickly spreads by root suckers. Typical habitats in the Czech Republic are railway corridors, where *Ailanthus* often forms thickets by clonal growth. It often grows along with *Robinia pseudoacacia* stands and together with other neophytes such as *Erigeron annuus*, *Lycium barbarum* and *Parthenocissus inserta*. It is occasionally planted in dry grasslands on steep slopes where it forms clonal scrub or shruby woodland with thermophilous species, vegetation similar to that found in South-eastern Europe. The invasive populations in the Czech Republic occur mainly in southern Moravia. The impact of *A. altissima* on native vegetation is through its formation of dense thickets and allelopathic effects inhibiting growth and germination of native species. The plant sap can induce dermatitis in humans. Another important aspect is that the root system of this tree often damages pavements, walls and buildings.

**Fig. 8.12** Examples of invasive plant species in the Czech Republic: (**a**) *Ailanthus altissima*, Prague-Troja; (**b**) *Amaranthus powellii*, Lednice, southern Moravia; (**c**) *A. retroflexus*, Lipůvka, southern Moravia; (**d**) *Asclepias syriaca*, Břeclav, southern Moravia. Photo credits: P. Pyšek (a, b), M. Chytrý (c) and Z. Lososová (d)

***Amaranthus powellii*** (*Amaranthaceae*, Figs. 8.11c and 8.12b) is an annual herbaceous plant native to tropical and subtropical South and Central America. Its invaded range includes temperate regions in North America, Europe, Asia, Africa and Australia, where it grows in open ruderal vegetation in urban areas and as a weed in fields and gardens. It was first recorded in the Czech Republic in a potato field near Mnichovo Hradiště, north-eastern Bohemia, in 1853. The earliest herbarium specimens are from 1931. Introduced into the Czech Republic as contaminant in grain, oil seed, ore, cotton and wool, it is common in warm lowland areas and is spreading into colder areas and higher altitudes. Until the 1960s it occurred only in ruderal habitats in villages and along roads. It started to spread rapidly in the 1980s, supported by frequent planting of maize and beet. In the 1990s it became common as a weed on arable land. It is resistant to some herbicides and in the last decade its invasion has been supported by their regular application in agriculture, along railways and in urban areas. Compared to the congeneric *A. retroflexus*, it is less widely distributed, confined to warmer areas, less drought-resistant and more nutrient-demanding, which is reflected in *A. powellii* being currently more invasive in arable land and vegetable gardens than in ruderal habitats. Due to its strong hybridization potential (two hybrids are listed by Pyšek et al. 2012a) it may affect closely related species, although the viability of some of the hybrids is low.

***Amaranthus retroflexus*** (*Amaranthaceae*, Figs. 8.11d and 8.12c) is an annual herbaceous plant native to North America, introduced into regions with temperate and warm climates on all continents. In its native range it is a component of pioneer riverine vegetation. The earliest records in the Czech Republic, where it was introduced as a contaminant in grain, oil seed, cotton and wool, are from 1818 in Prague and 1822 in Uherské Hradiště, south-eastern Moravia. Seed contamination of soils and commodities is linked to its extreme fecundity, with a single plant producing up to 500,000 seeds that remain viable in the soil for up to 5 years. It grows in relatively dry and nutrient-rich urban and agricultural habitats such as rubbish dumps, soil heaps after digging along roads, rivers and railroads, and as a weed on arable land; it is recorded from 10 habitat types. Its invasion has been supported by an increase in maize and beet planting, although currently it benefits mostly from its tolerance to herbicides, high salinity levels, polluted soil and eutrophication of the landscape. Further rapid spread is unlikely since the species is widespread and common and appears to have colonized most available habitats. However, further spread of herbicide-resistant populations in less suitable habitats as well as into higher altitudes is possible. The major impact of this species is a reduction of crop yield. It hybridizes with some species from the *A. hybridus* agg., including cultivated ornamentals or pseudocereals.

***Ambrosia artemisiifolia*** (*Asteraceae*, Fig. 8.11e) is an annual herbaceous plant that is native to North America, including the central and eastern USA, where it grows as a pioneer species in open semi-arid habitats. In its invaded range, which includes all continents (except Antarctica) as well as some islands (New Zealand, Hawaii, Madagascar and Mauritius), it is known from a wide range of open and nutrient-rich, disturbed ruderal habitats and arable land. This species was introduced into Europe in the second half of the nineteenth century as a contaminant of agricultural commodities, bird seed and with agricultural machinery. Several independent introductions are documented. Early introductions into Europe mostly resulted in short-lived casual occurrences. Established populations only developed in the first decades of the twentieth century, and commonly after World War II. Within the last few decades *A. artemisiifolia* has significantly increased its range and abundance in many European countries, and spread into a number of habitats. The invasion is supported by the production of a high amount of seed (1200–2500 seeds per plant) that forms a long-term persistent soil-seed bank, with seed remaining viable for up to 40 years. The first record in the Czech Republic is from 1883, in a clover field near Třeboň, southern Bohemia, and from a field near Plzeň-Doudlevce, western Bohemia. The seed was probably introduced with clover seed from North America. The next wave of introductions occurred in the second half of the twentieth century from different sources: with grain from Canada and with soybeans from North America, and with Ukrainian grain and Soviet ore. This species is confined to the warm parts of the Czech Republic and prefers open dry habitats on sandy or gravelly substrata with a low cover of vegetation. After rapid spread in the 1980s–1990s the invasion has decelerated. *Ambrosia artemisiifolia* is limited by requiring well-aerated soils and being a weak competitor. It rarely spreads in urban spaces, in sand pits, on coal mining heaps or in semi-natural grasslands. In southern Moravia, however, it has started recently to occur as a weed in maize fields. It grows mostly in ruderal habitats, in particular along railways where it is supported by the use of herbicides, to which some populations are resistant. This resistance is due in part to the fact that herbicides are applied at the end of spring when *Ambrosia* populations are only starting to germinate as their phenological optimum is in autumn. It is scattered and locally abundant along railways but does not yet form monodominant stands. The pollen of *A. artemisiifolia* is the most allergenic of all plant species occurring in Europe. Ragweed pollen peaks in August and September in the Czech Republic, especially in southern Moravia. It is spread by easterly winds, most probably from the Hungarian lowlands. Due to its so far rare occurrence in the Czech Republic it does not present a serious problem for the allergic population in this country, but its importance as an allergen is likely to increase in the future. The economic costs of ragweed invasion in Germany are estimated at 32 million EUR annually, nearly entirely incurred within the health sector. Furthermore, in other countries *A. artemisiifolia* significantly reduces crop yields, especially in spring-sown crops like sunflower, soybean, maize and vegetables. However, in most of Europe, the infestation of agricultural fields is a relatively recent phenomenon, thus impacts on crop yield are still minor.

***Asclepias syriaca*** (*Apocynaceae*, Figs. 8.11f and 8.12d) is a rhizomatous perennial native to eastern North America, with a distribution ranging from Canada to North Carolina. In its native range this species grows in prairies, along roads and railways and at disturbed sites. It has been introduced into many parts of the world, and the introduction into Europe dates back to 1629. The first record of its planting in the Bohemian lands is from the Lány chateau park, central Bohemia, in 1785, and it escaped from cultivation already before 1821. It was planted for multiple purposes in the past, including medicinal, textile, oil and honey production. It is currently cultivated mostly as an ornamental. In the Czech Republic it is abundant and invasive only in warm lowlands, especially in southern Moravia, where it spreads over long distances by seed. However, the spread by rhizomes is more important, such as with contaminated soil during railway construction. In Bohemia, it forms rather large but isolated stands, and long-distance dispersal is limited. Populations occur in open habitats along roads, railways, in vineyards and shruby margins, abandoned places in settlements and on dry banks of streams. The tall stands of *Asclepias syriaca* persist for decades and are relatively species-rich. The species is toxic to humans and herbivores. Some impact of *Asclepias* on species diversity is reported from Southern Europe where it colonizes nutrient-poor soils and sand dunes.

***Bassia scoparia*** (*Amaranthaceae*, Fig. 8.11g) is an annual plant up to 2 m tall, native to a large area from South-eastern Europe to East Asia, where it grows in dry open steppe habitats and as a weed in arable fields. It has been introduced as a contaminant of crop seed and an ornamental plant into rather warm parts of Europe, South Africa, both Americas and Australia. Within this extremely variable species, two poorly defined subspecies are traditionally distinguished in the Czech Republic, **subsp. *scoparia*** and **subsp. *densiflora***. The first report of cultivation of the subspecies *scoparia* in the Czech Republic dates back to 1811 and that of its escape to 1819. Plants referable to subspecies *densiflora* were first collected in Moravia in Popice in 1901 and in Bohemia in Praha-Zlíchov in 1930; later they were introduced with commodities from the former Soviet Union. The rapid spread in warm areas only started in the early 1990s. This species prefers sandy and gravelly soil poor in nutrients, and occurs in open dry habitats; it grows well on salty and polluted soils. *Bassia scoparia* colonizes ruderal habitats, namely along railways, but it also grows along roads, in sand pits and on sand heaps, forming dense closed stands, which persist for several years. It is wind-pollinated, thus maintaining a genetic link with populations in cultivation. In southern Moravia it is reported to occur as a weed in maize fields. Populations are herbicide-resistant and have their development peak in the autumn, therefore they are not harmed by the spring application of herbicides. Both subspecies occur together at the same localities, but the subspecies *scoparia* is more common.

***Beta vulgaris* Altissima Group** (*Amaranthaceae*) includes, in addition to sugar beet, also annual weedy types that were first recorded in the 1980s in the Czech Republic. These plants were introduced with beet seed from Southern and South-western Europe, mainly Italy and France, where they originated from the pollination of cultivated sugar beet (*B. vulgaris* Altissima Group) by the pollen of the wild *B. vulgaris* subsp. *maritima* or of weedy annual plants derived from some cultivars of the Altissima Group. The first records of this "weedy beet" come from the 1970s in the UK. It started to spread very fast only rather recently. A survey from 2006 revealed that weedy beet occurred on 70% of the farms in the Czech Republic growing sugar beet and on 4% of those its density exceeded 1000 plants/ha. Plants of weedy *B. vulgaris* occur in a variety of growth forms, from large to small roots and differing in root branching, total plant weight or in the number seeds they produce. They germinate early in the season before cultivated sugar beet. The seeds form a long-term persistent seed bank, which makes eradication difficult. Occurrence of weed beet in fields is a serious economic problem as it competes with sugar beet and makes it more difficult to harvest and process sugar beet. The chemical control of weedy beet is prevented by the impossibility of using herbicides in sugar beet fields. Alternative (mechanical) methods can partly reduce the seed set, but high regeneration capacity makes it necessary to repeat the cutting of flowering stems several times in a season. Hybridization with native species in the Czech Republic has not been reported, but transfer of genes from GMO sugar beet to weedy species was observed.

**Fig. 8.13** Examples of invasive plant species in the Czech Republic: (**a**) *Bidens frondosus*, Cep sandpit near Suchdol nad Lužnicí, southern Bohemia; (**b**) *Bunias orientalis*, Pouzdřany, southern Moravia; (**c**) *Conyza canadensis*, a railway bank near Všetaty, central Bohemia; (**d**) *Echinocystis lobata*, Sázava River bank in Sázava, central Bohemia; (**e**) *Echinops sphaerocephalus*, Kobylí hlava hill near Hluk, southern Moravia. Photo credits: M. Chytrý

***Bidens frondosus*** (*Asteraceae*, Figs. 8.11h and 8.13a) is an annual plant native to large areas in North America, from southern Canada to the southern USA, where it grows in riparian habitats, on lake shores, as well as along roads and railways. The invaded range covers Europe, Asia and New Zealand. It was introduced into Europe, possibly as a contaminant in wool, in the late eighteenth century, with the first record from 1777 in Poland. In the Czech Republic it was first reported in 1907 and became naturalized in the first half of the twentieth century. At present it occurs in most parts of this country and colonizes a wide range of moist and nutrient rich habitats. It is recorded in 21 habitat types, and is a dominant species in vegetation consisting of nitrophilous annual hygrophilous plants (Fig. 8.8). Its competitiveness with native *Bidens* species is enhanced by its markedly higher drought resistance. It occurs on riverbanks, shores of water bodies, in road ditches and moist waste places. This species is common in urban areas, industrial zones and mining areas. It forms dense but small populations, usually up to several $m^2$. *Bidens frondosus* has little impact on the diversity of invaded communities; early germination and tall stature enable it to suppress native species with which, nevertheless, it coexists following invasion. It appears to have colonized the majority of suitable habitats, and further spread thus depends on the frequency of available sites.

***Bunias orientalis*** (*Brassicaceae*, Figs. 8.13b and 8.14a) is a biennial or perennial herbaceous plant up to 1.7 m tall, reproducing by seed and root fragments. Its native range is in Siberia and Eastern and South-eastern Europe, although according to some authors it was originally restricted to Armenia. It grows at the edges of forests and along riverbanks. It has been introduced into North America and most of Europe, where it has been known since the seventeenth century. The first record in the Czech Republic is from 1856. It was introduced as a contaminant in maize seed and with horse fodder imported from Russia. Until the 1920s there were only a few occurrences reported in Europe, but it became more common in the Czech Republic after World War II. Recent spread is often due to unintentional transport with soil. It has spread very fast in the last two decades, mainly in grassland along roads and railways, in abandoned areas, on reclaimed mining areas in the lowlands and moderately warm mid-altitudinal regions. Only recently it started to extend its altitudinal range and is also spread by floods. It occurs in a range of ruderal, but also semi-natural habitats, assigned to nine habitat types. Optimum conditions include perennial thermophilous ruderal vegetation on deep and dry clayey soils rich in mineral nutrients. This species mostly invades secondary human-made habitats, where the impact on native diversity is relatively low; however, it may become a troublesome weed in some agroecosystems. It can reduce species diversity in grasslands, in which it has developed large and dense populations in the last 20 years.

***Cannabis sativa* var. *spontanea*** (*Cannabaceae*, Fig. 8.14b), an annual herbaceous plant up to 1.3 m tall, native to dry steppe areas in Central Asia, where it also grows at disturbed sites along roads and railways, and in cities and villages, i.e. in habitats similar to where it grows in its invaded range in Europe and North America. This taxon has been introduced into Europe for use as bird seed and as an agricultural crop used for fibre and was also unintentionally introduced as a crop contaminant. In the Czech Republic it was first recorded growing in the wild in 1868, at

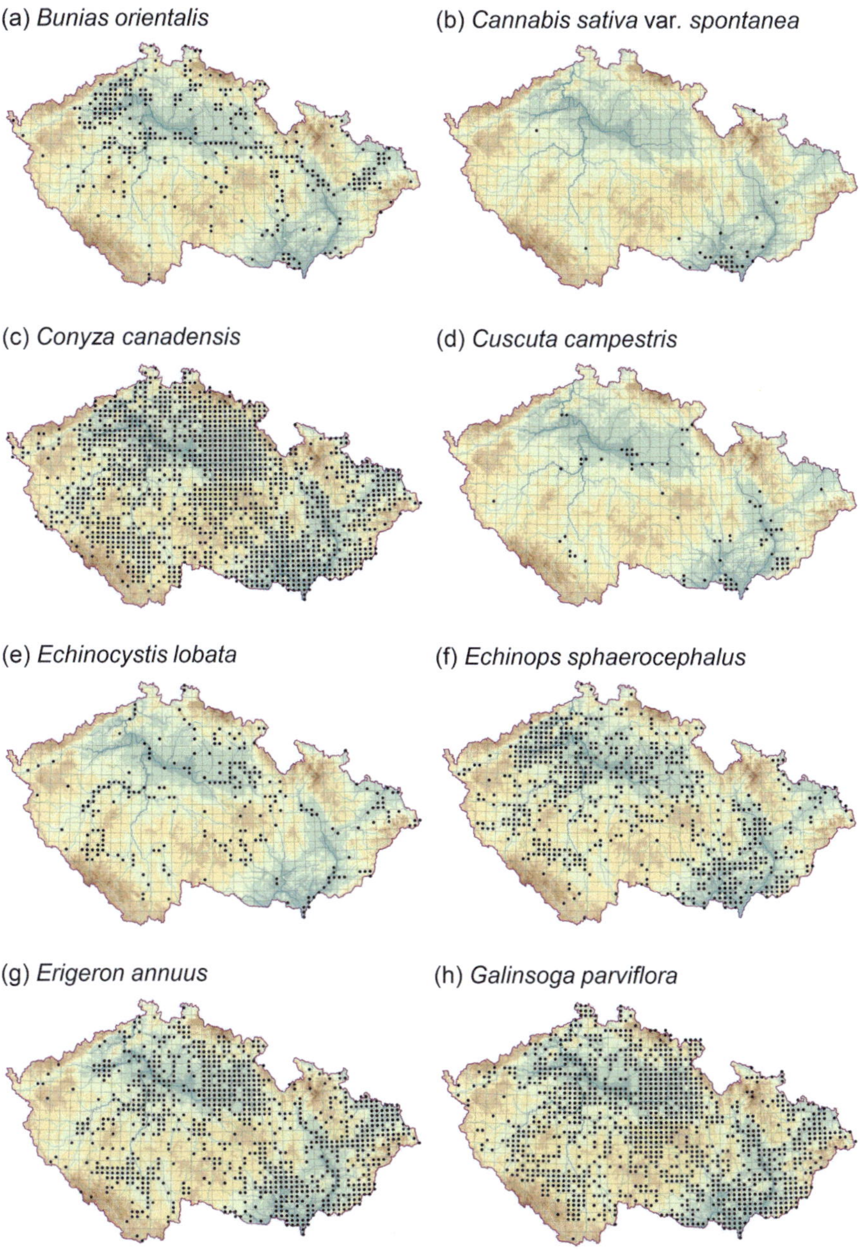


**Fig. 8.14** Distribution of selected invasive plant species in the Czech Republic

Hustopeče in southern Moravia. Its distribution in the Czech Republic is confined to the warm and dry areas in southern Moravia, where it spreads in agricultural landscapes and in the surroundings of villages on deep, dry soils rich in nitrogen and mineral nutrients. It invades annual vegetation in ruderal habitats and on arable land, and produces dense stands on rubbish dumps, waste places, along roads and paths, and at the margins of vineyards, maize and other fields. Hybridization between *C. sativa* var. *spontanea* and cultivated forms is reported.

***Conyza canadensis*** (*Asteraceae*, Figs. 8.13c and 8.14c) is an annual species native to North America where it grows at disturbed open patches in meadows or fields. A single plant produces up to 100,000 cypselae. The species has been unintentionally introduced into all other continents except Antarctica. It is common, especially in regions with temperate and subtropical climates. Typical habitats in the invaded range are dumps, ruderal and urban sites, road and railway verges or disturbed grasslands. The first record in the Czech Republic is from 1750. It is reported as naturalized and common in early floras from the late eighteenth and the early nineteenth century, while numerous records from the beginning of the eighteenth century exist for neighbouring Germany. It is widely distributed in the Czech Republic, and occurs in 33 habitat types; it is a dominant in annual ruderal vegetation and thrives in many other habitats (Fig. 8.8). Although it was introduced a long time ago, it is still spreading into higher altitudes and semi-natural vegetation, forest clearings and other habitats. Its invasion is supported by its resistance to triazine-based herbicides, which is reported to have developed in the 1970s, and which enables it to spread in urban areas or industrial zones. It grows as a weed in vineyards, hop fields, orchards and vegetable fields. In Europe, it is a significant weed in warmer areas and in fields where it may reach high densities and therefore negatively affect crop yield.

***Cuscuta campestris*** (*Cuscutaceae*, Fig. 8.14d) is an annual parasitic plant native to western North America where it grows in open grasslands and fields. It has been introduced into most of the world, often as a contaminant of seed. The first record in Europe is probably from Prague in 1883. The distribution of this species in the Czech Republic is scattered. It is common only in warm lowlands such as those of the Labe River and in southern Moravia, where it usually occurs in wasteland along roads and also as a weed in clover and alfalfa fields and gardens, or in southern Bohemia, where it grows on the bottoms of drained fishponds. It only occurs in four habitat types and as a dominant in annual and perennial nitrophilous ruderal vegetation; its recent spread is attributed to its high tolerance of salinity, a feature that enables it to spread along roads. Where it is abundant it has a major impact on agriculture in that it reduces yield. It was considered as rare in the 1960s; at present it is starting to extend its altitudinal range.

***Echinocystis lobata*** (*Cucurbitaceae*, Figs. 8.13d and 8.14e) is an annual vine native to North America where it grows at open sunny sites in floodplains and at the fringes of forests. This species was introduced into temperate and continental Europe in the early twentieth century (1906, Slovakia) and into the Bohemian lands in 1911. Its introduction into Europe is linked with botanical gardens, from where it spread across the continent. The frequency of planting has increased in the last 20 years, which accelerated the spread of this species. Its invasion in the Czech Republic was supported by seed transported by water, especially during floods, and it is still actively spreading via water in Eastern Europe. This mode of dispersal is reflected by its distribution in the Czech Republic, where it is confined to areas along large rivers. It invades semi-natural habitats, being assigned to seven habitat types, with its ecological optimum in lowland nitrophilous herbaceous fringes, willow galleries on riverbanks and perennial nitrophilous vegetation at mesic to wet sites. This species is demanding of light, nutrients and moisture, which limits its invasion to the vicinity of rivers. The impact of this species is through its ability to cover large areas and overgrow native vegetation. The whole plant contains substances that are toxic to humans and animals.

***Echinops sphaerocephalus*** (*Asteraceae*, Figs. 8.13e and 8.14f) is a tall herbaceous perennial that is native to Europe, with its native distribution ranging from Southern Europe to Southern Siberia. It occurs as an introduced species in the rest of Europe and North America. In both distribution ranges it grows at disturbed sites along roads and rivers, in urban areas and dry grasslands, preferring nutrient-rich soils and sunny places. In the Czech Republic it is grown as a garden ornamental and is still sown in the wild by bee keepers, e.g. on railway banks. The first record from the Czech Republic is from 1871. This species is relatively widely distributed, being more abundant in warm areas and occurs in nine habitat types and thrives best in perennial thermophilous ruderal vegetation. It has recently spread in dry, disturbed ruderal grasslands. In the past its occurrence was restricted to the surroundings of villages and stone quarries, while currently it is common in suburban areas and spreads in the landscape along roads and is extending its altitudinal range. It also invades mesic habitats such as road ditches. No clear impacts on biodiversity or human health are known.

***Erigeron annuus*** (*Asteraceae*, Figs. 8.14g and 8.15a) is native to North America (northern and eastern USA and south-eastern Canada) where it grows in dry forests, forest clearings, open rocky sites, grasslands, along roads and railways, and as a weed in gardens and fields. Its introduced range covers Europe, and parts of Asia and New Zealand. Introduced into Europe as an ornamental in the early eighteenth century, its later spread across the continent was mostly due to unintentional introductions as a seed and soil contaminant. The earliest record of this species in the Czech Republic is from 1884. There are two subspecies differing in their distribution in the Czech Republic: **subsp. *septentrionalis*** is more common, while **subsp. *annuus*** is scattered. However, its precise distribution and invasion history is difficult to outline as both subspecies are easy to confuse, making literature reports less reliable. It grows in 12 habitat types, with an ecological optimum in the herbaceous fringes of lowland rivers, forest clearings and perennial thermophilous ruderal vegetation (Fig. 8.8), and is occasionally planted in gardens. It prefers light and dry habitats, with its recent spread supported by large-scale disturbances, eutrophication and extension of suburban areas. In the second half of the twentieth century it was mostly an urban weed limited to gardens and waste places but currently it is present in open landscapes, where it occurs not only at disturbed sites but also in

**Fig. 8.15** Examples of invasive plant species in the Czech Republic: (**a**) *Erigeron annuus*, Klecany, central Bohemia; (**b**) *Galinsoga parviflora*, Brno; (**c**) *G. quadriradiata*, Ketkovice, southern Moravia; (**d**) *Helianthus tuberosus*, Prague; (**e**) *Heracleum mantegazzianum*, Františkovy Lázně, western Bohemia. Photo credits: P. Pyšek (a) and M. Chytrý (b–e)

semi-natural grassland. Its invasion is supported by disturbance of the soil surface by occasional mowing. The impact of this species is as a weed of arable land.

***Fraxinus pennsylvanica*** (*Oleaceae*) is a deciduous tree up to 25 m tall, native to eastern North America, where it is a light-demanding and early successional species in a wide range of environmental conditions, although it mostly occurs on moist and nutrient-rich soils along rivers. The first record of its planting in Europe is from 1783, and in the Czech Republic from 1835. This species is planted in towns and occasionally also in forests. It spreads into natural and semi-natural alluvial forests, where it thrives best. It also occurs in scrub and pioneer woodlands in forest clearings and spreads in suburbs and coal-mining areas, where it is planted for habitat reclamation. As a fast-growing and early-reproducing woody plant, it establishes in ruderal vegetation in waste places early in succession and later invades grassland, scrub and forest margins. Its distribution is scattered, confined mainly to large river floodplains. At present it is spreading fast but its invasion is in the initial phase.

***Galinsoga parviflora*** (*Asteraceae*, Figs. 8.14h and 8.15b) is an annual species native to the South American Andes, with secondary distribution in temperate and subtropical regions. It is recorded growing from the lowlands up to 3600 m a.s.l. As for its congener *G. quadriradiata*, its natural habitat in its native range are floodplains. In the invaded range it occurs as a weed in arable fields and gardens as well as in disturbed urban habitats. A single plant can produce up to 6000 highly germinable cypselae. The earliest record of *G. parviflora* in Europe is from 1785 at a botanical garden in Paris, France. In the Czech Republic it was first reported in 1880 and records started to accumulate rapidly; it spread very quickly from the 1920s to the 1940s, when this species became common. Following the initial introduction into botanical gardens and subsequent escape, the main pathway of further spread was as a contaminant of soil and crop seed. This species occurs in a wide range of habitats in the Czech Republic (13; Fig. 8.8), however, in most of them only occasionally. It requires well-aerated or loose, moist and nitrogen-rich soil, but cannot tolerate a high cover of other species and is sensitive to frost; these factors limit its colonization of sites at higher altitudes. It is abundant in gardens, on rubbish and compost heaps, along walls in cities and in root-crops. In the Czech Republic it is common and has colonized the majority of sites with suitable habitats. Recently, for example, it has started to appear in road verges. This species spreads both by seed and clonally by adventitious roots at the stem base; if weeded manually, rooting bases of stems remain in the soil and ripening of cypselae continues for a few days after removal. These features make it a serious weed in crops. It is a host plant of some agriculturally important viruses, insects and nematodes.

***Galinsoga quadriradiata*** (*Asteraceae*, Figs. 8.15c and 8.16a) is an annual plant native to South and Central America, where it occurs in periodically flooded sites along rivers. As for its previously described congener, it has been introduced into other continents where it occurs in similar human-made habitats. It was introduced into Europe in the nineteenth century. The earliest record of planting in the Czech Republic is from 1823 and that of escaped plants from Prague in 1890. Both species are very similar in terms of pathways of introduction and dispersal, ecological requirements and habitat affinities. The rapid spread of this species occurred later than that of its congener, during the second half of the twentieth century. Currently

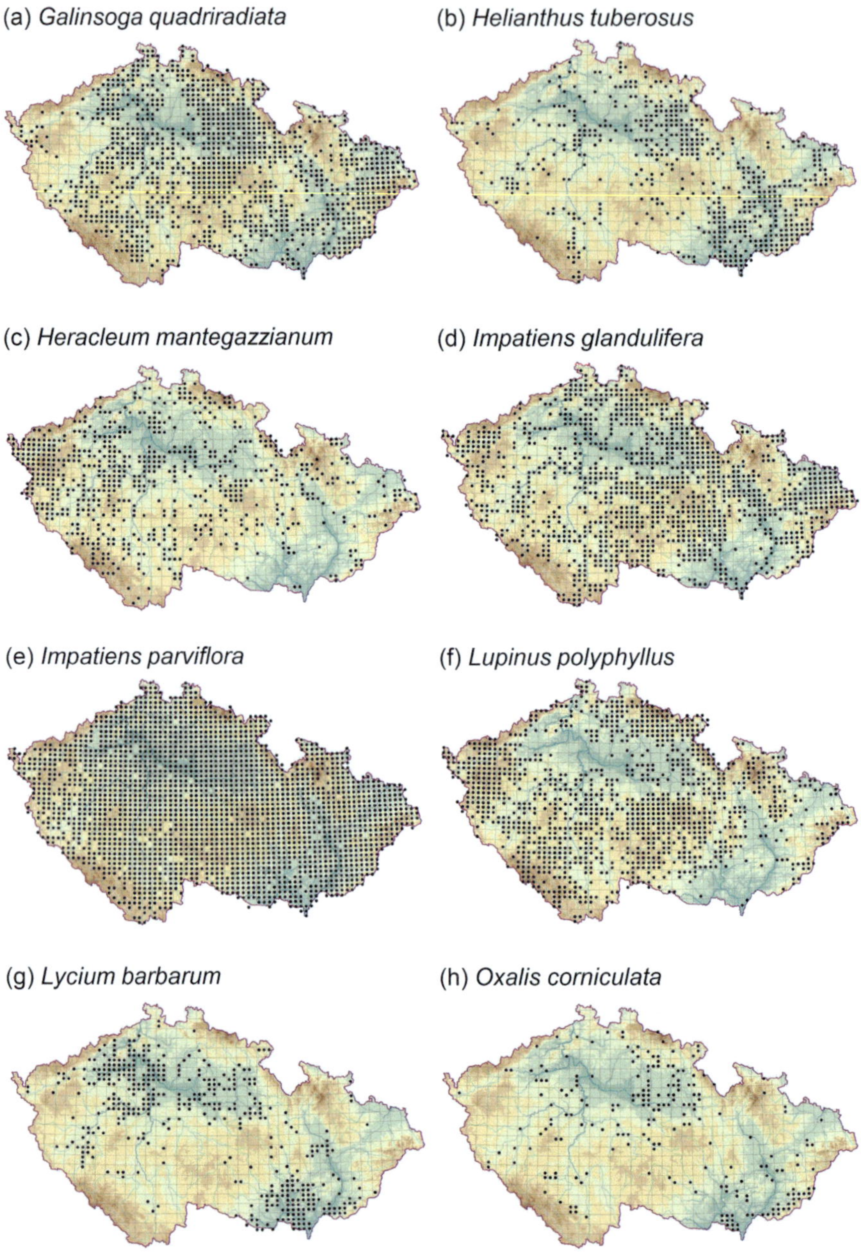


**Fig. 8.16** Distribution of selected invasive plant species in the Czech Republic

it is less common than *G. parviflora* but also occurs in most of this country. It is tolerant of high concentrations of salt and heavy-metals. Its impact is that of a serious weed in agricultural crops and as a host plant of some agriculturally important viruses, insects and nematodes.

***Helianthus tuberosus*** (*Asteraceae*, Figs. 8.15d and 8.16b) is a perennial herbaceous plant up to 3 m tall, reproducing by seeds and tubers. This species is a native to the central and eastern parts of the USA and south-eastern Canada, where it grows in wet meadows and abandoned fields. In its invaded range (parts of North and South America, northern Africa, Australasia, temperate Asia and Europe) it occupies open ruderal sites along roads and rivers as well as field edges and urban habitats with nutrient-rich soils. This species was introduced into Europe probably in the seventeenth century. The first record in the wild in the Czech Republic dates back to 1885. *Helianthus tuberosus* is planted for ornamental purposes and as food for wild animals, namely wild boar. Recently, the tubers started to be more popular as dietetic food. This species prefers clayey, humid, nutrient-rich soils and spreads locally near plantings, such as in villages and forest openings. It is, however, most invasive in floodplains where tubers are spread by floods; on fluvial sediments tubers can sprout from a depth of up to 1 m. It is invasive particularly along large rivers in Moravia and in the surroundings of lowland settlements. *Helianthus tuberosus* is assigned to seven habitat types and is dominant in nitrophilous herbaceous fringes of lowland rivers, but also thrives in perennial ruderal vegetation at warm and mesic sites. Primary (F1) hybrids with other species of the genus are not recorded. The probability of their occurrence is low due to the late flowering of *H. tuberosus*, which prevents the seed from ripening under local conditions. Impact on the species diversity of invaded communities is reported from the Czech Republic: stands reaching 50–100% cover reduce species richness by ~30%.

***Heracleum mantegazzianum*** (*Apiaceae*, Figs. 8.15e and 8.16c) is usually monocarpic and short-lived perennial species which can live up to 13 years, reproducing exclusively by seed, forming a short-term persistent seed bank lasting for at least 5 years and persisting in some localities for decades. A single plant produces on average 20,000 seeds of which up to 90% germinate. It is native to the Western Greater Caucasus, where it grows in tall-forb vegetation below the tree line, on forest clearings and along forest margins. In 1817 it was introduced into Europe as a garden ornamental and multiple introductions followed. Now it is considered as invasive or naturalized in many European countries and North America. The invasion of *H. mantegazzianum* in the Czech Republic is recorded in detail. The first record comes from 1862, when it was introduced into the park of the Kynžvart chateau in western Bohemia. Fifteen years later it was found to have escaped and naturalized in the close vicinity of this garden. The duration of the lag phase, or the time between its introduction and the start of its exponential spread, is estimated as ~60 to 70 years, with the rapid invasion starting in the 1940s. Rapid spread and increase in abundance of *H. mantegazzianum* was promoted by a radical change in land use and human-mediated disturbances after World War II, especially in the west of the country, where the species was originally introduced. *Heracleum mantegazzianum* invades nutrient-rich sites in semi-natural grasslands, forest edges and anthropogenic habitats. However, it is also able to establish in nutrient-poor habitats, such as fen meadows or forest clearings on acidic soils. Based on local conditions, it can form large populations of thousands of individuals; however, more often it is found in smaller populations of few individuals along linear landscape features such as roads and streams. In the Czech Republic, it occurs mainly in western Bohemia, from where it spread eastwards. In other parts of the country it forms mostly small stands. It rarely occurs in dry and warm lowlands. *Heracleum mantegazzianum* is reported to reduce species diversity of invaded communities; stands with 70–100% cover reduce species richness by 50–60%. This species is harmful to humans due to its phytotoxic sap causing blisters on the skin. It is difficult to eradicate due to the existence of a seed bank and high regeneration ability.

***Impatiens glandulifera*** (*Balsaminaceae*, Figs. 8.16d and 8.17a) is an annual species native to the Himalayas, reaching up to 2.5 m in height and one of the most invasive species in Europe. It was introduced as a garden ornamental into Europe (UK) in 1839 and first recorded as escaped in 1855. Currently it is recorded in 35 European countries. It has also been introduced into North America. In both distribution ranges it grows on disturbed riverbanks, in roadside ditches and at forest edges. In the Czech Republic, the first record of its planting as a garden ornamental comes from 1846 and that of an occurrence outside cultivation from 1896. However, rapid invasion only started in the mid-twentieth century and currently the species is common in this country but only rarely cultivated. *Impatiens glandulifera* is a dominant species in the nitrophilous herbaceous fringes of lowland rivers. It also finds optimum conditions in willow galleries on loamy and sandy river banks and in riverine reed vegetation, and it occurs in 16 habitat types. It also invades fresh soil heaps, forest clearings, forest edges on slopes and forest road verges. Until recently its invasion in the Czech Republic was restricted to the floodplains and surroundings of villages with nutrient-rich humus and permanently moist soils. However, in the last few decades this species has started to widen its habitat niche by spreading outside floodplains, such as in forest clearings and abandoned meadows. At present, it is beginning to colonize drier sites, which in some cases are less rich in nutrients and shaded by trees. This species produces a higher biomass than its congeners and is plastic in terms of its response to nutrient availability and shading, and also exhibits some genetically based population differentiation. Its competitive ability in the Czech Republic may be reduced by late frosts, to which seedlings are sensitive. It regenerates well after disturbance by means of adventitious roots growing from stem nodes. Due to its rapid spread and extensive populations in riparian habitats, it is considered a conservation problem. However, despite forming populations with a cover of up to 90%, it does not markedly reduce the numbers of species co-occurring in invaded stands, although invasion does alter species composition. *Impatiens glandulifera* is also known to reduce the availability of pollinators for co-occurring native species.

**Fig. 8.17** Examples of invasive plant species in the Czech Republic: (**a**) *Impatiens glandulifera*, Radošov, western Bohemia; (**b**) *Impatiens parviflora*, Aš, western Bohemia; (**c**) *Lupinus polyphyllus*, Slavkovský les Mts, western Bohemia; (**d**) *Lycium barbarum*, Výhon hill near Blučina, southern Moravia; (**e**) *Prunus serotina*, Hory, western Bohemia. Photo credits: M. Chytrý (a, b, d, e) and P. Pyšek (c)

***Impatiens parviflora*** (*Balsaminaceae*, Figs. 8.16e and 8.17b) is an annual herbaceous plant assumed to be native to the mountains in Central Asia (including the southern part of western Siberia, western Mongolia, the adjacent Turanian region and the western Himalayas). Its habitats in its native range include shaded stream-sides and stony mountain slopes. In its introduced range, it is widely distributed in Europe, Africa, North America and Asia; the first record in Europe is from 1831 from a botanical garden in Geneva. In the Czech Republic it was first recorded in 1844 in a botanical garden in Prague, and in the wild in about 1870. It became widely distributed in the Czech Republic after World War II, with the period of its most rapid spread occurring in the 1970s and 1980s, when it dominated the understory in both natural and cultivated forests. Subsequently, its invasion was slowed by an introduced monophagous aphid, *Impatientium asiaticum*. At present, *I. parviflora* is common all over the country except in treeless landscapes or nutrient-poor coniferous forests. It is, however, less dominant than three decades ago and appears to have already colonized the majority of suitable habitats. *Impatiens parviflora* is strongly confined to sites shaded by trees, preferring loose, moist humus- and nutrient-rich soils, and invading both disturbed ruderal habitats in and around settlements and more natural forest habitats. It is recorded from 45 habitat types and is a dominant in a number of them (perennial nitrophilous herbaceous vegetation at mesic sites, alluvial forests, oak-hornbeam forests and ravine forests) and a frequent dominant in *Robinia pseudoacacia* plantations (Fig. 8.8). In the Czech Republic *I. parviflora* is reported to be less plastic in terms of its response to nutrients and shading than its congener *I. glandulifera*, but exhibits stronger genetically based population differentiation. It is also highly sensitive to frosts. Its impact on native biodiversity in the invaded range is probably weak because of its poor competitive ability.

***Lupinus polyphyllus*** (*Fabaceae*, Figs. 8.16f and 8.17c) is a rhizomatous perennial up to 1.6 m tall, native to western North America, where it grows in wet mountain grasslands and along streams. It has been introduced and became naturalized in the eastern parts of North America and in Europe, including its northern part. In Europe it was first recorded in the UK in 1826 and in the Czech Republic in 1895. It is planted as a garden ornamental, but the invasion was more promoted by it being sown in the wild. From the late nineteenth century it was sown, as a nitrogen-fixing plant, in forests for enrichment and amelioration of acidic soils and as a pasture crop for animals. It is still used for stabilizing the soil of banks along roads and railways. It differs from most other invasive herbaceous plants in the Czech Republic in being confined to non-calcareous, slightly moist and nutrient-poor soils in cold hilly landscapes and foothills. A genetic link with ornamentals cultivated as *L.* ×*hybridus* is probable though not proven. This species is common in the Czech Republic, especially in the west. It is recorded in 14 habitat types and can be dominant in perennial thermophilous ruderal vegetation (Fig. 8.8). It invades unmanaged grasslands along road sides and railways, at ruderal and disturbed sites, in meadows and the margins of forests, or on sandy soils. However, at present *L. polyphyllus* exerts its main impact by dominating vegetation in game preserves, former military areas and slag heaps. Despite forming stands with a cover of 60–95% in invaded vegetation, its effect on native species richness is rather moderate, with ~20% reduction being reported.

***Lycium barbarum*** (*Solanaceae*, Figs. 8.16g and 8.17d) is a shrub up to 3 m tall. Some authors consider it to be a native to South-eastern Europe and Asian parts of the

Mediterranean area, but it is more likely a native to China. It has been introduced into other parts of Europe and Asia, northern Africa and North America. The first record of planting in the Czech Republic is from 1785 and that of its occurrence in the wild is from 1870. This species is rather common, being more abundant in warm areas. Its distribution to a large extent still corresponds to plantings along railways in the nineteenth and early twentieth century. In warm and dry regions it was also planted in hedgerows, as an ornamental shrub and for soil stabilization in the past, but currently is only rarely used. It is a deep-rooting, light-demanding species, colonizing deep, relatively dry, nutrient-rich basic soils at exposed or disturbed sites. Most stands originate from planted shrubs that spread by vigorous clonal growth; it does not reproduce by seed in the Czech Republic as it rarely bears fruit and seedlings do not establish. Long-distance dispersal is by root fragments and rooting branches that get dispersed following disturbance, e.g. during the remodelling of railway corridors. Populations last for a long time and are supported by cutting, which reduces competition from co-occurring trees and initiates new growth. It occurs as a dense stout shrub, a climber in stands of trees, or a small shrub on extremely dry substrates. It grows in 11 habitat types (Fig. 8.8). The invasion of the Czech Republic seems to have reached its peak: the populations persist and increase in size, but new stands rarely develop. The ability of *L. barbarum* to rapidly occupy new sites and exert a strong impact on native diversity is documented for other parts of Europe.

***Oxalis corniculata*** (*Oxalidaceae*, Fig. 8.16h) is an annual, biennial or short-lived creeping perennial herbaceous plant with a native distribution range in the tropical and subtropical belts of Asia and Africa, including the Mediterranean region. It has been introduced into the temperate zones of the Northern Hemisphere (Europe, North America and Asia) where it became naturalized in open disturbed sites along roads, in the edges of fields and urban gardens. It occurs as a serious weed in gardens, fields, lawns and glasshouses. In the Czech Republic it was first reported in 1852 and is most likely still being introduced with contaminated soil. It occurs in nine habitat types and thrives best, with an ecological optimum in annual vegetation on arable land and annual vegetation in trampled habitats. At present its occurrence is still mostly confined to cities and villages with its invasion being supported by the spread of suburban areas with garden allotments. It spreads, both by seed and stem fragments, colonizing loose well-aerated soils such as garden beds, in glasshouses or heaps of garden substrata. Its ability to grow fast makes it a noxious garden weed, especially early in the season when it profits from competitive superiority over seedlings of other species. It also occurs in joints between sections of pavement, in railway areas and along walls, but only rarely as a weed of agricultural crops.

***Oxalis dillenii*** (*Oxalidaceae*, Fig. 8.18a) is an annual, biennial or short-lived erect, decumbent or prostrate, rarely creeping perennial herbaceous plant that is native to eastern and central North America, where it grows in prairies and broad-leaved forests, but also in a variety of disturbed habitats. It was most likely introduced into Europe during the first half of nineteenth century as a garden ornamental. It became naturalized starting in the second half of the nineteenth century, but its spread was recorded only after World War II. The likely pathway of introduction

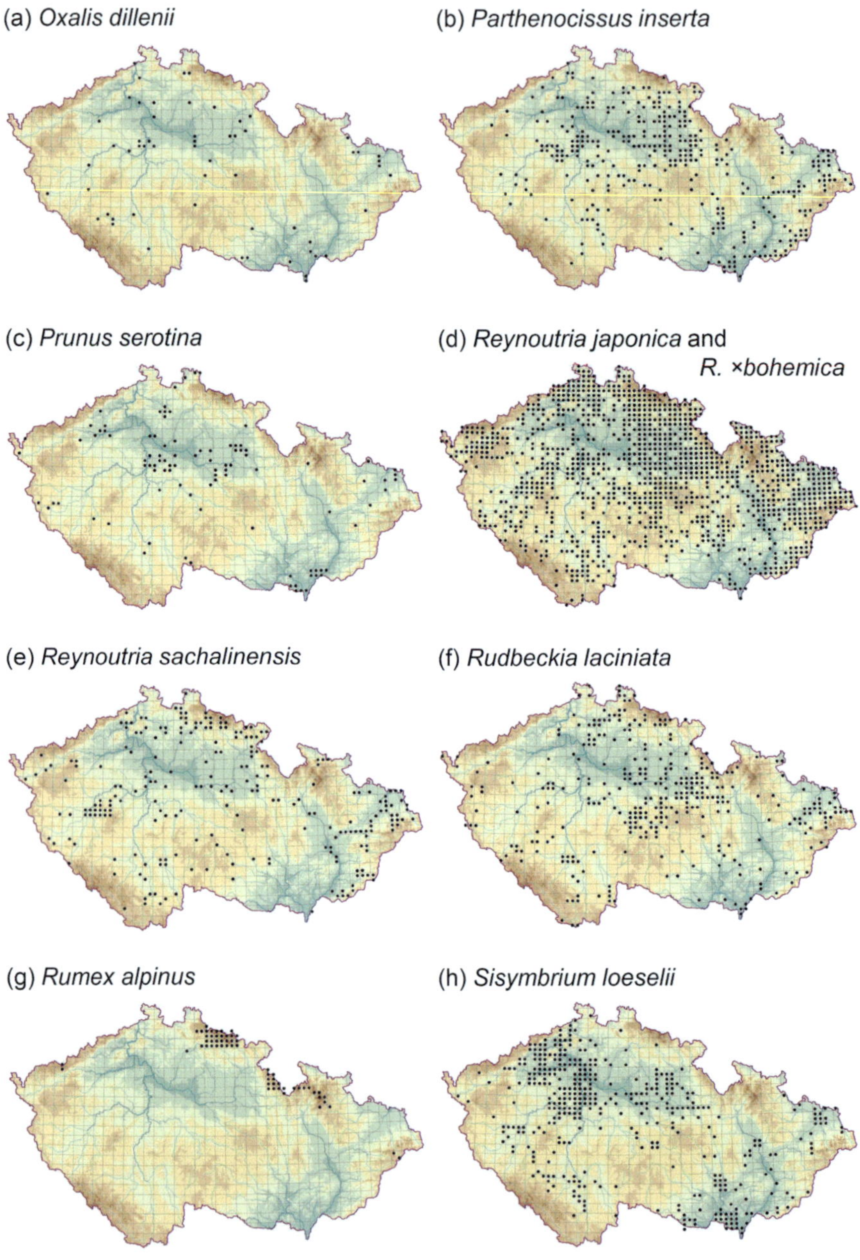


**Fig. 8.18** Distribution of selected invasive plant species in the Czech Republic. Note that *Oxalis dillenii* is probably under-recorded

into the Czech Republic is via contaminated soil. It is confined to warm lowland areas and occurs in nine habitats, mostly in settlements and suburban areas. The invasion started during the last decade when this species began to spread in a wide range of habitats such as waste places, rubbish dumps, along paths and railways, field margins, open grassland, root crops and fodder crops. It is likely to continue spreading, but because it is a weak competitor which does not dominate vegetation, it is unlikely to have a serious impact.

***Parthenocissus inserta*** (*Vitaceae*, Fig. 8.18b) is a woody vine native to North America. It was introduced into Europe about 1800 and the first record of an escaped population in the Czech Republic is from 1900. Planted as an ornamental species in parks and gardens, it escapes and spreads in villages and their surroundings, from where it invades alluvial forests in the warmer parts of this country (southern Moravia, central Bohemia). This species requires moist, nutrient-rich soils and spreads slowly, supported by succession of ruderal scrub towards tree-dominated stands. It is able to propagate both by seed and clonally, which enhances its invasive potential and impact on invaded vegetation.

***Pinus strobus*** (*Pinaceae*) is a coniferous tree with its native distribution range in eastern North America and a part of Central America. In its native range it is an important forestry species forming extensive, often monodominant, forests. *Pinus strobus* was introduced into Europe in 1705. The introduction to the Bohemian lands for forestry purposes, with the first record in 1784, predates that for ornamental purposes (1812). In the Czech Republic this species is locally invasive on sandy acidic soils in sandstone areas, especially in the Elbe Sandstones (Bohemian Switzerland), where it gradually outcompetes the native *P. sylvestris*. It was first planted there in 1798 and started to spread in the 1950s in mixed forests, but also in other habitat types, including open vegetation on rock outcrops. Its invasion and impact is restricted to extremely poor soils; the native pine regenerates poorly in *P. strobus*-dominated stands. The invasion in this sandstone area has accelerated since the 1990s and is continuing. This species exerts strong impacts on invaded communities by developing dense stands with a thick layer of slowly decomposing litter.

***Populus ×canadensis*** (*Salicaceae*) is a fast-growing tree up to 40 m tall. It originated as a product of both intentional and spontaneous hybridization of the European *P. nigra* with the North American *P. deltoides* around 1750. This hybrid was widely planted for forestry purposes, in wind-breaks and as an ornamental species in parks and gardens. The first record of cultivation of *P. ×canadensis* in the Czech Republic is from 1852. In the Czech Republic it reproduces both by seed and root suckers and can spread a great distance from plantings. *Populus ×canadensis* invades areas with disturbed soils along streams, which is reflected in its distribution in the Czech Republic being concentrated in floodplains, however, it is often found also in urban and suburban areas and sand pits. It is often planted at sites of former hardwood floodplain forests, but it also spontaneously occurs in more natural alluvial forests, willow galleries on loamy and sandy river banks and in other habitats (14 habitat types, Fig. 8.8). The spread is supported by it being more drought resistant, compared to *P. nigra*. The recorded impact of this hybrid is through further hybridization with the native *P. nigra*, which is at present exacerbated by utilization of *P.* ×*canadensis* as a biofuel plant.

***Prunus serotina*** (*Rosaceae*, Figs. 8.17e and 8.18c) is native to large areas in eastern North America and Central America, where it grows as a tree up to 35 m tall. It reproduces by seed dispersed by birds and small vertebrates, as well as vegetatively through the formation of dense polycorms. Its habitat in its native range is deciduous and pine forests. The species has been introduced into Europe and Asia. The first record of a European introduction is from 1623 in France. *Prunus serotina* was planted as a forest and park tree, and since it is tolerant of air pollution and poor soil, it was also widely used in urban areas, for soil amelioration and for reclamation in mining areas. Following its introduction this species was released from the effect of a parasitic fungus that is known to control its abundance in its native range, which might have contributed to its invasiveness. In the Czech Republic, the species was introduced into cultivation in 1811. In this country it is a small tree or shrub reaching 3–6 m in height. It occurs on various soils, but prefers moist, acidic, well-drained soils. Its niche in the Czech Republic is still rather narrow. It occurs in six habitat types, thriving best in alluvial forests and acidophilous oak forests. It also spreads in forest clearings and along forest paths and roads, reducing the abundance and richness of understorey species. It is also able to establish in abandoned fields on sandy soils and in sand pits. Invasion in the Czech Republic is currently markedly accelerating, especially in the lowlands, being supported by ongoing eutrophication of nutrient-poor forests, and is also beginning to colonize treeless landscapes. The distribution in the country is still rather restricted. *Prunus serotina* exerts an impact on native species by being a strong competitor and forming thickets. Impact on humans can be through bark and seeds that are toxic.

***Quercus rubra*** (*Fagaceae*, Fig. 8.19a) is a tree up to 40 m tall, native to a large area in eastern North America where it is an important source of timber. It grows in a wide range of dry-mesic to mesic sites and occurs in various habitats ranging from nutrient-rich soils to sandy plains and rock outcrops. Currently this species is widely cultivated in the temperate zone of Europe and Asia as a popular forestry and ornamental tree. It was introduced into Europe in 1691 and has been planted in the Czech Republic since 1799. In the Czech Republic, it was mostly used as a garden and park ornamental until several decades ago when it started to be introduced into forest plantations, often in monocultures; in the 2000s it was planted on more than 4000 ha. It is also used for reclaiming post-mining areas and for afforestation of arable land. It is now widespread and spreading mainly in central and eastern Bohemia in the Labe River lowland. It is recorded from 14 habitat types (Fig. 8.8). *Quercus rubra* has a short juvenile period and spreads into surrounding vegetation because it is more shade-tolerant than native oaks. It prefers open forest on light, nutrient-poor soils. Acorns are spread outside forests by birds, but populations are only able to establish in nutrient-poor mesic habitats. As a fast- and vigorously growing, shade-tolerant tree, it has an impact on forest understorey, exacerbated by its slowly decomposing leaf litter inhibiting succession.

***Reynoutria japonica* var. *japonica*** and ***R. sachalinensis*** (*Polygonaceae*, Figs. 8.18d and 8.19b) are rhizomatous perennials native to East Asia, from where

**Fig. 8.19** Examples of invasive plant species in the Czech Republic: (**a**) *Quercus rubra*, Chlum u Třeboně, southern Bohemia; (**b**) *Reynoutria japonica*, Prague; (**c**) *Robinia pseudoacacia*, Veverská Bítýška, southern Moravia; (**d**) *Rudbeckia laciniata*, Třeboň, southern Bohemia; (**e**) *Solidago canadensis*, Prague; (**f**) *S. gigantea*, Lednice, southern Moravia. Photo credits: M. Chytrý (a, c, d) and P. Pyšek (b, e, f)

they were introduced into Europe as garden ornamentals and fodder plants in the nineteenth century. In the Czech Republic the genus *Reynoutria* (syn. *Fallopia* p.p.) is represented also by the invasive hybrid ***R. ×bohemica***. This hybrid, first grown in gardens in the UK in 1872, is likely to have arisen on this continent several times independently and is also known from the native range of the parental species. In their native range, species of *Reynoutria* grow along rivers and in disturbed open habitats and at the edges of forests. *Reynoutria japonica* var. *japonica* is also known to colonize volcanic slopes and bare lava fields. The invaded range includes Europe, North America and Australasia. All three taxa have been introduced into the Czech Republic as garden ornamentals; the first record of *R. japonica* var. *japonica* in cultivation is from 1883, and outside cultivation from 1902. *Reynoutria sachalinensis* was first collected in 1921, and the earliest record of the hybrid *R. ×bohemica* is from 1950. The invasion occurred in the second half of the twentieth century when this species colonized most of this country. That of the hybrid lagged behind the two parental species but proceeded faster because it is competitively superior to its parents. In the early 2000s, *R. japonica* var. *japonica* was recorded from 1335 localities, *R. sachalinensis* from 261 and the hybrid from 382. Typical of the *Reynoutria* complex of taxa is the large intraspecific variation in ploidy in both the native and invaded ranges, and interspecific hybridization that is common especially in invaded regions. In the Czech Republic, all three taxa are still planted as garden ornamentals. Their dispersal is mainly vegetative by the regeneration from rhizome and stem segments transported in contaminated soil and water. Sexual reproduction rarely occurs, being constrained by the lack of pollen grains in some taxa or inefficient seedling establishment. *Reynoutria japonica* var. *japonica* has been introduced into Europe as a single female clone, which spread across the continent. *Reynoutria* taxa became widely naturalized and are now among the worst invasive plants in the Czech Republic in terms of their impact. They occur in a number of habitat types (12) and are dominant in perennial nitrophilous herbaceous vegetation growing at mesic to wet sites (Fig. 8.8). They are demanding in terms of moisture and nitrogen and invade settlements and floodplains, where the invasion is supported by periodical large-scale disturbances such as floods, during which rhizomes are spread and new habitats suitable for colonization created. The strongest impact is recorded in river floodplains in north-eastern Moravia or northern Bohemia. Due to their high competitive ability, high biomass production and efficient vegetative reproduction, knotweeds are classified as transformer species (sensu Richardson et al. 2000) that cause alterations in water regime and displace native plant species. The invasion by *Reynoutria* taxa has the most severe impact on species richness and diversity of any of the Central European alien plants, reducing the number of species present prior to invasion by 66–86%, depending on the taxon. *Reynoutria* taxa are not reported to affect human health, but do affect the infrastructure by damaging roads and flood-prevention structures, and increasing the erosion potential of rivers. For *R. japonica*, a biological control agent, the psyllid *Aphalara itadori*, was recently released in the UK.

***Robinia pseudoacacia*** (*Fabaceae*, Fig. 8.19c) is a deciduous tree up to 30 m tall. In its native range in central and eastern North America it grows as an early successional species in open and disturbed habitats. The tree has a good regeneration capacity, resprouting well from roots and stumps. The invaded range includes temperate areas. *Robinia pseudoacacia* was introduced into Europe in 1601 as an ornamental species. Later it was used for timber production and for erosion control. It has been cultivated in the Czech Republic since 1710, after which it was widely planted in parks. The first record of its spontaneous occurrence in the wild is from 1874. Since the 1760s it has been planted for timber, erosion control, to support soil eutrophication and as a bee plant. It grows up to ~500 m a.s.l., mostly in central Bohemia and southern Moravia, locally in extensive groves. It is tolerant of drought and air pollution, and grows on sandy, poorly drained and saline substrates. It is resistant to fire and trimming and does not suffer from attacks by pests and diseases. It reproduces by seed, but seedlings only establish following disturbance at warm sites. Spread is by root suckers, and root fragments also contribute to dispersal. This is why most populations are found close to original plantings, whereas long-distance dispersal happens only occasionally, and is mostly linked to transport of substrate from mines, quarries and sand pits. Its local spread is accelerated by attempts at eradication that mostly fail and only serve to stimulate resprouting. The species is recorded in 24 habitats (Fig. 8.8). The whole plant is toxic for humans and cattle, and produces nitrogene compounds and allelopathic substances that inhibit germination and growth of native species. However, stands of *R. pseudoacacia* harbour some scarce native species, especially vernal geophytes (e.g. of the genera *Allium* and *Gagea*), that are able to tolerate the effects of the allelopathic compounds. Its poor ability of spreading over long-distance and the difficulty in eradicating it are probably the reasons why old populations of *R. pseudoacacia* are often tolerated by nature conservation authorities, except when they invade steppe vegetation with a high conservation value.

***Rudbeckia laciniata*** (*Asteraceae*, Figs. 8.18f and 8.19d) is a stout perennial species reproducing both by seed and rhizome fragments. Its native range is in northeastern Canada and eastern and central USA, where it grows along streams and in wet habitats from the lowlands to the mountains. The invaded range includes Europe, spans central Russia and the Caucasus, China, Japan and New Zealand. It has been planted in Europe since the early seventeenth century and in the Czech Republic since the nineteenth century. At present a number of cultivated forms as well as hybrids with *R. nitida* are frequently planted in gardens, but only the one with the oldest history of planting has escaped from cultivation. The first record in the wild from the Czech Republic dates to 1895. Currently this species is common, namely in eastern, northern and south-western parts of this country. *Rudbeckia laciniata* efficiently spreads by rhizome fragments dispersed by water in riparian habitats, wet meadows, and along roads and railways with disturbed soil. It occurs in nine habitat types. About 80% of the populations persist at invaded sites for many decades, with a maximum of up to 135 years recorded for one clone. This species forms stands that may reach cover values of up to 80–100% and reduce species diversity compared to uninvaded stands by ~30%, depending on the measure used.

***Rumex alpinus*** (*Polygonaceae*, Fig. 8.18g) is a clonal herbaceous perennial up to 1.5 m tall, native to the mountains of Central and Southern Europe, and the Caucasus region and northern Turkey. The species was introduced into mountainous areas elsewhere in Europe, Great Britain, North America and South-east Asia. The first literature record from the Czech Republic is from 1819, but this species might have been introduced as early as the second half of the sixteenth or the early seventeenth century by woodcutters from the Alps coming to the Krkonoše and Orlické hory Mts. This species occurs in mountain ranges in the northern part of this country, with invasive populations concentrated in the Krkonoše Mts where rapid invasion started after World War II, supported by abandonment of mountain grasslands. Plants spread by rhizomes and invade abandoned nutrient-rich meadows or cattle pastures. This species also occurs in a variety of habitats including some semi-natural ones, e.g. gravel riverbanks and fringes of montane brooks. Current spread is supported by the ruderalization of mountain landscapes. Invasive populations may reach a cover of up to 75–100% and reduce species richness of invaded communities by ~50%.

***Rumex longifolius* subsp. *sourekii*** (*Polygonaceae*) is a rhizomatous perennial up to 1.8 m tall. This subspecies is described from the Czech Republic. *Rumex longifolius* is native to the Pyrenees and mountains in Scotland, Ireland and Scandinavia, where it grows in moist grasslands and along streams. The first record of invasive populations in the Czech Republic comes from 1961, and a rapid invasion followed towards the end of the twentieth century; in the 1990s *R. longifolius* subsp. *sourekii* was locally common in the Krkonoše, Krušné hory and Jizerské hory Mts, and it is still spreading. The populations are restricted to mountain areas where they occur at disturbed sites along water courses, in ruderalized or abandoned meadows and pastures, road verges and human settlements, finding their ecological optimum in perennial nitrophilous herbaceous vegetation. The subspecies *sourekii* is much more widespread than the rare type subspecies. Currently it also colonizes suitable habitats at lower altitudes. It is sensitive to mowing. Potential impact of this invasion results from the fact that *Rumex* species easily hybridize; it is recorded that this taxon has produced three hybrids with native species in the Czech Republic.

***Sisymbrium loeselii*** (*Brassicaceae*, Fig. 8.18h) is an annual plant whose native distribution ranges from Southern Europe, including the Mediterranean area, to Central Asia, where it grows on mountain slopes and in disturbed road verges and field edges. The invaded range includes the rest of Europe and North America. The first record in the Czech Republic dates to 1819. It is more common in warm areas. It occurs in a wide range of habitat types (16), being dominant in annual vegetation in ruderal habitats (Fig. 8.8). In the past populations occurred mainly in urban areas, but over the last 30 years it has started to spread into open landscapes. Currently it occurs in villages, quarries and mining areas, on reclaimed spoil heaps and in abandoned fields, and also colonizes disturbed steppe vegetation. It is a competitively weak, early successional thermophilous species colonizing newly created habitats,

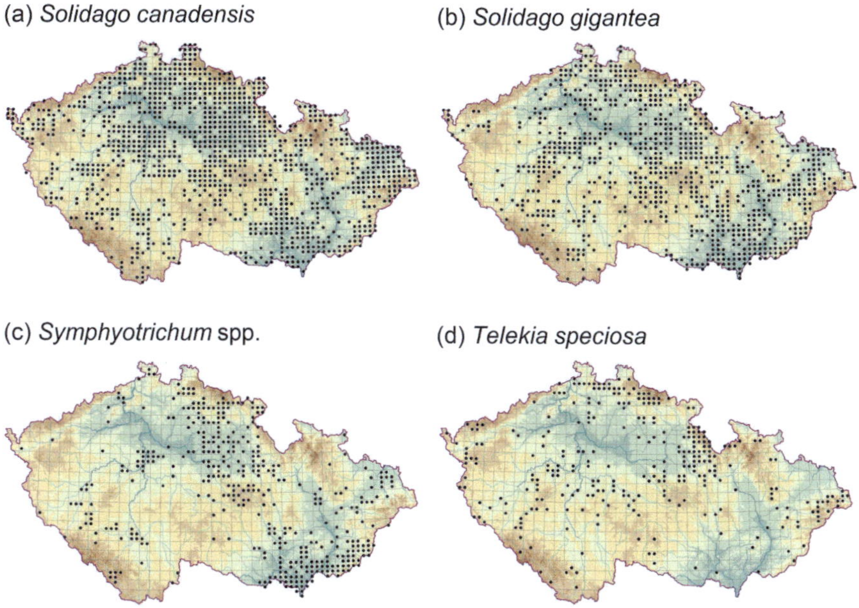


**Fig. 8.20** Distribution of selected invasive plant species in the Czech Republic. *Symphyotrichum* spp. includes mainly *S. lanceolatum* and *S. novi-belgii*

which easily becomes established following disturbance in grassland; it rarely occurs on arable land.

***Solidago canadensis*** (*Asteraceae*, Figs. 8.19e and 8.20a) is a rhizomatous herbaceous plant reproducing vegetatively and by seeds. Its native range includes the whole of North America, from Alaska and Labrador in the north to Mexico and Florida in the south. This species is naturalized in Europe, East Asia, Australia and New Zealand. In its native range it occurs at the edges of forests, along rivers and in a variety of disturbed habitats such as abandoned pastures, roadsides, abandoned fields, grasslands and urban areas. It occupies similar habitats in its invaded range. *Solidago canadensis* has been introduced into botanical gardens in Europe as an ornamental plant, with the first record on this continent being from 1645. It was introduced into the Czech Republic as a bee plant and a garden ornamental and first recorded in the wild in 1838. The invasion started as early as the first half of the nineteenth century. It is now common in most parts of this country, occurring in 13 habitat types, being dominant in perennial thermophilous ruderal vegetation (Fig. 8.8). Seeds and rhizomes are often dispersed as soil contaminants. It is suggested that this species supresses co-occurring species by means of allelopathic compounds released into the soil. A negative impact on human health due to pollen allergies is suspected, albeit without supporting data.

***Solidago gigantea*** (*Asteraceae*, Figs. 8.19f and 8.20b) is a rhizomatous perennial plant reproducing both by seeds and vigorous clonal growth of rhizomes. It is native to southern Canada and eastern USA, and has been introduced into Europe, East Asia and New Zealand. In Europe it was first planted as an ornamental plant in botanical gardens. The first record is from 1758 in London. In its native range it is found in grasslands and open forests, while in its invaded range it grows mainly in forests and in disturbed habitats along roads and railways. Introduced as a garden ornamental and important bee plant, it was first reported to occur in the wild in the Czech Republic in 1851. The invasion started in the second half of the nineteenth century, supported by seeds and rhizomes dispersed with contaminated soil. By the 1930s it formed stands along rivers and also started to spread into disturbed sites such as coal mining slag heaps in north-eastern Moravia. Compared to its congener *S. canadensis*, this species forms denser stands, prefers moister and more nutrient-rich soils and is less common. It is also more confined to riverbanks and floodplains of large rivers. It occurs in 13 habitat types and is dominant in perennial nitrophilous herbaceous vegetation at mesic and wet sites (Fig. 8.8). Impact of this species on native vegetation is similar to that of *S. canadensis*. It is known to reduce the species richness and diversity of invaded plant communities by ~25–30%.

***Symphoricarpos albus*** (*Caprifoliaceae*, Fig. 8.21a) is a shrub with a height of up to 2 m, native to the western part of North America. This species has been introduced outside its native range, including the Czech Republic, as an ornamental commonly planted in parks and gardens, along fences and roads, and in forests as shelters for pheasants. Its introduced range is almost cosmopolitan. The earliest record from Europe is probably from 1879. In the Czech Republic, it only occasionally reproduces by seed, relying on vegetative dispersal, and occurs scattered throughout the whole country, most frequently being planted and subsequently escaping in colline to submontane areas. Besides scrub vegetation it is also found in perennial nitrophilous herbaceous vegetation at mesic sites. It occurs in nine habitat types, including semi-natural vegetation such as riverine reed stands, fringes of montane brooks, willow galleries on river banks and alluvial forests. Populations persist at invaded sites for a long time and, to some extent, are resistant to shading by trees; very few other species co-occur with it in the understorey.

***Symphyotrichum lanceolatum*** (syn. *Aster lanceolatus*; *Asteraceae*, Figs. 8.20c and 8.21b) is a herbaceous perennial native to eastern North America. It was introduced into Europe in 1837 and became naturalized across most of the continent, from south-western France to southernmost Scandinavia, up to central Russia and with isolated occurrences on the Iberian Peninsula. It is frequently planted in the Czech Republic, escaping from cultivation and invading a wide range of habitat types (19), including semi-natural riparian habitats (Fig. 8.8). It occurs as a dominant in perennial nitrophilous herbaceous vegetation at mesic sites and in nitrophilous herbaceous fringes of lowland rivers, and is often found thriving in reed- and tall-sedge beds. The invasion is ongoing, particularly on river banks in southern Moravia, but also along rivers in northern Bohemia. It is known to exert significant impacts on invaded communities. Occasional mowing of *Symphyotrichum* spp.

**Fig. 8.21** Examples of invasive plant species in the Czech Republic: (**a**) *Symphoricarpos albus*, Žďár nad Sázavou, Bohemian-Moravian Highlands; (**b**) *Symphyotrichum lanceolatum*, Mikulov, southern Moravia; (**c**) *Telekia speciosa*, Průhonice, central Bohemia. Photo credits: M. Chytrý (a, b) and P. Pyšek (c)

stands (as for *Helianthus*, *Rumex* and *Solidago* mentioned above) causes their rapid regeneration resulting in emergence of more dense stands.

***Symphyotrichum novi-belgii*** (syn. *Aster novi-belgii*; *Asteraceae*, Fig. 8.20c) is a perennial species with a native distribution range covering a ~150-km-wide belt along the Atlantic coast of North America, from the Appalachian Mts to southern Canada. As for the previous species, natural habitats include riparian communities along rivers and lakes. It is naturalized in Europe, from northern Italy to southern Scandinavia, in the UK and France, with isolated occurrences in Romania and the central part of European Russia. It was introduced into Europe in 1710. The first record from the Czech Republic is from 1850. Planted as an ornamental, this species is less invasive than *S. lanceolatum*, occurring in 12 habitat types, but it also invades semi-natural habitats such as alluvial meadows of lowland rivers. It occurs as a dominant in perennial nitrophilous herbaceous vegetation at mesic and wet sites. It is known to reduce the species diversity of invaded communities by ~30–40%, depending on the measure and scale.

***Symphyotrichum ×salignum*** (syn. *Aster ×salignus*; *Asteraceae*, Fig. 8.20c) is an anecophyte resulting from the hybridization of the above two North American species, *S. lanceolatus* and *S. novi-belgii*, which most likely happened in a European garden. It is almost sterile, with less than 0.1% of cypselae ripening. At present it is cultivated and naturalized all over Europe and was first collected in the wild in 1872. The distribution in the Czech Republic is scattered. It invades mainly riparian scrub.

***Symphyotrichum versicolor*** (syn. *Aster ×versicolor*; *Asteraceae*, Fig. 8.20c) is considered to be a product of an artificial crossing between *S. laevis* and *S. novi-belgii*, probably in Europe. It is planted as an ornamental in most countries in Central and Western Europe, where it has escaped from cultivation and has become naturalized. In the Czech Republic it occurs in six habitat types, invading mainly perennial thermophilous ruderal vegetation and perennial nitrophilous herbaceous vegetation at mesic and wet sites.

***Telekia speciosa*** (*Asteraceae*, Figs. 8.20d and 8.21c) is a rhizomatous perennial up to 2 m tall, native to the mountains of Southern and Eastern Europe, northern Anatolia and the Caucasus. *Telekia speciosa* grows at the edges of mountain forests or clearings and along roads and brooks. It is naturalized in areas outside its native distribution in Europe and the European part of Russia, in similar habitats as in its native range. This nitrophilous, shade-tolerant species, requiring moist clayey soils, was introduced as a garden ornamental into the Czech Republic, with the first record in the wild dated around 1820. It is still commonly planted and escapes along streams, in the surroundings of parks and gardens where it is cultivated, as well as in other habitat types such as forest margins, old forest clearings or unmown road ditches. It thrives best in perennial nitrophilous herbaceous vegetation at mesic and wet sites. It occurs most commonly at middle altitudes, especially in north-eastern Bohemia. Impact on human health is documented: some people are allergic to its pollen and it can also cause skin irritation.

**Acknowledgements** We thank Jiří Danihelka, Zdeněk Kaplan and Marinus Werger for helpful comments on the manuscript, Ondřej Hájek for preparing the maps, Zdeňka Lososová for a photograph, Kryštof Chytrý for preparing photographic plates, Pavel Dřevojan and Zuzana Sixtová for editing the list of references, Tony Dixon for improving our English and Irena Axmanová for her help with proof correcting. This work was supported by project no. 14-36079G, Centre of Excellence Pladias (Czech Science Foundation), and long-term research development project no. RVO 67985939 (The Czech Academy of Sciences).

## References

Aikio S, Duncan RP, Hulme PE (2010) Lag-phases in alien plant invasions: separating the facts from the artefacts. Oikos 119:370–378

Alexander JM, Kuffer C, Daehler CC, Edwards PJ, Pauchard A, Seipel T (2011) Assembly of nonnative floras along elevational gradients explained by directional ecological filtering. Proceedings of the National Academy of Sciences of the USA 108:656–661

Balogh L, Dancza I, Király G (2004) A magyarországi neofitonok időszerű jegyzéke, és besorolásuk inváziós szempontból [A recent list of neophytes in Hungary and their classification according to their success]. In: Mihály B, Botta-Dukát Z (eds) Biológiai inváziók Magyarországon: Özönnövények [Biological invasions in Hungary: invasive plants]. Természet BÚVÁR Alapítvány Kiadó, Budapest, pp 61–92

Bastl M, Kočár P, Prach K, Pyšek P (1997) The effect of successional age and disturbance on the establishment of alien plants in man-made sites: an experimental approach. In: Brock JH, Wade M, Pyšek P, Green D (eds) Plant invasions: studies from North America and Europe. Backhuys Publishers, Leiden, pp 191–201

Becker T, Dietz H, Billeter R, Bush MB, Buschmann H, Edwards PJ (2005) Altitudinal distribution of alien plant species in the Swiss Alps. Perspectives in Plant Ecology, Evolution and Systematics 7:173–183

Blackburn TM, Pyšek P, Bacher S, Carlton JT, Duncan RP, Jarošík V, Wilson JRU, Richardson DM (2011) A proposed unified framework for biological invasions. Trends in Ecology and Evolution 26:333–339

Blackburn TM, Essl F, Evans T, Hulme PE, Jeschke JM, Kühn I, Kumschick S, Marková Z, Mrugała A, ... Bacher S (2014) A unified classification of alien species based on the magnitude of their environmental impacts. PLoS Biology 12:e1001850

Celesti-Grapow L, Alessandrini A, Arrigoni PV, Banfi E, Bernardo L, Bovio M, Brundu G, Cagiotti MR, Camarda I, ... Blasi C (2009) Inventory of the non-native flora of Italy. Plant Biosystems 143:386–430

Chytrý M (ed) (2007–2013) Vegetace České republiky 1–4 [Vegetation of the Czech Republic 1–4]. Academia, Praha

Chytrý M, Rafajová M (2003) Czech national phytosociological database: basic statistics of the available vegetation-plot data. Preslia 75:1–15

Chytrý M, Pyšek P, Tichý L, Knollová I, Danihelka J (2005) Invasions by alien plants in the Czech Republic: a quantitative assessment across habitats. Preslia 77:339–354

Chytrý M, Jarošík V, Pyšek P, Hájek O, Knollová I, Tichý L, Danihelka J (2008a) Separating habitat invasibility by alien plants from the actual level of invasion. Ecology 89:1541–1553

Chytrý M, Maskell LC, Pino J, Pyšek P, Vilà M, Font X, Smart SM (2008b) Habitat invasions by alien plants: a quantitative comparison among Mediterranean, subcontinental and oceanic regions of Europe. Journal of Applied Ecology 45:448–458

Chytrý M, Pyšek P, Wild J, Pino J, Maskell LC, Vilà M (2009a) European map of alien plant invasions, based on the quantitative assessment across habitats. Diversity and Distributions 15:98–107

Chytrý M, Wild J, Pyšek P, Tichý L, Danihelka J, Knollová I (2009b) Maps of the level of invasion of the Czech Republic by alien plants. Preslia 81:187–207

Chytrý M, Wild J, Pyšek P, Jarošík V, Dendoncker N, Reginster I, Pino J, Maskell L, Vilà M, ... Settele J (2012) Projecting trends in plant invasions in Europe under different scenarios of future land-use change. Global Ecology and Biogeography 21:75–87

Čuda J, Skálová H, Janovský Z, Pyšek P (2015) Competition among native and invasive *Impatiens* species: the roles of environmental factors, population density and life stage. AoB Plants 7:plv033

Čuda J, Skálová H, Janovský Z, Pyšek P (2016) Juvenile biological traits of *Impatiens* species are more strongly associated with naturalization in temperate climate than their adult traits. Perspectives in Plant Ecology, Evolution and Systematics 20:1–10

Danihelka J (2013) Botanické součty, rozdíly a podíly [Botanical counts]. Živa 61:69–72
Danihelka J, Chrtek J Jr, Kaplan Z (2012) Checklist of vascular plants of the Czech Republic. Preslia 84:647–811
Davis MA, Grime JP, Thompson K (2000) Fluctuating resources in plant communities: a general theory of invasibility. Journal of Ecology 88:528–534
di Castri F (1989) History of biological invasions with special emphasis on the Old World. In: Drake JA, Mooney HA, di Castri F, Groves RH, Kruger FJ, Rejmánek M, Williamson M (eds) Biological invasions: a global perspective. Wiley, Chichester, pp 1–30
Dostál P, Dawson W, van Kleunen M, Keser LH, Fischer M (2013) Central European plant species from more productive habitats are more invasive at a global scale. Global Ecology and Biogeography 22:64–72
Dostálek J, Frantík T, Šilarová V (2016) Changes in the distribution of alien plants along roadsides in relation to adjacent land use over the course of 40 years. Plant Biosystems 150:442–448
Essl F, Rabitsch W (eds) (2002) Neobiota in Österreich. Umweltbundesamt, Wien
Essl F, Dullinger D, Kleinbauer I (2009) Changes in the spatio-temporal patterns and habitat preferences of *Ambrosia artemisiifolia* during the invasion of Austria. Preslia 81:119–133
Essl F, Dullinger S, Rabitsch W, Hulme PE, Hülber K, Jarošík V, Kleinbauer I, Krausmann F, Kühn I, ... Pyšek P (2011) Socioeconomic legacy yields an invasion debt. Proceedings of the National Academy of Sciences of the USA 108:203–207
European Union (2014) Regulation (EU) No 1143/2014 of the European Parliament and of the Council of 22 October 2014 on the prevention and management of the introduction and spread of invasive alien species. Official Journal of the European Union L 317:35–55
Genovesi P, Carboneras C, Vilà M, Walton P (2014) EU adopts innovative legislation on invasive species: a step towards a global response to biological invasions? Biological Invasions 17:1307–1311
Hejda M (2013) Do species differ in their ability to coexist with the dominant alien *Lupinus polyphyllus*? A comparison between two distinct invaded ranges and a native range. NeoBiota 17:39–55
Hejda M, Pyšek P, Jarošík V (2009a) Impact of invasive plants on the species richness, diversity and composition of invaded communities. Journal of Ecology 97:393–403
Hejda M, Pyšek P, Pergl J, Sádlo J, Chytrý M, Jarošík V (2009b) Invasion success of alien plants: do habitat affinities in the native distribution range matter? Global Ecology and Biogeography 18:372–382
Hejda M, Chytrý M, Pergl J, Pyšek P (2015) Native-range habitats of invasive plants: are they similar to invaded-range habitats and do they differ according to the geographical direction of invasion? Diversity and Distributions 21:312–321
Hejda M, Štajerová K, Pyšek P (2017) Dominance has a biogeographical component: do plants tend to exert stronger impacts in their invaded rather than native range? Journal of Biogeography 44:18–27
Hejný S, Jehlík V, Kopecký K, Kropáč Z, Lhotská M (1973) Karanténní plevele Československa [Quarantine weeds of Czechoslovakia]. Studie ČSAV 1973(8):1–156
Hejný S, Kopecký K, Jehlík V, Krippelová T (1979) Přehled ruderálních rostlinných společenstev Československa [Survey of ruderal plant communities of Czechoslovakia]. Rozpravy Československé akademie věd, Řada matematických a přírodních věd 89(2):1–100
Hejný S, Slavík B, Chrtek J, Tomšovic P, Kovanda M (eds) (1988) Květena České socialistické republiky 1 [Flora of the Czech Socialist Republic 1]. Academia, Praha
Hejný S, Slavík B, Hrouda L, Skalický V (eds) (1990) Květena České republiky 2 [Flora of the Czech Republic 2]. Academia, Praha
Hejný S, Slavík B, Kirschner J, Křísa B (eds) (1992) Květena České republiky 3 [Flora of the Czech Republic 3]. Academia, Praha
Holub J, Jirásek V (1967) Zur Vereinheitlichung der Terminologie in der Phytogeographie. Folia Geobotanica et Phytotaxonomica 2:69–113
Horáčková J, Juřičková L, Jarošík V, Šizling A, Pyšek P (2014) Invasiveness does not predict impact: response of native land snail communities to plant invasions in riparian habitats. PLoS One 9:e108296

Hulme PE, Pyšek P, Nentwig W, Vilà M (2009) Will threat of biological invasions unite the European Union? Science 324:40–41
Jandová K, Klinerová T, Müllerová J, Pyšek P, Pergl J, Cajthaml T, Dostál P (2014) Long-term impact of *Heracleum mantegazzianum* invasion on soil chemical and biological characteristics. Soil Biology and Biochemistry 68:270–278
Jehlík V (ed) (1998) Cizí expanzivní plevele České republiky a Slovenské republiky [Alien expansive weeds of the Czech Republic and the Slovak Republic]. Academia, Praha
Jehlík V, Hejný S (1974) Main migration routes of adventitious plants in Czechoslovakia. Folia Geobotanica et Phytotaxonomica 9:241–248
Kalníková V (2012) Rozšíření invazních neofytů a sukcese na štěrkových náplavech na tocích Moravskoslezských Beskyd a jejich podhůří [Distribution of invasive neophytes and succession on gravel sediments along streams of the Moravskoslezské Beskydy Mts and their foothills]. MSc thesis, Masaryk University, Brno
Kalusová V (2009) Rostlinné invaze v aluviálních biotopech dolního toku Moravy a Dyje [Plant invasions in floodplain habitats of the lower Morava nad Dyje Rivers]. MSc thesis, Masaryk University, Brno
Kalusová V, Chytrý M, Kartesz JT, Nishino M, Pyšek P (2013) Where do they come from and where do they go? European habitats as donors of alien plants globally. Diversity and Distributions 19:199–214
Kalusová V, Chytrý M, Peet RK, Wentworth TR (2014) Alien species pool influences the level of habitat invasion in intercontinental exchange of alien plants. Global Ecology and Biogeography 23:1366–1375
Kalusová V, Chytrý M, Peet RK, Wentworth TR (2015) Intercontinental comparison of habitat levels of invasion between temperate North America and Europe. Ecology 96:3363–3373
Kaplan Z (2012) Flora and phytogeography of the Czech Republic. Preslia 84:505–574
Kowarik I (1995) Time lags in biological invasions with regard to the success and failure of alien species. In: Pyšek P, Prach K, Rejmánek M, Wade M (eds) Plant invasions: general aspects and special problems. SPB Academic Publishers, Amsterdam, pp 15–38
Křivánek M (2006) Biologické invaze a možnosti jejich předpovědi [Biological invasions and possibilities for their prediction]. Acta Průhoniciana 84:5–73
Křivánek M, Pyšek P (2006) Predicting invasions by woody species in a temperate zone: a test of three risk assessment schemes in the Czech Republic (Central Europe). Diversity and Distributions 12:319–327
Kubešová M, Moravcová L, Suda J, Jarošík V, Pyšek P (2010) Naturalized plants have smaller genomes than their non-invading relatives: a flow cytometric analysis of the Czech alien flora. Preslia 82:81–96
Kumschick S, Bacher S, Evans T, Marková Z, Pergl J, Pyšek P, Vaes-Petignat S, van der Veer G, Vilà M, Nentwig W (2015) Comparing impacts of alien plants and animals using a standard scoring system. Journal of Applied Ecology 52:552–561
Lambdon PW, Pyšek P, Basnou C, Hejda M, Arianoutsou M, Essl F, Jarošík V, Pergl J, Winter M, … Hulme PE (2008) Alien flora of Europe: species diversity, temporal trends, geographical patterns and research needs. Preslia 80:101–149
Linc O (2012) Efektivita likvidace invazních druhů v České republice na příkladu bolševníku velkolepého [The efficiency of eradication of invasive species in the Czech Republic exemplified by *Heracleum mantegazzianum*]. BSc thesis, University of Economics, Prague
Lockwood JL, Cassey P, Blackburn TM (2009) The more you introduce the more you get: the role of colonization pressure and propagule pressure in invasion ecology. Diversity and Distributions 15:904–910
Lonsdale WM (1999) Global patterns of plant invasions and the concept of invasibility. Ecology 80:1522–1536
Lososová Z, Cimalová Š (2009) Effects of different cultivation types on native and alien weed species richness and diversity in Moravia (Czech Republic). Basic and Applied Ecology 10:456–465

Lososová Z, Grulich V (2009) Chorological spectra of arable weed vegetation types in the Czech Republic. Phytocoenologia 39:235–252
Lososová Z, Simonová D (2008) Changes during the 20th century in species composition of synanthropic vegetation in Moravia (Czech Republic). Preslia 80:291–305
Lososová Z, Chytrý M, Cimalová Š, Kropáč Z, Otýpková Z, Pyšek P, Tichý L (2004) Weed vegetation of arable land in Central Europe: gradients of diversity and species composition. Journal of Vegetation Science 15:415–422
Lososová Z, de Bello F, Chytrý M, Kühn I, Pyšek P, Sádlo J, Winter M, Zelený D (2015) Alien plants invade more phylogenetically clustered communities and cause their even stronger clustering. Global Ecology and Biogeography 24:786–794
Matějček T (2008) Výskyt invazních druhů rostlin v břehové vegetaci vybraných vodních toků [Occurrence of invasive plant species in riverbank vegetation along selected water courses]. Zprávy České botanické společnosti 43, Materiály 23:169–182
McDougall KL, Alexander JM, Haider S, Pauchard A, Walsh NG, Kueffer C (2011) Alien flora of mountains: global comparisons for the development of local preventive measures against plant invasions. Diversity and Distributions 17:103–111
Medvecká J, Kliment J, Májeková J, Halada Ľ, Zaliberová M, Gojdičová E, Feráková V, Jarolímek I (2012) Inventory of the alien flora of Slovakia. Preslia 84:257–310
Mihulka S (1998) The effect of altitude on the pattern of plant invasions: a field test. In: Starfinger U, Edwards K, Kowarik I, Williamson M (eds) Plant invasions: ecological mechanisms and human responses. Backhuys, Leiden, pp 313–320
Moravcová L, Pyšek P, Jarošík V, Havlíčková V, Zákravský P (2010) Reproductive characteristics of neophytes in the Czech Republic: traits of invasive and non-invasive species. Preslia 82:365–390
Müllerová J, Pyšek P, Jarošík V, Pergl J (2005) Aerial photographs as a tool for assessing the regional dynamics of the invasive plant species *Heracleum mantegazzianum*. Journal of Applied Ecology 42:1042–1053
Nentwig W, Bacher S, Pyšek P, Vilà M, Kumschick S (2016) The Generic Impact Scoring System (GISS): a standardized tool to quantify the impacts of alien species. Environmental Monitoring and Assessment 188:315
Niklfeld H (1998) Mapping the flora of Austria and the eastern Alps. Revue Valdôtaine d'Histoire Naturelle, Suppl 51:53–62
Ööpik M, Kukk T, Kull K, Kull T (2008) The importance of human mediation in species establishment: analysis of the alien flora of Estonia. Boreal Environment Research 13:53–67
Otýpková Z, Chytrý M, Tichý L, Pechanec V, Jongepier JW, Hájek O (2011) Floristic diversity patterns in the White Carpathians Biosphere Reserve, Czech Republic. Biologia 66:266–274
Pergl J, Dušek J, Hošek M, Knapp M, Simon O, Berchová K, Bogdan V, Černá M, Poláková S, … Svobodová J (2016) Metodiky mapování a monitoringu invazních (vybraných nepůvodních) druhů [Guidelines for mapping and monitoring invasive (selected alien) species]. Botanický ústav AV ČR, Průhonice
Pergl J, Genovesi P, Pyšek P (2016b) Better management of alien species. Nature 531:173
Pergl J, Sádlo J, Petřík P, Danihelka J, Chrtek J Jr, Hejda M, Moravcová L, Perglová I, Štajerová K, Pyšek P (2016c) Dark side of the fence: ornamental plants as a source for spontaneous flora of the Czech Republic. Preslia 88:163–184
Pergl J, Sádlo J, Petrusek A, Laštůvka Z, Musil J, Perglová I, Šanda R, Šefrová H, Šíma J, … Pyšek P (2016d) Black, Grey and Watch Lists of alien species in the Czech Republic based on environmental impacts and management strategy. NeoBiota 28:1–37
Prach K, Řehounková K, Konvalinková P, Trnková R (2008) Invaze a sukcese [Invasion and succession]. Zprávy České botanické společnosti 43, Materiály 23:41–49
Pyšek P, Chytrý M (2014) Habitat invasion research: where vegetation science and invasion ecology meet. Journal of Vegetation Science 25:1181–1187
Pyšek P, Jarošík V (2005) Residence time determines the distribution of alien plants. In: Inderjit (ed) Invasive plants: ecological and agricultural aspects. Birkhäuser Verlag, Basel, pp 77–96

Pyšek P, Jarošík V, Kučera T (2002a) Patterns of invasion in temperate nature reserves. Biological Conservation 104:13–24
Pyšek P, Sádlo J, Mandák B (2002b) Catalogue of alien plants of the Czech Republic. Preslia 74:97–186
Pyšek P, Jarošík V, Kučera T (2003) Inclusion of native and alien species in temperate nature reserves: an historical study from Central Europe. Conservation Biology 17:1414–1424
Pyšek P, Sádlo J, Mandák B, Jarošík V (2003b) Czech alien flora and a historical pattern of its formation: what came first to Central Europe? Oecologia 135:122–130
Pyšek P, Chocholoušková Z, Pyšek A, Jarošík V, Chytrý M, Tichý L (2004) Trends in species diversity and composition of urban vegetation over three decades. Journal of Vegetation Science 15:781–788
Pyšek P, Richardson DM, Rejmánek M, Webster G, Williamson M, Kirschner J (2004b) Alien plants in checklists and floras: towards better communication between taxonomists and ecologists. Taxon 53:131–143
Pyšek P, Jarošík V, Chytrý M, Kropáč Z, Tichý L, Wild J (2005) Alien plants in temperate weed communities: prehistoric and recent invaders occupy different habitats. Ecology 86:772–785
Pyšek P, Křivánek M, Jarošík V (2009) Planting intensity, residence time, and species traits determine invasion success of alien woody species. Ecology 90:2734–2744
Pyšek P, Jarošík V, Hulme PE, Kühn I, Wild J, Arianoutsou M, Bacher S, Chiron F, Didžiulis V, ... Winter M (2010) Disentangling the role of environmental and human pressures on biological invasions across Europe. Proceedings of the National Academy of Sciences of the USA 107:12157–12162
Pyšek P, Jarošík V, Chytrý M, Danihelka J, Kühn I, Pergl J, Tichý L, Biesmeijer J, Ellis WN, ... Settele J (2011a) Successful invaders co-opt pollinators of native flora and accumulate insect pollinators with increasing residence time. Ecological Monographs 81:277–293
Pyšek P, Jarošík V, Pergl J (2011b) Alien plants introduced by different pathways differ in invasion success: unintentional introductions as greater threat to natural areas? PLoS One 6:e24890
Pyšek P, Jarošík V, Pergl J, Wild J (2011c) Colonization of high altitudes by alien plants over the last two centuries. Proceedings of the National Academy of Sciences of the USA 108:439–440
Pyšek P, Chytrý M, Pergl J, Sádlo J, Wild J (2012a) Plant invasions in the Czech Republic: current state, introduction dynamics, invasive species and invaded habitats. Preslia 84:575–630
Pyšek P, Danihelka J, Sádlo J, Chrtek J Jr, Chytrý M, Jarošík V, Kaplan Z, Krahulec F, Moravcová L, ... Tichý L (2012b) Catalogue of alien plants of the Czech Republic (2nd edition): checklist update, taxonomic diversity and invasion patterns. Preslia 84:155–255
Pyšek P, Jarošík V, Hulme PE, Pergl J, Hejda M, Schaffner U, Vilà M (2012) A global assessment of invasive plant impacts on resident species, communities and ecosystems: the interaction of impact measures, invading species' traits and environment. Global Ecology and Biogeography 18:1725–1737
Řehounková K, Prach K (2008) Spontaneous vegetation succession in gravel-sand pits: a potential for restoration. Restoration Ecology 16:305–312
Reif J, Hanzelka J, Kadlec T, Štrobl M, Hejda M (2016) Conservation implications of cascading effects among groups of organisms: the alien tree *Robinia pseudoacacia* in the Czech Republic as a case study. Biological Conservation 198:50–59
Rejmánek M (1989) Invasibility of plant communities. In: Drake JA, Mooney HA, di Castri F, Groves RH, Kruger FJ, Rejmánek M, Williamson M (eds) Biological invasions: a global perspective. Wiley, Chichester, pp 369–388
Rejmánek M, Richardson DM, Pyšek P (2005) Plant invasions and invasibility of plant communities. In: van der Maarel E (ed) Vegetation ecology. Blackwell, Oxford, pp 332–355
Richardson DM, Pyšek P (2012) Naturalization of introduced plants: ecological drivers of biogeographic patterns. New Phytologist 196:383–396
Richardson DM, Pyšek P, Rejmánek M, Barbour MG, Panetta FD, West CJ (2000) Naturalization and invasion of alien plants: concepts and definitions. Diversity and Distributions 6:93–107

Roy HE, Adriaens T, Aldridge DC, Bacher S, Bishop JDD, Blackburn TM, Branquart E, Brodie J, Carboneras C, … Zenetos A (2015) Invasive alien species – prioritising prevention efforts through horizon scanning. European Commission, Brussels

Rumlerová Z, Vilà M, Pergl J, Nentwig W, Pyšek P (2016) Scoring environmental and socioeconomic impacts of alien plants invasive in Europe. Biological Invasions 18:3697–3711

Sádlo J, Chytrý M, Pyšek P (2007) Regional species pools of vascular plants in habitats of the Czech Republic. Preslia 79:303–321

Schönfelder P (1999) Mapping the flora of Germany. Acta Botanica Fennica 162:43–53

Šenová V, Matějček T (2013) Zatížení biotopů geograficky nepůvodními druhy rostlin na příkladu povodí Ploučnice [Level of invasion by alien plants across habitats in the basin of the Ploučnice River]. Geografie 118:356–371

Šíma J (2008) Právní úprava problematiky nepůvodních druhů rostlin v České republice a ve světě [Legislation related to non-native species in the Czech Republic and the world]. Zprávy České botanické společnosti 43, Materiály 23:213–218

Simonová D, Lososová Z (2008) Which factors determine plant invasions in man-made habitats in the Czech Republic? Perspectives in Plant Ecology, Evolution and Systematics 10:89–100

Skálová H, Guo W-Y, Wild J, Pyšek P (2017) *Ambrosia artemisiifolia* in the Czech Republic: historic and current invasion pathways predict future spread. Preslia 89:1–16

Slavík B, Smejkal M, Dvořáková M, Grulich V (eds) (1995) Květena České republiky 4 [Flora of the Czech Republic 4]. Academia, Praha

Slavík B, Chrtek J Jr, Tomšovic P (eds) (1997) Květena České republiky 5 [Flora of the Czech Republic 5]. Academia, Praha

Slavík B, Chrtek J Jr, Štěpánková J (eds) (2000) Květena České republiky 6 [Flora of the Czech Republic 6]. Academia, Praha

Slavík B, Štěpánková J, Štěpánek J (eds) (2004) Květena České republiky 7 [Flora of the Czech Republic 7]. Academia, Praha

Sojneková M, Chytrý M (2015) From arable land to species-rich semi-natural grasslands: succession in abandoned fields in a dry region of central Europe. Ecological Engineering 77:373–381

Štěpánková J, Chrtek J Jr, Kaplan Z (eds) (2010) Květena České republiky 8 [Flora of the Czech Republic 8]. Academia, Praha

Strayer DL, D'Antonio C, Essl F, Fowler M, Geist J, Hilt S, Jarić I, Jöhnk K, Jones CG, ... Jeschke JM (2017) Boom-bust dynamics in biological invasions: definitions, causes, detection, and description. Ecology Letters 20:1337–1350

Tokarska-Guzik B (2005) The establishment and spread of alien plant species (kenophytes) in the flora of Poland. Wydawnictwo Uniwersytetu Śląskiego, Katowice

van Kleunen M, Dawson W, Essl F, Pergl J, Winter M, Weber E, Kreft H, Weigelt P, Kartesz J … Pyšek P (2015) Global exchange and accumulation of non-native plants. Nature 525:100–103

Verloove F (2006) Catalogue of neophytes in Belgium (1800–2005). Scripta Botanica Belgica 39:1–89

Vilà M, Pujadas J (2001) Land-use and socio-economic correlates of plant invasions in European and North African countries. Biological Conservation 100:397–401

Vilà M, Basnou C, Pyšek P, Josefsson M, Genovesi P, Gollasch S, Nentwig W, Olenin S, Roques A, ... DAISIE partners (2010) How well do we understand the impacts of alien species on ecosystem services? A pan-European, cross-taxa assessment. Frontiers in Ecology and the Environment 8:135–144

Vilà M, Espinar JL, Hejda M, Hulme PE, Jarošík V, Maron JL, Pergl J, Schaffner U, Sun Y, Pyšek P (2011) Ecological impacts of invasive alien plants: a meta-analysis of their effects on species, communities and ecosystems. Ecology Letters 14:702–708

Vítková M, Müllerová J, Sádlo J, Pergl J, Pyšek P (2017) Black locust (*Robinia pseudoacacia*) beloved and despised: a story of an invasive tree. Forest Ecology and Management 384:287–302

Vymyslický T (2001) Rozšíření vybraných invazních druhů rostlin na aluviích jihomoravských řek [Distribution of selected invasive plant species in the floodplains of the southern Moravian rivers]. MSc thesis, Masaryk University, Brno

Williamson M (1996) Biological invasions. Chapman & Hall, London
Williamson M (2002) Alien plants in the British Isles. In: Pimentel D (ed) Biological invasions: economic and environmental costs of alien plant, animal, and microbe species. CRC Press, Boca Raton, pp 91–112
Williamson M, Pyšek P, Jarošík V, Prach K (2005) On the rates and patterns of spread of alien plants in the Czech Republic, Britain and Ireland. Ecoscience 12:424–433
Winter M, Schweiger O, Klotz S, Nentwig W, Andriopoulos P, Arianoutsou M, Basnou C, Delipetrou P, Didžiulis V, … Kühn I (2009) Plant extinctions and introductions lead to phylogenetic and taxonomic homogenization of the European flora. Proceedings of the National Academy of Sciences of the USA 106:21721–21725

# Chapter 9
# Conservation of Botanical Diversity in the Czech Republic

**Handrij Härtel**

**Abstract** Nature conservation in the Czech Republic has a long history that started in the nineteenth century when the first protected areas were established. Modern conservation, based on the 1992 Nature and Landscape Conservation Act and integrating EU legislation (Natura 2000), has formed a robust framework for protecting habitat and species diversity in an extensive network of protected areas. Major threats to plant diversity are expressed in the Red List, which classifies half of the Czech native flora (including naturalized archaeophytes) as either threatened (critically threatened, endangered, vulnerable) or extinct. Present-day plant conservation in the Czech Republic profits from a good knowledge of the flora and vegetation, including invasive species, as well as from results of detailed habitat mapping. A growing attention is being paid to the application of restoration management in various habitat types, such as mires, species-rich meadows and post-mining sites, as well as in military training areas.

## 9.1 Introduction

The present-day richness of the Czech flora is a result of various ecological and historical factors, including the biogeographical position, geological and geomorphological diversity and the position of the Czech Republic between the northern (continental) and Alpine glaciers during the Pleistocene glaciations. The last mentioned factor is particularly obvious when comparing the Czech Republic with relatively species-poor landscapes in adjacent more northerly areas in Germany and Poland. Moreover, long-lasting human activities during the Holocene contributed to the preservation and further spread of many species of open habitats, including relicts and other nowadays rare species. Therefore, the territory of the Czech Republic comprises a heterogeneous mosaic of various types of landscapes, both

H. Härtel (✉)
Bohemian Switzerland (České Švýcarsko) National Park Administration,
Pražská 52, 407 46 Krásná Lípa, Czech Republic
e-mail: h.hartel@npcs.cz



M. Chytrý et al. (eds.), *Flora and Vegetation of the Czech Republic*, Plant and Vegetation 14, DOI 10.1007/978-3-319-63181-3_9

(relatively) natural and cultural, harbouring a large variety of habitats and considerable biodiversity.

Moreover, the long history of conservation in the present-day Czech Republic has contributed to the preservation of many valuable parts of nature, particularly within a dense network of protected areas, such as national parks, protected landscape areas and nature reserves or monuments. Most of these protected areas include the local flora and vegetation as one of the conservation objectives. The recent integration of Czech conservation activities into the EU network of protected areas Natura 2000 was an important milestone in terms of a long-term perspective towards the sustainable maintenance of species and habitat diversity in this country.

## 9.2 Legal and Historical Background of Conservation

The basis for the legal framework of nature conservation has been formed, as in other European countries, since the Middle Ages. At that time, however, the motivations for nature protection were mainly economic, cultural or aesthetic. A remarkable attempt was made by Emperor Charles IV in the fourteenth century in his *Maiestas Carolina*, a legal proposal suggesting strict rules for the protection of forests. However, this legal act never came into force due to strong resistance from the kingdom's aristocracy. Old-growth forests have remained preserved mainly in large domains such as the Křivoklát king's domain in central Bohemia. Truly modern conservation efforts were made during the period of Romanticism in the nineteenth century. The first protected area in the current Czech territory, the Žofín Old-growth Forest, was established in 1838 by Count Georg August Longueval-Buquoy on his Nové Hrady estate in southern Bohemia, followed by the Hojná voda Old-growth Forest in the same year. In 1858 Prince Jan Schwarzenberg established the Boubín Primeval Forest reserve, probably the best-known old-growth forest within the Czech Republic.

The involvement of the state in nature conservation started after the establishment of Czechoslovakia (1918). This period is closely connected with the founder of modern nature and landscape conservation in Czechoslovakia, the general authority for state nature conservation Rudolf Maximovič (1886–1963), who served as the highest nature conservation official from 1922 to 1948. He also participated in the founding of the International Union for Conservation of Nature (IUCN). His work was continued after World War II by Jaroslav Veselý (1906–1985), the first director of the State Institute for the Care of Historic Monuments and Nature Protection, an organization managing at that time both cultural and natural heritage in the whole of Czechoslovakia. In this period the first large-scale protected areas in present-day Czech Republic were established: the first Protected Landscape Area (Bohemian Paradise) in 1955 and the first National Park (Krkonoše Mts) in 1963.

Towards the end of the twentieth century, the adoption of the key legislative instrument, Act no. 114/1992, on the Conservation of Nature and Landscape, marks the most important milestone in modern conservation history. Unlike the previous nature conservation legislation from 1956, this act integrated new approaches and principles, such as prevention, sustainable use of natural resources, ecosystem approach and ecosystem networking. Moreover, after the accession of the Czech Republic to the European Union in 2004, this act transposed the EU nature conservation legislation (Directive 79/409/EEC on the conservation of wild birds, also referred to as Bird Directive, and Directive 92/43/EEC on the conservation of natural habitats and of wild fauna and flora, also referred to as Habitats Directive) in the Czech legal system. Nowadays, this act integrates a whole array of legal instruments and conservation issues, ranging from general conservation to special protection of both sites and species. While the relevance of the special protection tools (specially protected species and specially protected areas, including Natura 2000) for conserving biodiversity within the Czech Republic is obvious and will be discussed in more detail below, the potential of the general conservation tools included in the act, such as the territorial systems of ecological stability (ecological network), important landscape elements and general protection of all species, is often underestimated by conservation practitioners. Species conservation is directly implemented by Decree no. 395/1992 of the Ministry of the Environment with the annexes including the list of specially protected species of animals, plants and fungi. Further acts or decrees adopted by the Czech Parliament, Government or Ministry of the Environment ensure protection of particular protected areas or deal with particular conservation issues.

Recently, EU Regulation no. 1143/2014 on Invasive Alien Species has affected nature conservation in the Czech Republic. However, so far the list of invasive alien species of Union concern included in this regulation contains a very limited group of species, most of which do not occur in Central Europe, whilst other important invasive plants, such as species of *Reynoutria*, are not included (see Chap. 8, this book).

After the political changes in 1989, the Czech Republic also became party to several international conventions. Of these the Convention on the Conservation of European Wildlife and Natural Habitats (Bern Convention), Convention on International Trade in Endangered Species of Wild Fauna and Flora (CITES), Convention on Wetlands of International Importance especially as Waterfowl Habitat (Ramsar Convention) and the Convention on Biological Diversity are all at least partly related to plant species protection. The last one is implemented in the Czech Republic by the National Biodiversity Strategy of the Czech Republic (Ministry of the Environment of the Czech Republic 2005). This document sets priorities for both in-situ and ex-situ nature conservation in order to implement the three main objectives of this Convention: (i) conservation of biological diversity, (ii) sustainable use of components of biological diversity and (iii) fair and equitable sharing of the benefits arising out of the utilization of genetic resources. There are a whole range of further legal instruments directly or indirectly influencing biodiversity conservation in the Czech Republic, such as the Act on Forests or the Act on Water.

Botanists have always played an important role throughout the whole modern history of Czechoslovak and Czech nature conservation. Many conservationists, who significantly influenced the development of nature conservation during the twentieth century, started their professional careers as botanists, including Jaroslav Veselý (1906–1985) and Jan Čeřovský (1930–2017). Similarly, several Czech botanists working at research institutes and universities have been very actively involved in nature conservation both conceptually (legislation, proposals for new protected areas, red lists, etc.) and in terms of field work (mapping flora and vegetation in protected areas, research on the biology of threatened species, etc.). They contributed significantly to the existing efficient legislation and high-quality conservation work, including the dense network of protected areas in the whole of former Czechoslovakia including Slovakia and Carpathian Ruthenia (now Transcarpathian Region of Ukraine). Examples include prominent Czech botanists, such as Karel Domin (1882–1953), Jaromír Klika (1888–1957), Emil Hadač (1914–2003), Jan Jeník (*1929), Josef Holub (1930–1999), Jan Květ (*1933) and František Procházka (1939–2004).

Many principal works of the Czech botanical school became an important base for applied conservation science and conservation planning. This applies, for example, to several vegetation mapping projects, especially the Map of Natural Vegetation of Europe. This project, launched and led for many years by Robert Neuhäusl (1930–1991) from the Institute of Botany of the Czechoslovak Academy of Sciences in Průhonice (Bohn et al. 2000–2003), is a major achievement that became an essential basis for habitat mapping across Europe and for the European map of Biogeographical Regions used in the EU Habitats Directive and the Emerald Network under the Bern Convention. Within the Czech Republic, the Geobotanical Map of 'reconstructed natural vegetation' at a scale of 1:200,000 (Mikyška et al. 1968–1972) and the Map of Potential Natural Vegetation of the Czech Republic at a scale of 1:500,000 (Neuhäuslová et al. 1997, 1998) have served for many years as principle sources of scientific information for nature conservation applications. At the regional and local levels, maps of potential natural vegetation have been made for several large-scale protected areas including the Podyjí National Park (Chytrý and Vicherek 1995), Křivoklátsko UNESCO Biosphere Reserve and Protected Landscape Area (Kolbek and Moravec 1995) and Šumava National Park (Neuhäuslová 2001).

Thanks to the botanists of both academic and conservation institutions there are comprehensive surveys of the flora or vegetation of several large-scale protected areas, such as the Podyjí National Park (Chytrý and Vicherek 1995; Grulich 1997) and the UNESCO Biosphere Reserves and Protected Landscape Areas Křivoklátsko (Kolbek et al. 1997, 1999a, b, 2001a, b, 2003) and Bílé Karpaty (White Carpathians; Jongepier and Pechanec 2006). The flora of Saxon-Bohemian Switzerland, a transboundary German-Czech National Park and Protected Landscape Area, is currently under preparation (Härtel et al. 2001, 2013).

The Habitat Catalogue for the Czech Republic (Chytrý et al. 2001, 2010) has been a key instrument in the habitat mapping not only for Natura 2000 but for any field work on habitats and vegetation in the Czech Republic; the same author has also edited a four-volume monograph Vegetation of the Czech Republic (Chytrý 2007–2013). Floristic data and vegetation classification also played a key role in the classification of the biogeographical regions in the Czech Republic (Culek 1996; Culek et al. 2005).

## 9.3 Major Threats to Plant Diversity

Holub (2000) analyzed the list of extinct and missing taxa and the main factors causing the extinction and the loss of plant diversity in the Czech and Slovak Republics. The greatest loss of species was recorded for arable weed communities (e.g. *Camelina alyssum*, *Cuscuta epilinum*, *Lolium remotum* and *L. temulentum*), followed by wetlands, dry grasslands and herbaceous fringe vegetation, thermophilous forests, salt marshes and heaths. By contrast, forest vegetation was significantly less affected by the species loss. Holub (2000) suggests that sub-Mediterranean and continental species are more endangered than Central European, sub-Atlantic, boreal or even alpine species. The main causes of the decline in plant diversity recognized by Holub (2000) still remain valid, including intensive agriculture, such as the use of fertilizers and general eutrophication of the landscape. However, some factors that caused the extinctions or retreat of many species until the 1980s, such as acid rain and other kinds of pollution or extensive hydrological changes in the landscape (regulated and channelized streams, large-scale drainage, etc.) no longer rank among the key drivers of contemporary change. Today plant conservation in the Czech Republic faces threats similar to those faced in other post-communist countries in Central Europe, which are related to dynamic changes in the structure of the landscape. A general trend in recent decades is an increasing polarization between the rapidly developing metropolitan areas especially around the cities of Prague, Brno, Ostrava and Plzeň, and peripheral areas with a declining human population and infrastructure. Although the total agricultural production has fallen significantly since 1989, the productivity has gradually increased (Czech Statistical Office, www.czso.cz). As a result, on one hand, there is growing land-use intensification, while on the other hand, land is being abandoned in other parts of the country, with both of these trends leading to a change from the previous fine-scale landscape mosaic to a more homogeneous landscape. These trends are mainly reflected in the changes in habitats in agricultural landscapes whereas forest areas are less affected. This situation corresponds to the fact that most of the critically threatened taxa listed in the Czech Red List (Grulich 2012) are non-forest species.

The political changes in 1989 and subsequent changes in the economy brought not only new challenges but also many new opportunities for the conservation of biodiversity. In submontane and montane landscapes, large areas of arable land were converted into permanent grassland with predominantly positive effects on habitat and species diversity. By contrast, similar conversion from grassland to forest, mainly at high altitudes, has had a rather negative effect on biodiversity. Large areas in the Czech Republic were abandoned or left to secondary succession with ambivalent results. While in some habitats, such as former mining sites, the natural succession might have positive effects on biodiversity, in many other habitats, mainly in grasslands and some colline woodlands, natural succession is currently a major threat to species diversity due to the decline in the abundance of low-competitive and light-demanding species, such as *Adenophora liliifolia* (Fig. 9.1a) and *Lathyrus pisiformis*. Natural processes, mainly succession and competition, are

**Fig. 9.1** Examples of threatened species protected within the Natura 2000 network (species listed in the annexes of the Habitats Directive) with an indication of their category in the 2012 edition of the Czech Red List: (**a**) *Adenophora liliifolia* (C1b); (**b**) *Angelica palustris* (C1t); (**c**) *Coleanthus subtilis* (C3); (**d**) *Cypripedium calceolus* (C2b); (**e**) *Dracocephalum austriacum* (C1r); (**f**) *Gentianella praecox* subsp. *bohemica* (C1t). See Sect. 9.4.2 for explanation of Red List categories. Photo credits: V. Kalníková (a), J. Navrátilová (b), K. Šumberová (c), P. Bauer (d), V. Nejeschleba (e) and A. Zvára (f)

also evaluated as a serious threat to the species protected in the Czech Republic by the European Union (Habitats Directive, Natura 2000; Turoňová 2016). These processes are currently more important than direct habitat destruction or fragmentation. Several species protected by the Habitats Directive have an "unfavourable-bad" conservation status according to assessment reports, such as *Gentianella praecox* subsp. *bohemica*, *Himantoglossum adriaticum*, *Jurinea cyanoides*, *Lindernia procumbens* and *Luronium natans* (Holá and Turoňová 2014). A specific threat, mainly to endemics and other species of high conservation priority, is interspecific hybridization, such as in the case of *Viola lutea* subsp. *sudetica* in the Krkonoše Mts (hybridization with *V. tricolor*; Krahulcová et al. 1996).

From the perspective of the winner–loser replacement paradigm, the long list of losers in the Czech flora particularly reflects the decline of oligotrophic habitats and the cessation of traditional land-use, such as low-intensity grazing, different combinations of mowing and grazing, the limited application of fertilizers to grasslands and coppicing in woodlands. The winners can be found not only among the successful invasive species (see below) but also among native species. In the Czech Republic they include several herbaceous species such as *Calamagrostis epigejos*, *C. villosa* and *Urtica dioica*, and also woody species such as *Fraxinus excelsior* and *Sambucus nigra*.

Growing attention is being paid to invasive alien species and their impact on native biodiversity. The Czech Republic is a country in which there is a very good knowledge of this topic (Pyšek et al. 2002a, 2003b, 2009, 2012a, b; Chytrý et al. 2005; Jarošík et al. 2011b; Foxcroft et al. 2013; see also Chap. 8, this book) and the only country in Europe with a comprehensive study dealing in detail with the patterns and drivers of plant invasions in protected areas. These analyses indicate that protected areas serve as a barrier against invasions (Pyšek et al. 2003a). For example, the percentage of alien species in the flora of small-scale nature reserves is significantly smaller if they are located within a large protected area (Pyšek et al. 2002a). However, although in protected areas the percentage of alien flora is significantly smaller than in the surrounding landscape, the picture is less optimistic when looking at the presence of widespread invasive species in protected areas (Pyšek et al. 2013). This can be demonstrated by the example of the large-scale colonization of the Bohemian Switzerland National Park by *Pinus strobus* (see Sect. 9.7.3), where the devastating impact on the natural habitats has been caused by a large-scale invasion of a single alien tree species (Härtel and Hadincová 1998; Wild et al. 2013). There are many examples of local eradication projects in the Czech Republic, of for example *Heracleum mantegazzianum*, *Pinus strobus*, *Reynoutria japonica*, *R. sachalinensis*, *R. ×bohemica* and *Robinia pseudoacacia*. However, a clear, consistent and realistic strategy for managing invasive plant species at the national level is still lacking. The recently published Black, Grey and Watch Lists of Alien Species in the Czech Republic (Pergl et al. 2016; see also Chap. 8, Sect. 8.9, this book), for both plants and animals, is an important step towards such a strategy. These lists provide a robust means of assessing invasive alien species, primarily for nature conservation purposes. It includes criteria for classifying invasive species, based on their mode of spread and impact, followed by assessment of management options and related recommendations. There are currently 78 plant species on the Black List (see Chap. 8, this book). Sometimes, different approaches to managing invasive species applied along the state border may result in serious challenges for conservation, mainly in transboundary protected areas (Härtel et al. 2015), especially in the case of invasive species distributed by streams along the state border (Schiffleithner and Essl 2016).

## 9.4 Protected Plant Species, Red List and Red Data Book

### 9.4.1 *Protected Plant Species*

The Nature and Landscape Conservation Act no. 114/1992 (see Sect. 9.2) distinguishes between general protection and special protection of species. General protection protects plant (and animal) species from general impacts that could result in a significant reduction in their abundance or extinction over a large area. By contrast, special protection is the legal tool for protecting all individuals of a species including all stages of its development and their habitats. Specially protected species, a category including rare species, threatened species and species significant from a cultural or scientific point of view, are listed in the annexes of the Decree of the Ministry of the Environment no. 395/1992. They comprise 487 taxa of vascular plants, further divided into three categories according to their vulnerability: 246 critically threatened, 149 endangered and 92 vulnerable taxa. These taxa make up less than 20% of the 2642 native taxa and naturalized archaeophytes in the Czech flora (Danihelka 2013). The list of specially protected species (Decree no. 395/1992) also implements the protection of species listed in the EU Habitats Directive (41 vascular plant species occurring in the Czech Republic, Table 9.1, Figs. 9.1 and 9.2), as well as species listed in the annexes of the Council of Europe's Convention on the Conservation of European Wildlife and Natural Habitats (also referred to as Bern Convention). Exemptions from the provisions protecting these species can only be granted under conditions specified in the Nature and Landscape Conservation Act no. 114/1992.

### 9.4.2 *Red List and Red Data Book*

The Czech Republic ranks among the countries with an excellent knowledge of its flora and understanding of the extent to which it is endangered. The first efforts aimed at elaborating the Czechoslovak Red List for Vascular Plants occurred in the 1970s and were stimulated by the 12th International Botanical Congress in Saint Petersburg (then Leningrad) in 1975 and the IUCN European Red List of Plants (Lucas and Walters 1976). The conference of the Czechoslovak Botanical Society

**Table 9.1** Threatened species of the Czech Republic protected in the European Union by the Directive 92/43/EEC on the conservation of natural habitats and of wild fauna and flora, also referred to as Habitats Directive (in annexes II, IV and V). An asterisk indicates that the species is listed as priority species in the Annex II of the Habitats Directive. The synonyms in brackets refer to names of taxa used in the original legislation in case they differ from the Checklist of vascular plants of the Czech Republic (Danihelka et al. 2012)

| |
|---|
| *Aconitum firmum* subsp. *moravicum* |
| *Adenophora liliifolia* |
| *Aldrovanda vesiculosa* |
| *Angelica palustris* |
| *Arnica montana* |
| *Artemisia pancicii** |
| *Asplenium adulterinum* |
| *Campanula bohemica** |
| *Campanula gelida** |
| *Cerastium alsinifolium** |
| *Cirsium brachycephalum* |
| *Coleanthus subtilis* |
| *Crambe tataria* |
| *Cypripedium calceolus* |
| *Dianthus arenarius* subsp. *bohemicus** |
| *Dianthus lumnitzeri** |
| *Dianthus moravicus** |
| *Dracocephalum austriacum* |
| *Echium maculatum* (*Echium russicum*) |
| *Galanthus nivalis* |
| *Galium sudeticum** |
| *Gentianella praecox* subsp. *bohemica* (*Gentianella bohemica*)* |
| *Gladiolus palustris* |
| *Himantoglossum adriaticum* |
| *Iris arenaria* (*Iris humilis* subsp. *arenaria*) |
| *Jurinea cyanoides** |
| *Klasea lycopifolia* (*Serratula lycopifolia*)* |
| *Ligularia sibirica* |
| *Lindernia procumbens* |
| *Liparis loeselii* |
| *Luronium natans* |
| *Lycopodium* spp. |
| *Minuartia smejkalii** |
| *Pedicularis sudetica** |
| *Poa riphaea** |
| *Pulsatilla grandis* |
| *Pulsatilla patens* |
| *Stipa glabrata* (*Stipa zalesskii*)* |
| *Tephroseris longifolia* subsp. *moravica* |
| *Thesium ebracteatum* |
| *Trichomanes speciosum* |

**Fig. 9.2** Examples of threatened species protected within the Natura 2000 network (species listed in the annexes of the Habitats Directive) with an indication of their category in the 2012 edition of the Czech Red List: (**a**) *Ligularia sibirica* (C1b); (**b**) *Liparis loeselii* (C1t); (**c**) *Luronium natans* (C1b); (**d**) *Pulsatilla grandis* subsp. *grandis* (C2b); (**e**) *Pulsatilla patens* (C1t); (**f**) *Trichomanes speciosum* (C2r). See Fig. 9.1 for details. Photo credits: V. Kalníková (a), V. Nejeschleba (b), P. Bauer (c), Š. Koval (d), A. Zvára (e) and V. Sojka (f)

held in Prague in 1976 decided to prepare the first Red List of Czech vascular plants. The *List of extinct, endemic and threatened taxa of vascular plants in the flora of the Czech Socialistic Republic (first draft)* was prepared by Josef Holub, František Procházka and Jan Čeřovský (Holub et al. 1979) and included three main categories: extinct taxa (A), endemic taxa (B) and threatened taxa (C), each subdivided into several sub-categories. This first version of the national Red List raised a lot of interest among both botanists and conservationists, stimulated further floristic research and conservation activities, and inspired efforts to prepare local Red Lists or Red Data Books for administrative entities or natural landscape units in this country, for example northern Bohemia (Kubát 1986), southern Bohemia (Chán 1999; Lepší et al. 2013), the Šumava Mts (Procházka and Štech 2002), the Krkonoše Mts (Štursa et al. 2009) and the Jeseníky Protected Landscape Area (Bureš 2013). In 1999 the Red Data Book of Higher Plants in the Czech and Slovak Republics was published (Čeřovský et al. 1999), covering a total of 400 vascular plant taxa. The second version of the Czech Red List of vascular plants was published by Holub and Procházka (2000) and accompanied by the *Black List of taxa that had disappeared from the floras of the Czech and Slovak Republics* (Holub 2000). The latest (third) version of the Czech Red List was published by Grulich (2012). Unlike the 1979 Red List, the 2000 and 2012 versions no longer include lists of endemics and subendemics as a special category (B). In all three versions the classification criteria are based on expert judgement, because the quantitative approach used by IUCN (2001) is not always fully applicable at a regional scale (Gärdenfors et al. 2001; Miller et al. 2007; IUCN 2012). However, the rigorous application of the same criteria in all three editions of the Red List facilitates an analysis of temporal trends in the endangerment of particular species. The principles of species categorization used in the Czech Red List are described in detail by Grulich (2012). In spite of some differences, the Czech Red List categories are very close to and basically compatible with the IUCN approach. To achieve a full compatibility and better international comparison, the next Red List will use a parallel system assessing the categories according to both the Czech and IUCN Red List criteria (Grulich in prep.). Temporal trends in the Czech threatened flora are assessed based on comparison of the 1979, 2000 and 2012 Red Lists by Grulich (2012). The most recent Red List (Grulich 2012) includes a total of 1720 taxa (for examples of distribution maps of several Red List species see Figs. 9.3 and 9.4), of which 74 taxa are classified as extinct or vanished (A1), 53 as missing (A2) with 29 additional uncertain cases (A3), 471 critically threatened (C1), 357 endangered (C2), 356 vulnerable (C3), 233 lower risk – near threatened (C4a) and 147 lower risk – data deficient (C4b). The total number of 1720 (including the C4 category, i.e. lower risk and data deficient categories) Red List taxa make up 64.3% of the Czech native flora including naturalized archaeophytes (Danihelka 2013), or 50.7% if the C4 category is excluded. This means that half of the Czech native flora is either threatened (in categories C1, C2, C3) or extinct.

In the 2012 Red List, an additional measure is used for the critically threatened (C1) and endangered (C2) taxa with the aim of distinguishing between different causes of endangerment, namely species being: in serious decline (C1t, C2t), just rare (C1r, C2r), or a combination of both (C1b, C2b). For instance, *Rubus*

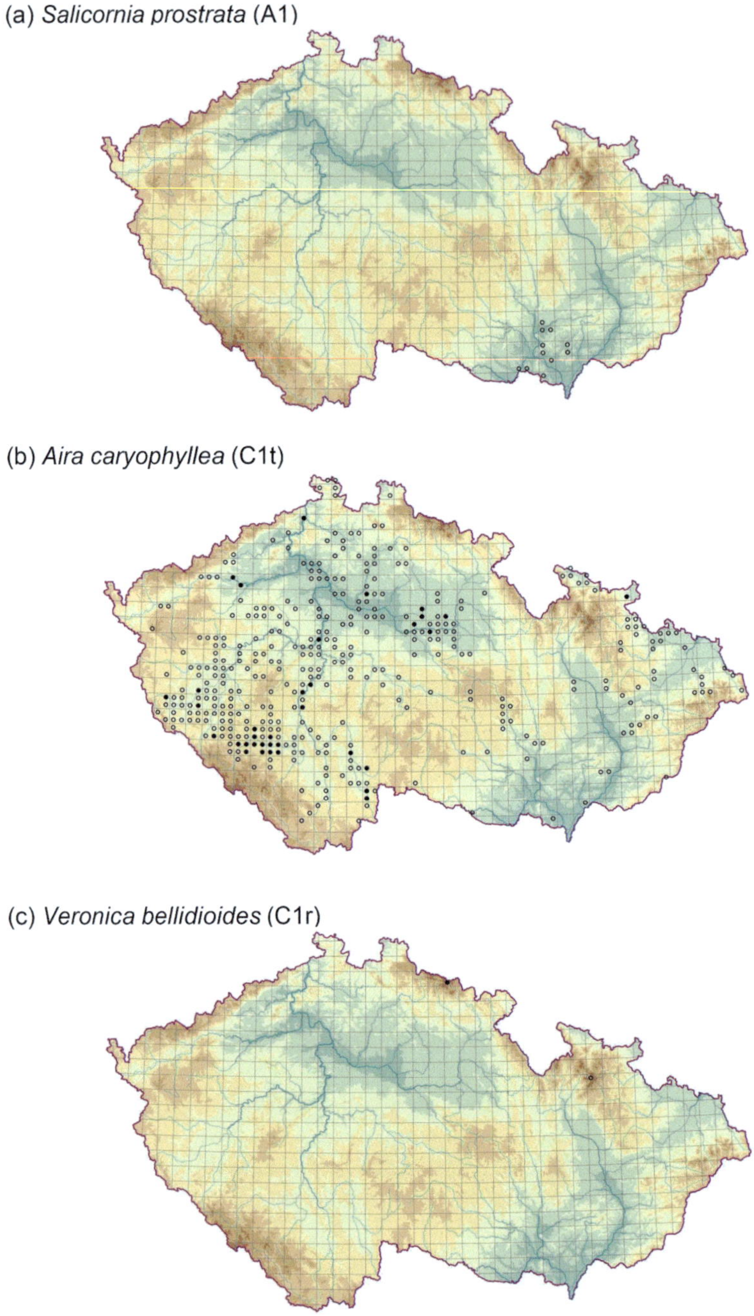


**Fig. 9.3** Examples of distributions of extinct and critically threatened species classified according to the 2012 edition of the Czech Red List. *Full circles* indicate records after the year 2000, *empty circles* only records until 2000. See Sect. 9.4.2 for explanation of Red List categories (data obtained from the Pladias database (www.pladias.cz); all the maps in this chapter were prepared by O. Hájek)

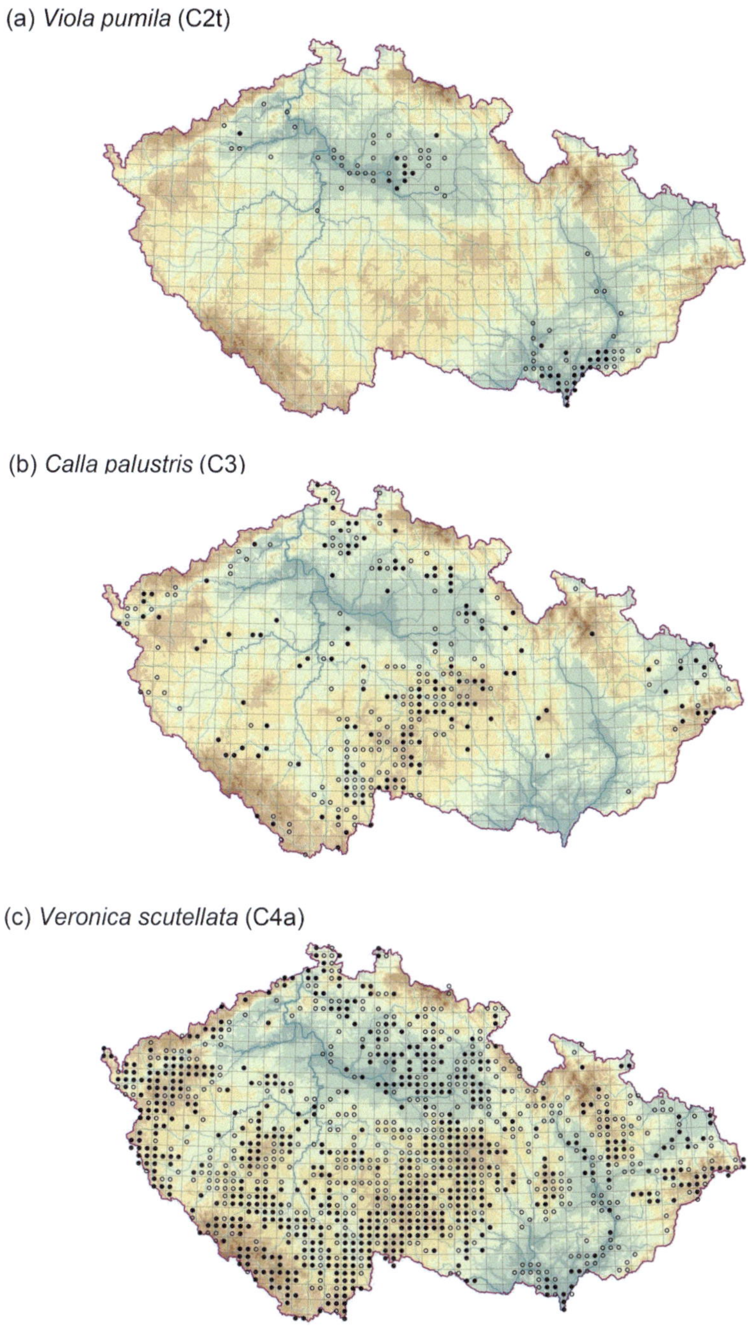


**Fig. 9.4** Examples of distributions of threatened species classified to various categories of the 2012 edition of the Czech Red List. See Fig. 9.3 for details

*chamaemorus* (listed as C1r), an Arctic species, extremely rare in Central Europe, occurs as a glacial relict in the Krkonoše Mts both on the Czech and Polish side and is not directly threatened in the Czech Republic since all of its sites are protected within the core zone of the Krkonoše National Park. By contrast, *Pedicularis palustris* subsp. *palustris* (C1t) is a typical example of a formerly rather widespread species that has undergone a dramatic decline in recent decades.

Most of the taxa listed in categories A1 or A2 are extinct or missing nationally; however, seven of them were endemics and, therefore, they are extinct globally (see Chap. 3, this book). In contrast, there are several examples of species evaluated in previous Red List editions as extinct or missing in the Czech Republic but recently rediscovered after a long time, such as *Luronium natans* (Suda et al. 2000; Härtel and Bauer 2002) and *Pilularia globulifera* (Ekrtová et al. 2008).

If we compare the last Red List (Grulich 2012) with the list of species protected by law, the number of legally protected species is less than half of the total number of 1184 species included in the Red List as critically threatened, endangered and vulnerable. Of the critically threatened species classified as such because of a decline in their abundance (C1t), more than one third (depending on different taxonomic approaches used in the Red List and List of Protected Species, respectively) are not protected by law. In many cases the absence from the protected species list is due to difficulties in active protection of the particular species, such as in the case of some taxa occurring nowadays as agricultural weeds or as taxonomically critical taxa (e.g. *Taraxacum* spp.). However, there are also several easy-to-identify critically threatened species that are not protected, such as *Aira caryophyllea*, *Arnoseris minima* and *Pyrola chlorantha*.

Comparison of national Red Lists among countries is rather difficult due to the different systems used for categorizing endangerment, different years of publication and differences in the areas of the countries, since the risk of extinction is dependent on scale (Hartley and Kunin 2003). However, several attempts have been made to compare national red lists within Europe (e.g. Gabrielová et al. 2011) or to compile Red List of Vascular Plants for Europe (Bilz et al. 2011) or Central Europe (Schnittler and Günther 1999). Holub (2000) identifies extinct and missing taxa shared by a group of several Central European countries (Austria, Czech Republic, Germany, Poland and Slovakia). Regardless of the afore-mentioned limitations of international comparisons, the Czech Republic, with 50% of its taxa threatened or extinct, ranks among the European countries with the highest share of these taxa in the country's native flora.

There have been very few attempts to assess the regional endangerment of species across state borders. In the Czech Republic, the best example is probably the Red List for the Krkonoše/Karkonosze Mts (Štursa et al. 2009), which assigns categories for the whole Czech-Polish mountain region regardless of the state border and the Red List categories in the respective countries. Such examples are particularly valuable from the conservation perspective as they facilitate a coordinated conservation/management approach in transboundary protected areas, since the category of protection or category of endangerment sometimes differs significantly between neighbouring countries. Transboundary studies based on national Red List categories, without the assignment of the Red List category for the whole transboundary area, are more frequent, including a recent study in the Erzgebirge/Krušné hory Mts (Germany/Czech Republic; Müller and Kubát 2013).

Besides vascular plants, Red Lists for bryophytes (Kučera et al. 2012) and lichens (Liška et al. 2008) are also available (see Chaps. 4 and 5, this book). Endangered plant communities were listed by Moravec et al. (1995) and an attempt to assess the endangerment of habitats in the Czech Republic was made by Kučera (2009).

## 9.5 National System of Protected Areas and Their Botanical Value

In April 2017, there were 2610 specially protected areas (hereinafter referred to as "protected areas") in the Czech Republic, covering 1,363,922 ha, i.e. 17.3% of this country's territory (according to http://drusop.nature.cz; Fig. 9.5). There are six categories of protected areas. Large-scale protected areas include national parks (4) and protected landscape areas (26), and small-scale protected areas include national nature reserves (107), nature reserves (809), national nature monuments (120) and nature monuments (1544). The large-scale protected areas have their own administrative authorities and can also include small-scale protected areas nested within

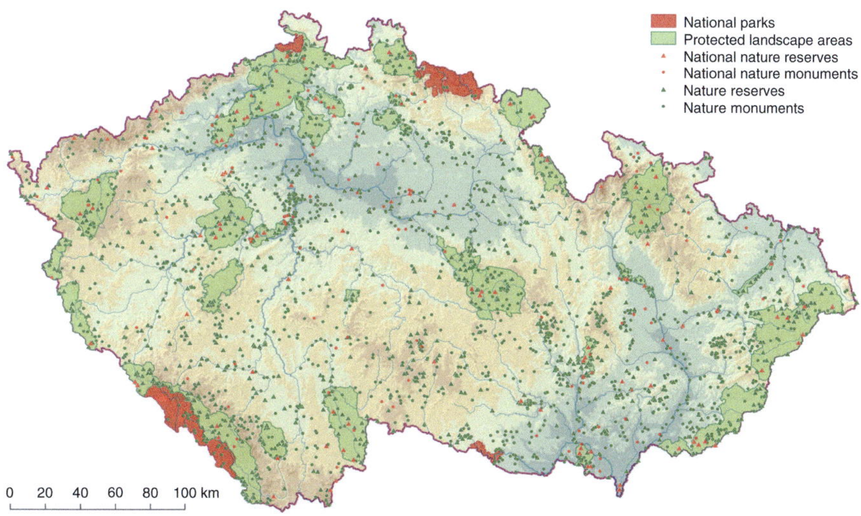


**Fig. 9.5** Protected areas in the Czech Republic. Most of the large-scale protected areas (national parks and protected landscape areas) are situated in uplands and mountains rather than in intensively managed lowland landscapes. All four Czech national parks are situated in peripheral areas on the national borders. Small-scale protected areas are distributed throughout the Czech Republic including its densely populated areas such as around the city of Prague (data on protected areas were provided by © Nature Conservation Agency of the Czech Republic)

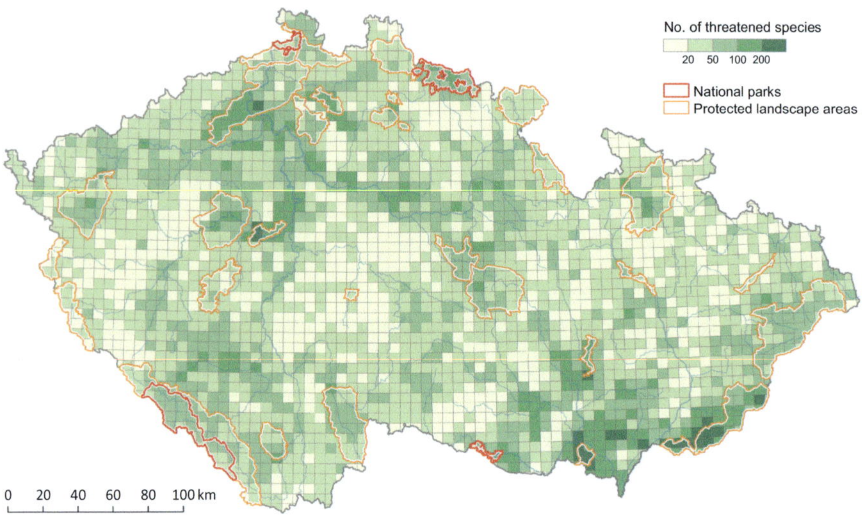


**Fig. 9.6** Overlap of the regions harbouring a high number of threatened plant species with large-scale protected areas (national parks and landscape protected areas). The map shows the richness of threatened plants (categories C1, C2 and C3 in the 2012 edition of the Czech Red List) in quadrants of 5 × 3 min (approximately 5.5 × 5.9 km). National parks and protected landscape areas only partly overlap the regions rich in threatened species, which occur mainly at low altitudes, whereas the large-scale protected areas are situated predominantly at middle and high altitudes (species distribution data were obtained from the Pladias database (www.pladias.cz) and data on protected areas were provided by © Nature Conservation Agency of the Czech Republic)

them. Additional sites are protected within the EU protected area network Natura 2000 (see Sect. 9.6). The protected areas in the Czech Republic are described in a 14-volume monograph including detailed description of their flora and vegetation (Mackovčin and Sedláček 1999–2009).

The overlap of the regions harbouring a high number of threatened plant species with the protected areas is obvious from Fig. 9.6. Not surprisingly, most of the protected areas, mainly national parks and protected landscape areas, are situated in uplands and on mountains, not in intensively managed landscapes. All four Czech national parks are situated in peripheral areas of this country, on the borders with Germany, Poland and Austria, and they are part of transboundary national parks of international importance. Rejzková (2013) reports that 38% of the species in the Czech flora that are threatened (categories C1, C2 and C3) are recorded in Czech national parks, although national parks cover only 1.5% of the territory of the country. However, there are significant differences in the species diversity among these parks. For instance, the Podyjí National Park, an outstanding regional biodiversity hotspot (Grulich 1997; Michalcová et al. 2014), has a higher number of threatened vascular plant species than the ten times larger Šumava National Park. Podyjí hosts many rare and threatened species, such as *Actaea europaea*, *Bupleurum affine*, *Buphthalmum salicifolium*, *Daphne cneorum*, *Tordylium maximum*, *Veratrum nigrum* and *Verbascum speciosum*, and also two endemic species of *Sorbus*

(*S. cucullifera* and *S. thayensis*; Lepší et al. 2015). In contrast, the Bohemian Switzerland National Park, established mainly for protecting geodiversity (Härtel et al. 2007a), is characterized by a low species diversity of phanerogams but a high diversity of cryptogams (Härtel et al. 2007b; Sádlo et al. 2007). Finally, the Krkonoše National Park (Fig. 9.7a) is unique among the four national parks in harbouring the highest number of threatened plant species, including several glacial relicts (e.g. *Luzula spicata*, *Pedicularis sudetica* and *Rubus chamaemorus*) and the highest concentration of endemic plant taxa within the Czech Republic, mainly species of the genus *Hieracium*; see Chap. 3, Sect. 3.4, this book). This is mainly due to the biogeograhic position of this highest mountain group of all Hercynian mountain systems. The Krkonoše Mts form an isolated high-mountain island where the northern and alpine vegetation met during the Quaternary glaciations and after climatic amelioration during the Pleistocene/Holocene transition an isolated island of the arctic-alpine tundra remained there (Jeník 1961).

Unlike national parks, established in remote places with large areas of natural (and semi-natural) habitats and supporting primarily natural processes, the protected landscape areas focus on protection of harmonic cultural landscapes that have been affected and managed by humans for a long time. Their main conservation objective is the active management and development of an environmentally sustainable and spatially differentiated use of the landscape. Among all the protected area categories, the protected landscape areas cover the largest percentage of the Czech territory (14.4%) and are distributed throughout the country. Therefore, together with the national parks, protected landscape areas contribute significantly to the conservation of Czech plant diversity, some of them having attracted the attention of botanists for a long time. Examples of protected landscape areas of outstanding botanical value include: (i) the České středohoří Protected Landscape Area, a species-rich volcanic landscape with a high number of threatened species of vascular plants (Kubát 1970, 1986), including four endemic species of the genus *Sorbus* (*S. albensis*, *S. bohemica*, *S. milensis*, and *S. portae-bohemicae*, Kaplan et al. 2016), with some species reaching their western distribution limit there, such as the continental grasses *Helictotrichon desertorum* and *Stipa glabrata* (syn. *S. zalesskii*) – this area is also very rich in orchids (38 taxa, including regionally extinct and missing ones; Nepraš et al. 2008); (ii) the Bohemian Karst (Český kras) Protected Landscape Area, hosting a range of rare and threatened species (Skalický and Jeník 1974); critically threatened taxa (C1 category) include *Adenophora liliifolia*, *Anacamptis pyramidalis*, *Dracocephalum austriacum* (Fig. 9.1e) and recently discovered *Notholaena marantae* (Špryňar 2004); (iii) the Pálava Protected Landscape Area, including the Pavlov Hills (Fig. 9.7b) and adjacent areas, a unique botanical hotspot in the Pannonian part of the Czech Republic containing a number of species of the Pontic-Pannonian geoelement and including a range of critically threatened taxa, e.g. *Arabis nemorensis*, *Arenaria grandiflora*, *Leucojum aestivum*, *Salvia aethiopis*, *Scorzonera parviflora*, *Stipa eriocaulis*, *Thesium dollineri*, *Tripolium pannonicum* subsp. *pannonicum* and *Viola kitaibeliana* (some of them known in the Czech Republic only from Pálava); (iv) the Bílé Karpaty Protected Landscape Area,

**Fig. 9.7** Examples of protected areas in the categories of national park, protected landscape area, national nature reserve and national nature monument, all of which are also registered as Important Plant Areas: (**a**) Krkonoše National Park, the oldest national park in the Czech Republic (established 1963), conserving a unique island of arctic-alpine tundra in Central Europe, with strong impact of tourism even in the core zone and high level of fragmentation of the *Pinus mugo* krummholz in the subalpine belt by a dense network of mountain paths; (**b**) Pálava Protected Landscape Area, a cultural landscape in the warmest part of the Czech Republic, protecting an exceptionally rich diversity of Pannonian habitat types and species: a limestone ridge of the Pavlov Hills; (**c**) Boubínský prales National Nature Reserve, established in 1858 as one of the best examples of old-growth forests in Bohemia; (**d**) Swamp National Nature Monument protecting wide range of threatened species and types of fen vegetation. Photo credits: K. Antošová (Krkonoše National Park archive, a), J. Kmet (b), T. Vrška (c), D. Turoňová (d)

a flysch landscape in the White Carpathian Mts on the Czech-Slovak border, characterized by extraordinary species-rich meadows (see Sect. 9.7.3; Jongepierová 2008; Michalcová et al. 2014; Chytrý et al. 2015).

The large number of small-scale protected areas is also of high importance for the conservation of plant diversity, mainly for targeted conservation of particular species or habitats. According to Pyšek et al. (2002b), 2152 vascular plant taxa at the specific and subspecific levels (including 92 hybrids) are recorded in Czech nature reserves including approximately 80% of the native species of the Czech flora (Pyšek et al. 2002a). These small-scale nature reserves and nature monuments are distributed throughout the Czech Republic including its densely populated parts. In some cases they represent the last refugia for threatened and rare species and habitats within the human-modified landscape. For instance, even in the city of Prague, there are almost a hundred protected areas predominantly conserving dry grasslands (Jarošík et al. 2011a) and harbouring many threatened species, such as *Allium strictum*, *Anemone sylvestris*, *Gagea bohemica*, *Helianthemum canum*, *Iris aphylla*, *Linum flavum*, *Muscari tenuiflorum*, *Prunus fruticosa*, *Pulsatilla pratensis* subsp. *bohemica*, *Thalictrum foetidum* and *Thesium bavarum* (e.g. Špryňar and Münzbergová 1998).

Many Czech protected areas or their significant parts are included in various international networks. For instance, several protected areas are identified as Important Plant Areas (Čeřovský et al. 2007; see Sect. 9.7.1). Furthermore, there are 14 sites protected within the Ramsar Convention in this country (www.ramsar.org/wetland/czech-republic). They not only include natural wetlands, such as mires, springs and mountain streams (Springs and mires of the Slavkov Forest, Krkonošská rašeliniště, Jizera Headwaters, Třeboňská rašeliniště, Šumavská rašeliniště and Krušnohorská rašeliniště) and floodplains (Liběchovka and Pšovka brooks, Poodří, Mokřady dolního Podyjí and Litovelské Pomoraví), but also important man-made wetlands, such as fishponds (Novozámecký and Břehyňský rybník, Lednické rybníky and Třeboňské rybníky). A very unique Ramsar Site is the Punkva subterranean stream in the Moravian Karst. A comprehensive list of the Czech Ramsar Sites and other nationally or regionally important wetlands was compiled by Chytil et al. (1999). Another international network is represented by Biosphere Reserves designated within the UNESCO Man and Biosphere Programme. Currently, there are six Biosphere Reserves in the Czech Republic: Křivoklátsko, Třeboňsko, Dolní Morava (Lower Morava), Šumava, Bílé Karpaty (White Carpathians) and Krkonoše. They cover several representative landscape types of the Czech Republic, as regards their exceptional plant and habitat diversity.

Finally, military training areas, both active and abandoned, are of key importance for conserving biodiversity, including that of vascular plants, in the Czech Republic. Their military use, which imposed regular disturbance combined with the no-access regime for the public, made them important biodiversity hotspots within the Czech Republic. They require a very specific management (see Sect. 9.7.3) and some of them are part of the protected area network (including Natura 2000 sites).

Administration and management of the Czech network of protected areas is challenging for several reasons. First, a typical feature of the Czech protected area network is a very high number (relative to the country's size) of individual sites, the majority of which are small. This is basically due to the highly fragmented landscape resulting in a high number of small sites suitable for conservation. It is also due to the very detailed knowledge of nature and its biodiversity in the Czech Republic. However, such a system of protected areas puts high demands on conservation planning and management. Furthermore, some protected areas face difficult dilemmas. For example, the bark beetle outbreak in the Šumava National Park is a serious challenge for nature conservation authorities. Despite the fact that there is evidence that these insect outbreaks in montane spruce forests have a positive effect on their structure and diversity (Jonášová and Prach 2004; Müller et al. 2008; Lehnert et al. 2013), it is still difficult to win the acceptance of the key stakeholder groups for the idea of protection of large-scale processes of natural disturbances and regeneration in this national park (Křenová and Kiener 2012). In contrast, protected areas at low altitudes, typically consisting of various types of lowland forest, are facing the dilemma of whether management should be passive or active. There are several studies demonstrating how the biodiversity in lowland (mainly oak) woodlands historically depended on active management, whereas successional changes during the twentieth century resulted in a significant decrease in the herb-layer species diversity in these woodlands (Hédl et al. 2010; Kopecký et al. 2013). Application of traditional forest management is an effective means of reversing the decrease in diversity in these woodlands and bringing back some rare plant species (Douda et al. 2017). However, in some strictly protected areas at low altitudes, such as the Podyjí National Park and some national nature reserves, it is sometimes challenging to achieve a balance between conservation of biodiversity and protection of natural processes.

## 9.6 Habitat Mapping and the Natura 2000 Network in the Czech Republic

The Czech Republic hosts a wide range of natural habitats, and vegetation and habitat mapping has a long tradition in this country (see Sect. 9.2). Botanists have also paid attention to assessing the extent to which plant communities (Moravec et al. 1995; Chytrý 2007–2013) and habitats (Kučera 2009) are endangered. However, the decision to undertake a detailed mapping of habitats, based on extensive field work carried out throughout the whole country, was initiated by the accession of the Czech Republic to the EU and the transposition of EU legislation, including the Habitats Directive. An ambitious project of habitat mapping was started in 2000, and between 2001 and 2004, the basic mapping was carried out with the aim of collecting data for identifying Sites of Community Importance, followed by regular updating of the mapping results. The mapping methodology (Guth and Kučera 2005) is based on the habitat classification described in the Habitat Catalogue of the Czech Republic (Chytrý et al. 2001, 2010). This Catalogue not only contains the Natura 2000 habitat types but also other natural habitats occurring within the Czech Republic. In its second edition it distinguishes 140 units or subunits, which are fully compatible with the habitat types listed in Annex I of the Habitats Directive.

The results of the first phase of the habitat mapping project are summarized by Härtel et al. (2009). This project gathered data on all the natural habitat types and their distribution in the Czech Republic. The value of this data goes far beyond the purpose of Natura 2000 and is of great significance for Czech nature conservation as it is an up-to-date source of quantitative and qualitative information on habitats in the whole country. More than 750 experts were involved in the mapping. At the beginning, training was organized to standardize the views on habitat classification and its application in mapping. The habitat mapping project has been continuing since 2006 by permanent updating with approximately one twelfth of the country remapped annually in the field. This updating is carried out in about 3500 mapping districts of 1000–3000 ha (Králová and Tomášek 2009).

The outcomes of the mapping in the form of habitat maps were published by Härtel et al. (2009) and more recently by Chobot (2016) (see examples in Figs. 9.8 and 9.9). There are 60 Natura 2000 habitats in the Czech Republic (Table 9.2), of which 58 are recorded in the Continental Biogeographical Region and 35 in the Pannonian Biogeographical Region, the latter region covering only ~4% of the Czech territory in the south-east of the country (for EU biogeographical regions see http://bit.ly/2nPdhQA).

The Natura 2000 habitat types recorded in the Czech Republic vary significantly in terms of rarity, distribution and endangerment. There are several Natura 2000 habitat types that are widespread in this country, such as 6510 Lowland hay meadows (*Alopecurus pratensis*, *Sanguisorba officinalis*), 9170 *Galio-Carpinetum* oak-hornbeam forests, 9110 *Luzulo-Fagetum* beech forests and 9130 *Asperulo-Fagetum* beech forests (the names follow Annex I of the Habitats Directive; see Table 9.2). In contrast, many other habitat types cover very small areas within this country and are often restricted to particular areas with specific natural conditions, such as several alpine or Pannonian types. Some of them occur in the Czech Republic at their distribution limit. Examples include habitats: 7210 Calcareous fens with *Cladium mariscus* and species of the *Caricion davallianae*, 3220 Alpine rivers and the herbaceous vegetation along their banks, 8160 Medio-European calcareous scree of hill and montane levels, and 3230 Alpine rivers and their ligneous vegetation with *Myricaria germanica*.

Among the 60 Natura 2000 habitat types in the Czech Republic, there are 19 priority habitats with a high level of protection under EU legislation. They include a wide range of habitat types considered as particularly rare or endangered within the European Union; some of them are, however, locally relatively common in the Czech Republic, such as 9180 *Tilio-Acerion* forests of slopes, screes and ravines or 91E0 Alluvial forests with *Alnus glutinosa* and *Fraxinus excelsior* (*Alno-Padion*, *Alnion incanae*, *Salicion albae*) (Fig. 9.9a).

The results of the habitat mapping together with those of mapping the Natura 2000 species served as a basis for identifying the potential Sites of Community Importance (pSCIs); this identification was carried out separately for both biogeographical regions occurring in the country. After the recent extension in 2016, the proposal of the final national list of Sites of Community Importance in the Czech Republic comprises 1112 sites (http://drusop.nature.cz) covering 795,108 ha, i.e. ~10% of the area in this country (for the map of Sites of Community Importance see

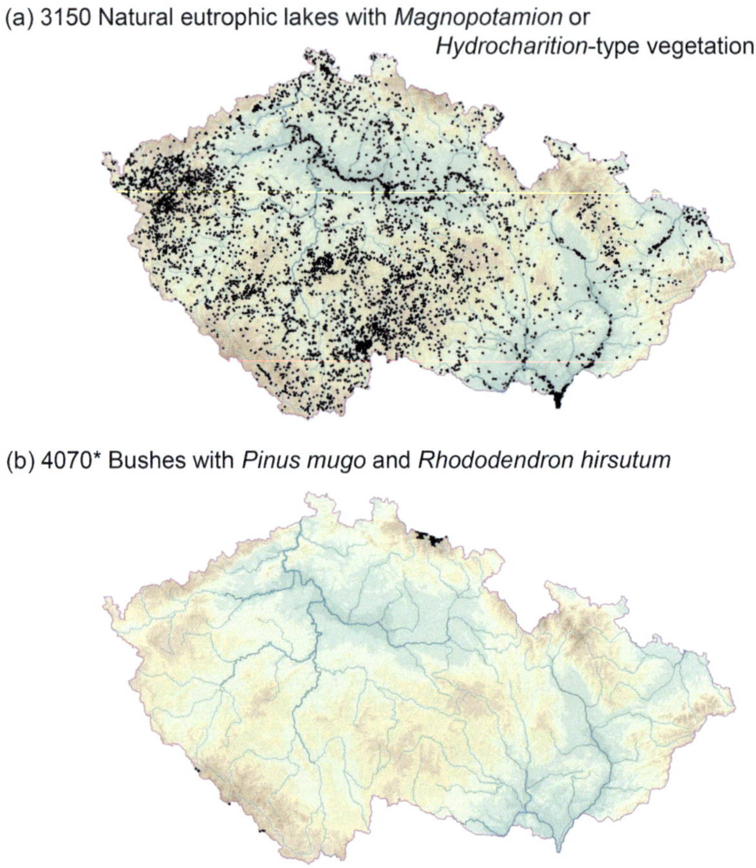


**Fig. 9.8** Selected results of the national project of habitat mapping showing the distribution of two natural habitat types occurring in the Czech Republic that are listed in Annex I of the Habitats Directive. The codes correspond to the Natura 2000 codes. Asterisks indicate priority habitat types. (**a**) 3150 Natural eutrophic lakes with *Magnopotamion* or *Hydrocharition*-type vegetation: this habitat type is widespread in naturally eutrophic and mesotrophic still water bodies throughout the Czech Republic, however generally endangered mainly by eutrophication; (**b**) 4070* Bushes with *Pinus mugo* and *Rhododendron hirsutum* (*Mugo-Rhododendretum hirsuti*): a rare habitat type in the Czech Republic, only represented by *Pinus mugo* scrub, which is native to the Krkonoše and Šumava Mts; in contrast, *Rhododendron hirsutum* does not occur in the Czech Republic (data on the distribution of habitats were provided by © Nature Conservation Agency of the Czech Republic)

Fig. 9.10). The area of individual sites varies substantially. The largest sites are situated mainly in the national parks (Šumava and Krkonoše), protected landscape areas (Beskydy, Bílé Karpaty and Blanský les) or military training areas (Hradiště and Boletice). An important issue is the cross-border connectivity of Sites of Community Importance between EU countries. The map of natural habitats at Natura 2000 sites in the Bavarian Forest and Šumava National Parks is an example of a successful

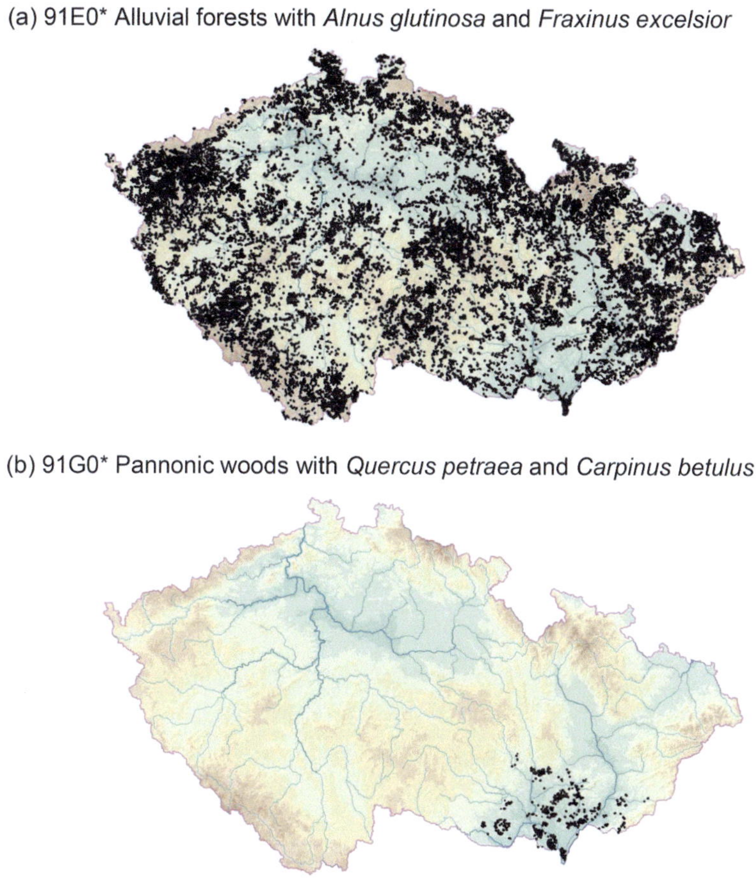


**Fig. 9.9** Selected results of the national project of habitat mapping showing the distribution of two natural habitat types occurring in the Czech Republic that are listed in Annex I of the Habitats Directive. See Fig. 9.8 for details. (**a**) 91E0* Alluvial forests with *Alnus glutinosa* and *Fraxinus excelsior* (*Alno-Padion*, *Alnion incanae*, *Salicion albae*): an example of a priority Natura 2000 habitat type that is common in the Czech Republic; the map illustrates the extent and detail of the habitat mapping carried out throughout the country; (**b**) 91G0* Pannonic woods with *Quercus petraea* and *Carpinus betulus*: an example of a habitat type that in the Czech Republic is strongly related with the Pannonian Biogeographical Region (data on the distribution of habitats were provided by © Nature Conservation Agency of the Czech Republic)

German-Czech cross-border cooperation based on collaboration in the habitat mapping (Hußlein and Kiener 2007).

As an EU member state, the Czech Republic is obliged to provide assessment reports on the conservation status of species and natural habitats of the European Union, listed in annexes of the Habitats Directive, every 6 years; the most recent report was provided in 2013 (Chobot 2016). The basic messages resulting from the habitat mapping are that most of the area in the Czech Republic is covered by non-natural habitats, such as arable land, forest plantations, urban areas and forests with

**Table 9.2** Natural habitat types occurring in the Czech Republic and listed in the Annex I of Directive 92/43/EEC on the conservation of natural habitats and of wild fauna and flora, also referred to as Habitats Directive. These natural habitat types require the designation of Special Areas of Conservation. An asterisk indicates priority habitat types. The code corresponds to the Natura 2000 code

| |
|---|
| **Coastal and halophytic habitats** |
| 1340* Inland salt meadows |
| **Coastal sand dunes and inland dunes** |
| 2330 Inland dunes with open *Corynephorus* and *Agrostis* grasslands |
| **Freshwater habitats** |
| 3130 Oligotrophic to mesotrophic standing waters with vegetation of the *Littorelletea uniflorae* and/or of the *Isoëto-Nanojuncetea* |
| 3140 Hard oligo-mesotrophic waters with benthic vegetation of *Chara* spp. |
| 3150 Natural eutrophic lakes with *Magnopotamion* or *Hydrocharition*-type vegetation |
| 3160 Natural dystrophic lakes and ponds |
| 3220 Alpine rivers and the herbaceous vegetation along their banks |
| 3230 Alpine rivers and their ligneous vegetation with *Myricaria germanica* |
| 3240 Alpine rivers and their ligneous vegetation with *Salix elaeagnos* |
| 3260 Water courses of plain to montane levels with the *Ranunculion fluitantis and Callitricho-Batrachion* vegetation |
| 3270 Rivers with muddy banks with *Chenopodion rubri* p.p. and *Bidention* p.p. vegetation |
| **Temperate heath and scrub** |
| 4030 European dry heaths |
| 4060 Alpine and Boreal heaths |
| 4070* Bushes with *Pinus mugo* and *Rhododendron hirsutum* (*Mugo-Rhododendretum hirsuti*) |
| 4080 Sub-Arctic *Salix* spp. scrub |
| 40A0* Subcontinental peri-Pannonic scrub |
| **Sclerophyllous scrub (matorral)** |
| 5130 *Juniperus communis* formations on heaths or calcareous grasslands |
| **Natural and semi-natural grassland formations** |
| 6110* Rupicolous calcareous or basophilic grasslands of the *Alysso-Sedion albi* |
| 6150 Siliceous alpine and boreal grasslands |
| 6190 Rupicolous pannonic grasslands (*Stipo-Festucetalia pallentis*) |
| 6210 Semi-natural dry grasslands and scrubland facies on calcareous substrates (*Festuco-Brometalia*) (* important orchid sites) |
| 6230* Species-rich *Nardus* grasslands, on siliceous substrates in mountain areas (and submountain areas in Continental Europe) |
| 6240* Sub-Pannonic steppic grasslands |
| 6250* Pannonic loess steppic grasslands |
| 6260* Pannonic sand steppes |
| 6410 *Molinia* meadows on calcareous, peaty or clayey-silt-laden soils (*Molinion caeruleae*) |
| 6430 Hydrophilous tall herb fringe communities of plains and of the montane to alpine levels |
| 6440 Alluvial meadows of river valleys of the *Cnidion dubii* |
| 6510 Lowland hay meadows (*Alopecurus pratensis*, *Sanguisorba officinalis*) |
| 6520 Mountain hay meadows |

(continued)

**Table 9.2** (continued)

| |
|---|
| **Raised bogs and mires and fens** |
| 7110* Active raised bogs |
| 7120 Degraded raised bogs still capable of natural regeneration |
| 7140 Transition mires and quaking bogs |
| 7150 Depressions on peat substrates of the *Rhynchosporion* |
| 7210* Calcareous fens with *Cladium mariscus* and species of the *Caricion davallianae* |
| 7220* Petrifying springs with tufa formation (*Cratoneurion*) |
| 7230 Alkaline fens |
| **Rocky habitats and caves** |
| 8110 Siliceous scree of the montane to snow levels (*Androsacetalia alpinae* and *Galeopsietalia ladani*) |
| 8150 Medio-European upland siliceous screes |
| 8160* Medio-European calcareous scree of hill and montane levels |
| 8210 Calcareous rocky slopes with chasmophytic vegetation |
| 8220 Siliceous rocky slopes with chasmophytic vegetation |
| 8230 Siliceous rock with pioneer vegetation of the *Sedo-Scleranthion* or of the *Sedo albi-Veronicion dillenii* |
| 8310 Caves not open to the public |
| **Forests** |
| 9110 *Luzulo-Fagetum* beech forests |
| 9130 *Asperulo-Fagetum* beech forests |
| 9140 Medio-European subalpine beech woods with *Acer* and *Rumex arifolius* |
| 9150 Medio-European limestone beech forests of the *Cephalanthero-Fagion* |
| 9170 *Galio-Carpinetum* oak-hornbeam forests |
| 9180* *Tilio-Acerion* forests of slopes, screes and ravines |
| 9190 Old acidophilous oak woods with *Quercus robur* on sandy plains |
| 91D0* Bog woodland |
| 91E0* Alluvial forests with *Alnus glutinosa* and *Fraxinus excelsior* (*Alno-Padion*, *Alnion incanae*, *Salicion albae*) |
| 91F0 Riparian mixed forests of *Quercus robur*, *Ulmus laevis* and *Ulmus minor*, *Fraxinus excelsior* or *Fraxinus angustifolia*, along the great rivers (*Ulmenion minoris*) |
| 91G0* Pannonic woods with *Quercus petraea* and *Carpinus betulus* |
| 91H0* Pannonian woods with *Quercus pubescens* |
| 91I0* Euro-Siberian steppic woods with *Quercus* spp. |
| 91 T0 Central European lichen Scots pine forests |
| 91 U0 Sarmatic steppe pine forest |
| 9410 Acidophilous Picea forests of the montane to alpine levels (*Vaccinio-Piceetea*) |

a low natural value, whereas natural habitats cover 17% of the Czech territory (Miko and Hošek 2010) including mainly near-natural forests and grasslands; other types of natural habitats are of marginal importance in terms of the area covered. The natural habitat types are considerably more frequent in the peripheral parts of this country, in less productive areas and partly overlap with protected areas.

Forests (including forest plantations) are the most widespread habitat type in the Czech Republic, covering more than one third of this country. Despite the fact that

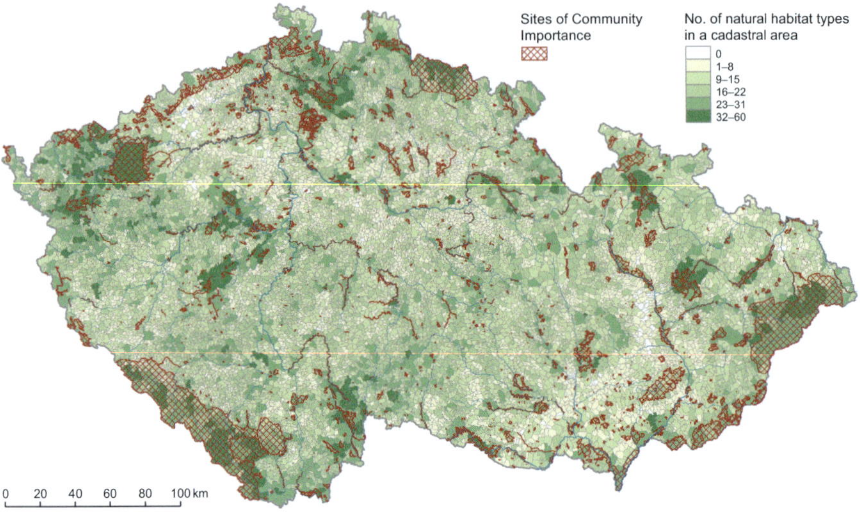


**Fig. 9.10** Sites of Community Importance. The results of the national mapping of habitat types and Natura 2000 species were used as a basis for identifying the potential Sites of Community Importance, therefore, there is a significant overlap of the Sites of Community Importance with the areas harbouring a high number of natural habitat types (counted within a cadastral area). These sites are considerably more frequent in less productive areas in the peripheral parts of this country. After the update in 2016, there are 1112 Sites of Community Importance in the Czech Republic covering 795,108 ha, i.e. ~10% of the country's area. They vary substantially in their area. The largest sites are situated mainly in the national parks, protected landscape areas or military training areas (data on natural habitats and Sites of Community Importance were provided by © Nature Conservation Agency of the Czech Republic)

forests rank among the stabilizing landscape components in the Czech Republic and their biodiversity is far less endangered than that of various open vegetation types, Czech forests are still subjected to unsustainable forestry practices, including clear-cutting, plantations (mainly *Picea abies* and *Pinus sylvestris* monocultures) and spread of alien species (*Pinus strobus*, *Quercus rubra* and others). Pristine forests cover only ~0.1% of the total forest area in the Czech Republic (http://bit.ly/2pjoMQs, 2012 data) and are predominantly found in large-scale protected areas, mainly in national parks.

Grasslands are the second most widespread group of habitats in the Czech Republic. Natural and semi-natural grasslands make up less than one third of the total area of grassland in this country and are very sensitive to all kinds of changes in their management. The recent changes in the Czech landscape include both intensification and lack of management. Among the Natura 2000 habitat types, the habitats 6410 *Molinia* meadows on calcareous, peaty or clayey-silt-laden soils (*Molinion caeruleae*) and 1340 Inland salt meadows, rank among the most endangered (Lustyk et al. 2014).

The conservation status of nearly all types of wetland habitats is currently evaluated as unfavourable despite many individual conservation and restoration projects (see Sect. 9.7.3). The main reason for the degradation of wetlands is still eutrophication and intensive fish production in the case of still waters, and stream regulation and pollution in the case of running water. In contrast, various types of alpine habitats or rocks and rocky slopes have a considerably better conservation status and are often not directly endangered.

## 9.7 Conservation and Management of Threatened Plant Species and Their Habitats

Despite the many successes and huge effort put into nature conservation in the Czech Republic, the conservation still faces a continuing biodiversity decline and degradation of natural habitats. This is also true for the diversity of vascular plants and their habitats. Therefore, it is essential for practical plant conservation to set priorities in terms of both species and sites. Furthermore, it is necessary to elaborate action plans for selected priority species in order to reverse negative trends in the population dynamics of the most threatened species by direct management interventions combining ex-situ and in-situ conservation approaches. Finally, there is an indisputable need for extensive restoration not only of selected habitats but also of landscapes (Prach et al. 2012).

### *9.7.1 Prioritization*

There have been several attempts to set conservation priorities for threatened taxa. Gabrielová et al. (2011) compared the endangerment of critically threatened species in the Czech flora (C1 category, based on the Red List valid at that time; Holub and Procházka 2000) with their endangerment in other European countries, based on their national Red Lists. This study concludes that ~12% of the critically threatened species in the Czech flora are not common in any other European country. Approximately half of these species are endemics (see Chap. 3, Sect. 3.4, this book) and the rest are very rare and threatened species across the whole of Europe, such as *Dracocephalum austriacum*, *Ligularia sibirica*, *Liparis loeselii* and *Pedicularis exaltata* (Gabrielová et al. 2011). The Czech Republic therefore has (or shares with other countries) the responsibility for their conservation. Another important study is that of Schnittler and Günther (1999), which focuses on Central-European vascular plants requiring priority conservation measures based on an analysis of national Red Lists of European countries and species distribution maps.

An area-based prioritization approach aimed at identifying sites of key importance for plant conservation is used in the international project Important Plant Areas (IPAs; Anderson et al. 2005). In 2002, the Czech Republic was among the first seven countries in Central and Eastern Europe that started to identify the best sites for wild plants, fungi and their habitats, using agreed standard criteria for the selection of IPAs. A total of 75 IPAs were identified in the Czech Republic, covering an area of 151,975 ha (Čeřovský et al. 2007; Fig. 9.11). Almost all Czech IPAs are

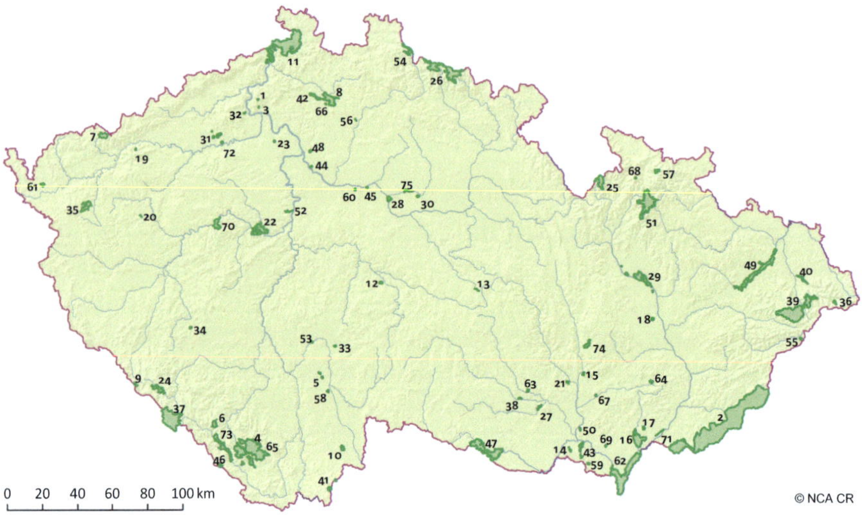


**Fig. 9.11** Important Plant Areas (IPAs) in the Czech Republic. The IPAs are sites of key importance for plant conservation based on international criteria. A total of 75 IPAs were identified in the Czech Republic, covering an area of 151,975 ha. The map is based on Čeřovský et al. (2007), with site numbers corresponding to this source (data were provided by © Nature Conservation Agency of the Czech Republic)

protected areas or Natura 2000 sites. This showed that the protected area network provided good coverage of botanically important sites in the Czech Republic. The Czech IPAs include not only the well-known botanical nature reserves, such as the Lovoš National Nature Reserve, which protects species-rich thermophilous vegetation, including the endemic species *Sorbus bohemica* and the Praděd National Nature Reserve in the Hrubý Jeseník Mts, which protects several endemic species, but also several large-scale protected areas, such as the White Carpathians Protected Landscape Area, Krkonoše National Park and Bohemian Switzerland National Park (the last because of its high richness of cryptogams).

## 9.7.2 Action Plans for Threatened Plant Species

Action plans (also referred to as species recovery programmes) for threatened species are prepared for plant species with a high extinction risks in the country, for which usual conservation tools, such as protection by law, creating a protected area and management of sites, are not sufficient. The responsible body for the preparation and implementation of action plans in the Czech Republic is the Nature Conservation Agency of the Czech Republic. The implementation of the action plans is limited by personnel and financial resources. As a consequence, action plans have so far only been approved and implemented for four taxa (Mináriková et al. 2011): *Angelica palustris*, *Dianthus arenarius* subsp. *bohemicus*, *Gentianella praecox* subsp. *bohemica* and *Potamogeton praelongus*. Action plans for *Adenophora liliifolia*, *Gentianella amarella* subsp. *amarella*, *G. obtusifolia* subsp. *sturmiana*, *Ornithogalum pyrenaicum* subsp. *sphaerocarpum* and *Pulsatilla patens* are currently under preparation (for more information see www.zachranneprogramy.cz; for selection criteria see Kostiuková and Čepelová 2017). The action plan for *Gentiana verna* subsp. *verna* was terminated as it appeared to be extremely difficult to realize, mainly due to the low genetic diversity of the remnant population in southern Bohemia (Kirschner et al. 2011; Kirschnerová et al. 2011).

An example of an intensively investigated threatened taxon for which there is an actively implemented action plan, is *Gentianella praecox* subsp. *bohemica* (Fig. 9.1f), which is endemic to the Bohemian Massif occurring only in the Czech Republic and adjacent regions of Germany, Poland and Austria. This strictly biennial plant, formerly rather abundant in a wide range of grasslands, became critically threatened within less than a century. It grew in pastures, short-grass meadows and at various disturbed sites. Its population decline during the twentieth century was caused mainly by eutrophication and abandonment of grazing, and consequently was overgrown by more competitive herbaceous plants. Some sites where this species occurred were afforested. Nowadays *Gentianella praecox* subsp. *bohemica* is protected by law in the Czech Republic and by the Habitats Directive (Natura 2000) in the European Union. Recently, there were only 59 localities of this species in the Czech Republic (Martinec 2015). According to the action plan the conservation objective is to stop the decline in the number of populations and population sizes by optimizing management, to continue monitoring of all recent sites and raise public awareness, mainly among farmers, of the species' conservation and management needs. Recent research indicates that the key for effective management of this taxon is to ensure that there is a sufficient level of disturbance, together with the removal of biomass (mowing or grazing, preferably by sheep and goats) and the creation of patches for seed germination by turf disruption (Milberg 1994; Huhta et al. 2000; Brabec 2005; Brabec et al. 2011). Bucharová et al. (2012) show that even small populations (10–15 flowering individuals) are able to survive if managed appropriately. However, there are currently both viable and steadily declining populations in the Czech Republic (Martinec 2015).

### 9.7.3 Ecological Restoration

Restoration ecology as a field of applied research was introduced to the Czech Republic in the mid-1990s (Prach et al. 2007). Ecological restoration in the broad sense, including various management activities aimed at improving the conservation status of habitats or species, has been part of nature conservation since the 1980s, even though often limited to voluntary conservation work. A particularly important impetus for undertaking restoration was the extreme landscape destruction and exploitation in north-western Bohemia caused by opencast coal mining in the 1970–1980s, which at that time was at an unprecedented scale within the whole of Czechoslovakia. However, it became obvious in the 1990s that the extent of the landscape and habitats requiring intensive restoration after four decades of communist regime and its harmful environmental impacts would be significantly broader than just the post-mining and post-industrial sites.

As elsewhere in Central Europe, wetlands and streams are habitats that need restoring in many cases (Dudgeon et al. 2006). Although rivers and wetlands have been affected by human activity since the Middle Ages, serious engineering of rivers started only in the nineteenth century and continued with growing intensity during the twentieth century. At that time not only large rivers but also the majority of small streams were at least partly regulated or, in some cases, almost completely channelized. Less regulated large rivers are rather exceptional, such as the Labe (Elbe) between the city of Ústí nad Labem and its estuary in Germany (Fig. 9.12a). Similarly, wet meadows, mires, fens and other types of wetland were subject to drastic measures, mainly large-scale drainage aimed at expanding the area of arable land in former Czechoslovakia. These measures not only resulted in serious consequences for landscape hydrology and biodiversity but also brought negative economic consequences, such as an increase in the impact of flood events. Therefore, very soon after the political change in 1989, the Czech Ministry of the Environment launched a River System Restoration Programme. Later, further options for restoration measures in wetlands came with EU funding. During the 25 years of wetland restoration in the Czech Republic a large number of projects have been carried out, mainly at a local scale. These projects predominantly focused on restoration of small streams, sometimes linked with measures to improve passability for migrating organisms or to enhance flood control. Restoration activities on large rivers are still rare (Just et al. 2012). For the protection of plant diversity, restoration of wet grasslands and especially mires are of particular importance. The latter have been successful in several places across the country mainly in the mountains, such as the Šumava Mts (Bufková et al. 2008), Krušné hory Mts and Krkonoše Mts (Lanta et al. 2006). The large-scale restoration of drained mires in the Šumava National Park (UNESCO Biosphere Reserve and Šumavská rašeliniště (Šumava Peatlands) Ramsar Site) is probably the best example in the Czech Republic. Some mires in this area were strongly disturbed, mainly during the twentieth century, by peat extraction, drainage and forestry. These mires, however, are of outstanding importance in terms of their relictness and being unique habitats with rare and threatened taxa such as *Andromeda polifolia*, *Betula nana*, *Carex chordorrhiza*, *C. dioica*, *Corallorhiza trifida*, *Listera cordata*, *Pinus uncinata* subsp. *uliginosa* and *Scheuchzeria palustris* (Spitzer and Bufková 2013). Therefore, soon after the establishment of the Šumava National Park (1991), the first activities in the monitoring of mires started, which resulted in the ambitious Mire Restoration Programme (Fig. 9.12c). Nineteen sites with a total area of 500 ha were restored by 2010 (Bufková et al. 2010) and monitored thereafter in a project that also included volunteers and NGOs (Bufková and Stíbal 2012).

The restoration and recreation of species-rich grasslands in the White Carpathians (UNESCO Biosphere Reserve and Protected Landscape Area) is one of the best

**Fig. 9.12** Examples of ecological restoration: (**a**) Little regulated large rivers are very rare in the Czech Republic: the Labe River in the Elbe Sandstones Protected Landscape Area with a rare habitat of muddy river banks and sandstone rocks in the Kaňon Labe National Nature Reserve (in the background); (**b**) The banks of the Labe River are the last remaining sites in this country of the critically threatened species *Corrigiola litoralis*; (**c**) Březnická slať mire in the Šumava National Park in 2012, 6 years after restoration; (**d**) Management of species-rich hay meadows in the Javorůvky Nature Reserve, Bílé Karpaty (White Carpathians) Protected Landscape Area; (**e**) Sheep grazing as part of the steppe restoration programme, Raná National Nature Reserve in the České středohoří Protected Landscape Area. Photo credits: P. Bauer (a, b), I. Bufková (c), J. W. Jongepier (d) and R. Hamerský (e)

examples of a successful restoration project, not only in the Czech Republic but also beyond. The White Carpathians is a large (1150 km$^2$) transboundary Czech-Slovak protected landscape area hosting unique biodiversity, mainly in grasslands; some sites in the Čertoryje National Nature Reserve even hold world records in local species richness of vascular plants (Merunková et al. 2012; Wilson et al. 2012; Chytrý et al. 2015; see also Chap. 7, Sect. 7.6, this book). These White Carpathian grasslands and fens also host many rare and critically threatened species, such as *Campanula cervicaria*, *Catabrosa aquatica*, *Gentianella amarella* subsp. *amarella*, *G. lutescens* subsp. *lutescens*, *Liparis loeselii*, *Ornithogalum pyrenaicum* subsp. *sphaerocarpum* and *Tephroseris longifolia* subsp. *moravica* (Jongepierová 2008). Since the 1990s, there has been a long-term project aimed at the restoration of large areas of species-rich grasslands that were previously converted to arable land or to grassland seeded with commercial clover-grass seed mixtures. This project, undertaken in close cooperation between the state nature conservancy and NGOs, succeeded in restoring and re-creating species-rich grasslands across a total area of 600 ha at ~40 sites, by developing a regional seed mixtures suitable for the local species-rich hay meadows and in increasing and restoring the species diversity at the regrassed sites, including those with occurrence of rare and threatened species (Jongepierová et al. 2007, 2012; Jongepierová 2008; Prach et al. 2013; Fig. 9.12d).

Other examples of restoration include the management by grazing of dry grasslands in the Bohemian Karst Protected Landscape Area in central Bohemia, the aim of which was to restore these dry grasslands and the landscape mosaic by using small-scale rotational grazing (Dostálek and Frantík 2012) by mixed herds of sheep and goats, which has a positive effect on vegetation and species diversity (Mayerová et al. 2010, 2012).

Currently, an ambitious project focused on the conservation of rare steppe habitats is underway in the western part of the České středohoří Mts. This area has a continental climate and hosts a wide range of rare and threatened steppic species, including *Adonis vernalis*, *Astragalus exscapus*, *Helictotrichon desertorum* subsp. *basalticum*, *Stipa pennata* and *S. pulcherrima*. The objective of this project, funded by the LIFE+ programme (EU financial instrument), is to maintain or restore local steppe habitats by using traditional farming practices (goat and sheep grazing; Fig. 9.12e) to maintain their unique biodiversity and raise public awareness about the values of steppes and the importance of their management (Vlačiha 2016).

Among the forest habitats, probably the current most ambitious restoration project is being carried out in the Bohemian Switzerland National Park. A long-term goal in this national park, covering an area of 7929 ha, is to restore the wilderness conditions by (i) converting the extensive spruce (*Picea abies*) plantations occupying more than half of the area of this park into near-natural mixed forests dominated by European beech (*Fagus sylvatica*) and (ii) eliminating invasive alien species, mainly the white pine (*Pinus strobus*), along with further invasive plant species, such as *Impatiens glandulifera*, *Reynoutria japonica* and *R. sachalinensis* (Fig. 9.13a, b). White pine, native to eastern North America, was introduced into Bohemia at the end of the eighteenth century. It was recorded as an invasive species in Bohemian Switzerland in the second half of the twentieth century. It is able to suppress the native species by forming dense uniform stands and by building up a thick layer of needle litter (Härtel and Hadincová 1998). It also outcompetes the native *P. sylvestris* by crown competition (Mácová 2001). Immediately after the establishment of the national park in 2000, a long-term project aimed at eradicating *P. strobus* was launched, accompanied by intensive research (Hadincová et al. 2007, 2008; Mácová and Tichý 2007; Mácová 2008; Münzbergová et al. 2010, 2013; Wild et al. 2013). By 2016, almost 90% of the *Pinus strobus* stands in the national park had been eradicated, representing ~66,000 $m^3$ of timber, with the total eradication costs amounting to almost 1.5 million €. The focus of current and future management is on the elimination of natural seeding and on monitoring of the already cleaned parts of the national park. Due to the rugged terrain of this sandstone pseudokarst area it is thought that this phase will be the most time-consuming and expensive part of the project.

In the same national park the secondary succession after a fire in 2006, when 18 ha of burnt forest were subsequently left to spontaneous succession, provided a valuable experience (Fig. 9.13c, d). Results based on the research and monitoring of this area documented a very dynamic succession accompanied by a significant increase in species diversity at all successional stages compared to unburnt plots (Adámek et al. 2016). Moreover, this fire helped to suppress the locally abundant invasive *Pinus strobus*. Based on this positive effect of post-fire succession, fire management is considered as an easy and cheap alternative to the common practice of restoration management in forests (Král et al. 2012). Due to the existing legislation and risks, however, such practices have not so far been used in forests in the Czech Republic.

Recently, there have been experiments involving grazing in the forests in the Bohemian Karst Protected Landscape Area and in the Podyjí National Park. Other experiments aim at restoring coppice systems (e.g. at Křtiny Training Forest Enterprise of Mendel University Brno and in Podyjí National Park; Vrška 2012).

The post-mining and post-industrial sites are a special target for restoration, both in terms of the level of human impact and restoration potential. Contrary to widespread public opinion, these sites are often of exceptional importance for the diversity of species and habitats. As a consequence of the heterogeneous environment, often combined with their extremely low productivity and the presence of early successional stages, quarries, sand and gravel pits (Řehounková et al. 2012), and even former open-cast mines and mining spoil heaps can host many rare and threatened species. Because of their very specific conditions, such sites are often neglected by state conservation authorities. Despite many studies advocating the advantages of spontaneous succession as a cheaper and environmentally more friendly alternative to technical reclamation, including positive effects on species richness and habitat diversity, priority is unfortunately still given to the technical approaches, such as artificial afforestation (Prach and Pyšek 2001; Hodačová and Prach 2003; Tropek et al. 2010; Prach et al. 2011; Prach 2012; Tropek and Prach 2012).

Finally, abandoned military training areas are very specific landscapes, mainly in post-communist countries where extensive areas were formerly dedicated to military training activities. Today, both abandoned and still used military training areas belong, from a conservation point of view, to the most important natural areas in the Czech Republic. Some of them are protected, mainly within the Natura 2000 net-

**Fig. 9.13** Examples of ecological restoration: (**a, b**) Repeated photographs from 1994 and 2015 show successful eradication of the invasive *Reynoutria sachalinensis* along the Křinice stream in the Bohemian Switzerland National Park; (**c, d**) Repeated photographs of an 18 ha plot of burnt forest (2006) in the Bohemian Switzerland National Park shortly after the fire and after 10 years of a spontaneous secondary succession (2016); (**e**) a tank road in the largest Czech military training area of Hradiště near the former village of Mětikalov (Meckl) in the Doupovské hory Mts; a rich mosaic of patches with various disturbance regimes resulting in different successional stages, together with volcanic bedrock, created high habitat and species diversity. Photo credits: H. Härtel (a), P. Bauer (b), V. Sojka (c, d) and J. Matějů (e)

work. The exceptional values which make them biodiversity hotspots, at least at a country level, result from several key factors including (i) their large extent, (ii) the absence of permanent residents, (iii) their origin in a formerly heterogeneous cultural landscape, rich in both species and habitats, (iv) the permanent disturbances creating fine-scale habitat mosaics of different successional stages, and (v) the absence of intensive agricultural and forestry practices, helping to preserve an oligotrophic environment and, as a consequence, preventing or minimizing negative effects such as eutrophication, invasion by alien species and biotic homogenization. Several military training areas were established in former Czechoslovakia after World War II when the country became a part of the Soviet bloc. After the Soviet Army left Czechoslovakia in the early 1990s, some military areas were abandoned and elsewhere the frequency of military training decreased significantly. Several areas were designated as nature reserves or Natura 2000 sites. In 2016, the large military training area in the Brdy Mts in western and central Bohemia was dissolved and converted into the Brdy Protected Landscape Area. All abandoned military areas face similar challenges in terms of management, above all the lack of active military use resulting in large-scale succession of woody species, reduction in early successional stages and the loss of the fine-scale landscape mosaic. This is valid also for some military training areas where nowadays only small parts are still used by the army, such as in the Hradiště Military Training Area in the Doupovské hory Mts (Fig. 9.13e). This largest military training area in this country, established in 1953, currently covers 280 km$^2$. The residents of the area were moved out by 1954 and the town of Doupov and 67 villages were completely demolished. Currently this military area hosts a unique diversity of species, habitats and landscape features as a result of its base-rich volcanic bedrock (basalt) and specific natural and cultural history (Matějů et al. 2016). The list of rare and threatened vascular plant species recorded in this military area includes several critically threatened taxa of the Czech flora such as *Androsace septentrionalis*, *Gentianella obtusifolia* subsp. *sturmiana* and *Pulsatilla patens*. At present, besides the lack of active management, this military training area is also endangered by a large-scale invasion of *Lupinus polyphyllus*. As a consequence, alternative management practices have been discussed including the introduction of large herbivores, especially the European bison (*Bison bonasus*). The first rewilding attempts were made recently in the Milovice-Mladá SCI (a Natura 2000 site in a lowland of central Bohemia), a part of a former military training area, where European bison, Exmoor pony (*Equus ferus caballus*; a horse breed native to the British Isles) and auroch-like cattle (a breed of *Bos taurus* similar to the extinct *Bos primigenius*) were introduced for low-cost maintenance of open vegetation and related threatened and rare species, such as *Gentiana cruciata*. Other management options were discussed or tested in some abandoned military training areas such as off-roading, quad-biking, horse-riding and other leisure-time activities, but their effects tend to be concentrated in particular places and paths, leaving most of the area to spontaneous succession (Marhoul and Zámečník 2012).

To conclude, ecological restoration in the Czech Republic is based on good knowledge and practical experience and its major advantage is a targeted approach implemented by local authorities, including state nature conservation authorities and NGOs in close collaboration with research institutions. There is a huge potential for using natural processes as sometimes the best and the cheapest solution, particularly when compared with expensive technical reclamations. However, there are many areas in urgent need of regular management in order to maintain species and habitat diversity. They include not only meadows, pastures and other types of agricultural land but also various types of species-rich lowland forest. Therefore, ensuring the sustainable management of these sites, including the restoration of traditional management practices, as well as combating new threats, such as the growing impact caused by the invasion by alien species, will rank among the crucial tasks of nature conservation in the Czech Republic in the future.

**Acknowledgements** I thank Milan Chytrý, Jiří Danihelka, Petr Pyšek, Karel Prach, Ivana Jongepierová, Vít Grulich and Marinus Werger for critical reading, valuable comments and text improvements. Tony Dixon kindly edited the English. I am also grateful to Ondřej Hájek and Jan Wild for preparing the maps, to Kamila Antošová, Petr Bauer, Iva Bufková, Roman Hamerský, Jan W. Jongepier, Veronika Kalníková, Jiří Kmet, Štěpán Koval, Jan Matějů, Jana Navrátilová, Vladimír Nejeschleba, Václav Sojka, Kateřina Šumberová, Dana Turoňová, Tomáš Vrška and Aleš Zvára for photographs and to the Centre of Excellence Pladias and the Nature Conservation Agency of the Czech Republic for providing data on the distributions of species and habitats.

## References

Adámek M, Hadincová V, Wild J (2016) Long-term effect of wildfires on temperate *Pinus sylvestris* forests: vegetation dynamics and ecosystem resilience. Forest Ecology and Management 380:285–295

Anderson S, Kušík T, Radford E (eds) (2005) Important plant areas in Central and Eastern Europe. Plantlife International, Salisbury

Bilz M, Kell SP, Maxted N, Lansdown RV (2011) European Red List of vascular plants. Publications Office of the European Union, Luxembourg

Bohn U, Neuhäusl R, Gollub G, Hettwer C, Neuhäuslová Z, Raus T, Schlüter H, Weber H (eds) (2000–2003) Karte der natürlichen Vegetation Europas. Maßstab 1: 2 500 000. Teil 1–3/Map of the natural vegetation of Europe. Scale 1: 2 500 000. Part 1–3. Landwirtschaftsverlag, Münster

Brabec J (2005) Současný stav rozšíření hořečku mnohotvarého českého (*Gentianella praecox* subsp. *bohemica*) v ČR [Recent distribution of *Gentianella praecox* subsp. *bohemica* in the Czech Republic]. Zprávy České botanické společnosti 40:1–44

Brabec J, Bucharová A, Štefánek M (2011) Vliv obhospodařování na životní cyklus hořečku mnohotvarého českého (*Gentianella praecox* subsp. *bohemica*) [Effect of management on life cycle of *Gentianella praecox* subsp. *bohemica*]. Příroda 31:85–109

Bucharová A, Brabec J, Münzbergová Z (2012) Effect of land use and climate change on future fate of populations of an endemic species of central Europe. Biological Conservation 145:39–47

Bufková I, Stíbal F (2012) Restoration of drained mires in the Šumava National Park. In: Jongepierová I, Pešout P, Jongepier JW, Prach K (eds) Ecological restoration in the Czech Republic. Nature Conservation Agency of the Czech Republic, Prague, pp 80–82

Bufková I, Stíbal F, Loskotová E (2008) Ecology and restoration of drained mires in the Šumava National Park (Czech Republic). In: Farrell C, Feehan J (eds) After wise use – the future of peatlands. Proceedings of the 13th International Peat Congress, Tullamore 2008. International Peat Society, Jyväskylä, pp 380–384

Bufková I, Stíbal F, Mikulášková E (2010) Restoration of drained mires in the Šumava National Park, Czech Republic. In: Eiseltová M (ed) Restoration of lakes, streams, floodplains, and bogs in Europe: principles and case studies. Springer, Dordrecht, pp 331–354
Bureš L (2013) Chráněné a ohrožené rostliny Chráněné krajinné oblasti Jeseníky [Protected and endangered plants of the Jeseníky Protected Landscape Area]. Agentura Rubico, Olomouc
Čeřovský J, Feráková V, Holub J, Maglocký Š, Procházka F (eds) (1999) Červená kniha ohrožených a vzácných druhů rostlin a živočichů ČR a SR. Vol. 5 Vyšší rostliny [Red Data Book of threatened and rare species of plants and animals in the Czech Republic and Slovak Republic. Vol. 5. Higher plants]. Príroda, Bratislava
Čeřovský J, Podhajská Z, Turoňová D (eds) (2007) Botanicky významná území České republiky [Important plant areas in the Czech Republic]. Agentura ochrany přírody a krajiny ČR, Praha
Chán V (ed) (1999) Komentovaný červený seznam květeny jižní části Čech [Annotated Red List of the southern Bohemian flora]. Příroda 16:1–284
Chobot K (ed) (2016) Druhy a přírodní stanoviště. Hodnotící zprávy o stavu v České republice 2013 [Species and natural habitats. Assessment reports on the conservation status in the Czech Republic 2013]. Agentura ochrany přírody a krajiny ČR & Ministerstvo životního prostředí ČR, Praha
Chytil J, Hakrová P, Hudec K, Husák Š, Jandová J, Pellantová J (eds) (1999) Mokřady České republiky – přehled vodních a mokřadních lokalit ČR [Wetlands of the Czech Republic – a survey of the water and wetland sites in the Czech Republic]. Český ramsarský výbor, Mikulov
Chytrý M (ed) (2007–2013) Vegetace České republiky 1–4 [Vegetation of the Czech Republic 1–4]. Academia, Praha
Chytrý M, Vicherek J (1995) Lesní vegetace Národního parku Podyjí/Thayatal. Die Waldvegetation des Nationalparks Podyjí/Thayatal. Academia, Praha
Chytrý M, Kučera T, Kočí M (eds) (2001) Katalog biotopů České republiky [Habitat catalogue of the Czech Republic]. Agentura ochrany přírody a krajiny ČR, Praha
Chytrý M, Pyšek P, Tichý L, Knollová I, Danihelka J (2005) Invasions by alien plants in the Czech Republic: a quantitative assessment across habitats. Preslia 77:339–354
Chytrý M, Kučera T, Kočí M, Grulich V, Lustyk P (eds) (2010) Katalog biotopů České republiky [Habitat catalogue of the Czech Republic]. Ed. 2. Agentura ochrany přírody a krajiny ČR, Praha
Chytrý M, Dražil T, Hájek M, Kalníková V, Preislerová Z, Šibík J, Ujházy K, Axmanová I, Bernátová D, ... Vymazalová M (2015) The most species-rich plant communities in the Czech Republic and Slovakia (with new world records). Preslia 87:217–278
Culek M (ed) (1996) Biogeografické členění České republiky [Biogeographical division of the Czech Republic]. Enigma, Praha
Culek M, Buček A, Grulich V, Hartl P, Hrabica A, Kocián J, Kyjovský Š, Lacina J (2005) Biogeografické členění České republiky [Biogeographical division of the Czech Republic]. Ed 2. Agentura ochrany přírody a krajiny ČR, Praha
Danihelka J (2013) Botanické součty, rozdíly a podíly [Botanical counts]. Živa 61:69–72
Danihelka J, Chrtek J Jr, Kaplan Z (2012) Checklist of vascular plants of the Czech Republic. Preslia 84:647–811
Dostálek J, Frantík T (2012) The impact of different grazing periods in dry grasslands on the expansive grass *Arrhenatherum elatius* L. and on woody species. Environmental Management 49:855–861
Douda J, Boublík K, Doudová J, Kyncl M (2017) Traditional forest management practices stop forest succession and bring back rare plant species. Journal of Applied Ecology 54:761–771
Dudgeon D, Arthington AH, Gessner MO, Kawabata Z-I, Knowler DJ, Lévêque C, Naiman RJ, Prieur-Richard A-H, Soto D, ... Sullivan CA (2006) Freshwater biodiversity: importance, threats, status and conservation challenges. Biological Reviews 81:163–182
Ekrtová E, Ekrt L, Košnar J, Zapomělová E, Čejková A (2008) Míčovka kulkonosná (*Pilularia globulifera*) znovu objevena v České republice [*Pilularia globulifera* rediscovered in the Czech Republic]. Zprávy České botanické společnosti 43:193–208

Foxcroft LC, Pyšek P, Richardson DM, Genovesi P (eds) (2013) Plant invasions in protected areas: patterns, problems and challenges. Springer, Dordrecht

Gabrielová J, Fialová T, Münzbergová Z (2011) Jak je to s ohrožením kriticky ohrožených druhů rostlin České republiky v jiných evropských zemích? [Critically endangered plant species of the Czech Republic: what is the situation of their endangerment in other European countries?]. Příroda 31:299–345

Gärdenfors U, Hilton-Taylor C, Mace GM, Rodríguez JP (2001) The applications of IUCN Red List criteria at regional level. Conservation Biology 15:1206–1212

Grulich V (1997) Atlas rozšíření cévnatých rostlin Národního parku Podyjí/Thayatal. Verbreitungsatlas der Gefässpflanzen des Nationalparks Podyjí/Thayatal. Masarykova univerzita, Brno

Grulich V (2012) Red List of vascular plants of the Czech Republic: 3rd edition. Preslia 84:631–645

Guth J, Kučera T (2005) Natura 2000 habitat mapping in the Czech Republic: methods and general results. Ekológia 24 (Suppl):39–51

Hadincová V, Köhnleinová I, Marešová J (2007) Invasive behaviour of white pine (*Pinus strobus*) in sandstone areas in the Czech Republic. In: Härtel H, Cílek V, Herben T, Jackson A, Williams R (eds) Sandstone landscapes. Academia, Praha, pp 219–224

Hadincová V, Münzbergová Z, Wild J, Šajtar L, Marešová J (2008) Dispersal of invasive *Pinus strobus* in sandstone areas of the Czech Republic. In: Tokarska-Guzik B, Brock JH, Brundu G, Child LE, Daehler CC, Pyšek P (eds) Plant invasions: human perception, ecological impacts and management. Backhuys Publishers, Leiden, pp 117–132

Härtel H, Bauer P (2002) Das Vorkommen von *Luronium natans* (L.) Raf. im Elbsandsteingebirge. Sächsische Floristische Mitteilungen 7:20–25

Härtel H, Hadincová V (1998) Invasion of white pine (*Pinus strobus* L.) into the vegetation of the Elbsandsteingebirge (Czech Republic/Germany). In: Synge H, Akeroyd J (eds) Planta Europa proceedings. The Swedish Threatened Species Unit & Plantlife, Uppsala, pp 251–255

Härtel H, Bauer P, Wild J (2001) Botanický výzkum národního parku České Švýcarsko a chráněné krajinné oblasti Labské pískovce: principy, výsledky a perspektivy [Botanical research of the České Švýcarsko (Bohemian Switzerland) National Park and Labské pískovce (Elbe Sandstones) Protected Landscape Area: principles, results and perspectives (Northern Bohemia, Czech Republic)]. Příroda 19:59–65

Härtel H, Cílek V, Herben T, Jackson A, Williams R (eds) (2007a) Sandstone landscapes. Academia, Praha

Härtel H, Sádlo J, Świerkosz K, Marková I (2007b) Phytogeography of the sandstone areas in the Bohemian Cretaceous Basin (Czech Republic/Germany/Poland). In: Härtel H, Cílek V, Herben T, Jackson A, Williams R (eds) Sandstone landscapes. Academia, Praha, pp 177–189

Härtel H, Lončáková J, Hošek M (eds) (2009) Mapování biotopů v České republice. Východiska, výsledky, perspektivy [Habitat mapping in the Czech Republic. Background, results, perspectives]. Agentura ochrany přírody a krajiny ČR, Praha

Härtel H, Bauer P, Benda P, Blažej L, Holešinský O, Klitsch M, Marková I, Phoenix J, Riebe H, Šteflová D (2013) Biological research and monitoring in the Elbe Sandstones (Saxon-Bohemian Switzerland). In: Migoń P, Kasprzak M (eds) Sandstone landscapes, diversity, ecology and conservation. Proceedings of the 3rd International Conference on Sandstone Landscapes, Kudowa-Zdrój (Poland), 25–28 April 2012. University of Wrocław, Wrocław, pp 38–43

Härtel H, Pergl J, Šíma J, Bauer P (2015) Invazní rostliny v chráněných územích [Invasive plants in protected areas]. Fórum ochrany přírody 2015(3):16–24

Hartley S, Kunin WE (2003) Scale dependency of rarity, extinction risk and conservation priority. Conservation Biology 17:1559–1570

Hédl R, Kopecký M, Komárek J (2010) Half a century of succession in a temperate oakwood: from species-rich community to mesic forest. Diversity and Distributions 16:267–276

Hodačová D, Prach K (2003) Spoil heaps from brown coal mining: technical reclamation versus spontaneous revegetation. Restoration Ecology 11:385–391

Holá E, Turoňová D (2014) Hodnotící zprávy o stavu mechorostů a cévnatých rostlin – podruhé [Assessment reports on the conservation status of bryophytes and vascular plants – for the second time]. Ochrana přírody 2014(4):22–25

Holub J (2000) Černá listina vymizelých taxonů květeny České republiky a Slovenské republiky [Black List of taxa disappeared from the floras of the Czech and Slovak Republics]. Preslia 72:167–186

Holub J, Procházka F (2000) Red List of vascular plants of the Czech Republic – 2000. Preslia 72:187–230

Holub J, Procházka F, Čeřovský J (1979) Seznam vyhynulých, endemických a ohrožených taxonů vyšších rostlin květeny ČSR (1. verze) [The list of extinct, endemic and threatened taxa of vascular plants of the Czech Socialistic Republic (1st version)]. Preslia 51:213–237

Huhta A-P, Lennartsson T, Tuomi J, Rautio P, Laine K (2000) Tolerance of *Gentianella campestris* in relation to damage intensity: an interplay between apical dominance and herbivory. Evolutionary Ecology 14:373–392

Hußlein M, Kiener H (eds) (2007) Natura 2000. Divoké srdce Evropy. Europas Wildes Herz, Grafenau/Vimperk

IUCN (2001) IUCN Red List categories and criteria: version 3.1. IUCN, Gland/Cambridge

IUCN (2012) Guidelines for application of IUCN Red List criteria at regional and national levels: version 4.0. IUCN, Gland/Cambridge

Jarošík V, Konvička M, Pyšek P, Kadlec T, Beneš J (2011a) Conservation in a city: do the same principles apply to different taxa? Biological Conservation 144:490–499

Jarošík V, Pyšek P, Kadlec T (2011b) Alien plants in urban nature reserves: from red-list species to future invaders? NeoBiota 10:27–46

Jeník J (1961) Alpinská vegetace Krkonoš, Králického Sněžníku a Hrubého Jeseníku [Alpine vegetation of the Krkonoše, Králický Sněžník and Hrubý Jeseník]. Nakladatelství Československé akademie věd, Praha

Jonášová M, Prach K (2004) Central-European mountain spruce (*Picea abies* (L.) Karst.) forests: regeneration of tree species after bark beetle outbreak. Ecological Engineering 23:15–27

Jongepier J, Pechanec V (2006) Atlas rozšíření cévnatých rostlin CHKO Bílé Karpaty [Distribution atlas of vascular plants of the White Carpathians Protected Landscape Area]. ZO ČSOP Bílé Karpaty, Veselí nad Moravou

Jongepierová I (ed) (2008) Louky Bílých Karpat [Grasslands of the White Carpathian Mountains]. ZO ČSOP Bílé Karpaty, Veselí nad Moravou

Jongepierová I, Mitchley J, Tzanopoulos J (2007) A field experiment to recreate species rich hay meadows using regional seed mixtures. Biological Conservation 139:297–305

Jongepierová I, Prach K, Řehounková K (2012) Recreation of species rich grasslands in the Bílé Karpaty Mountains. In: Jongepierová I, Pešout P, Jongepier JW, Prach K (eds) Ecological restoration in the Czech Republic. Nature Conservation Agency of the Czech Republic, Prague, pp 45–46

Just T, Pithart D, Bufková I (2012) Wetlands and streams. Introduction. In: Jongepierová I, Pešout P, Jongepier JW, Prach K (eds) Ecological restoration in the Czech Republic. Nature Conservation Agency of the Czech Republic, Prague, pp 67–73

Kaplan Z, Danihelka J, Lepší M, Lepší P, Ekrt L, Chrtek J Jr, Kocián J, Prančl J, Kobrlová L, Hroneš M, Šulc V (2016) Distributions of vascular plants in the Czech Republic. Part 3. Preslia 88:459–544

Kirschner J, Kirschnerová L, Bartish I (2011) Conservation status of two isolated populations of *Gentiana verna* (*Gentianaceae*) in the Czech Republic: insights from an allozyme analysis. Phyton (Horn) 51:177–199

Kirschnerová L, Moravcová L, Rauch O (2011) Dlouhodobé sledování změn populace *Gentiana verna* v národní přírodní památce Rovná [Monitoring of long-term changes in the isolated population of *Gentiana verna* in the Rovná Nature Reserve]. Příroda 30:91–114

Kolbek J, Moravec J (eds) (1995) Mapa potenciální přirozené vegetace Biosférické rezervace Křivoklátsko [Map of potential natural vegetation of the Biosphere Reserve Křivoklátsko]. Botanický ústav AV ČR & Ministerstvo životního prostředí ČR, Průhonice

Kolbek J, Blažková D, Husová M, Moravec J, Neuhäuslová Z, Sádlo J (1997) Potenciální přirozená vegetace Biosférické rezervace Křivoklátsko [Potential natural vegetation of the Biosphere Reserve Křivoklátsko]. Academia, Praha

Kolbek J, Blažková D, Břízová E, Kučera T, Ložek V, Rybníček K, Rybníčková E, Rydlo J (1999a) Vegetace Chráněné krajinné oblasti a Biosférické rezervace Křivoklátsko 1. Vývoj krajiny a vegetace, vodní, pobřežní a luční společenstva [Vegetation of the Protected Landscape Area and Biosphere Reserve Křivoklátsko 1. Development of the landscape and vegetation, aquatic, river bank and meadow communities]. Agentura ochrany přírody a krajiny ČR & Botanický ústav AV ČR, Praha

Kolbek J, Mladý F, Petříček V et al. (1999b) Květena Chráněné krajinné oblasti a Biosférické rezervace Křivoklátsko 1. Mapy rozšíření cévnatých rostlin [Flora of the Protected Landscape Area and Biosphere Reserve Křivoklátsko 1. Distribution maps of vascular plants]. Agentura ochrany přírody a krajiny ČR & Botanický ústav AV ČR, Praha

Kolbek J, Mladý F, Brabec E, Hroudová Z, Kučera T, Vítková M (2001a) Květena Chráněné krajinné oblasti a Biosférické rezervace Křivoklátsko 2. Rozbor a syntéza [Flora of the Protected Landscape Area and Biosphere Reserve Křivoklátsko 2. Analysis and synthesis]. Botanický ústav AV ČR & Správa chráněných krajinných oblastí ČR, Praha

Kolbek J, Neuhäuslová Z, Sádlo J, Dostálek J, Havlíček P, Husáková J, Kučera T, Kropáč Z, Lecjaksová S (2001b) Vegetace Chráněné krajinné oblasti a Biosférické rezervace Křivoklátsko 2. Společenstva skal, strání, sutí, primitivních půd, vřesovišť, termofilních lemů a synantropní vegetace [Vegetation of the Protected Landscape Area and Biosphere Reserve Křivoklátsko 2. Plant communities of the rocks, screes, primitive soils, heathlands, thermophilous margins and ruderal sites]. Academia, Praha

Kolbek J, Bílek O, Boublík K, Brabec J, Černý T, Härtel H, Husová M, Jelínek J, Kučera T, .... Vítková M (2003) Vegetace Chráněné krajinné oblasti a Biosférické rezervace Křivoklátsko 3. Společenstva lesů, křovin, pramenišť, balvanišť a acidofilních lemů [Vegetation of the Protected Landscape Area and Biosphere Reserve Křivoklátsko 3. Plant communities of the forests, shrubs, forest springs, boulder screes and acidophilous fringes]. Academia, Praha

Kopecký M, Hédl R, Szabó P (2013) Non-random extinctions dominate plant community changes in abandoned coppices. Journal of Applied Ecology 50:79–87

Kostiuková J, Čepelová B (2017) Jak probíhal výběr druhů pro záchranné programy ohrožených druhů rostlin [How the selection of species for the action plans for the endangered plant species took place]. Fórum ochrany přírody 2017(1):13–16

Krahulcová A, Krahulec F, Kirschner J (1996) Introgressive hybridization between a native and an introduced species: *Viola lutea* subsp. *sudetica* × *V. tricolor*. Folia Geobotanica et Phytotaxonomica 31:219–244

Král K, Trochta J, Vrška T (2012) Can fire and secondary succession assist in the regeneration of forests in a national park? In: Jongepierová I, Pešout P, Jongepier JW, Prach K (eds) Ecological restoration in the Czech Republic. Nature Conservation Agency of the Czech Republic, Prague, pp 24–26

Králová T, Tomášek M (2009) Metodika aktualizace vrstvy mapování biotopů [Methods of updating the habitat mapping layer]. In: Härtel H, Lončáková J, Hošek M (eds) Mapování biotopů v České republice. Východiska, výsledky, perspektivy [Habitat mapping in the Czech Republic. Background, results, perspectives]. Agentura ochrany přírody a krajiny ČR, Praha, pp 19–25

Křenová Z, Kiener H (2012) Europe's Wild Heart – still beating? Experiences from a new transboundary wilderness area in the middle of the Old Continent. European Journal of Environmental Sciences 2:115–124

Kubát K (1970) Rozšíření některých druhů rostlin v Českém středohoří. Fytogeografická studie [Distribution of some plant species in the České středohoří Mts. A phytogeographic study]. Okresní vlastivědné muzeum, Litoměřice

Kubát K (1986) Červená kniha vyšších rostlin Severočeského kraje [Red Data Book of higher plants of the Northern Bohemian region]. Okresní vlastivědné muzeum, Litoměřice

Kučera T (2009) Červená kniha biotopů – shrnutí [Red Data Book of habitats – a summary]. In: Härtel H, Lončáková J, Hošek M (eds) Mapování biotopů v České republice. Východiska, výsledky, perspektivy [Habitat mapping in the Czech Republic. Background, results, perspectives]. Agentura ochrany přírody a krajiny ČR, Praha, pp 66–71

Kučera J, Váňa J, Hradílek Z (2012) Bryophyte flora of the Czech Republic: updated checklist and Red List (2012) and a brief analysis. Preslia 84:813–850

Lanta V, Mach J, Holcová V (2006) The effect of dam construction on the restoration succession of spruce mires in the Giant Mountains (Czech Republic). Annales Botanici Fennici 43:260–268

Lehnert LW, Bässler C, Brandl R, Burton PJ, Müller J (2013) Conservation value of forests attacked by bark beetles: highest number of indicator species is found in early successional stages. Journal of Nature Conservation 21:97–104

Lepší P, Lepší M, Boublík K, Štech M, Hans V (eds) (2013) Červená kniha květeny jižní části Čech [The Red Data Book of South Bohemian flora]. Jihočeské muzeum, České Budějovice

Lepší M, Lepší P, Koutecký P, Bílá J, Vít P (2015) Taxonomic revision of *Sorbus* subgenus *Aria* occurring in the Czech Republic. Preslia 87:109–162

Liška J, Palice Z, Slavíková Š (2008) Checklist and Red List of lichens of the Czech Republic. Preslia 80:151–182

Lucas GL, Walter SM (eds) (1976) List of rare, threatened & endemic plants for the countries of Europe. International Union for Conservation of Nature and Natural Resources, Morges

Lustyk P, Oušková V, Kratochvílová L, Chobot K (2014) Zpráva o stavu biotopů 2013. Hodnocení stavu a trendů evropsky významných typů přírodních stanovišť [Report on the conservation status of natural habitats. Assessment of state and trends in natural habitats of European importance in the Czech Republic]. Ochrana přírody 2014(3):30–33

Mackovčin P, Sedláček M (eds) (1999–2009) Chráněná území ČR [Protected areas of the Czech Republic] I–XIV. Agentura ochrany přírody a krajiny ČR, Praha

Mácová M (2001) Tree-ring analysis of crown competition and climatic sensitivity in *Pinus strobus* and *P. sylvestris* stands in the Elbe River Sandstone Mountains. Dendrochronologia 19:103–113

Mácová M (2008) Dendroclimatological comparison of native *Pinus sylvestris* and invasive *Pinus strobus* in different habitats in the Czech Republic. Preslia 80:277–289

Mácová M, Tichý T (2007) Dendroclimatological comparison of invasive *Pinus strobus* and native *Pinus sylvestris* growing in the Czech Republic on ecological and geographical gradients. In: Härtel H, Cílek V, Herben T, Jackson A, Williams R (eds) Sandstone landscapes. Academia, Praha, pp 225–229

Marhoul P, Zámečník J (2012) Abandoned military areas. Introduction. In: Jongepierová I, Pešout P, Jongepier JW, Prach K (eds) Ecological restoration in the Czech Republic. Nature Conservation Agency of the Czech Republic, Prague, pp 111–113

Martinec P (2015) Hořeček mnohotvarý český [The early gentian]. Ochrana přírody 70(1):18–23

Matějů J, Hradecký P, Melichar V (eds) (2016) Doupovské hory [Doupovské hory Mts]. Česká geologická služba, Praha

Mayerová H, Čiháková K, Florová K, Kladivová A, Šlechtová A, Trnková E, Münzbergová Z (2010) Vliv pastvy ovcí a koz na vegetaci suchých trávníku v CHKO Český kras [Impact of sheep and goat grazing on vegetation of dry grasslands in the Bohemian Karst PLA]. Příroda 27:53–74

Mayerová H, Tichý T, Heřman P, Čiháková K, Münzbergová Z (2012) Grazing of dry grasslands in the Bohemian Karst. In: Jongepierová I, Pešout P, Jongepier JW, Prach K (eds) Ecological restoration in the Czech Republic. Nature Conservation Agency of the Czech Republic, Prague, pp 47–50

Merunková K, Preislerová Z, Chytrý M (2012) White Carpathian grasslands: can local ecological factors explain their extraordinary species richness? Preslia 84:311–325

Michalcová D, Chytrý M, Pechanec V, Hájek O, Jongepier JW, Danihelka J, Grulich V, Šumberová K, Preislerová Z, ... Zelený D (2014) High plant diversity of grasslands in landscape context: a comparison of contrasting regions in Central Europe. Folia Geobotanica 49:117–135

Miko L, Hošek M (eds) (2010) State of nature and landscape in the Czech Republic. Report 2009. Nature Conservation Agency of the Czech Republic, Prague
Mikyška R, Deyl M, Holub J, Husová M, Moravec J, Neuhäusl R, Neuhäuslová-Novotná Z (1968–1972) Geobotanická mapa ČSSR 1. České země [Geobotanical map of the CSSR 1. Bohemian lands]. Vegetace ČSSR, Ser A, 2:1–204. Academia, Praha [Explanatory text 1968, map sheets 1968–1972]
Milberg P (1994) Germination ecology of the endangered grassland biennial *Gentianella campestris*. Biological Conservation 70:287–290
Miller RM, Rodríguez JP, Aniskowicz-Fowler T, Bambaradeniya C, Boles R, Eaton MA, Gärdenfors U, Keller V, Molur S, Walker S, Pollock C (2007) National threatened species listing based on IUCN criteria and regional guidelines: current status and future perspectives. Conservation Biology 21:684–696
Mináriková T, Zmeškalová J, Větrovcová J, Bína P, Šlechtová A (2011) Záchranné programy ohrožených druhů v České republice [Action plans for endangered species in the Czech Republic]. Životné prostredie 45:249–255
Ministry of the Environment of the Czech Republic (2005) National biodiversity strategy of the Czech Republic. Ministry of the Environment of the Czech Republic, Prague
Moravec J, Balátová-Tuláčková E, Blažková D, Hadač E, Hejný S, Husák Š, Jeník J, Kolbek J, Krahulec F, ... Vicherek J (1995) Rostlinná společenstva České republiky a jejich ohrožení [Red list of plant communities of the Czech Republic and their endangerment]. Severočeskou přírodou Suppl:1–206
Müller F, Kubát K (eds) (2013) Seltene Pflanzen im Erzgebirge. Situation in Böhmen und in Sachsen. CDSM.cz, Ústí nad Labem
Müller J, Bußler H, Goßner M, Rettetbach T, Duelli P (2008) The European spruce bark beetle *Ips typographus* (L.) in a national park – from pest to keystone species. Biodiversity and Conservation 17:2979–3001
Münzbergová Z, Hadincová V, Wild J, Herben T, Marešová J (2010) Spatial and temporal variation in dispersal pattern of an invasive pine. Biological Invasions 12:2471–2486
Münzbergová Z, Hadincová V, Wild J, Kindlmannová J (2013) Variability in the contribution of different life stages to population growth as a key factor in the invasion success of *Pinus strobus*. PLoS One 8:e56953
Nepraš K, Kroufek R, Kubát K, Vlačiha V (2008) Orchideje Českého středohoří [Orchids of the České středohoří Mts]. Oblastní muzeum v Litoměřicích, Litoměřice
Neuhäuslová Z (ed) (2001) Mapa potenciální přirozené vegetace Národního parku Šumava [The map of potential natural vegetation of the Šumava National Park]. Silva Gabreta Suppl 1:1–189
Neuhäuslová Z, Moravec J, Chytrý M, Sádlo J, Rybníček K, Kolbek J, Jirásek J (1997) Mapa potenciální přirozené vegetace České republiky 1:500 000 [Map of potential natural vegetation of the Czech Republic 1:500 000]. Botanický ústav AV ČR, Průhonice
Neuhäuslová Z, Blažková D, Grulich V, Husová M, Chytrý M, Jeník J, Jirásek J, Kolbek J, Kropáč Z, ... Sádlo J (1998) Mapa potenciální přirozené vegetace České republiky. Textová část [Map of potential natural vegetation of the Czech Republic. Explanatory text]. Academia, Praha
Pergl J, Sádlo J, Petrusek A, Laštůvka Z, Musil J, Perglová I, Šanda R, Šefrová H, Šíma J, Vohralík V, Pyšek P (2016) Black, grey and watch lists of alien species in the Czech Republic based on environmental impacts and management strategy. NeoBiota 28:1–37
Prach K (2012) Coal mining spoil heaps in the Most region: restoration potential of spontaneous succession. In: Jongepierová I, Pešout P, Jongepier JW, Prach K (eds) Ecological restoration in the Czech Republic. Nature Conservation Agency of the Czech Republic, Prague, pp 97–98
Prach K, Pyšek P (2001) Using spontaneous succession for restoration of human-disturbed habitats: experience from Central Europe. Ecological Engineering 17:55–62
Prach K, Mars R, Pyšek P, van Diggelen R (2007) Manipulation of succession. In: Walker LR, Walker J, Hobbs RJ (eds) Linking restoration and ecological succession. Springer, New York, pp 121–149
Prach K, Řehounková K, Řehounek J, Konvalinková P (2011) Ecological restoration of central European mining sites: a summary of a multi-site analysis. Landscape Research 36:263–268

Prach K, Pešout P, Jongepierová I (2012) Restoration ecology and ecological restoration in the Czech Republic. In: Jongepierová I, Pešout P, Jongepier JW, Prach K (eds) Ecological restoration in the Czech Republic. Nature Conservation Agency of the Czech Republic, Prague, pp 9–10

Prach K, Jongepierová I, Řehounková K (2013) Large-scale restoration of dry grasslands on ex-arable land using a regional seed mixture: establishment of target species. Restoration Ecology 21:33–39

Procházka F, Štech M (2002) Komentovaný černý a červený seznam cévnatých rostlin české Šumavy [Annotated Black and Red List of vascular plants of the Czech part of the Šumava Mts]. Správa NP a CHKO Šumava & EkoAgency KOPR, Vimperk

Pyšek P, Jarošík V, Kučera T (2002a) Patterns of invasion in temperate nature reserves. Biological Conservation 104:13–24

Pyšek P, Kučera T, Jarošík V (2002b) Plant species richness of nature reserves: the interplay of area, climate and habitat in a Central European landscape. Global Ecology and Biogeography 11:279–289

Pyšek P, Jarošík V, Kučera T (2003a) Inclusion of native and alien species in temperate nature reserves: an historical study from Central Europe. Conservation Biology 17:1414–1424

Pyšek P, Sádlo J, Mandák B, Jarošík V (2003b) Czech alien flora and a historical pattern of its formation: what came first to Central Europe? Oecologia 135:122–130

Pyšek P, Křivánek M, Jarošík V (2009) Planting intensity, residence time, and species traits determine invasion success of alien woody species. Ecology 90:2734–2744

Pyšek P, Chytrý M, Pergl J, Sádlo J, Wild J (2012a) Plant invasions in the Czech Republic: current state, introduction dynamics, invasive species and invaded habitats. Preslia 84:575–629

Pyšek P, Danihelka J, Sádlo J, Chrtek J Jr, Chytrý M, Jarošík V, Kaplan Z, Krahulec F, Moravcová L, ... Tichý L (2012b) Catalogue of alien plants of the Czech Republic (2nd edition): checklist update, taxonomic diversity and invasion patterns. Preslia 84:155–255

Pyšek P, Genovesi P, Pergl J, Monaco A, Wild J (2013) Plant invasions of protected areas in Europe: an old continent facing new problems. In: Foxcroft LC, Pyšek P, Richardson DM, Genovesi P (eds) Plant invasions in protected areas: patterns, problems and challenges. Springer, Dordrecht, pp 209–240

Řehounková K, Řehounek J, Prach K (eds) (2012) Near-natural restoration vs. technical reclamation of mining sites in the Czech Republic. Faculty of Science, University of South Bohemia, České Budějovice

Rejzková T (2013) Ohrožené druhy rostlin v národních parcích ČR: srovnání z pohledu chráněných fenoménů a managementových přístupů [Threatened plant species of the national parks in the Czech Republic: a comparison from the protected phenomena and management approaches point of view]. BSc thesis, Charles University, Praha

Sádlo J, Härtel H, Marková I (2007) Diversity of flora and vegetation of the sandstone areas in the Bohemian Cretaceous Basin (Czech Republic/Germany/Poland). In: Härtel H, Cílek V, Herben T, Jackson A, Williams R (eds) Sandstone landscapes. Academia, Praha, pp 161–176

Schiffleithner V, Essl F (2016) Is it worth the effort? Spread and management success of invasive alien plant species in a Central European National Park. NeoBiota 31:43–61

Schnittler M, Günther K-F (1999) Central European vascular plants requiring priority conservation measures – an analysis from national Red Lists and distribution maps. Biodiversity and Conservation 8:891–925

Skalický V, Jeník J (1974) Květena a vegetační poměry Českého krasu z hlediska ochrany přírody [Flora and vegetation of the Bohemian Karst in view of the nature conservation]. Bohemia Centralis 3:101–140

Spitzer K, Bufková I (2013) Peatlands of Šumava. Administration of the Šumava National Park and Protected Landscape Area, Vimperk

Špryňar P (2004) Poznámky k překvapivému výskytu podmrvky jižní (*Notholaena marantae*) a sleziníku hadcového (*Asplenium cuneifolium*) na ultrabazickém pikritu v Českém krasu [Notes on the surprising occurrence of *Notholaena marantae* and *Asplenium cuneifolium* on the ultrabasic picrite in Bohemian Karst (Central Bohemia)]. Zprávy České botanické společnosti 39:321–338

Špryňar P, Münzbergová Z (1998) Prodromus pražské květeny [Prodrome of the Prague Flora]. Muzeum a současnost, Řada přírodovědná 12:129–222
Štursa J, Kwiatkowski P, Harčarik J, Zahradníková J, Krahulec F (2009) Černý a červený seznam cévnatých rostlin Krkonoš [Black and red list of vascular plants of the Krkonoše/Karkonosze Mts. (Western Sudetes CZ, PL)]. Opera Corcontica 46:67–104
Suda J, Bauer P, Brabec J, Hadinec J (2000) Znovunalezené druhy naší květeny – žabníček vzplývavý [Re-discovered species of our flora – *Luronium natans*]. Živa 48:205–207
Tropek R, Prach K (2012) Mining and post-industrial sites. Introduction. In: Jongepierová I, Pešout P, Jongepier JW, Prach K (eds) Ecological restoration in the Czech Republic. Nature Conservation Agency of the Czech Republic, Prague, pp 89–93
Tropek R, Kadlec T, Karesová P, Spitzer L, Kocárek P, Malenovsky I, Banar P, Tuf IH, Hejda M, Konvička M (2010) Spontaneous succession in limestone quarries as an effective restoration tool for endangered arthropods and plants. Journal of Applied Ecology 47:39–147
Turoňová D (2016) Cévnaté rostliny [Vascular plants]. In: Chobot K (ed) Druhy a přírodní stanoviště. Hodnotící zprávy o stavu v České republice 2013 [Species and natural habitats. Assessment reports on the conservation status in the Czech Republic 2013]. Agentura ochrany přírody a krajiny ČR & Ministerstvo životního prostředí ČR, Praha, pp 86–109
Vlačiha V (2016) 40 let chráněné krajinné oblasti České středohoří [The České středohoří/ Bohemian Mittelgebirge hills protected landscape area celebrates forty years]. Ochrana přírody 2016(2):10–12
Vrška T (2012) Forests. Introduction. In: Jongepierová I, Pešout P, Jongepier JW, Prach K (eds) Ecological restoration in the Czech Republic. Nature Conservation Agency of the Czech Republic, Prague, pp 11–16
Wild J, Hadincová V, Münzbergová Z, Härtel H (2013) Modell der räumlichen Ausbreitung der Weymouth-Kiefer (*Pinus strobus* L.) im Nationalpark Böhmische Schweiz. In: Seiler U, Wild J, Csaplovics E (eds) Historische Waldentwicklung in der Sächsisch-Böhmischen Schweiz, Fernerkundung und angewandte Geoinformatik (TU Dresden), vol 10, pp 41–61
Wilson JB, Peet RK, Dengler J, Pärtel M (2012) Plant species richness: the world records. Journal of Vegetation Science 23:796–802

# Index of Plant and Syntaxon Names





M. Chytrý et al. (eds.), *Flora and Vegetation of the Czech Republic*, Plant and Vegetation 14, DOI 10.1007/978-3-319-63181-3

**F**

## N

MIX
Papier aus verantwortungsvollen Quellen
Paper from responsible sources
FSC® C105338

If you have any concerns about our products,
you can contact us on
**ProductSafety@springernature.com**

In case Publisher is established outside the EU,
the EU authorized representative is:
**Springer Nature Customer Service Center GmbH**
**Europaplatz 3, 69115 Heidelberg, Germany**

Printed by Libri Plureos GmbH
in Hamburg, Germany